MW00966555

Soil Biology

Volume 31

Series Editor
Ajit Varma, Amity Institute of Microbial Technology,
Amity University Uttar Pradesh, Noida, UP, India

For further volumes:
http://www.springer.com/series/5138

Erika Kothe • Ajit Varma
Editors

Bio-Geo Interactions in Metal-Contaminated Soils

Editors
Erika Kothe
Friedrich Schiller Universität Jena
Institute of Microbiology
Neugasse 25
07745 Jena
Germany
erika.kothe@uni-jena.de

Dr. Ajit Varma
Amity University Uttar Pradesh
Amity Institute of Microbial Sciences
Noida Uttar Pradesh
Block A, Ground Floor, Sector 125
India
ajitvarma@aihmr.amity.edu

ISSN 1613-3382
ISBN 978-3-642-23326-5 e-ISBN 978-3-642-23327-2
DOI 10.1007/978-3-642-23327-2
Springer Heidelberg Dordrecht London New York

Library of Congress Control Number: 2011943312

Printed on acid-free paper

Springer is part of Springer Science+Business Media (www.springer.com)

Preface

The book aims at interdisciplinary research between geosciences and biology with major input from microbiology. The view on geoactive principles of bacteria and fungi in soil comprises the ecological understanding of habitat formation as well as modeling of matter fluxes. The combination of geological and biological methodology brings forward a general understanding of processes at the interfaces of the microbial cell and the plant root at soil particles in specifically metal-contaminated land. Since metal contamination is an increasing ecological and ecotoxicological risk, the basic understanding of processes involved in metal mobilization as well as sorption and mineralization are the key features for remediation actions in the field of heavy metal-contaminated land management. This is also reflected in the growing interest in bio–geo-interactions by a number of new periodicals as well as study courses, like biogeosciences, geomicrobiology, or geoecology.

Researchers and graduate students working in the field of biogeosciences and ecology, consulting companies which now become aware of the potential lying in the use of biology and look for biological means to manage schemes of soil remediation, or small- and medium-sized enterprises involved in bioremediation are the groups targeted with this book. We intended to provide state-of-the-art knowledge of phytostabilization and phytoextraction for their use. Since many of the sites in question are former mining sites, the federal and state governments might be interested in finding new solutions to the risk management of heavy metal-contaminated sites. For their use, specifically, the chapters on bioremediation or phytoremediation were added.

The chapters of the book start from a general overview on contaminated soil with a description of the role of physical, chemical, and biological compartments and their interaction with organic and inorganic contaminants which is followed by an introduction to biogeosciences in heavy metal-contaminated soils. After introducing the different mechanisms at play in the interactions of soil, microorganisms, plants, and the water phase necessary to transfer metals to biological systems, the mineralogy and geochemistry basics in assessing potential hazards at mining sites are addressed.

As mineralogy might also contain natural tracer elements for following processes between biotic and abiotic systems, rare earth elements are introduced as a means to follow the patterns of distribution between bios and geos in acidic systems. The direct impact of microorganisms on the substrate is specifically shown in an example of geomicrobial manganese redox reactions for metal release from manganese (hydr)oxides forming a biogeochemical barrier in a contaminated soil substrate. A potential for microbially induced biomineralization to natural attenuation is presented in the contaminated system at an abandoned mining site in Sardinia. Biomineralization investigated, in particular, at post-mining area is discussed regarding its potential to reduce the bioavailability of toxic heavy metals such as cadmium or lead. Since metal speciation plays a major role in element behavior, uranium speciation is investigated in seepage water and pore water of heavy metal-contaminated soil. In the context of uranium speciation and complexation, sophisticated methods like time-resolved laser-induced fluorescence spectroscopy or X-ray absorption spectroscopy are explained as suitable tools to study uranium in acidic and metal-rich waters.

The next chapters then move to plant–microbe interactions and their role in sustaining plant growth at heavy metal-contaminated sites, with heavy metal resistance of soil bacteria, the role of mycorrhiza in reforestation programs, and the occurrence of adapted plants at historical copper spoil heaps in Austria. Occurrence and capabilities of heavy metal-resistant plant growth promoting bacteria dwelling in the rhizosphere are presented with regard to bioaugmentation strategies. The examination of natural vegetation on heavy metal rich soils is necessary to find hyperaccumulator plants, which leads to the topic of phytoremediation of metal-contaminated soils. One such example is the nickel hyperaccumulating plant *Alyssum bertolonii*, which is presented as a model system for studying biogeochemical interactions.

The role of soil organic matter on metal mobility as well as the influence of symbiotic associations like different types of mycorrhiza for phytoremediation at different sites is presented. Again, use of metal-resistant bacterial strains as inocula for bioremediation of metal containing soils is presented. The investigation of experiments at laboratory, lysimeter, and field scale leads to the final chapter in which theoretical foundations of integrated modeling approaches in biogeochemistry are presented.

Different tools are discussed and a molecular understanding of dominant and relevant processes in contaminated soils is introduced. The hydrogeology and microbiology of former mining sites, specific microbial communities, and plant–microbe interactions are described with respect to state-of-the-art research.

The presentation of this book within the soil biology series would not have been possible without the help and collaboration of all our partners, specifically from UMRELLA, to whom we are deeply indebted.

Jena, Germany — Erika Kothe and Götz Haferburg
Noida, Uttar Pradesh, India — Ajit Varma

Contents

Contributors

Abate C. M. Universidad Nacional de Tucumán, Tucumán, Argentina

Adlassnig W. Core Facility Cell Imaging and Ultrastructure Research, University of Vienna, Vienna, Austria, wolfram.adlassnig@univie.ac.at

Amoroso M. J. Planta Piloto de Procesos Industriales y Microbiológicos (PROIMI), Avenida Belgrano y Pasaje, Caseros, Tucuman, Argentina

Arnold T. Helmholtz-Zentrum Dresden-Rossendorf, Institut fur Radiochemie, Dresden, Germany

Banasova V. Institute of Botany, Slovak Academy of Sciences, Bratislava, Slovakia

Baumann N. Helmholtz-Zentrum Dresden-Rossendorf, Institut fur Radiochemie, Dresden, Germany, n.baumann@hzdr.de

Behl R. K. Department of Plant Breeding, CCS Haryana Agricultural University, Hisar, Haryana, India, rkbehlprof@gmail.com

Bodescu F. Research Centre for Ecological Services (CESEC), Faculty of Biology, University of Bucharest, Bucharest, Romania

Buchel G. Institute of Earth Sciences, Friedrich-Schiller-University Jena, Jena, Germany

Cecchi L. Department of Evolutionary Biology, University of Firenze, Firenze, Italy

Ciamporova M. Institute of Botany, Slovak Academy of Sciences, Bratislava, Slovakia, milada.ciamporova@savba.sk

Cidu R. Department of Earth Sciences, University of Cagliari, Cagliari, Italy, cidur@unica.it

Cobzaru I. Institute of Biology Bucharest, Romanian Academy, Bucharest, Romania

Das A. Amity Institute of Microbial Technology (AIMT), Amity University Uttar Pradesh (AUUP), Noida, UP, India, adas@amity.edu

Giudici G. De Department of Earth Sciences, University of Cagliari, Cagliari, Italy, gbgiudic@unica.it

Durisova E. Institute of Botany, Slovak Academy of Sciences, Bratislava, Slovakia

Farcasanu I. C. University of Bucharest, Bucharest, Romania, ileana.farcasanu@g.unibuc.ro

Gawronski S. Institute of Botany, Jagiellonian University, Kraków, Poland, stefangaw@gmail.com

Gherghel Felicia Department of Mycology, Philipps-University of Marburg, Marburg, Germany

Gonnelli C. Department of Evolutionary Biology, University of Firenze, Firenze, Italy

Grawunder Anja Institute of Geosciences, Friedrich Schiller University Jena, Jena, Germany, anja.grawunder@uni-jena.de

Gube Matthias Institute of Microbiology, Friedrich-Schiller-Universitat, Jena, Germany, matthias.gube@uni-jena.de

Gurinova E. Institute of Botany, Slovak Academy of Sciences, Bratislava, Slovakia

Haferburg Gotz Institut fur Mikrobiologie, Friedrich-Schiller Universitat, Jena, Germany, goetz.haferburg@uni-jena.de

Iacob C. Lythos Research Center, Faculty of Geology, University of Bucharest, Bucharest, Romania

Ion S. Institute of Mathematical Statistics and Applied Mathematics, Romanian Academy, Bucharest, Romania

Iordache Virgil Research Centre for Ecological Services (CESEC), Faculty of Biology, University of Bucharest, Bucharest, Romania, virgil.iordache@g.unibuc.ro

Jianu D. Lythos Research Center, Faculty of Geology, University of Bucharest, Bucharest, Romania, denisa0301@yahoo.com

Kothe Erika Institute of Microbiology, Friedrich-Schiller-Universitat, Jena, Germany, erika.kothe@uni-jena.de

Krause Katrin Institute of Microbiology, Friedrich Schiller University, Jena, Germany, katrin.krause@uni-jena.de

Lacatusu R. Research Institute for Soil Science and Agrochemistry (ICPA), Bucharest, Romania

Lattanzi P. Department of Earth Sciences, University of Cagliari, Cagliari, Italy, lattanzp@unica.it

Lichtscheidl Irene K. Core facility Cell Imaging and Ultrastructure Research, University of Vienna, Vienna, Austria

Lonschinski M. Institute of Geosciences, Friedrich-Schiller University, Jena, Germany

Lorenz C. Chair of Environmental Geology, Brandenburg Technical University, Cottbus, Germany, christian.lorenz@tu-cottbus.de

Matache Mihaela University of Bucharest, Bucharest, Romania, mihaela.matache@gmail.com

Medas D. Department of Earth Sciences, University of Cagliari, Cagliari, Italy, dmedas@unica.it

Mengoni A. Department of Evolutionary Biology, University of Firenze, Firenze, Italy, alessio.mengoni@unifi.it

Merten Dirk Institute of Geosciences, Friedrich Schiller University Jena, Jena, Germany, dirk.merten@uni-jena.de

Nadubinská M. Institute of Botany, Slovak Academy of Sciences, Bratislava, Slovakia

Narula N. Department of Microbiology, CCS Haryana Agricultural University, Hisar, Haryana, India, neeru_narula@yahoo.com

Neagoe Aurora Research Centre for Ecological Services (CESEC), Faculty of Biology, University of Bucharest, Bucharest, Romania, aurora.neagoe@unibuc.eu

Onete M. Institute of Biology Bucharest, Romanian Academy, Bucharest, Romania

Orza R. Lythos Research Center, Faculty of Geology, University of Bucharest, Bucharest, Romania

Petrescu L. Laboratory of Geochemistry, Faculty of Geology, University of Bucharest, Bucharest, Romania

Podda F. Department of Earth Sciences, University of Cagliari, Cagliari, Italy, fpodda@unica.it

Purice D. Institute of Biology Bucharest, Romanian Academy, Bucharest, Romania

Reinicke Martin Institut fur Mikrobiologie, Friedrich-Schiller Universitat, Jena, Germany, martin.reincke@uni-jena.de

Ryszka P. Institute of Environmental Sciences, Jagiellonian University, Kraków, Poland

Schindler Frank Institute of Microbiology, Friedrich-Schiller-Universitat, Jena, Germany, schindler@uni-jena.de

Schutze Eileen Institute of Microbiology, Microbial Phytopathology, Friedrich Schiller University Jena, Jena, Germany, eileen-schuetze@web.de

Scradeanu D. Laboratory of Hydrogeology, Faculty of Geology and Geophysics, University of Bucharest, Bucharest, Romania

Sherameti I. Institute of General Botany and Plant Physiology, Friedrich-Schiller-University Jena, Jena, Germany, irenasherameti@yahoo.de

Soare B. Lythos Research Center, Faculty of Geology, University of Bucharest, Bucharest, Romania

Turnau K. Institute of Environmental Sciences, Jagiellonian University, Kraków, Poland, katarzyna.turnau@uj.edu.pl

Varma Ajit Amity Institute of Microbial Technology (AIMT), Amity University Uttar Pradesh (AUUP), Noida, UP, India, ajitvarma@aihmr.amity.edu

Wernitznig S. Core facility Cell Imaging and Ultrastructure Research, University of Vienna, Vienna, Austria

Zook D. Boston University, Boston, MA, USA, dzook@bu.edu

Chapter 1
Contaminated Soil: Physical, Chemical and Biological Components

Aparajita Das, Irena Sherameti, and Ajit Varma

1.1 Introduction

Soil may be defined as a thin layer of earth's crust which serves as a natural medium for growth of plants. Soil contains organic and inorganic materials that serve as a reservoir of nutrients as well as water for crops and provides mechanical anchorage. The organic portion, which is derived from the decayed remains of plants and animals, is concentrated in the dark uppermost topsoil. The inorganic portion was formed over thousands of years by physical, chemical and microbiological weathering of underlying bedrock. Unpolluted and fertile soils are required for agriculture to supply the mankind with sufficient food and other necessary requirements. Due to extensive range of industrial and agricultural activities, a high number of chemical contaminants get released into the ecosystem. As a result of this, the soil gets contaminated with the build-up of persistent toxic compounds, chemicals, salts, radioactive materials or disease causing agents in soils, which have adverse effects on plant growth and animal health. Depending on soil type, such agents may be distributed with the water phase in soil and influence the functions adversely. Biological remediation using microorganisms and plants is generally considered a safe and a less expensive method for the removal of hazardous contaminants from the environment.

This chapter provides an overview of the contaminated soils, their physical, chemical and biological components and briefly discusses the importance of heavy

A. Das • A. Varma (✉)
Amity Institute of Microbial Technology (AIMT), Amity University Uttar Pradesh (AUUP), Sec-125, Expressway, Noida, UP, India
e-mail: adas@amity.edu; ajitvarma@aihmr.amity.edu

I. Sherameti
Institute of General Botany and Plant Physiology, Friedrich-Schiller-University Jena, Dornburger Strasse 159, 07743 Jena, Germany
e-mail: irenasherameti@yahoo.de

E. Kothe and A. Varma (eds.), *Bio-Geo Interactions in Metal-Contaminated Soils*, Soil Biology 31, DOI 10.1007/978-3-642-23327-2_1,

metal tolerant AM fungi and heavy metal tolerant plants for reclamation of degraded soils.

1.2 Physical Properties of Soil

1.2.1 Soil Texture

Soil texture refers to the size distribution of soil particles and the relative percentage of sand, silt and clay particles present in a mass of soil with material sizes less than 2 mm. Materials of more than 2 mm are coarse fragments (gravel, stone, boulder, etc.). Soil texture includes only mineral particles that vary in size from those easily seen with the unaided eye to those below the range of a high-powered microscopy and it does not include any organic matter. Soils with the best structure for most crops are some types of loams, namely sandy loam, silty loam, sandy clay and clay loams which best retain water and supply dissolved minerals for nutrition. The available water holding capacity of soil is related to soil texture. Clayey soils show high water holding capacity, high plasticity and stickiness as well as swelling capacity whereas sandy soils are conspicuous by the absence of these properties. The most important way in which soil texture affects plant growth is the provision of water and, with it, nutrient supply (Pani 2007).

1.2.2 Soil Structure and Composition

Structure of soil refers to the arrangement of soil particles into groups that help in water and nutrient supplying ability of the soil, and air supply to plants' roots. Soil structure is the product of processes that aggregate cement and compact or unconsolidated soil materials. In essence, soil structure is a physical condition that is distinct from that of the initial material from which it has been formed, and can be related to processes of soil formation. The pads (natural aggregates) are separated from adjoining pads by surfaces of weakness. Soil structure influences plant growth rather indirectly. The pores sizes are the controlling factors governing water, air and temperature in soil, which in turn govern plant growth. One of the best examples of the effect of soil structure on plant growth is the emergence of seedlings in the seedbed. The seedlings are very sensitive to soil physical condition so that there should not be any hindrance to the emergence of tender seedlings and there should be optimum soil water and soil aeration. The soil in the seedbed should have a crumb structure so that the pads are soft and porous and roots of the seedling can penetrate it easily. A hard compact layer in contrast would hamper root growth. Soil structure can be divided into three groups as depicted in Table 1.1 (cf. Pani 2007). Most soils consist of four major components, namely mineral materials, organic

Table 1.1 Classification of soil structure (cf. Pani 2007)

Types (form)	Class (size)	Grade (degree of aggregation)
Platy	Very fine	Structure less
Prism	Fine	Weak
Blocky and nuciform	Medium	Strong
Spheroid	Coarse or thick	Moderate
	Very coarse or very weak	

matter, water and air. Of the total volume, about half is covered by mineral materials (45%) and organic matter (5%). The rest is contributed by air and water together.

1.2.3 *Soil Water*

Water is crucial for plant growth. Soil is capable of being a storehouse of water and becoming the main source of water for land plants. Mineral soil consists of 25% water and only half of it is available to plants (Pani 2007). If adequate water supply is unavailable in soil then normal plant functions are disturbed, and the plant gradually wilts, stops growing and dies. Thus, water is necessary in maintaining the turgidity of plants. Plants are most prone to damage from water deficiency during the vegetative and reproductive stages of growth. In addition, many plants are most sensitive to salinity during the germination and seedling growth stages.

If a cubic foot of typical silt loam topsoil was separated into its component parts, about 45% of the volume would be mineral matter (soil particles), organic residue would occupy about 5% of the volume, and the rest would be pore space. The pore space is the voids between soil particles and is occupied by either air or water. The quantity and size of the pore spaces are determined by the soil's texture, bulk density and structure. When rain or irrigation water is supplied to a field, it seeps into the soil. This process is called infiltration. Water is held in soil in two ways: as a thin coating on the outside of soil particles and in the pore spaces. Soil water in the pore spaces can be divided into two different forms: gravitational water and capillary water. As water infiltrates into a soil, the pore spaces fill with water. As the pores are filled, water moves through the soil by gravity and capillary forces. Water movement continues downward until a balance is reached between the capillary forces and the force of gravity. Water is pulled around soil particles and through small pore spaces in any direction by capillary forces. When capillary forces move water from shallow water table upward, salts may precipitate and concentrate in the soil as water is removed by plants and evaporation (according to http://www.ag.ndsu.edu, dated 16 May 2011). Most of the water that enters the plant roots does not stay in the plant. Less than 1% of the water withdrawn by the plant is actually used in photosynthesis (i.e., assimilated by the plant). The rest of the water moves to the leaf surfaces where it transpires (evaporates) to the atmosphere. The rate at which a plant takes up water is controlled by its physical characteristics, the atmosphere and soil environment. Plants can extract only the soil water that is in contact with their roots (according to http://www.ag.ndsu.edu, dated 16 May 2011). Soil is a medium that stores and moves water. Thus, the amount of water present in the soil plays a significant role in several natural processes such as evaporation, infiltration and drainage of water, diffusion of gases, conduction of heat, and movement of salts and nutrients. Plants meet their water requirement from water stored in soil.

1.2.4 Soil Aeration and Plant Growth

Oxygen is required by microbes and plants for respiration (oxybiont). As the plant's roots depend on beneficial mycorrhizospheric microbes for optimal growth and development, both partners of this symbiotic/associative relationship need to be supported by sustaining soil structure and composition. Oxygen taken up and carbon dioxide evolved are stoichiometric. Under anaerobic conditions, gaseous carbon compounds in addition to carbon dioxide are released. Root elongation is particularly sensitive to aeration. Oxygen deficiency disturbs metabolic processes in plants, resulting in the accumulation of toxic substances in plants and low uptake of nutrients. Certain plants such as rice are adapted to grow under submerged conditions. These have large internal air spaces, which facilitate oxygen transport to the roots, under water saturated conditions.

1.3 Soil Contamination

Industrial revolution in the twentieth century has resulted in the generation of vast array of chemical products which has resulted in contamination of the environment. Since soil and ground water are preferred sinks for complex contamination, various chemical and biological soil properties are profoundly altered, which affects biodiversity and soil function (Singh et al. 2009).

1.3.1 Common Chemicals Involved in Soil Contamination

Soil pollution is caused by chemicals of both organic and inorganic contaminants. The most prominent chemical groups of organic contaminants are fuel hydrocarbons, polynuclear aromatic hydrocarbons (PAHs), polychlorinated biphenyls (PCBs), chlorinated aromatic compounds, detergents and pesticides. Inorganic species include nitrates, phosphates and heavy metals such as cadmium, chromium and lead; inorganic acids; and radionuclides (radioactive substances). Among the sources of these contaminants are agricultural runoffs, acidic precipitates, industrial waste materials and radioactive fallout. Some of the prominent chemicals involved in soil contamination are discussed below:

(a) Petroleum hydrocarbons

In petrol polluted soil, benzene, toluene, ethylbenzene and xylene (BTEX) isomers are present in the water soluble fraction, causing pollution The most infamous class of hazardous compounds found in petrol, diesel, hydrocarbon-based oil as well as coal tar and its derivatives are the polycyclic aromatic hydrocarbons (PAHs). PAHs are hydrophobic, chemical compounds consisting of fused aromatic rings, not containing heteroatoms or carrying substituents, e.g., naphthalene, anthracene, coronene, pyrene, etc. (Surridge et al. 2009). They are detected in air, soil and sediment, surface water, groundwater, and road runoff and are dispersed from the atmosphere to vegetation and contaminate foods. As a result of their natural and anthropogenic sources, in combination with global transport phenomena, PAHs can be found in all regions of the world, including extreme environments with minimal human impact, such as Antarctica. Depending on the source of contamination, soils can contain PAH concentrations ranging between 1 μg/kg and 300 g/kg total PAHs. Further, due to incomplete combustion of materials such as coal and wood, atmospheric levels of PAHs ranging from 60 to 3 mg/m^3 air have been reported. Therefore, the development of practical bioremediation strategies for heavily impacted sites is urgently needed (Loick et al. 2009).

(b) Heavy metals

Heavy metals such as lead, arsenic and many more represent a highly abundant group of toxic compounds in the soil environment. Although their natural presence is limited to a few soil habitats, they are locally present as a consequence of human activities including mining, processing or the extensive use of the metal. Thus heavy metal pollution in soils constitutes a highly complex disruption of ecological equilibrium. Several heavy metals such as copper, zinc and iron are essential for the physiological functioning of living organisms, but they all become toxic at high concentrations. The toxicity of a metal depends on the metal itself, its total concentration, the availability of the metal to the organism and the organism itself (Kamal et al. 2010).

(c) Pesticides

A pesticide may be a chemical substance, biological agent (such as a virus or bacterium), antimicrobial, disinfectant used against any pest. Prominent chemical pesticide families include organochlorines, organophosphates and carbamates. Organochlorine hydrocarbons (e.g., DDT) could be separated into dichlorodiphenylethanes, cyclodiene compounds and other related compounds. Insecticides DDT and gammaxene were widely used at the end of World War II. Insects soon became resistant to DDT and as the chemical did not decompose readily, it persisted in the environment. Since it was soluble in fat rather than water, it biomagnified up the food chain and disrupted calcium metabolism in birds, causing eggshells to be thin and fragile. As a result, large birds of prey such as the brown pelican, ospreys, falcons and eagles became endangered (according to http://www.pollutionissues.com, dated 26.5.11).

(d) Solvents that are used in industries

Toluene is an aromatic hydrocarbon with a methyl side chain, widely used as an industrial feedstock, octane booster in fuel, solvent in paints, rubber, printing, adhesives, lacquers, disinfectants and in production of phenol (Surridge et al. 2009).

1.3.2 Sources of Soil Contamination

Ecosystems have been polluted with diverse kinds of chemicals which are released by various human and natural activities. Excessive levels of inorganic fertilizer related chemicals introduced into soil, such as ammonia, nitrates, phosphates which accumulate there or lead to contamination of water courses and air, have resulted in noteworthy environmental damage. Metals such as lead, arsenic, cadmium, copper, zinc, nickel and mercury are continuously being added to our soils through various agricultural activities such as agrochemicals usage and long-term application of urban sewage sludge in agricultural soils, industrial activities such as waste disposal, waste incineration and vehicle exhausts. All these sources cause accumulation of metals and metalloids in our agricultural soils and pose serious threat for biota and human health (Forstner 1995). Sites contaminated by heavy metal include battery disposal areas, burn pits, chemical disposal areas, contaminated marine sediments, electroplating/metal finishing shops and fire fighting training areas as well as landfills and burial pits. Metals such as lead, arsenic, etc. are commonly called heavy metals, although this term strictly refers to metallic elements with a specific mass higher than 5 g/cm^3 which are able to form sulfides (Gaur and Adholeya 2004; Singh et al. 2009). The other sources of metals in the soil are diverse, including burning of fossil fuels, mining and smelting of metalliferous ores, municipal wastes, sewage sludge amendments and the use of pigments. All these can also cause a considerable detrimental effect on soil ecosystems, environment and human health due to their motilities and solubility which determine their speciation (Khan 2005). At low concentrations, some metals such as copper, chromium, molybdenum, nickel, selenium and zinc are vital to healthy functioning and reproduction of microorganism, plants and animals (Marschner and Romheld 1995; Gaur and Adholeya 2004). However, at high concentrations, these essential elements may cause direct toxicity or reproductive effects. Some elements such as arsenic, cadmium, lead and mercury are not required by living components of ecosystem as these elements have no beneficial effect. These elements even at low concentrations can cause toxicity to plants and animals when they are present in the environment and their accumulation over time in the bodies of animals can cause serious illness (Alloway 1995; Gaur and Adholeya 2004; Adamo and Zampella 2008; Mohapatra 2008). In some cases, the soil may be contaminated to such an extent that it may be classified as a hazardous waste (Berti and Jacob 1996; Khan 2005). Migration of these contaminants into noncontaminated areas as dust or leachates through the soil and spreading of heavy metal containing sewage sludge are examples of events that contribute towards contamination of our ecosystems.

Soil contamination with heavy metal mixtures is receiving increasing attention from the public as well as governmental bodies, particularly in developing countries (Khan 2005). The remediation of such soils is important because these usually cover large areas that are rendered unsuitable for agricultural and other human use.

Trace metal contamination of soils can occur naturally from geological sources, for example Cu and Ni contamination of basaltic soils from the basalt parent material (Pal et al. 2010), or as a result of a wide range of industrial and agricultural activities. Metals and metalloids enter soils and waters because of many processes including atmospheric deposition from industrial activities or power generation; disposal of wastes such as sewage sludge, animal manures, ash, domestic and industrial wastes or byproducts; irrigation and flood or seepage waters and the utilization of fertilizers, lime or agrochemicals. Radionuclides are building up in some areas due to deliberate or accidental releases related to their use in energy production or for military purposes. It has been found that sewage sludge contains maximum amount of metals among different sources (Singh et al. 2009; Pal et al. 2010).

Heavy metals are deposited in soils by atmospheric input and the use of mineral fertilizers or compost, and sewage sludge disposal. It is well known that heavy metals are not biodegraded in contrast to organic pollutants and thus need to be physically removed or be immobilized (Gaur and Adholeya 2004; Mohapatra 2008). To some extent, microorganisms can modify the toxicity of metals by altering the bioavailability through oxidation and reduction. Traditionally, remediation of heavy metals contaminated soils involves either on-site management or excavation, and subsequent disposal to a landfill site. However, this method of disposal merely shifts the contamination problem elsewhere along with the hazards associated with transportation of contaminated soil and migration of contaminants from landfill into adjacent environments. Soil washing for removing contaminants from soil is an alternative to disposal to landfill. This method is, however, costly and produces a residue rich in heavy metals, which will require further treatment or burial. Conventional remediation methods usually involve excavation and removal of contaminated soil layer, physical stabilization (mixing of soil with cement, lime, apatite, etc.) and washing of contaminated soils with strong acids or heavy metal chelators (Pal et al. 2010). To prevent decrease in available arable land for cultivation from heavy metal contamination, remediation action is required.

1.4 Chemical Fate and Mobility of Heavy Metals

The fate and transport of a metal in soil and groundwater depend significantly on the chemical form and speciation of the metal. The mobility of metals in ground-water systems is hindered by reactions that cause metals to adsorb or precipitate, or chemistry that tends to keep metals associated with the solid phase and prevents them from dissolving. These mechanisms can retard the movement of metals and also provide a long-term source of metal contaminants (Evanko and Dzombak 1997).

1.4.1 Lead

The primary industrial sources of lead (Pb) contamination include metal smelter and processing, secondary metals production, lead battery manufacturing, pigment and chemical manufacturing, and lead-contaminated wastes. Widespread contamination due to the former use of lead in gasoline is also of concern. Lead released to groundwater, surface water and land is usually in the form of elemental lead, lead oxides and hydroxides, and lead metal oxyanion complexes. Most lead released to the environment is retained in the soil. The primary processes influencing the fate of lead in soil include adsorption, ion exchange, precipitation and complexation with absorbed organic matter. These processes limit the amount of lead that can be transported into the surface water or groundwater (Evanko and Dzombak 1997).

1.4.2 Chromium

Chromium (Cr) is one of the less common elements and does not occur naturally in elemental form, but usually as chromates. Chromium is mined as a primary ore product in the form of the mineral chromite, $FeCr_2O_4$. Hexavalent chromium is widely used in many industrial processes such as electroplating and wood preservation. Major sources of Cr contamination include releases from electroplating processes and the disposal of chromium containing wastes. Most of the chromium released into natural waters is particle associated, however, and is ultimately deposited into the sediment (Evanko and Dzombak 1997). Commercially available forms of hexavalent chromium (Cr (VI)) are potassium chromate and potassium dichromate. The chromium manufacturing industry produces a large quantity of solid and liquid waste containing hexavalent chromium. The treatment of these wastes is essential before discharging them to the environment. Cr (VI) compounds are highly water soluble, toxic and carcinogenic in mammals (Jeyasingh and Philip 2005). As chromium is widely used in many industries of which leather industries are the biggest consumers, wastes from tanneries pose a serious threat to the environment.

1.4.3 Arsenic

Arsenic (As) is a semimetallic element that occurs in a wide variety of minerals, mainly as As_2O_3, and can be recovered from processing of ores containing mostly copper, lead, zinc, silver and gold. It is also present in ashes from coal combustion. Many arsenic compounds absorb strongly to soils and are therefore transported only over short distances in groundwater and surface water (Evanko and Dzombak 1997).

1.4.4 Zinc

Zinc (Zn) does not occur naturally in elemental form. It is usually extracted from mineral ores to form zinc oxide (ZnO). The primary industrial use for Zinc is as a corrosion-resistant coating for iron or steel. Zinc is one of the most mobile heavy metals in surface waters and groundwater because it is present as soluble compounds at neutral and acidic pH values. At higher pH values, zinc can form carbonate and hydroxide complexes which control zinc solubility. Zinc readily precipitates under reducing conditions and in highly polluted systems when it is present at very high concentrations, and may co-precipitate with hydrous oxides of iron or manganese (Evanko and Dzombak 1997).

1.4.5 Cadmium

Cadmium (Cd) occurs naturally in the form of CdS or $CdCO_3$. Cadmium is recovered as a byproduct from the mining of sulfide ores of lead, zinc and copper. Sources of cadmium contamination include plating operations and the disposal of cadmium-containing wastes.

Cadmium is relatively mobile in surface water and ground-water systems and exists primarily as hydrated ions or as complexes with humic acids and other organic acids. Under acidic conditions, cadmium may also form complexes with chloride and sulfate. Cadmium is removed from natural waters by precipitation and sorption to mineral surfaces, especially oxide minerals, at higher pH values ($pH > 6$). Removal by these mechanisms increases as pH increases (Evanko and Dzombak 1997).

1.4.6 Copper

Copper (Cu) is mined as a primary ore product from copper sulfide and oxide ores. Mining activities are the major source of copper contamination in groundwater and surface waters. Other sources of copper include algicides, chromated copper arsenate (CCA) pressure treated lumber and copper pipes. Solution and soil chemistry strongly influence the speciation of copper in ground-water systems (Evanko and Dzombak 1997).

1.4.7 Mercury

The primary source of mercury is the sulfide ore cinnabar. Mercury (Hg) is usually recovered as a byproduct of ore processing. Release of mercury from coal combustion is a major source of mercury contamination. Releases from manometers at pressure measuring stations along gas/oil pipelines also contribute to mercury contamination (Evanko and Dzombak 1997). In contrast to the other metals and metalloids, microbial reduction of mercury can lead to volatilization of elemental mercury which has a high vapor pressure.

1.5 Soil Health and Need for Biological Remediation

Contaminated soils around the world have limited value for farming purposes. Affected soil that are excessively polluted become relatively sterile to maximum life forms. Many technologies are currently used to clean up heavy metal contaminated soils. The most commonly used ones are soil removal and land filling stabilization/solidification, physico-chemical extraction, soil washing, flushing, bioremediation and phytoremediation. None of above mentioned techniques are completely accepted as best treatment option because either they offer a temporary solution, or simply immobilize the contaminant or are costly when applied to large areas (Jeyasingh and Philips 2005). Productivity of conventional agricultural systems largely depends on the functional process of soil microbial communities. The structure and diversity of microbial communities are influenced by the soil structure and spatial distribution, as well as by the relationship between abiotic and biotic factors of the microbial communities (Jeyasingh and Philips 2005; Surridge et al. 2009). Biological remediation using microorganisms and plants is generally considered a safe and a less expensive method for the removal of hazardous contaminants from the environment. Certain indigenous microorganisms including bacteria and fungi are able to degrade PAHs in soil, leading to in situ rehabilitation of contaminated soils. Specific plants are also adapted to grow on contaminated sites. Arbuscular mycorrhizal fungi (AM fungi) can colonize the plants growing in contaminated soil and thus AM fungi can be exploited for phytoremediation (Bothe et al. 2010). Bioremediation, i.e., the use of living organisms to manage or remediate pollute soils, is an emerging technology. It is defined as the elimination, attenuation or transformation of polluting or contaminating substances by the use of biological processes. Phytoremediation, the use of plants to remediate or clean-up contaminated soils, is another promising method to remove and/or stabilize soils contaminated with heavy metals (Gaur and Adholeya 2004). Therefore, sustainable on-site techniques for remediation of heavy metal contaminated sites need to be developed.

1.6 Possible Role of AM Fungi in Remediation

Plants growing on contaminated soil are poor in health. Not all plants can grow on metal containing soil. The introduction of an AM fungal inoculum into such areas might be one of the strategies for establishing of mycorrhizal herbaceous plant species. AM fungal isolates differ in their effect on heavy metal uptake by plants. Some reports indicate higher concentrations of heavy metals in plants due to AM, whereas others have found a reduced plant concentration, e.g., for Zn and Cu in mycorrhizal plants. Thus, selection of appropriate isolates could be of importance for a given phytoremediation strategy. AM fungal species can be isolated from areas which are either naturally enriched by heavy metals or old mine/industry waste sites in origin. In this context, AM fungi constitute an important functional component of the soil–plant system that is critical for sustainable productivity in degraded soils. Thus, AM fungi may also play a role in the protection of roots from heavy metal toxicity by mediating interactions between metals and plant roots (Khan 2005).

AM fungi are of importance as they play a vital role in metal tolerance and accumulation. External fungal mycelium of AM fungi provides a wider exploration of soil volumes by spreading beyond the root exploration zone which otherwise is unavailable for uptake by roots alone and thus provide access to a greater volume of heavy metals present in the root zone. A greater volume of metals is also stored in the mycorrhizal structures in the root and spores. For example, concentrations of over 1,200 mg/kg of Zn have been reported in fungal tissues of *Glomus mosseae* and over 600 mg/kg in *Glomus versiforme*. Another important feature of this symbiosis is that AM fungi can increase plant establishment and growth despite high levels of soil heavy metals, due to better nutrition water availability and soil aggregation properties associated with this symbiosis. AM fungus is significant in the ecological improvement of rhizosphere (Gaur and Adholeya 2004). Several of the heavy metal tolerant AM fungi isolated from polluted soils can be useful for reclamation of degraded soils as they are found to be associated with a large number of plant species in heavy metal polluted soil. A detailed review on role and prospects of arbuscular mycorrhizal fungi in phytoremediation of heavy metal contaminated soils is available in Gaur and Adholeya (2004).

Gildon and Tinker (1981, 1983) isolated a mycorrhizal strain which tolerated 100 mg/kg of Zn in the soil. Considerable amount of AM fungal colonization was also reported in an extremely polluted metal mining area with an HCl-extractable Cd soil concentration of more than 300 mg/kg. Similarly, Weissenhorn et al. (1993) isolated mycorrhizal fungi from two heavy metal-polluted soils, which were found to be more resistant to Cd than a reference strain. Sambandan et al. (1992) reported 15AM fungal species from heavy metal contaminated soils from India. Of the 15AM species isolated, *Glomus geosporum* was encountered at all the sites studied. The percentage colonization ranged from 22 to 71% and spore count was as high as 622 per 100 g of soil. Weissenhorn et al. (1995) suggested a high tolerance of indigenous AM fungal population to elevated metal concentrations in soil and inside the roots (Gaur and Adholeya 2004).

Turnau et al. (2001) analyzed the community of AM fungi in the roots of *Fragaria vesca* growing in Zn-contaminated soil. Seventy percent of the root samples containing positively stained fungal hyphae were found to be colonized by *G. mosseae*. Another unique AM fungal species, *Scutellospora dipurpurascens* has been reported by Griffioen et al. (1994) from the rhizosphere of *Agrostis*

Table 1.2 AM fungi in metal polluted habitat (Mathur et al. 2007)

Glomus claroideum
Glomus mosseae
Gigaspora sp.
Glomus tenue
Glomus ambisporum
Glomus fasiculatum
Glomus intraradices
Glomus macrocarpum
Glomus etunicatum

capillaris growing in contaminated surroundings of a zinc refinery in the Netherlands. This indicates that these fungi have evolved Zn and Cd tolerance and that they might play an important role in conferring Zn or Cd tolerance in plants. Mycorrhizal fungi have also been shown to be associated with metallophyte plants on highly polluted soils, where only adapted plants such as *Viola calaminaria* (violet) can grow (Gaur and Adholeya 2004). A *Glomus* sp. isolated from the roots of the violet plant improved maize growth in a polluted soil (Hildebrandt et al. 1999; Gaur and Adholeya 2004; Bothe et al. 2010) and reduced root and shoot heavy metal concentrations in comparison to a common *Glomus* isolate or non-colonized controls. A list of AMF fungi present in contaminated habitat is given in Table 1.2.

1.7 Possible Role of Heavy Metal Tolerant Plants in Remediation

A metallophyte is a plant that can tolerate high levels of heavy metals such as lead, zinc and others. Plants that do well on heavy metal enriched soils can be used in the phytoremediation of polluted areas. The ability of metallophytes to tolerate extreme metal concentrations commends them as the perfect option for ecological restoration of metal-contaminated sites. Metallophyte can be used to stabilize soils against the erosion of the surface soil by wind and rainfall, and they can colonize disturbed areas. Metallophytes can be used to extract heavy metals (Bothe et al. 2010). As discussed above AM Fungi might contribute to metal tolerance. Plant species belonging to plant families Chenopodiaceae, Cruciferaceae, Plumbaginaceae, Juncaceae, Juncaginaceae, Amaranthaceae and few members of Fabaceae are believed not to form a symbiosis with AM fungi (Khan 2005). A detailed review on the role of soil microbes in the rhizospheres of plants growing on trace metal contaminated soils in phytoremediation was done by Khan (2005).

Many hyperaccumulators belong to the family nonmycorrhized family Brassicaceae but there are conflicting reports regarding their mycotrophic status. Hirrel et al. (1978) reported them to be nonmycorrhizal but 1–5% AM fungal root colonization occurred in seven species of crucifers when grown in the presence of a mycorrhizal companion plant. DeMars and Boerner (1996) made an extensive literature survey of crucifers and revealed that roots of 18.9% of the 946 members investigated were found to be colonized with internal hyphae, occasional vesicles but no arbuscules. Since arbuscules are the major site of nutrient exchange, including metals, between the plant and the AM fungi; these associations are expected to be non-functional (Bago 2000). However, a function as such for metals, e.g., in cell walls of dead hyphae, is still feasible. Pawlowska et al. (1996) found roots of *Biscutella laevigata*, a Brassicaceous plant colonizing the calamine mounds in Poland, to be mycorrhizal but without arbuscules. However, a latter study by Orlowska et al. (2002), who re-examined the mycorrhizal status of *B. laevigata* and the role of restoration of zinc-wastes on mycorrhization of this cruciferous plant species, observed AM hyphae, vesicles, as well as arbuscules in roots collected prior to seed maturity. *Minuarita verna* can endure the highest concentrations of heavy metals on any Central European metallophyte. *Armeria maritima* ssp. *halleri* occurs in coastal salt marshes and contains 20-fold and 88-fold greater concentrations of lead and copper, respectively, in its roots than in its leaves, indicating that the metals are immobilized in its roots. The genus *Thlaspi* comprises several closely related species

Table 1.3 Metal tolerant plants

Alyssum argenteum	*Alyssum constellatum*	*Alyssum euboeum*
Alyssum obovatum	*Ariadne shaferi*	*Armeria denticulate*
Austromyrtus bidwilliia	*Berkheya coddii*	*Berkheya zeyheri*
Blepharis acuminate	*Clerodendrum infortunatum*	*Cnidoscolus bahianus*
Cochlearia aucheri	*Cochlearia sempervivum*	*Crotalaria trifoliastrum*
Croton bonplandianus	*Croton campestris*	*Dichapetalum gelonioides*
Dicoma niccolifera	*Dodonaea microzyga*	*Elsholtzia haichowensis*
Elsholtzia patrini	*Eremophila weldii*	*Eriachne mucronata*
Euphorbia selloi	*Evovulvus alsinoides*	*Garcinia polyneura*
Garcinia reäoluta	*Geniosporum tenuiflorum*	*Gochnatia recuräa*
Grevillea acuaria	*Heliotropium salicoides*	*Hybanthus epacroides* sp. bilobus
Hybanthus floribundus	*Merremia xanthophylla*	*Mosiera araneosa*
Myristica laurifolia	*Ouratea striata*	*Pearsonia metallifera*
Pentacalia cristalensis	*Pentacalia moaensis*	*Pentacalia trichotoma*
Pentacalia trineura	*Phyllanthus balgooyi*	*Phyllanthus chryseus*
Phyllanthus cinctus	*Phyllanthus comptus*	*Phyllanthus cristalensis*
Phyllanthus discolour	*Phyllanthus formosus*	*Phyllanthus incrustatus*
Phyllanthus microdictyus	*Phyllanthus mirificus*	*Phyllanthus myrtilloides*
Planchonella oxyhedra	*Polygonum posumbu*	*Psidium araneosum*
Ptilotus obovatum	*Rhus wildii*	*Richardia grandiflora*
Rinorea bengalensis	*Rinorea javanica*	*Senecio lydenburgensis*
Senecio pauperculus	*Shorea tenuiramulosa*	*Sida linifolia*
Solidago hispida	*Streptanthus polygaloides*	*Tephrosia polyzyga*
Tephrosia villosa	*Thlaspi montanum* var *siskiyouense*	*Trichospermum kjelbergii*
Trymalium myrtillus	*Turnera subnuda*	*Turnera trigona*
Walsura monophylla	*Waltheria indica*	*Westringia rigida*

According to http://www.metallophytes.com, dated 06.05.2011

that grow well on heavy metal containing soils, e.g., *Thlaspi praecox* and *T. caerulescens* growing in heavy metal soils. *Thlaspi* species are generally short living annual plants and they have developed means to keep their seeds free of toxic heavy metal concentrations. *Viola lutea* ssp. *calaminaria* (yellow zinc violet) and *Viola lutea* ssp. *westfalica* (blue zinc violet) occupy heavy metal heaps with endemic occurrences (Tonin et al. 2001; Bothe et al. 2010). A list of metal tolerant plants is given in Table 1.3.

1.8 Conclusions

The present chapter discusses soil which consists of a complex mixture of particulate materials derived from abiotic parent minerals, living biota, organic detritus and humic substances. Soil can have naturally high concentrations of heavy metals as a result of the weathering of parental material with high amounts of heavy metal minerals or due to contamination associated with several human activities such as mining. Heavy metal contaminations of soils affect the population of living components present in it. Microorganisms have the capability to interact in a variety of specialized ways with metals. Specifically mycorrhizal associates may contribute to metal tolerance of plants. Such metalliferous plants that flourish on heavy metal rich soils can be used in phytoremediation of polluted areas.

References

Adamo P, Zampella M (2008) Chemical speciation to assess potentially toxic metals (PTMs) bioavailability and geochemical forms in polluted soils. In: De Vivo B, Belkin HE, Lima A (eds) Environmental geochemistry. Elsevier, Amsterdam, pp 175–203

Alloway BJ (1995) Soil processes and the behaviour of metals. In: Alloway BJ (ed) Heavy metals in soils, 2nd edn. Blackie Academic and Professional, London, pp 11–37

Bago B (2000) Putative sites for nutrient uptake in arbuscular mycorrhizal fungi. Plant Soil 226:263–274

Berti WR, Jacob LW (1996) Chemistry and phytotoxicity of soil trace elements from repeated sewage sludge application. J Environ Qual 25:1025–1032

Bothe H, Regvar M, Turnau K (2010) Arbuscular mycorrhiza, heavy metal and salt tolerance. In: Sherameti I, Varma A (eds) Soil heavy metals. Springer, Heidelberg, pp 87–107

DeMars BG, Boerner REJ (1996) Vesicular arbuscular mycorrhizal development in the Brassicaceae in relation to plant life span. Flora 191:179–189

Evanko CR, Dzombak DA (1997) Remediation of metals contaminated soils and groundwater. Ground Water Remediation Technologies Analysis Center. E series: TE-97-01

Forstner U (1995) Land contamination by metals – globe scope and magnitude of problem. In: Allen HE, Huang CP, Bailey GW, Bowers AR (eds) Metal speciation and contamination of soil. Lewis, Boca Raton, FL, pp 1–3

Gaur A, Adholeya A (2004) Prospects of arbuscular mycorrhizal fungi in phytoremediation of heavy metal contaminated soils. Curr Sci 86:528–534

Gildon A, Tinker PB (1981) A heavy metal-tolerant strain of a mycorrhizal fungus. New Phytol 95:263–268

Gildon A, Tinker PB (1983) Interactions of vesicular-arbuscular mycorrhiza infections and heavy metals in plants II. The effects of infection on uptake of copper. Trans Br Mycol Soc 77: 648–649

Griffioen WAJ, Iestwaart JH, Ernst WHO (1994) Mycorrhizal infection of *Agrostis capillaris* population on a copper contaminated soil. Plant Soil 158:83–89

Hildebrandt U, Kaldorf M, Bothe H (1999) The zinc violet and its colonization by arbuscular mycorrhizal fungi. J Plant Physiol 154:709–717

Hirrel MC, Mehravaran H, Gerdemann JW (1978) Vesicular arbuscular mycorrhizae in the Chenopodiaceae and Cruciferae: do they occur? Can J Bot 56:2813–2817

Loick N, Hobbs PJ, Hale MDC, Jones DL (2009) Bioremediation of Poly aromatic Hydrocarbon (PAH) contaminated soil by composting. Crit Rev Environ Sci Technol 39:271–332

Jeyasingh J, Philip L (2005) Bioremediation of chromium contaminated soil: optimization of operating parameters under laboratory conditions. J Hazard Mater B118:113–120

Kamal S, Prasad R, Varma A (2010) Soil microbial diversity in relation to heavy metals. In: Sherameti I, Varma A (eds) Soil heavy metals. Springer, Heidelberg, pp 31–63

Khan AG (2005) Role of soil microbes in the rhizospheres of plants growing on trace metal contaminated soils in phytoremediation. J Trace Elem Med Biol 18:355–364

Marschner H, Romheld V (1995) Strategies of plants for acquisition of iron. Plant Soil 165: 262–274

Mathur N, Singh J, Bohra S, Quaizi A, Vyas A (2007) Arbuscular mycorrhizal fungi: a potential tool for phytoremediation. J Plant Sci 2:127–140

Mohapatra PK (2008) Textbook of environmental microbiology. IK International, New Delhi, pp 411–418

Orlowska E, Sz Z, Jurkiewicz A, Szarek-Lukaszewska G, Turnau K (2002) Influence of restoration of arbuscular mycorrhiza of *Biscutella laevigata* L. (Brassicaceae) and *Plantago lanceolata* L. (Plantaginaceae) from calamine spoil mounds. Mycorrhiza 12:153–160

Pal S, Patra AK, Reza SK, Wildi W, Pote J (2010) Use of bio-resources for remediation of soil pollution. Nat Resour 1:110–125

Pani B (2007) Textbook of environmental chemistry. IK International, New Delhi, pp 365–373

Pawlowska TE, Blaszkowski J, Ruhling A (1996) The mycorrhizal status of plants colonizing a calamine spoil mound in southern Poland. Mycorrhiza 6:499–505

Sambandan K, Kannan K, Raman N (1992) Distribution of vesicular-arbuscular mycorrhizal fungi in heavy metal polluted soils of Tamil Nadu. J Environ Biol 13:159–167

Singh A, Kuhad RC, Ward OP (2009) Biological remediation of soil: an overview of global market and available technologies. In: Singh A, Kuhad RC, Ward OP (eds) Advances in applied bioremediation. Springer, Heidelberg, pp 1–18

Surridge AKJ, Wehner FC, Cloete TE (2009) Bioremediation of polluted soil. In: Singh A, Kuhad RC, Ward OP (eds) Advances in applied bioremediation. Springer, Heidelberg, pp 103–116

Tonin C, Vandenkoornhuyse P, Joner EJ, Straczek J, Leyval C (2001) Assessment of arbuscular mycorrhizal fungi diversity in the rhizosphere of *Viola calaminaria* and effect of these fungi on heavy metal uptake by clover. Mycorrhiza 10:161–168

Turnau K, Ryszka P, Gianinazzi-Pearson V, van Tuinen D (2001) Identification of arbuscular mycorrhizal fungi in soils and roots of plants colonizing zinc wastes in southern Poland. Mycorrhiza 10:169–174

Weissenhorn I, Leyval C, Berthelin J (1993) Cd-tolerant arbuscular mycorrhizal (AM) fungi from heavy-metal polluted soils. Plant Soil 157:247–256

Weissenhorn I, Leyval C, Berthelin J (1995) Bioavailability of heavy metals and abundance of arbuscular mycorrhiza in a soil polluted by atmospheric deposition from a smelter. Biol Fertil Soil 19:22–28

Chapter 2
Biogeosciences in Heavy Metal-Contaminated Soils

Götz Haferburg and Erika Kothe

2.1 Introduction

The development of the strongly interdisciplinary and highly applied research area of biogeosciences has been carried on to many fields, specifically those related to climate change or anthropogenic pollutions, where only the understanding of the interrelationships between both compartments allows prediction of changes introduced by mankind. This comparably young field of research integrates many different disciplines including hydrochemistry, plant physiology or microbiology and bacterial genetics, generally aiming at integration of effects of life (*βιός*) on Earth (*γεος*). Facing global change phenomena, we can make the general statement that the interference of humans with biogeochemical cycles leads to a dangerous imbalance in the overall mass balance of nature. The outcome of this disordered matter cycling for agricultural use of soils in Europe and subsequently for food security has recently been outlined (Miraglia et al. 2009). Biogeoscience arose as an answer to the alteration of our environment and is thought to deliver the scientific approach for measures to be taken to alleviate these deleterious effects of disequilibrium.

Aside from atmospheric changes and global climate change, the effects of pollution and their remediation are a key subject within biogeosciences. While organic pollution, including for example oil spills, may be accessible to microbial degradation as a remediation measure, the pollution, especially of soil, with (heavy) metals is not easily remediated since the pollutant cannot be degraded. Thus, one aim of biogeochemistry as a center piece within the biogeosciences is the understanding of the processes by which microbes influence pedogenesis and movement of metals in soil and water (Borch et al. 2010). This is important in any mining operation, where alteration and dissolution of bedrock and low-grade ore material

G. Haferburg (✉) • E. Kothe
Microbial Phytopathology, Friedrich-Schiller-University Jena, Neugasse 25, Jena, Germany
e-mail: goetz.haferburg@uni-jena.de

E. Kothe and A. Varma (eds.), *Bio-Geo Interactions in Metal-Contaminated Soils*,
Soil Biology 31, DOI 10.1007/978-3-642-23327-2_2,

are seen. The mechanisms by which bacteria and fungi impact the fate and transport of contaminants are yet to be completely understood. This is necessary to allow for modeling approaches of contaminant fate and distribution (Wiatrowski and Barkay 2005). Only with a sound scientific basis of contaminant transport and metal introduction into food chains, remediation strategies can be developed. Therefore biogeochemistry could also be seen as conception of an open-ended system: the metabolic diversity of (micro)organisms is extended by the network of material transport trails and especially metal trafficking and distribution pathways in any kind of habitat. A successful application of knowledge gained in biogeochemistry has been realized in biorecovery (also known as bioleaching or biomining), the exploitation of microbial metabolism in order to mobilize and enrich metals from low-grade ores and sewage sludge (Pathak et al. 2009). For example, approximately 25% of copper originates from bioleaching processes (Stabnikova et al. 2010). This may demonstrate the impact microbial metabolism can have on metal mobility.

Bioremediation is at least of the same importance as biorecovery in applied biogeochemistry. In bioremediation, single organisms – mainly plants, bacteria and fungi – or organisms in their interaction are adopted to convert contaminated soil and water to a condition which is not deleterious to plants, animals or humans. The synergistic effects of interacting organisms may involve, for example, iron oxidizing microorganisms in the rhizosphere of wetland plants to precipitate the excess iron (Laanbroek 2010). Essentially, every type and method of bioremediation benefit from the huge metabolic diversity of organisms. Making use of the potential of very specific metabolic adaptations of adequately niched organisms, the main advantages of bioremediation lies in the comparatively moderate investment necessary, low energy demand, inherent safety of biological processes to the environment, low waste production and, in optimal cases, self-sustainability (Haferburg and Kothe 2010). However, the process needs time, and thus a substantial amount of monitoring is required to allow for optimized bioremediation. Instruments for field applicable biological and chemical monitoring such as microarray analysis systems or reporter gene assays become more and more common (Chandler et al. 2010; Alkorta et al. 2006). In order to decide on the optimal strategy and the best possible monitoring of success, a large set of data from both biological and (hydro) geochemical/physico-chemical parameters is necessary. This initially high investment of site-specific research is then balanced in bioremediation by the long-term low-input strategy which often can be achieved in enhanced natural attenuation,[1] if an optimized strategy is developed. Such a strategy needs to be site specific since climatic condition, hydrogeochemical

[1] The term "natural attenuation" often causes misunderstanding. Here it reflects on the whole of all physical, chemical and biological processes which are active in a particular postindustrial area without any anthropogenic intrusion. Under particular conditions, the interplay of all these processes results in a decrease of mass, toxicity, mobility, volume or concentration of pollutants in soil and groundwater. "Enhanced natural attenuation" also comprises approaches of biostimulation and bioaugmentation.

settings and metal contaminations are different for every site necessitating different (micro)biological amendments for bioremediation. In contrast to conventional soil remediation standards (so-called Dutch standards) there are no harmonized standards in Europe for the potential application of bioremediation strategies yet. It will be both a challenge and a virtue to work on standardization within and for the European Union.

2.2 Metal Contaminated Soil

After estimation from 1995, a total amount of over 700 million kg of metals is being dumped in mine tailings worldwide annually (Warhurst 2002). Depending on the metal (As, Cd, Cu, Ni, Pb and Zn), the volume of tailing material ranges from 10,000 to 600,000 metric tons (ib.), illustrating the negative consequences of ore processing. When large volumes of geogenic substrate are excavated, waste rock material is often still rich in metals after the extraction process. The reallocated geogenic material is prone to weathering and source of continuous metal release. Usually, the leached residues are dumped onto waste piles. Under irrigated and aerobic conditions, acid mine drainage ensues, often seen as seepage effluent with high-metal load and low pH. This contamination of the water path (often running through arable land) leads to soils with an increasing amount of metal and, subsequently, to a slow and continuous toxification of plants and animals, thus allowing for introduction in food chains and intoxication of humans through food or drinking water. In addition, the dilution leads to three-dimensional expansion of contamination which makes re-concentration and removal of metals impossible, resulting in both losses of metals and arable land.

In 2008, 1.4 billion tons of metals was produced globally which is a production rate sevenfold higher than in 1950. In 1950, metal consumption was 77 kg per person and year, which increased to 213 kg in 2008, varying tremendously among countries. While the benefits of metal production are easy to recognize, the negative impact is less obvious. Global mining occupies a territory of approximately 37,000 km^2 which equals approximately the area of Belgium or 0.2% of the world's land surface (Dudka and Adriano 1997). In addition, approximately 240,000 km^2 (approximately the size of the UK) is influenced by metals released from waste dumps and open mines (Furrer et al. 2002). Estimates of the European Environment Agency listed 1.4 million contaminated sites (Prasad et al. 2010). Since metal contamination cannot be detoxified by degradation, metal contaminated soils have to be either remediated by removal of the metals from the arable land with subsequent safe deposition, or by changing land use after metals have been immobilized on the spot.

An issue closely linked to the health hazards of metal contaminated land is soil erosion and land degradation. Estimations of the annual loss of farming land predominantly by industrialization, contamination, urbanization and desertification range between 70 and 140,000 km^2. 4.3 million km^2 of arable land became abandoned during the last 40 years. Globally, 100 billion tons of topsoil is lost every year (Döös 2002). Natural pedogenesis proceeds five times slower than devastation of soil. Especially, scarcely vegetated, metalliferrous soils are prone to whatsoever mechanism of erosion. With the given numbers, it seems evident that soil protection, soil remediation and soil recovery are of ultimate importance, especially when relating this to the growing world's population. Biogeosciences is meant to deliver a scientific understanding of the environmental changes and elaborate proposals how to counteract.

2.3 Soil Analyses

All bioremediation strategies depend on the geological and hydrogeochemical description of the site to be remediated. Both the contaminated soil as well as the bedrock need to be analyzed in terms of occurrence and distribution of mineral phases (petrology and lithology), concentration and bioavailability (using sequential extraction to determine the mobile and easily mobilized fractions of metals), adsorptive capacity of the soil (e.g., for clay minerals and humic acids), metal complexion and oxidation state (to determine factors for reactive transport) and mobility (in ground- and surface waters including hydrogeology and capillary flow). For reactive transport in acid mine drainage impacted landscapes, the use of rare earth elements has proven helpful to determine source and sink relationships (Haferburg et al. 2007). Stable isotopes have been used for some time now to follow distribution paths of elements (Miljević and Golobocanin 2007). Since biological system shows a strong selection for lighter isotopes in some cases, depending on the transport proteins and enzymes involved in uptake and metabolism, this methods seems specifically well suited to show the role of (micro)biology in changes of soil types and metal transport over time.

Pedogenetic parameters such as soil type, soil texture and particle size, organic matter content, humus layer, soil density and soil morphology thus need to be determined to provide input parameters for prognoses of development of a site. Pedogenesis connects the geological site description with the inventory of (micro) biological data. In order to evaluate the fertility of the soil substrate present, it is inevitable to analyze the microbial colonization, composition of microflora, soil respiration and microbial activity with respect to metal solubilization. For example, organic carbon content strongly influences the mobility of metals. The analysis of the organic carbon content thus helps to decide on the use of organic amendments to stabilize metals in the soil and to decrease metal mobility.

Finally, evaluation of ecotoxicological hazards can follow from an integration of all site-specific compiled data. An important step forward is provided by modeling approaches which address the expansion of contaminants at landscape level and provide a prognosis for the success of specific remediation measure. Since modeling can be used to determine the amount of monitoring necessary, the potential of bioremediation depends not only on the geological and pedological character of the site but also on data processing.

2.4 Microbial Communities

For the use of microbes in bioremediation actions, it is mandatory to isolate, cultivate and select strains. Soil bacteria in general can be described as metabolically very heterogenous. At least 150 diverse metabolic pathways and 900 different reactions of bacteria have been revealed (Scheffer and Schachtschabel 2010). Hence, it seems rather a problem of the appropriate screening assay to derive isolates for the desired bioremediation application than to find the corresponding biochemical feature. Bioaugmentation is the insertion of living microbial biomass into soil in order to make use of the strain(s) properties in a clean-up attempt. As a prerequisite for a bioaugmentation application, these bacterial strains need to be resistant against the metals in bioavailable concentrations present at the site of interest. The combination of soluble and easily remobilizable metals varies at each site which may indicate that the isolation of strains from the site investigated is the optimal strategy. However, strains resistant to metals may also be derived from nonpolluted environment, albeit with lower efficiency. Nevertheless, strains originating from other sites may be used in bioaugmentation if they are capable of withstanding the multimetal stress prevalent at the current site. There is a long-lasting discussion on advantages and disadvantages of biostimulation versus bioaugmentation. In biostimulation the remediation progress is triggered by external supply of nutrients whereas in bioaugmentation microbial biomass is introduced into the site to remediate. Biostimulation is based on the assumption that the remediation effect can be performed with the well-adapted microorganisms already present in the ground but they depend in their metabolic activity on an external supply of nutrients. Bioaugmentation in contrast seems indicated if the autochthonous microflora cannot exert the necessary remediation function alone. In general, this discussion shows how much more research is needed on the theory of "everything is everywhere, but, the environment selects" (Becking 1934; de Wit and Bouvier 2006).

Until now bioaugmentation plays a far greater role in bioremediation of soils contaminated with aliphatic, aromatic and halogenated hydrocarbons. Introduction of microorganisms in organically polluted soils results ideally in the overall metabolisation of the pollutant by the microbial cell. The microbially driven processes of soil decontamination from hydrocarbons are entirely different from those essential in bioremediation of metal contaminated soils. In general, two types of organisms are applied to treat soils with a metal burden: plant growth promoting (rhizo) bacteria (PGPR) and metal (im) mobilizing bacteria (MB). The combination of both groups seems to be the key for phytoextraction or phytostabilization.

Meanwhile production and trade of PGPR became an important market. Plenty of small biofertilizer enterprises were founded especially in India and China long before the application of PGPR (often combined with mycorrhiza inocula) reached the market for agriculture and gardening elsewhere in the rest of the world. Interestingly, the first biofertilizers (also named bioinoculants) were already applied in the nineteenth century, most notably with nitrogen-fixing *Rhizobium* species. Owing to the continuous research and development on plant growth promoting properties of microorganisms the biofertilizer market has amazingly grown and was estimated with a volume of $690 million for the USA alone in 2001(de Freitas 2002).

Primarily PGPR were applied to strengthen plant health and to increase crop yield. Subsequently strategies on PGPR supported phytoremediation of organically or inorganically polluted soils were developed from the knowledge gained in sustainable agriculture. Originally the group of PGPR has been described for *Pseudomonas* strains active as biocontrol organisms (Kloepper and Schroth 1978). Biocontrol has huge power to contain plant pathogens that threaten crop yield in disturbed but agriculturally used land. Especially in environments that provoke stress in plants, the use of biocontrol microorganisms seems to be an advantageous alternative to the application of pesticides. Intensified pest infestation on stressed plants and subsided symptom development due to PGPR activity are well investigated (Han et al. 2005; Babalola 2010). Nevertheless the term PGPR, coined by Kloepper and Schroth, was later extended to be applied for any bacterial isolate actively or passively promoting plant growth. Table 2.1 shows examples of PGPR.

Table 2.1 Characteristics and examples of plant growth promoting bacteria applied in bioremediation strategies

Characteristics of PGPR	Bacteria	Plant partner	References
Nitrogen fixation			
Freely associated bacteria	*Azotobacter chroococcum*	*Brassica juncea*	Wu et al. (2006)
Symbiotic bacteria	*Sinorhizobium meliloti*	*Medicago truncatula*	Bianco and Defez (2009)
Phosphate mobilization			
Inorganic P source	*Pseudomonas aeruginosa*	*Vigna mungo*	Ganesan (2008)
Organic P source	*Bacillus amyloliquefaciens*	*Zea mays*	Idriss et al. (2002)
Siderophore release			
Hydroxamates	*Streptomyces acidiscabies*	*Cicer arietinum*	Dimkpa et al. (2008)
Phenol catecholates	*Rhizobium* sp.	*Sesbania procumbens*	Sridevi et al. (2008)
Carboxylates	*Pseudomonas fluorescens*	*Arachis hypogaea*	Dey et al. (2004)
Salicylic acid	*Arthrobacter oxidans*	*Pinus* sp.	Barriuso et al. (2008)
Auxin production			
Indole acetic acid	*Enterobacter chloacae*	*Oryza sativa*	Mehnaz et al. (2001)
Cytokinin	*Pseudomonas fluorescens*	*Glycine max*	de Salamone et al. (2005)
Gibberellin	*Bacillus pumilus*	*Alnus glutinosa*	Gutierrez-Mañero et al. (2001)
Influence on metal toxicity			
Increased Ni accumulation	*Bacillus subtilis*	*Brassica juncea*	Zaidi et al. (2006)
Increased Cd accumulation	*Xanthomonas* sp.	*Brassica napus*	Sheng and Xia (2006)
Reduction of Cr(VI) to Cr(III)	*Ochrobactrum intermedium*	*Helianthus annuus*	Faisal and Hasnain (2005)

From an ecological perspective, in order to reach a stable colonization of plant roots it seems more reasonable to work with consortia instead of single strain inocula. However, it remains difficult and challenging to monitor the metabolic activity of the different inoculated strains in situ. Fluorescence in situ hybridization (FISH) is a method becoming more and more popular in soil microbiology. It combines microscopy of habitat samples as, e.g., root hairs and soil particles with the specific detection of particular microbial populations. The detection of the cells of interest in a cell mixture is based on fluorescence signals of labeled oligonucleotides priming only sequence-specific targets within the DNA (Dubey et al. 2006). This method was, for instance, successfully applied to discover the physiologically active bacteria in heavy metal contaminated sites (Margesin et al. 2011). Application of the BIOLOG identification system is an indirect method to profile changes in carbon metabolism of a microbial community during a remediation operation (Miller and Rhoden 1991; Alisi et al. 2009). This type of profiling metabolic activity of rhizo-bacterial communities could help to estimate the success and stability of root colonization but it hardly answers questions on competition among single strains or establishment/survival of particular strains. Nevertheless indirect methods of pheno- and genotyping of inoculated consortia are favorable over direct plating methods due to the loss of cells in any (re-)isolation campaign, first described as great plate count anomaly (Staley and Konopka 1985).

The isolation and screening procedure for PGPR follows the classical scheme (for an extended survey, see Steele and Stowers 1991). The screening for the most common PGBR comprises assays on phosphate solubilization, nitrogen fixation, siderophore release and phytohormone production. More specific tests are performed on the reduction of metal toxicity for the plant in combination with a higher metal accumulation in the above ground plant tissue. Usually PGPR are isolated from the rhizosphere or even more specifically from the rhizoplane. Bacteria of the rhizosphere seem to be highly adapted to the conditions of that particular microhabitat, especially as a result of the continuous flux of nutrients from the root, as, e.g., sugars, amino acids, lipids, phenolics and even vitamins or enzymes. It has been hypothesized that the strongly regulated secretory system of plants supports, restricts or terminates bacterial colonization in the rhizosphere (Bednarek et al. 2010). Nevertheless, plant growth promoting bacteria can also easily be retrieved from bulk soil in many cases.

It is impossible to completely describe the entire microflora of any soil in terms of biodiversity, quantity, ecological functions and microbe–soil, microbe– microbe, microbe–plant interrelationships. Nevertheless, the typical composition of the microbial community is used to derive conclusions on soil fertility and plant health. Without its microbial content, soil would be the simple anchoring matrix for plants. Even the nutrient supply grinds to a halt if replenishment of nutrients by microbial activity would not continue. Biological parameters seem to be the more reliable indicators to follow the progress of soil treatment (Yakovchenko et al. 1996). There are at least two reasons why evaluation of the microflora is the pivot of soil status monitoring: (1) Microbial communities respond fast to environmental changes and possess a high potential for adaptation. (2) Microbial communities act as bottleneck of metals on the their way from the soil matrix into the plant biomass. Thereby the microbial cell ties the abiotic soil sphere onto higher organisms (Hargreaves et al. 2003).

Both, autochthonous microorganisms of the remediation site as well as allochthonous strains that originate from nonmetal contaminated habitats can be adopted for a remediation treatment. It was found that several isolates of actinobacteria originating from undisturbed nonmetalliferrous soils can exhibit the same remarkable resistance patterns towards a range of heavy metals such as strains derived from mining areas (Haferburg et al. 2009). The more interesting question is if isolates that follow r-selection or K-selection are more likely to support the remediation progress. r-strategists (in which "r" stands for growth rate) or synonymously zymogenous bacteria exhibit a high growth rate and occur rather in unstable environments disturbed only for a short period of time. (Micro) organisms of the r-strategy are characterized by occupying a small ecological niche due to a highly specialized metabolism. They are less well adapted to cope successfully with a broad range of widely varying soil factors such as, e.g., carbon content, nitrogen source, pH or temperature. Zymogenous bacteria have been reported to be very efficient in bioremediation of land contaminated with hydrocarbons (Beškoski et al. 2011). Autochthonous microorganisms, in contrast, are thought to be the better candidate for application in a remediation strategy to immobilize or extract metals from soil. The most important criterion for their selection seems adaptability to the occurring soil conditions. Metal resistance and adaptation to plant promoting life in the rhizosphere are essential for phytoremediation. For the characterization of well-adapted organisms the term K-selection has been coined.[2] In terms of nutrition, they are considered as oligotrophic with a broad range of accepted carbon and energy sources. It has been found that populations of K-strategists are more likely to be stable in metal contaminated soils (Kozdrój 1995). The widespread confusion in the pairs of concepts "autochthonous/zymogenous", "r- and K-selection" and "oligotrophic/copiotrophic" is nicely dissolved by Langer et al. (2004).

Turning away from ecologically important concepts in a bioremediation approach one further application-related aspect of biogeoscience should be addressed. Biogeoscience can also be seen as the science of the soil/cell interphase. The investigation of microbe–mineral interaction gives an insight into metal (im) mobilization in rhizosphere habitats and oxic/anoxic interphases of soils. Elements such as As, Cr, Mn, Se and U can be used by anaerobic organisms as terminal electron acceptor (Borch et al. 2010). Some of these elements are toxic to the majority of soil microorganisms and plants. The toxicity always depends on mobility and availability of the metal. As, for example, in the case of the soluble and highly poisonous U^{6+} some microorganisms can use it as electron acceptor. The reduced form (U^{4+}) is insoluble, precipitates and is barely bioavailable. Those

[2] K stands for the carrying capacity in the Verhulst equation to calculate population dynamics.

microorganisms using U to gain energy, for example, members of the genera *Desulfosporosinus* and *Clostridium* should possess resistance mechanisms to avoid intoxication. In general, U precipitation due to microbial energy metabolism leads to a shift of the U-rich liquid phase to the solid (Wall and Krumholz 2006). The decrease of metal-rich ground waters is related to an increase of Uranium in the sediment.

Besides oxidation/reduction reactions there are further routes for the immobilization of metal contaminants using physiology and biochemistry of soil microorganisms. Sorption benefits from the metal binding capacity of biomass. The cell envelope of the microbial cell is commonly negatively charged, also depending on the pH, and delivers binding properties to metal cations. In a so-called biocurtain, growing microbial biomass is applied as a filter in the migration way of the contaminant. This filter supports mitigation of the metal expansion. The metal load in the water path could be both concentrated and located for a subsequent removal. Microbial geotechnology (also called biogeotechnology) is the field of research within the biogeosciences which deals with the adoption and optimization of remediation strategies such as bioclogging (sealing of soil pores by microbial means) and biocementation (soil particle consolidation and precipitation due to microbial metabolism). These microbiological soil installations could improve the mechanical properties of soil in situ. Thus, they can replace the more energy demanding and eco-unfriendly mechanical and chemical methods (Ivanov and Chu 2008).

In contrast to microbial soil technologies for metal immobilization, microbial metabolites are investigated on their influence on metal mobility in soil. The release of microbial chelators can lead to a significant metal solubilization. Siderophores are secondary metabolites produced by many very different microorganisms under iron depleting conditions for a specific sequestration and uptake of ferric iron. Structurally they comprise at least three groups: hydroxamates, catecholates and carboxylates. The biosynthesis of these compounds is considered as part of the secondary metabolism, which means they are not indispensable to life. Nevertheless, iron deficiency stimulates synthesis and release of siderophores in many bacteria and also fungi. It has been shown that some siderophores can sequester other metals than Fe^{3+} as well. If for instance the soil actinomycete *Streptomyces pilosus*, the yeast *Rhodotorula mucilaginosa* or the fungus *Ustilago sphaerogena* release their siderophores, lead becomes mobilized from a mineral absorbent (Dubbin and Ander 2003). Siderophores of streptomycetes have also been shown to bind Ni and to promote plant growth in Cd contaminated soil (Dimkpa et al. 2008, 2009). The beneficial role siderophores can play in plant growth promotion and protection from metal toxicity in combination with enhanced metal accumulation during phytoremediation has been shown in a few studies. Nevertheless, an important but currently still unanswered question that remains is how inoculated siderophore producing microorganisms colonize plant roots and survive at metal contaminated sites. The detailed mechanisms by which these bacteria contribute to plant metal acquisition is yet unsolved (Rajkumar et al. 2010). Another group of soil microorganisms fulfills an ecological function opposed to mobilizing reactions. These bacteria and fungi are prone to degrade organic compounds that bind metals and keep them soluble; unbound metals often precipitate. Degradation of chelators that keep metals in solution has been shown, for example, for phytosiderophores (Zhang 1993). Gathering together all these often antagonizing biological and geological processes in soil biogeosciences could be described as an attempt to develop an overall understanding of matter cycle at different scales from micro-habitat, soil layer, cropland to globally influenced landscapes.

2.5 Plant–Microbe Interactions

The best illustration for the perfectly tuned interplay of plant and microbe is the mucilage layer of the root hair. 20–40% of the photosynthetically derived assimilates leave the plant as exudates through the root. This corresponds approximately to a release of 5–21% of the fixed carbon (Marschner 1995). Root exudates provide microorganisms with essential nutrients. Hiltner coined the term "rhizosphere effect" in 1904 therewith expressing the phenomenon of a largely increased microbial colonization on or in close vicinity of the root hair. Cell numbers can be 10–1,000 times higher compared to bulk soil (Hiltner 1904; Lugtenberg and Kamilova 2009). Between 60 and 80% of the exudates are taken up by the microflora of the rhizosphere as carbon and nitrogen source. The microflora in turn supports and protects plants by a huge number of growth promoting effects. This sum of interactions between microbial cells and plant roots in a hostile environment such as metal contaminated soils of mining areas may demonstrate the complexity of biogeosciences.

There seems to be a strict separation in phytostabilization and phytoextraction for most of the publications dealing with bioremediation of metal stressed soils. Chaney first described the potential that lies in the use of metal accumulating plants for the extraction of metals from agrable soils (Chaney 1983). It is an average calculation of 2–5% of accumulated toxic elements in the dry weight of plants exposed to the contaminated soil to make the plant suitable for a phytoremediation approach (Brown 1995). However, there is still ongoing discussion on the reasonable use of criteria to select the plants most suitable for phytoremediation.

Why does the combination of microbially supported phytostabilization and phytoextraction appear such appealing to some soil scientists? Soils of metal rich postmining areas are commonly scarce in nutrients, poor in carbon and infertile concerning basic plant growth conditions. Thus, before starting with a large-scale phytoextraction the basis of vegetation needs to be established. Phytostabilization means progressing soil formation and improvement of plant growth conditions while metal seepage with the water path is prevented. Phytostabilization mechanisms comprise precipitation of metals by bacterial and root surfaces, precipitation of metals by bacterial and root exudates, bacterial uptake and sequestration of metals, and root uptake of metals (Mendez and Maier 2008). The extrusion of immobilized metals can be prevented during nonvegetation period in well-rooted soils. Plants for phytostabilization are not chosen according to the potential of metal accumulation but rather referring to fast and huge biomass production. The well-developed plant cover results in minimization of the metal efflux thus being comparable to plants used for rhizofiltration. Pedogenesis directly depends on the functional interplay of soil microorganisms and vegetation. Subsequently, if stabilization has reached an advanced stage, a modified set of plants (accumulator plants) can be used to extract metals from soil. Plant growth promoting bacteria and metal mobilizing microorganisms are essential to accomplish metal removal from the ground. For a functional metal extraction it is evidently the "rhizo-environment" which needs to be developed. The ideal phytoremediation should combine phytostabilization and phytoextraction even if this concept seems conflictive since the first corresponds to accumulation and the latter rather to exclusion.

The governmental organization "Environment Canada" is involved in many programs for environmental protection and has developed the database PHYTOREM to collect knowledge on plants with capabilities to accumulate or hyperaccumulate metals. Until now the database comprises 775 plants from 76 families (McIntyre 2003). Only for a fractional part of these plants the synergistic interplay with PGPB in a potentially powerful bioremediation operation was studied. But in various reports it was shown how and to what extent PGPB can enhance the capacity of plants to extract metals from soils and sediments. Studies on nickel uptake by different *Brassica* species showed an increase in the mobile metal fraction and metal accumulation in the plant without yield loss after treatment with PGPB (Ma et al. 2009, Rajkumar and Freitas 2008a). Comparable observations were reported for the uptake and accumulation of copper and zinc by *Ricinus communis* after soil inoculation with different positively plant growth promoting *Pseudomonas* strains (Rajkumar and Freitas 2008b). Often the accumulated amount of metals reaches the multiple of the non-PGPB treated plant. This small selection of examples should illustrate: (1) The interaction of PGPB and plant can be very effective for metal removal from the ground, (2) The dependency of the extraction yield from the chosen plant and applied microorganisms and (3) There is no overall uptake for all metals present in the contaminated soil.

The interest in phytoremediation as alternative remediation technique grew tremendously during the last two decades of research on phytoextraction. Nevertheless, obviously much more is known on the plant part in the microbe–plant interaction during plant growth in contaminated land. Databases such as PHYTOREM deliver an indispensable basis for a systematic research on the potential microbial boost of metal mobilization and removal at concurrent support of plant nutrition and stress reduction. The plant physiological potential of metal extraction and accumulation in terms of element specificity and concentration are not only species dependent as has been shown by Lai and colleagues in a screening of more than 30 plants but also very much ecotype related and even highly variable among different clones derived from the same mother plant (Lai et al. 2010; Lombi et al. 2000; Nehnevajova et al. 2007).

Phytoremediation does not only benefit from the application of plant growth promoting bacteria the usage of mycorrhiza in the remediation of metal contaminated land is also valuable. Mycorrhiza plays a crucial role in plant nutrition especially phosphorous supply and protection from metal toxification. Some mycorrhiza fungi such as *Gigaspora margarita* help to keep the metal concentration in the above ground plant material low (Andrade et al. 2010). This would support the erection of a pioneering vegetation cover in hostile postmining land. Other fungal partners in this symbiosis seem to "feed" the plant associate with an extra portion of the metal load of the soil. Some strains of several *Glomus* species increase the metal content in plant tissue (Citterio et al. 2005).

2.6 Characteristics of Plant Growth Promotion in the Context of Biogeosciences

Plant growth is often limited by phosphate availability. With up to 80% of the total soil phosphorous the main resources are organic (Condron et al. 1985). Nevertheless, for accessing both the organic and the inorganic phosphorous pool microbial mineralization is a prerequisite. After the continuous application of inorganic P fertilizer for decades, many soils contain comparably high but inaccessible P sources. P immobilization in soils or washout into the waterpath is a rapid process resulting in both an ecological threat and an increase in fertilizer expenses. P mobilizing bacteria seem to afford the necessary alternative for the remobilization of the unavailable phosphorous pool of soils. The microbiological P fertilization becomes an important issue especially when realizing resources of rock phosphate which delivers the basis for P fertilization are declining. The global reserves would be exhausted within the next 80–250 years after estimation based on the current extraction rate (Smil 2000). Globally, there is no complete cycling of phosphorous as for sulfur or nitrogen.

Nitrogen fixing bacteria are other important associates in biofertilization. Costs for nitrogen fertilizers can be lowered by exploiting the potential of bacteria such as *Azotobacter*, *Azospirillum* and *Rhizobium* to fix nitrogen from air. The main advantage of this "biological nitrogen fertilization" in the rhizosphere is the supply of nitrogen sources at exactly the right place. Sixty-five percentage of the synthetically derived mineral nitrogen is lost from the plant–soil system by degassing and leaching (Bhattacharjee et al. 2008). Thus, application of nitrogen fixing bacteria helps to reduce costs for fertilizers and keeps the N supply of plants at a balanced level.

2.7 Tool Box Strategy: Optimizing the Microbiologically Supported Remediation of Heavy Metal Contaminated Soils

Fifty-two million hectares land in the European Union was estimated to be affected by soil degradation (Peuke and Rennenberg 2005). This is more than 16% of the total land area. The costs to remediate contaminated sites within the EU have been calculated to be between €59 and €109 billion (ib.). These numbers emphasize the demand for soil preserving clean-up technologies. The development of tool boxes could be a big step forward in the development of powerful phytoremediation strategies.

The tool box in bioremediation operates similar to the letter case in a printing press. The basic principle is the combination of three collections: (1) site description of postmining areas with a major consideration of pedologic characteristics, (2) collection of metal resistant and plant growth promoting microbial strains, including their metal resistance patterns and (3) seed collection of metal accumulating or excluding plants. Microorganisms as well as plants are adapted to their specific habitats or niches. This adaptation together with particular traits of the "metal metabolism" of microorganisms is exploited to shift metalliferous habitats into nonmetal-determined ones. By the strength of the microbe–plant interplay postmining land could be changed to fertile ground (again) which is attractive to agriculture or any other soil use regime. The selection of plants depends on the factors climate, soil type, extent and kind of contamination. The plants in turn determine (in conjunction with above-mentioned factors) the choice of microorganisms. The innovation is the usage of the many possibilities microorganisms possess to solubilize/mobilize metals for phytoextraction and contrary to prevent mobilization for the phytostabilization approach. The tool box system would offer a huge choice of combinations from a pool of accumulator or excluder plants and plant growth promoting or metal mobilizing microorganisms for the realization of a remediation plan. Chapter 11 within this volume reflects on the development and usage of such tool box systems. The basis for any screening of soil strain collections on PGP properties is their metal tolerance or resistance. This characteristic is considered to be most important to keep the ecological fitness after reintroduction for remediation purposes. To cover a broad spectrum of PGP properties, the screening should comprise assays on (1) nitrogen fixation, (2) phosphate mobilization, (3) siderophore release and (4) phytohormone production. The cumulative effect of the various PGP traits which almost every PGPB strain exhibits makes the evaluation of the contribution of single biochemical properties difficult. Nevertheless, the above-mentioned BIOLOG system provides an option for biochemical activity estimation in pure and mixed cultures. For the caskets of the tool box to be filled with plants active in the remediation process one should consider (1) genera and species known and characterized as hyperaccumulators or excluders, (2) ecotype adaptation of single species, (3) metal enrichment specificity, (4) transfer factors and (5) biomass production.

From an ecological point of view, it is not very clear if the application of a strain multitude as initial bioinoculum leads to a more successful and persistent colonization of the plant root compared to a minimal bioaugmentation consisting of only few strains with the same PGP properties. Nevertheless, following results of Kloepper and Schroth it seems advantageous to inoculate the rhizosphere with a high number of PGPB strains since only about 2–5% of the reintroduced rhizobacteria survive and exert beneficial activities when soil already contains a particularly adapted microflora which easily outcompetes new intruders (Kloepper and Schroth 1978).

2.8 Conclusions

Biogeosciences delivers the fundamental for the analysis of both natural processes and anthropogenically induced imbalances in the environment from the vantage point of the bio-geo interphase. Examples for the interphase of inanimate matter and living, metabolically active organisms are microbially colonized mineral surfaces (e.g., pyrite oxidation by *Acidithiobacillus ferrooxidans*), root hairs in sediment (e.g., rhizofiltration of arsenic by *Phragmites* rhizofiltration) or mineral formation due to CO_2 reduction (e.g., calcareous shale of coralline algae). The main focus in biogeosciences is laid on the cycle of matter. When different elements are investigated in the cycle of matter microorganisms as destruents, mineralizers, sorbents and mobilizers are of special interest with respect to remediation. Most of the reactions within the cycle can be performed by bacteria and fungi but no other organism. Especially for the state-of-the-art means of environmental protection, the metabolism of microorganisms is of great importance. The increase of worldwide occurring environmental disasters is likely to be one reason of the ascent of the biogeosciences we observe.

Biogeosciences in metal contaminated soil gives a paramount example for the interdisciplinary of research. The knowledge gained thanks to the pronounced interdisciplinary was implemented into many bioremediation projects for the decontamination of soils. Phytoremediation nicely demonstrates the exploitable interplay of soil, microorganism and plant for metal removal. The development of tool box systems indicates a new and much stronger habitat-oriented access in remediation practices. The tripartite remediation tool box system consists of (1) A hydrogeochemical site description, (2) A set of metal excluder or accumulator plants and (3) Metal-resistant and well-adapted plant growth promoting bacteria. The combination of plants and microorganisms is thought to result in an optimized phytoremediation of postmining sites greatly varying in soil parameters. Therewith the tool box strategy is an application of the accumulated knowledge from biogeosciences and could ideally be used for soil resource protection.

Acknowledgement We gratefully acknowledge support from EU (UMBRELLA, 7th framework program) and of JSMC.

References

Alisi C, Musella R, Tasso F, Ubaldi C, Manzo S, Cremisini C, Sprocati AR (2009) Bioremediation of diesel oil in a co-contaminated soil by bioaugmentation with a microbial formula tailored with native strains selected for heavy metals resistance. Sci Total Environ 407:3024–3032

Alkorta I, Epelde L, Mijangos I, Amezaga I, Garbisu C (2006) Bioluminescent bacterial biosensors for the assessment of metal toxicity and bioavailability in soils. Rev Environ Health 21: 139–152

Andrade SA, Silveira AP, Mazzafera P (2010) Arbuscular mycorrhiza alters metal uptake and the physiological response of *Coffea arabica* seedlings to increasing Zn and Cu concentrations in soil. Sci Total Environ 408:5381–5391

Baas Becking LGM (1934) Geobiologie of inleiding tot de milieukunde. Van Stockum WP and Zoon, The Hague, the Netherlands (in Dutch)

Babalola OO (2010) Beneficial bacteria of agricultural importance. Biotechnol Lett 32:1559–1570

Barriuso J, Solano BR, Gutiérrez Mañero FJ (2008) Protection against pathogen and salt stress by four plant growth-promoting rhizobacteria isolated from *Pinus* sp. on *Arabidopsis thaliana*. Phytopathology 98:666–672

Bednarek P, Kwon C, Schulze-Lefert P (2010) Not a peripheral issue: secretion in plant-microbe interactions. Curr Opin Plant Biol 13:378–387

Beškoski VP, Gojgić-Cvijović G, Milić J, Ilić M, Miletić S, Solević T, Vrvić MM (2011) Ex situ bioremediation of a soil contaminated by mazut (heavy residual fuel oil) – a field experiment. Chemosphere 83:34–40

Bhattacharjee RB, Singh A, Mukhopadhyay SN (2008) Use of nitrogen-fixing bacteria as biofertiliser for non-legumes: prospects and challenges. Appl Microbiol Biotechnol 80:199–209

Bianco C, Defez R (2009) *Medicago truncatula* improves salt tolerance when nodulated by an indole-3-acetic acid-overproducing *Sinorhizobium meliloti* strain. J Exp Bot 60:3097–3107

Borch T, Kretzschmar R, Kappler A, Cappellen PV, Ginder-Vogel M, Voegelin A, Campbell K (2010) Biogeochemical redox processes and their impact on contaminant dynamics. Environ Sci Technol 44:15–23

Brown KS (1995) The green clean: the emerging field of phytoremediation takes root. Bioscience 45:579–582

Chandler DP, Kukhtin A, Mokhiber R, Knickerbocker C, Ogles D, Rudy G, Golova J, Long P, Peacock A (2010) Monitoring microbial community structure and dynamics during in situ U(VI) bioremediation with a field-portable microarray analysis system. Environ Sci Technol 44:5516–5522

Chaney RL (1983) Plant uptake of inorganic waste constituents. In: Parr JF, Marsh PD, Kla JM (eds) Land treatment of hazadous wastes. Noyes Data Corporation, Park Ridge, NJ, pp 50–76

Citterio S, Prato N, Fumagalli P, Aina R, Massa N, Santagostino A, Sgorbati S, Berta G (2005) The arbuscular mycorrhizal fungus *Glomus mosseae* induces growth and metal accumulation changes in *Cannabis sativa* L. Chemosphere 59:21–29

Condron LM, Goh KM, Newman RH (1985) Nature and distribution of soil phosphorus as revealed by a sequential extraction method followed by 31P nuclear magnetic resonance analysis. Soil Sci 36:199–207

de Freitas JR (2002) Biofertilizers. In: Pimentel D (ed) Encyclopedia of pest management. CRC, 56, p 54

de Wit R, Bouvier T (2006) Everything is everywhere, but, the environment selects; what did Baas Becking and Beijerinck really say? Environ Microbiol 8:755–758

Dey R, Pal KK, Bhatt DM, Chauhan SM (2004) Growth promotion and yield enhancement of peanut (*Arachis hypogaea* L.) by application of plant growth-promoting rhizobacteria. Microbiol Res 159:371–394

Dimkpa C, Svatos A, Merten D, Büchel G, Kothe E (2008) Hydroxamate siderophores produced by *Streptomyces acidiscabies* E13 bind nickel and promote growth in cowpea (*Vigna unguiculata* L.) under nickel stress. Can J Microbiol 54:163–172

Dimkpa CO, Merten D, Svatos A, Büchel G, Kothe E (2009) Siderophores mediate reduced and increased uptake of cadmium by *Streptomyces tendae* F4 and sunflower (*Helianthus annuus*), respectively. J Appl Microbiol 107:1687–1696

Döös BR (2002) Population growth and loss of arable land. Global Environ Change 12:303–311

Dubbin WE, Ander EL (2003) Influence of microbial hydroxamate siderophores on Pb(II) desorption from α-FeOOH. Appl Geochem 18:1751–1756

Dubey SK, Tripathi AK, Upadhyay SN (2006) Exploration of soil bacterial communities for their potential as bioresource. Bioresour Technol 97:2217–2224

Dudka S, Adriano DC (1997) Environmental impacts of metal ore mining and processing: a review. J Environ Qual 26:590–602
Faisal M, Hasnain S (2005) Bacterial Cr (VI) reduction concurrently improves sunflower (*Helianthus annuus* L.) growth. Biotechnol Lett 27:943–947
Furrer G, Phillips BL, Ulrich KU, Pöthig R, Casey WH (2002) The origin of aluminum flocs in polluted streams. Science 297:2245–2247
Ganesan V (2008) Rhizoremediation of cadmium soil using a cadmium-resistant plant growth-promoting rhizopseudomonad. Curr Microbiol 56:403–407
García de Salamone IE, Hynes RK, Nelson LM (2005) Role of cytokinins in plant growth promotion by rhizosphere bacteria. In: Siddiqui ZA (ed) PGPR: biocontrol and biofertilization, 2nd edn. Springer, Dordrecht, pp 173–195
Gutierrez-Mañero FJ, Ramos-Solano B, Probanza A, Mehouachi J, Tadeo FR, Talon M (2001) The plant growth-promoting rhizobacteria *Bacillus pumilus* and *Bacillus licheniformis* produce high amounts of physiologically active gibberellins. Physiol Plant 111:1–7
Haferburg G, Groth I, Möllmann U, Kothe E, Sattler I (2009) Arousing sleeping genes: shifts in secondary metabolism of metal tolerant actinobacteria under conditions of heavy metal stress. Biometals 22:225–234
Haferburg G, Kothe E (2010) Metallomics: lessons for metalliferous soil remediation. Appl Microbiol Biotechnol 87:1271–1280
Haferburg G, Merten D, Büchel G, Kothe E (2007) Biosorption of metal and salt tolerant microbial isolates from a former uranium mining area. Their impact on changes in rare earth element patterns in acid mine drainage. J Basic Microbiol 47:474–484
Han J, Sun L, Dong X, Cai Z, Sun X, Yang H, Wang Y, Song W (2005) Characterization of a novel plant growth-promoting bacteria strain *Delftia tsuruhatensis* HR4 both as a diazotroph and a potential biocontrol agent against various plant pathogens. Syst Appl Microbiol 28:66–76
Hargreaves PR, Brookes PC, Ross GJS, Poulton PR (2003) Evaluating soil microbial carbon as indicator of long-term environmental change. Soil Biol Biochem 35:401–407
Hiltner L (1904) Über neuere Erfahrungen und Probleme auf dem Gebiete der Bodenbakteriologie unter besonderer Berücksichtigung der Gründüngung und Brache. Arbeiten der Deutschen Landwirtschaftlichen Gesellschaft 98:59–78 (in German)
Idriss EE, Makarewicz O, Farouk A, Rosner K, Greiner R, Bochow H, Richter T, Borris R (2002) Extracellular phytase activity of *Bacillus amyloliquefaciens* FZB45 contributes to its plant-growth-promoting effect. Microbiology 148:2097–2109
Ivanov V, Chu J (2008) Applications of microorganisms to geotechnical engineering for bioclogging and biocementation of soil in situ. Rev Environ Sci Biotechnol 7:139–153
Kloepper JW, Schroth MN (1978) Plant growth-promoting rhizobacteria on radishes. In: Proceedings of the 4th international conference on plant pathogenic bacteria, vol 2. Station de Pathologie Végétale et de Phytobactériologie, INRA, Angers, France, pp 879–882
Kozdrój J (1995) Microbial responses to single or successive soil contamination with Cd or Cu. Soil Biol Biochem 11:1459–1465
Laanbroek HJ (2010) Methane emission from natural wetlands: interplay between emergent macrophytes and soil microbial processes. A mini-review. Ann Bot 105:141–153
Lai HY, Juang KW, Chen ZS (2010) Large-area experiment on uptake of metals by twelve plants growing in soils contaminated with multiple metals. Int J Phytoremediation12:785–797
Langer U, Böhme L, Böhme F (2004) Classification of soil microorganisms based on growth properties: a critical view of some commonly used terms. J Plant Nutr Soil Sci 167:267–269
Lombi E, Zhao FJ, Dunham SJ, McGrath SP (2000) Cadmium accumulation in populations of *Thlaspi caerulescens* and *Thlaspi goesingense*. New Phytol 145:11–20
Lugtenberg B, Kamilova F (2009) Plant-growth-promoting rhizobacteria. Annu Rev Microbiol 63:541–556
Ma Y, Rajkumar M, Freitas H (2009) Improvement of plant growth and nickel uptake by nickel resistant-plant-growth promoting bacteria. J Hazard Mater 166:1154–1161

Margesin R, Płaza GA, Kasenbacher S (2011) Characterization of bacterial communities at heavy-metal-contaminated sites. Chemosphere 82:1583–1588

Marschner H (1995) Mineral nutrition of higher plants, 2nd edn. Academic, London

McIntyre T (2003) Phytoremediation of heavy metals from soils. Adv Biochem Eng Biotechnol 78:97–123

Mehnaz S, Mirza MS, Haurat J, Bally R, Normand P, Bano A, Malik KA (2001) Isolation and 16S rRNA sequence analysis of the beneficial bacteria from the rhizosphere of rice. Can J Microbiol 47:110–117

Miljević N, Golobocanin D (2007) Potential use of environmental isotopes in pollutant migration studies. Arh Hig Rada Toksikol 58:251–262

Miller JM, Rhoden DL (1991) Preliminary evaluation of Biolog, a carbon source utilization method for bacterial identification. J Clin Microbiol 29:1143–1147

Miraglia M, Marvin HJ, Kleter GA, Battilani P, Brera C, Coni E, Cubadda F, Croci L, De Santis B, Dekkers S, Filippi L, Hutjes RW, Noordam MY, Pisante M, Piva G, Prandini A, Toti L, van den Born GJ, Vespermann A (2009) Climate change and food safety: an emerging issue with special focus on Europe. Food Chem Toxicol 47:1009–1021

Mendez MO, Maier RM (2008) Phytostabilization of mine tailings in arid and semiarid environments-an emerging remediation technology. Environ Health Perspect 116:278–283

Nehnevajova E, Herzig R, Federer G, Erismann KH, Schwitzguébel JP (2007) Chemical mutagenesis – a promising technique to increase metal concentration and extraction in sunflowers. Int J Phytoremediation 9:149–165

Pathak A, Dastidar MG, Sreekrishnan TR (2009) Bioleaching of heavy metals from sewage sludge: a review. J Environ Manage 90:2343–2353

Peuke AD, Rennenberg H (2005) Phytoremediation. EMBO Rep 6:497–501

Prasad MN, Freitas H, Fraenzle S, Wuenschmann S, Markert B (2010) Knowledge explosion in phytotechnologies for environmental solutions. Environ Pollut 158:18–23

Rajkumar M, Freitas H (2008a) Effects of inoculation of plant-growth promoting bacteria on Ni uptake by Indian mustard. Bioresour Technol 99:3491–3498

Rajkumar M, Freitas H (2008b) Influence of metal resistant-plant growth-promoting bacteria on the growth of *Ricinus communis* in soil contaminated with heavy metals. Chemosphere 71:834–842

Rajkumar M, Ae N, Prasad MN, Freitas H (2010) Potential of siderophore-producing bacteria for improving heavy metal phytoextraction. Trends Biotechnol 28:142–149

Scheffer F, Schachtschabel P (2010) Bodenmikroflora. In: Scheffer F, Schachtschabel (eds) Lehrbuch der Bodenkunde. Spektrum Akademischer Verlag, Heidelberg, pp 84–99 (in German)

Sheng XF, Xia JJ (2006) Improvement of rape (*Brassica napus*) plant growth and cadmium uptake by cadmium-resistant bacteria. Chemosphere 64:1036–1042

Smil V (2000) Phosphorous in the environment: natural flows and human interferences. Annu Rev Energ Environ 25:53–88

Sridevi M, Kumar MG, Mallaiah KV (2008) Production of catechol-type of siderophores by *Rhizobium* sp. isolated from stem nodules of *Sesbania procumbens* (Roxb.) W and A. Res J Microbiol 3:282–287

Stabnikova O, Wang J-Y, Ivanov V (2010) Value-added biotechnological products from organic wastes. In: Wang LK, Ivanov V, Tay J-H (eds) Environ Biotechnol, 1st edn. Springer, Berlin, pp 343–394

Staley JT, Konopka A (1985) Measurements of in situ activities of nonphotosynthetic microorganisms in aquatic and terrestrial habitats. Annu Rev Microbiol 39:321–346

Steele DB, Stowers MD (1991) Techniques for selection of industrially important microorganisms. Annu Rev Microbiol 45:89–106

Wall JD, Krumholz LR (2006) Uranium reduction. Annu Rev Microbiol 60:149–166

Warhurst A (2002) Mining, mineral processing, and extractive metallurgy: an overview of the technologies and their impact on the physical environment. In: Warhurst A, Noronha L (eds) Environmental policy in mining: corporate strategy and planning for closure. CRC, Boca Raton, FL

Wiatrowski HA, Barkay T (2005) Monitoring of microbial metal transformations in the environment. Curr Opin Biotechnol 16:261–268

Wu SC, Cheung KC, Luo YM, Wong MH (2006) Effects of inoculation of plant growth promoting rhizobacteria on metal uptake by *Brassica juncea*. Environ Pollut 140:124–135

Yakovchenko V, Sikora LJ, Kaufman DD (1996) A biologically based indicator of soil quality. Biol Fertil Soils 21:245–251

Zaidi S, Usmani S, Singh BR, Musarrat J (2006) Significance of *Bacillus subtilis* strain SJ 101 as a bioinoculant for concurrent plant growth promotion and nickel accumulation in *Brassica juncea*. Chemosphere 64:991–997

Zhang FS (1993) Mobilisation of iron and manganese by plant-borne and synthetic metal chelators. Plant Soil 155–156:111–114

Chapter 3
The Role of Mineralogy and Geochemistry in Hazard Potential Assessment of Mining Areas

D. Jianu, V. Iordache, B. Soare, L. Petrescu, A. Neagoe, C. Iacob, and R. Orza

3.1 Introduction

Minerals are the original source and the most important pool of metals on Earth. Having a structured image on the role of mineralogy in the mobility of metals is crucial for understanding their biogeochemistry and designing the management of contaminated areas.

We define "mining areas" as territories with scales from local (contaminated sites) to regional size (contaminated basins or regions; see Iordache et al. 2012 for details about the span of scales and associated processes). At local scale, mining areas include primary and secondary natural and anthropic sources of metals. In a comprehensive review about the element distribution during primary metal production, Broadhurst et al. (2007) describe the types of solid waste resulting from mining/ore extraction, hydrometallurgical extraction, and pyrometallurgical extraction in terms of generic characteristics and process-related factors influencing the elements distribution: waste rock (mining dumps), concentration tailings, vat leach residues, effluent treatment residues, metal recovery slimes/sludges, smelter slags, smelter flue dusts, and smelter refining effluent treatment residues and slimes. In addition to the sources associated with primary metal production, it is not

The contributions of D. Jianu and V. Iordache to the writing of this chapter were equal.

D. Jianu (✉) • B. Soare • C. Iacob • R. Orza
Lythos Research Center, Faculty of Geology, University of Bucharest, Bd. Nicolae Balcescu Nr. 1 Sector 1, 010041 Bucharest, Romania
e-mail: denisa0301@yahoo.com

V. Iordache • A. Neagoe
Research Centre for Ecological Services (CESEC), Faculty of Biology, University of Bucharest, Bucharest, Romania

L. Petrescu
Laboratory of Geochemistry, Faculty of Geology, University of Bucharest, Bucharest, Romania

E. Kothe and A. Varma (eds.), *Bio-Geo Interactions in Metal-Contaminated Soils*, Soil Biology 31, DOI 10.1007/978-3-642-23327-2_3,

uncommon to locate secondary processing industries, with their particular solid wastes in mining regions. The dimension of the waste particles range from <μm (smelter dusts) to cm scale (mining dumps) and the surface of the contaminated sites associated with them range from less than a hectare (mining dumps) to tenths of km^2 (soil contaminated by smelters). The lifetime of the metal containing mineral particles is highly variable and environmental dependent. But in many cases in the surface atmospheric and hydrological conditions, it is within the scale window relevant for short-term (years) and long-term (decades) management. The enrichment factor of elements' concentrations in the solid wastes compared to the original mineral varies largely, although for many cases in tailings is up to hundreds of times larger (Broadhurst et al. 2007), but in all cases the reactive surface relevant for transfer to water phases and air is much larger in the solid wastes than in the original minerals. For that reason, the management of contamination in mining areas is dominated by the problem of sources associated with industrial activities.

The implementation of environmental policies for the management of contaminated areas is a complex problem for both scientific and socioeconomic reasons and requires scientifically based decision support (Iordache 2009; Rodrigues et al. 2009). Risk assessment is part of the decision support system for the management of contaminated sites (Carlon et al. 2004). The hazard of contaminated areas is defined in the context of risk assessment (Table 3.1). The key points from Table 3.1 are: (1) hazard is associated with the mobility analyses of metals, (2) because mobility covers many scales, the hazard is at multiscales; (3) exposure refers to the availability of metals for the target system; (4) risk is associated with the potential interception by the target system (about the coupling between mobility analyses and receptors analyses); (5) elements relevant for exposure analysis can result from mobility analysis because the mobility of metals at large scale and in food-chains depends on organisms; (6) the control of the hazard is the control of the mobility of metals at all scales. In the current risk assessment

Table 3.1 Position of hazard assessment in the risk assessment procedure (adapted from DEFRA 2002; Carlon et al. 2008; Gay and Korre 2006, 2009)

Nr	Step	Content
1	Hazard assessment	Characterization of the concentrations (total and fractions different mobility) of contaminants and of their distribution in space [*concentration in abiotic or biotic compartment, mg* kg^{-1}]
2	Exposure assessment	Correlation between the spatial distribution of contaminants concentrations, the spatial distribution of potential receptors, and the spatial distribution of control variables of metals mobility and intake [*concentration available for intake, mg* kg^{-1}]
3	Risk characterization	Estimation and spatial distribution (mapping) of risk indicators, e.g., intake of the amount of a contaminant in the population of potential receptors [*ratio between flux and receptor mass, mg* kg^{-1} *BW* d^{-1}]
4	Uncertainty analyses	Characterization of the spatial distribution of the overall uncertainty of the variables used in steps 1–3 in order to identify the areas with high risk and high uncertainty (deserving for this reason extra investigations) [*dimensionless*]

view, only the local scale (contamination site) is considered relevant, although there is ample evidence for the existence of hotspots at greater distance (see Iordache et al. 2012).

We differentiate between hazards with different timescales. Short-term hazard of a contaminated area exists when the stocks of metals are small and the intensity of metals' export is high. Management of the short-term hazard implies rapid and operational measures. Long-term hazard depends on the stock of metals and on the dynamics of internal and external conditions. A site with long-term hazard may also have a short-term one if the intensity of the exports is large. What is specific and usually not covered by the regulations about risk assessment is the case of a small hazard in the short term which may change if the time is long enough due to the large stock of metals. In such examples, the retention time of metals at the current export fluxes is a key variable in modeling the hazard on the long term and devising the management strategies of the contaminated areas. In a true contaminated basin, the local situations are of multiple types, thus requiring short-term measures organized under a long-term strategy in order to control the consequences of the hazard.

The short-term hazard of a contaminated area and its future hazards in different environmental scenarios depend on the stocks of metals, on the fluxes of outgoing elements, and on the retention time of the elements (ratio between stock and sum of fluxes). The types of fluxes directly relevant for this discussion are those driven by wind and by water. The hydrological fluxes can be manifested at surface (runoff) or underground (by groundwater). There are two variables influencing the intensity of the fluxes: the intensity of the carrier flux (hydrological, atmospheric), and the mobilization of metals by the carrier flux. Table 3.2 summarizes the hazard potential possibilities. The analyses of long-term hazard can relocate a contaminated site from one hazard situation to another because of changes in the intensity of the carrier flux or/and of the mobility of metals. Mineralogical variables can play a role in both of these potential long-term changes.

Assigning a hazard to a specific anthropogenic cause is an important issue, especially for the contaminated soils and sediment. Normalizing the concentrations of metals (e.g., by NASC in Hamilton 2000, reference concentrations in Gromet et al. 1984) is desirable, but should be region dependent. In some cases (e.g., serpentine soils) though, the natural background may be so high that the identification of contaminated areas would be difficult (Bonifacio et al. 2010).

In this chapter, we review the role of mineralogy in the mobility of metals from mining areas from a hazard assessment perspective. Although because of the small dimension of the particles and of the chemical and microbiological nature of the

Table 3.2 Matrix of situations for short-term hazard of a contaminated area

		Mobilization of metals by the carrier flux	
		Large	Small
Intensity of the carrier flux	Large	Situation H1: large hazard	Situation H2a: average hazard
	Small	Situation H2b: average hazard	Situation H3: small hazard

processes involved the scale specific to mineralogical role is very small, the discussion will be directed to the possible extent to aggregated roles at the contaminated site scale and to indirect roles at the scale of river systems ["direct" and "indirect" in the same sense introduced in Iordache et al. (2012) and used for organic carbon in Neagoe et al. (2012)]. After introducing several concepts about the stocks of potentially mobile metals, we review the role of mineralogy for the hazard situations introduced in Table 3.1 by type of carrier flux. We then illustrate the general considerations with a discussion about the specific hazard associated with tailing dams in Romania.

3.2 Stocks of Metals and Their Internal Heterogeneity

Key points in the evaluation of stocks are the correct estimation of the volumes of material and of the internal heterogeneity of metals distribution in the volume of interest. This internal heterogeneity is directly correlated with the mineralogy of the material. Geophysical characterization of waste stacks allows the development of structural and then hydrogeological models needed for the evaluation of metals export (part 4). The surface heterogeneity in terms of heavy metals distribution and availability is essential for devising appropriate phytostabilization methods (part 6).

Methodological literature for the management of contaminated sites is well developed. Campbell et al. (1999) review the use of geophysical methods for tailings/mine waste rocks and concludes that geoelectrical methods are most helpful. Commonly, the waste rock was found to be more conductive than surrounding areas as a result of high porosity, acidic and salty porewater and high concentrations of conductive minerals (sulfides, clays) (Campbell et al. 1999). However, electromagnetic methods fail in correct plume delineation in the case of preferential flows by geologic fractures and the presence of conductive clay lenses Campbell and Fitterman (2000). These authors (also Smith et al. 2000, 2002) suggest the most appropriate use of each geophysical method in mining areas: frequency domain electromagnetic for tracing acid mine drainage (AMD) plumes, direct current resistivity for identifying shallow (<10 m) water tables in waste piles and bottoms of shallow (<20 m) was piles, time-domain electromagnetic for deeper (10–30) water tables in mine piles, controlled source audiomagnetotellurics for deeper bottoms of mine waste piles, induced polarization for concentrations of sulfides in mine piles, ground-penetrating radar for monitoring plume remediation, magnetic methods for ferrous objects in mine piles, and seismic methods for tracing bottoms and edges of waste piles. For example, Poisson et al. (2009) used resistivity imaging (ERI) and ground-penetrating radar (GPR) on a mining dump and identified layers with different granularity and oxidization status. Iacob and Orza (2008) showed that integrated geoelectric and geomagnetic investigations on mine wastes gives good results for delineating the tailings in surface and in depth (Fig. 3.1). Using also ERI, electromagnetic conductivity, and GPR, Anterrieu et al. (2010) identified areas with different resistivity and shapes of anomalies

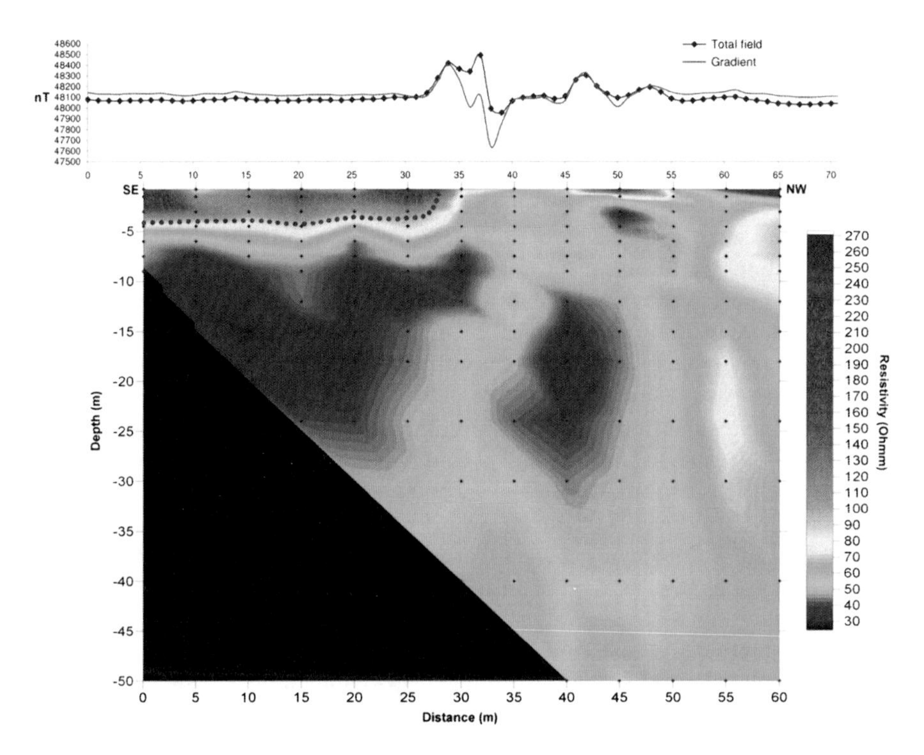

Fig. 3.1 Hanes mining dump, Romania – Depth and surface delineation of the tailing body using resistivity and geomagnetic (total field and gradient) surveys (Iacob and Orza 2008)

correlated with grain size, grain size distribution, minerals distribution and water content, with the overall effect of large acid mine drainage (AMD) generation, the most conductive parts of the pile. Iacob et al. (2009), Iacob (2011) showed that geophysical investigations can quantify valuable information regarding the acid-producing geochemical processes' status and distribution inside tailing ponds, creating premises for quantitative time modeling for acid leachings.

Additional information for mining areas comes from the observation that in many cases (although not usually in mining dumps and tailing dams) heavy metals are associated with minerals having strong magnetic properties because of their common genesis, weather of mining or industrial origin (see Hunt et al. 1995, for a review of the magnetic properties of rocks and minerals). Petrovsky et al. (2001) analyzed the magnetic properties of alluvial soil contaminated with ashes from a lead ore smelter and found strong correlations between magnetic minerals and the concentrations of Pb, Zn, and Cd. Hanesch and Scholger (2002) found that magnetic susceptibility anomalies in Austrian soils were correlated (in simple or logarithmic transformed form) either with geogenic anomalies (bedrock influences identified by similar anomalies in soil at depth and in surface) or with anthropogenic inputs of contaminants (as dispersed natural mineral and/or phases derived from industrial processes). Kukier et al. (2003) analyzed the mineralogy of magnetic ashes finding that magnetic fractions contained magnetite (Fe_3O_4), hematite (Fe_2O_3(III)), and some quartz (SiO_2) and mullite ($Al_8[(O,OH,F)|(Si,Al)O_4]_4$). The magnetic fractions had 10 times higher concentrations of Fe, and 2–4 times higher concentrations of Co, Ni, and Mn than nonmagnetic fly ashes (Kukier et al. 2003), and the concentrations in magnetic fly ashes were negatively correlated with elements associated with aluminosilicate matrix (Si, Al, K, Na). Jordanova et al. (2004) analyzed magnetic phases in the Danube River sediments and found a complex internal structure of the particles, including magnetite (Fe_3O_4) and to a lesser extent, oxides of other metals (including Cr, Ni, Zn, and Cu). Desenfant et al. (2004) demonstrated the association of magnetic anomalies with the distribution of lead, zinc, iron, and chromium in river sediments. The responsible particles were spherules of tenths of μm diameter and rich in iron oxides, in that cases derived probably from combustion processes (Desenfant and Petrovsky 2004). Recently, Lu et al. (2009) investigated the mineralogy and the leachability of magnetic fractions of fly ashes. They found these fractions relatively richer not only in Fe in particular, but also in Mn, Cr, Cu, Cd, and Pb, with leachability in magnetic compared to nonmagnetic fractions larger only in the case of Pb. The coupling between large distance atmospheric dispersion of ferromagnetic particles rich in heavy metals and geomorphological and vegetation dependent local runoff processes was also important in generating the distribution patterns of surface soil magnetic processes (Sangode et al. 2010). One can observe that while within mining dumps and tailing dams the magnetic methods may not be very useful because the heavy metals are in phases not associated with magnetic minerals, in areas resulted from dispersion of waste material in hydrosystems and also in areas contaminated by smelter atmospheric dispersion, the magnetic methods may be useful for extrapolating the concentrations of metals determined by geochemical methods. Such areas can cover large surfaces in mining-dominated catchments.

Another use of geophysical methods in our area of interest is for characterizing the depth of erodible layers. Usually, transport models of contaminated soil and particulate waste material work with an assumption of homogenous depth of contaminated soil over the catchment (Coulthard and Macklin 2003). Taking into consideration the additional complexity of variable depth of the erodible layer is an improvement of such model and may be operational over large areas at affordable costs using geophysical methods.

The extrapolating technique of local measurements (weather, geochemical or using correlations between metals' concentrations and other soil physical, chemical or geophysical variables) over the contaminated sites is another aspect controlling the final result of stock computation, or of hazard assessment (McGrath et al. 2004). The distribution patterns of metals concentration at impacted sites are highly complex because of the overlapping of natural and industrial processes; conventional geostatistical techniques (Silva et al. 2004) may not be reliable for appropriate description (Modis and Komnitsas 2008). Optimization of the mapping techniques of metals contamination in mining areas is an open problem (Modis et al 2008).

In conclusion, the distribution of minerals constrains the success of both geoelectric and geomagnetic methods in characterizing the extent and internal structure of contaminated areas, as well as their heterogeneity. The use of such geophysical methods is essential for the hazard assessment of mining areas.

3.3 Mobilization of Metals by Atmospheric Fluxes

In this part, we refer only to mobilization form solid wastes and particles that were already deposited, neglecting the patterns of metals' spreading from the chimney of smelters as dependent on the types of particles. Although the dispersal from solid wastes may be minor compared to the dispersal by hydrological fluxes from a flux quantification point of view, it has consequences for human health in mining areas (Bradshaw and Chadwick 1980; McDonald et al. 1980; Borgegard and Rydin 1989). Contaminated soils in mining areas can be a more important source of contaminated dust (for instance with Pb) than the solid wastes (Murgueytio et al. 1980). Interestingly, a source of contaminated dust can be also the floodplains of arid river systems (Taylor and Hudson-Edwards 2008). For a review of the medical geochemistry of dusts (influence of mineral type, crystal morphology, grain size, degree of encapsulation, and trace element content on bioaccessibility), see Plumlee and Ziegler (2007). These authors review also the situation of mine tailings and smelters from a human health perspective. Smith and Lee (2003) review the effects of soil dust on human health, with indirect relevance for the effects of the dust originating in soil contaminated by smelters or by mining dumps. For example, Mihalík et al. (2010) found that the impact of a uranium mining dump on the surrounding agricultural land was correlated with the predominant wind direction. Sampling no farther than 600 m from the dumps, they found large concentration both near the dumps and at distance, with no linear decrease as a result of complex deposition patterns [see also barrier effects at distance for atmospheric dispersal of metals in Iordache et al. (2012)]. The contamination of agricultural fields from smelters is common, although its extent in mining areas is relatively lower due to the hilly/mountain geomorphology of these areas.

In general, the sources of metals available for dispersion by wind can be the silty sand-like material resulting from the flotation-separation of polymetallic ores and efflorescence salts formed by evaporation. From a methodological point of view, the resulted dust is separated in fine particles (2.5 μm or less) and coarse particles (10 μm down to 2.5 μm) (Fig. 3.2). The crucial factors controlling the dispersal patterns are the dimensions and form of the mineral particles determining the aerodynamic parameters, and the specific weight of the mineral. For a discussion of these issues, see Chen and Fryrear (2001). For the influence of the specific weight of the particles on the transportation by wind, see Andreotti (2004) and Almeida et al. (2007). It is clear that the transport of metallic minerals with different specific gravity (e.g., Sphalerite (ZnS) – 4.1 g/cm^3, pyrite (FeS_2) – 5.1 g/cm^3, galena (PbS) – 7.5 g/cm^3, etc.) will be different, but detailed studies in mining areas are lacking. In addition to the properties of mineral particles, the wind erosion is also influenced by vegetation cover, soil moisture, and the structure of the surface (e.g., organic

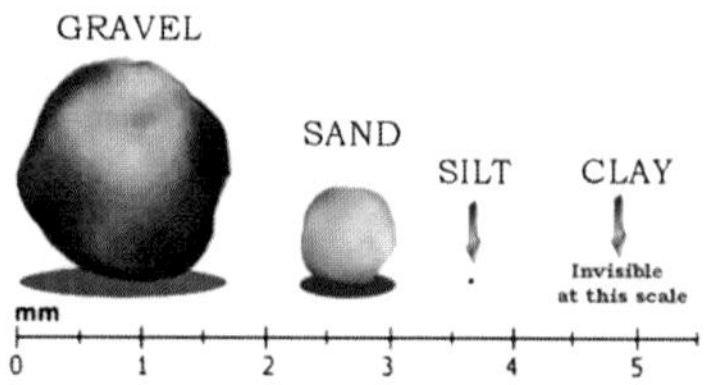

Fig. 3.2 Mobilization of particles by atmospheric fluxes is mostly controlled by the material particle size – sand and silt are the dominant particle sizes for tailing dams

material resulted from microbial and plant activity reduces the erosion). Neuman et al. (2009) analyzed experimentally in a wind tunnel the effects of shrinkage and cracking, spigotting, particle settling, and rewetting on the fugitive dust from mine tailings.

There are few detailed studies about the dust export (fluxes, mineralogy, and distance) from mining areas. Archibold (1985) used moss bag technique in order to monitor the dispersal of dust from uranium tailings with sandy texture (silt fraction between 3.3 and 44.3%). The impact was only at local scale, with maximum registered distance of 1 km from the tailing, usually between 300 and 500 m. There was no clear correlation of the distance with the wind force because of the topographical situation, and shorter distance of dispersion usually was associated with precipitations. The proximity of a sampling site from a point source may not have as result the dominance of the dust particles specific to that source, this depending on the properties of the dispersed particles. For instance, Chow and Watson (1998) notice that in the vicinity of a smelter the dominant particles were from more distant copper tailings. In another early study, Morin et al. (1999) documented the presence in soils around a smelter of Pb-bearing glassy particles transported by wind from the factory wastes.

Monna et al. (2006) use Pb isotopes to determine that the dust from numerous mine tailing dumps surrounding a city played a local role in the pollution with Pb, with automotive exhausts playing the dominant role.

One of the few studies characterizing the mixed mineralogy of tailings' and smelting dust is Teper (2009). Using pine needles as passive samplers, he identified three types of particles in function of their origin: particle from tailing ponds (Fe, Zn, Pb sulfides, Fe sulfates, Pb and Zn carbonates), particles from local soil and rock erosion (Ca–Mg and Ca carbonates, aluminosilicates and quartz), and particles from regional air pollution sources (Fe oxides, Fe–Zn oxides, and aluminosilicate glass). In sites no farther than 500 m from the tailings, the first type of particles were dominant (50%), but at the site located 2.5 km from the tailings particles were only just 9%. The vast majority of particle retained on needles were $<10\ \mu m$ and were trapped in stomata and epidermal tissue furrows (Teper 2009), so this information cannot be extrapolated to the deposition on soil. As one goes to larger distances from the mining area, tracing the sources of metals in atmospheric particles becomes a complex problem and discussion of the exact sources is usually not done (e.g., Tolocka et al. 2004). The only reasonable approach is coupled monitoring of depositions and dispersal modeling at different scales in space–time. Based on the current information tailings will not affect through wind areas beyond kilometers in distance, but smelters will affect up to tenths of kilometers depending on chimney characteristics, wind conditions, and geomorphology of the area.

3.4 Mobilization of Metals by Groundwater Fluxes

Groundwater plays an important role in the export of metals both at local (site) and regions levels when it has a low pH. This is almost always the case in mining areas, primarily because of mineralogical causes. At and around neutral pH, the export of metals by groundwater is of smaller significance at contaminated site catchment level and is mainly controlled by the colloidal fractions, with organic carbon playing an important role (Neagoe et al. 2012).

There are several types of conditions for the mobilization of metals by groundwater fluxes. Although these conditions controlling the vertical fluxes are in principle known, there heterogeneity in situ and the diversity of coupled processes bring difficulties in mine wastes leaching modeling (Hansen et al. 2008). Technological conditions control the specific surface of the waste particles and separate, for instance, the mining dumps from the tailing dams. Chemical reagents used in the extraction process may also influence the mobility of metals (for a review of the chemical reagents used in the mineral processing, see Pearse (2005), and for a review of synthetic reagent used in sulfide mineral flotation, see Zhang et al. (2009). Mineralogical conditions control the geochemical and biogeochemical processes, and will be detailed next. Exogenous abiotic conditions (pH, eH, concentration of oxygen, organic carbon, etc.) influence the kinetics and direction of the biogeochemical reactions. Exogenous biological conditions, mainly the microbiological ones in the case of contaminated groundwater generation, are crucial for the outcome and speed of the biogeochemical processes.

3.4.1 *Mineralogical Conditions*

When acid contaminated groundwater reaches the surface, the term used for it is "acid mine drainage" when the point is at or in the vicinity of the source contaminated site and is involved after that in mineralogical processes specific to

Fig. 3.3 Acid mine drainage from a closed mine gallery (*left*) and from closed tailing dam (*right*)

surface water and floodplains (part 5). But acid contaminated ground waters can extend far beyond such points, directly by groundwater flow, and even indirectly at large distance by contaminated rivers (Iordache et al. 2012). In mining areas, acid mine drainage is produced mainly by closed flooded mine and by tailings (Fig. 3.3). Mining dumps are not a major source of acid seepage water because of the comparatively lower reactive surface of the minerals involved (due to their bigger particle size), coupled with shorter transit time of the water.

A recent comprehensive source for the involved processes is Lottermoser (2007). We will summarize below some relevant conclusions from that manuscript.

From the potential of acid generating point of view, the weathering of mineral or poly-mineral association (dissolution, hydrolysis, oxidation and hydration) processes can be classified as acid-producing (i.e., generation of H^+), acid-buffering (i.e., consumption of H^+), or nonacid-generating or consuming reaction (i.e., no generation or consumption of H^+). For example, the degradation of pyrite (FeS_2) is an acid-producing reaction, whereas the weathering of calcite ($Ca[CO_3]$) is acid buffering, and the dissolution of quartz (SiO_2) does not consume or generate any acid. The balance of all chemical reactions, occurring within a particular waste at any time, will determine whether the material will "turn acid" and produce acid solutions.

Mine wastes are complex polyphasic systems comprising mineral associations containing aggregates from the original ore deposit, mechanical disaggregating phases, and anthropic added substances, from the ore concentrating processes such as cyanidation. The original mineral associations contain, apart from sulfides, a wide range of possible minerals including silicates, oxides, hydroxides, phosphates, halides, and carbonates. Silicates are the most common gangue

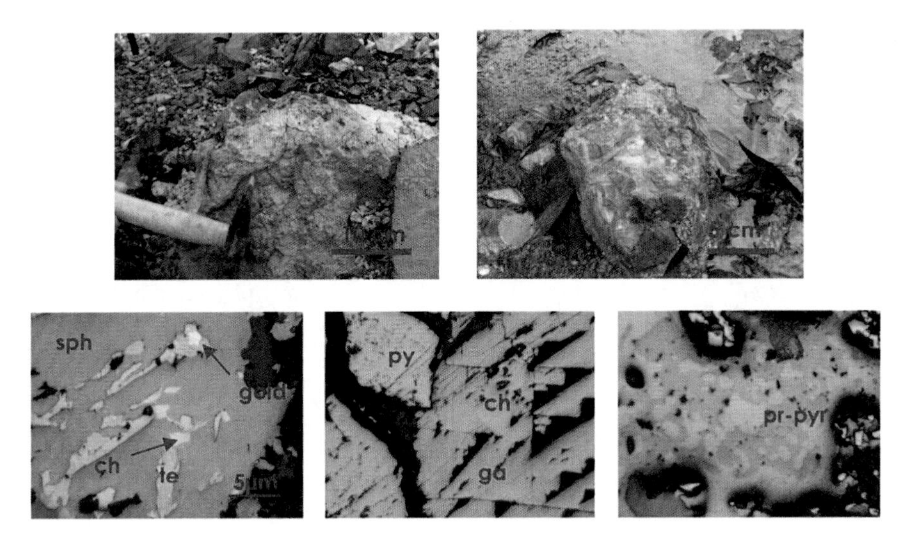

Fig. 3.4 Macroscopic and microscopic aspects of the mineralization processed in Apuseni Mountains, Romania (py-pyrite, ch-chalcopyrite, sph-sphalerite, te-tetraedrite, ga-galena, pr-proustite, pyr-pyrargyrite in a gangue of quartz)

minerals, and the sulfides may represent ore or gangue phases. Thus, the mineralogy of sulfidic wastes and ores is highly heterogeneous (Fig. 3.4).

When mining exposes sulfidic materials to an oxidizing environment, the materials become chemically unstable. A series of complex chemical weathering reactions are spontaneously initiated. This occurs because the mineral assemblages contained in the waste are not in equilibrium with the oxidizing environment. Weathering of the minerals proceeds with the help of atmospheric gases, meteoric water, and microorganisms.

We consider below the sulfidic tailing systems as it represents the most important acid generating system due to the small grain size of the material and, thus, the grate surface area, and active solution circulation, tailings being by definition solid–liquid systems.

The equilibrium of individual minerals in sulfidic wastes is influenced by: (a) the mineral's composition, crystal size, crystal shape, surface area, and crystal perfection; (b) the pH and dissolved carbon dioxide content of the solutions; (c) temperature; (d) redox conditions in system; (e) porosity of the system and (f) the elements mobility in system. For example, there is a large difference in dissolution rates between fine-grained waste and larger waste rock particles (diameters >0.25 mm). Smaller particles (diameters <0.25 mm) with their large surface areas contribute to the great majority of sulfide oxidation as well as silicate and carbonate dissolution (Stromberg and Banwart 1999; Lottermoser 2007).

Different minerals reacting with acidic solutions have different resistance to weathering. Minerals such as olivine ($(Mg,Fe)_2SiO_4$) and anorthite ($CaAl_2Si_2O_8$) are more reactive and less stable in the near-surface environment than K-feldspar

($KAlSi_3O_8$), biotite ($K(Mg,Fe^{2+},Mn^{2+})_3$ $[(OH,F)_2|(Al,Fe^{3+},Ti^{3+})Si_3O_{10}]$), muscovite ($KAl_2(AlSi_3O_{10})(F,OH)_2$), and albite ($NaAlSi_3O_8$). The rates of the different acid-buffering reactions are highly variable, and the major rock-forming minerals have been classified according to their relative pH-dependent reactivity. Compared with the weathering rates of even the most reactive silicate minerals, the reaction rates of carbonates are relatively rapid, particularly that of calcite ($Ca[CO_3]$) (Lottermoser 2007). Carbonates can rapidly neutralize acid. In an extreme case, calcite may even be dissolved at a faster rate than pyrite. As a consequence, drainage from a calcite-bearing waste may have a neutral pH, yet the quality of the mine drainage can eventually deteriorate and turn acid as the calcite dissolves faster than the pyrite (Lottermoser 2007).

Silicate minerals are abundant in sulfidic wastes, and their abundance may suggest that a waste rich in silicates has a significant buffering capacity. However, silicates do not necessarily dissolve completely, and the chemical weathering rate of silicates is very slow relative to the production rate of acid by pyrite oxidation. Therefore, rock-forming silicates do not buffer acid to a significant degree, and they only contribute token amounts of additional long-term buffering capacity to sulfidic wastes (Lottermoser 2007). Nonetheless, silicate mineral dissolution can maintain neutral conditions if the rate of acid production is quite slow and if abundant fine-grained, fast weathering silicates are present.

Sulfides are stable under strongly reducing conditions but the exposure of these minerals to oxidizing conditions will make them very unstable (Fig. 3.5). Pyrite (FeS_2) is the dominant metal sulfide phase in many ore deposits and its dissolution is the most important process in the formation of acid solutions. However, other sulfide minerals such as galena, sphalerite, and chalcopyrite, commonly occur with pyrite, and their evolution also influences the chemistry of the tailing system.

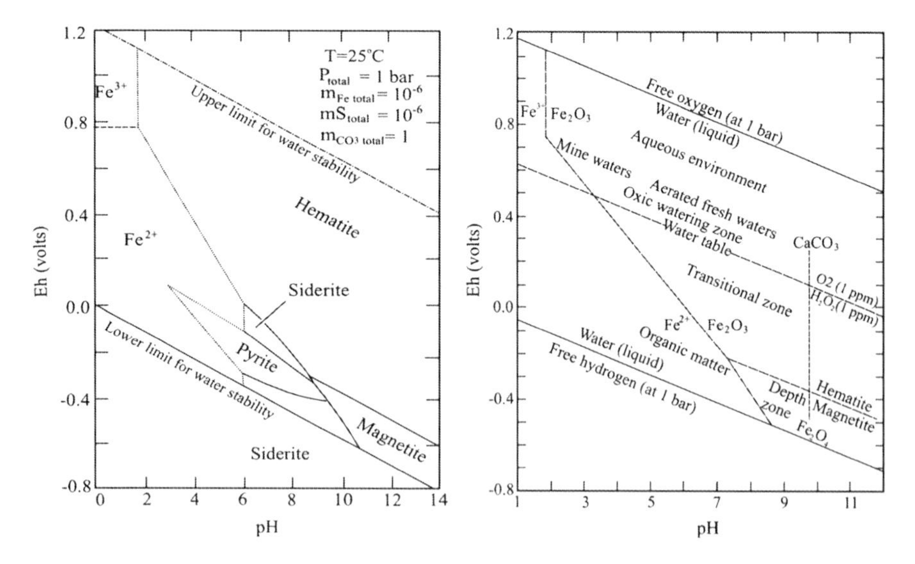

Fig. 3.5 Eh–pH diagrams showing fields of common iron minerals (Garrels and Christ 1965)

The weathering of various sulfides has been wide documented experimentally and through case studies (Lottermoser 2007). The major conclusion is that the sulfide minerals are different in what it concern acid-producing solutions, reactivity and dissolution products. Pyrite, marcasite (FeS_2), pyrrhotite ($Fe_{1-x}S$) and mackinawite [$(Fe,Ni)_{1-1.07}S$] appear to be the most reactive sulfides and their oxidation generates low pH waters. Other sulfides such as covellite (CuS), millerite (NiS), and galena (PbS) are generally far less reactive than pyrite. This is partly due to: (a) the greater stability of their crystal structure; (b) the lack of iron released; and (c) the formation of low solubility minerals such as cerussite ($PbCO_3$) or anglesite ($PbSO_4$), which may encapsulate sulfides such as galena preventing further oxidation. Other minerals, such as cinnabar (HgS) or molybdenite (MoS_2), are stabile in oxidizing environments. These sulfides are most resistant to oxidation and do not generate acidity.

The presence of iron in sulfide minerals or in solutions in contact with sulfides appears to be important for sulfide oxidation and acid solutions producing. High Fe^{2+} sulfides containing waste (e.g., pyrite, marcasite, pyrrhotite), or sulfides having iron as a major constituent (e.g., chalcopyrite ($CuFeS_2$), Fe-rich sphalerite) generate significantly more acidity than wastes with low percentages of iron sulfides or sulfides containing little iron (e.g., galena, Fe-poor sphalerite). Moreover, Fe^{2+} is oxidized to Fe^{3+}, which generates acid solutions through hydrolysis. Hence, sulfide minerals that do not contain iron in their structure (e.g., covellite, galena or iron-poor sphalerite) generate low acid solutions (Lottermoser 2007).

The (M)metal/(S)sulfur ratio in sulfides influences the amount of sulfuric acid produced by oxidation. For example, pyrite and marcasite have an M/S ratio of 1:2 and wile galena and sphalerite have an M/S ratio of 1:1.

While some sulfides can produce significant amounts of acid and other sulfides do not, there are nonsulfide minerals whose weathering or precipitation will also release hydrogen ions. First, the precipitation of Fe^{3+} hydroxides and aluminum hydroxides generates acid. Second, the dissolution of soluble Fe^{2+}, Mn^{2+}, Fe^{3+} and some sulfate salts such as jarosite ($KFe_3(SO_4(OH)_6)$), alunite ($KAl_3(SO_4(OH)_6)$), halotrichite ($FeAl_2(SO_4)_4{\cdot}22H_2O$), and coquimbite ($Fe_2SO_4{\cdot}9H_2O$), releases hydrogen. Soluble Fe^{2+} sulfate salts are particularly common in sulfidic wastes and a source of indirect acidity. For example, the dissolution of melanterite ($FeSO_4{\cdot}7H_2O$) results in the release of Fe^{2+}, which can be oxidized to Fe^{3+}. This resultant Fe^{3+} may precipitate as ferric hydroxide $Fe(OH)_3$ and generate hydrogen ions, or it may oxidize any pyrite present.

In general, increased hydrogen concentrations and acid production in mine wastes can be the result of: oxidation of Fe-rich sulfides; precipitation of Fe^{3+} and Al^{3+} hydroxides; and dissolution of soluble Fe^{2+}, Mn^{2+}, Fe^{3+}, and Al^{3+} sulfate salts (Lottermoser 2007).

Much of the buffering of the generated acidity by oxidation of pyrite, precipitation of Fe and Al hydroxides, and the dissolution of some secondary minerals is achieved through the reaction of the acid solution with rock-forming minerals in the sulfidic wastes. These gangue minerals have the capacity to buffer acid; that is, the minerals will react with and consume the hydrogen ions. Acid buffering is largely caused by the weathering of carbonates and silicates.

The buffering reactions occur under the same oxidizing conditions, which cause the weathering of sulfide minerals. However, unlike sulfide oxidation reactions, acid-buffering reactions are independent of the oxygen concentration of the gas phase or water in which the weathering reactions take place. The individual gangue minerals dissolve at different pH values, and buffering of the solution pH by individual minerals occurs within certain pH regions. As a consequence, depending on the type and abundance of gangue minerals within the waste (i.e., the buffering capacity of the material), not all sulfide wastes produce acidic leachates and the same environmental concerns.

Carbonate minerals play an important role in acid-buffering reactions. Minerals such as calcite ($CaCO_3$), dolomite [$CaMg(CO_3)$], ankerite [$Ca(Fe,Mg)(CO_3)$], or magnesite ($MgCO_3$) neutralize acid generated from sulfide oxidation. Calcite is the most important neutralizing agent, because of its common occurrence in a wide range of geological environments and its rapid rate of reaction compared to dolomite. Similar to pyrite weathering, grain size, texture and the presence of trace elements in the crystal lattice of carbonates may increase or decrease their resistance to weathering (Lottermoser 2007).

The major reservoir with buffering capacity in the environment is the silicate minerals, which make up the majority of the minerals in the Earth's crust. Chemical weathering of silicate minerals consumes hydrogen ions and occurs via congruent or incongruent weathering. Congruent weathering involves the complete dissolution of the silicate mineral and the production of only soluble components. Incongruent weathering is the more common form of silicate weathering whereby the silicate mineral is altered to another phase. The chemical composition of most silicates such as olivines, pyroxenes, amphiboles, garnets, feldspars, feldspathoids, clays, and micas is restricted to a range of elements.

From the export of metals perspective, the presence of sulfide minerals is important because these minerals are the primary source of acidic drainage associated with many mining and wastes. Lower acidities allow other metals that are associated with mining, such as aluminum, arsenic, cadmium, cobalt, copper, mercury, nickel, lead, and zinc, to enter the solution phase and be transported from the system. Drahota and Filippi (2009) review the secondary minerals of As in contaminated soils and waste systems.

Sulfide oxidation results also in the formation of secondary minerals occurring as cements and masses within the waste and as crust at or near the water's surface – efflorescences (Lottermoser 2007). The initial minerals that precipitate tend to be poorly crystalline, metastable phases that may transform to more stable phases over time (Lottermoser 2007). Secondary minerals can be grouped into sulfates, oxides, hydroxides and arsenates, carbonates, silicates and native elements. The type of secondary minerals formed in mine wastes is primarily controlled by the composition of the waste. Some of the secondary sulfates and carbonates are poorly soluble such as barite ($BaSO_4$), anglesite ($PbSO_4$), celestite ($SrSO_4$), and cerrusite ($PbCO_3$). As a result, once these minerals are formed, they will effectively immobilize alkali earth elements as well as lead. Their minerals act as sinks for sulfate, barium, strontium, and lead in oxidizing sulfidic wastes, and their precipitation controls the amount of sulfate, barium, strontium, and lead in AMD solutions (Lottermoser 2007). Secondary hydrous sulfates may redissolve in water and release their ions back into solution or may dehydrate to less hydrous or anhydrous compositions. Also, the hydrous Fe^{2+} sulfates may oxidize to Fe^{2+}–Fe^{3+} or Fe^{3+} sulfate salts. The newly formed secondary minerals are more stable and resistant to redissolution compared to their precursors. Secondary mineral possess large surface areas and adsorb or coprecipitate significant quantities of trace elements, including metals and metalloids. The precipitates effectively immobilize elements in acid mine waters and hence provide an important natural attenuation and detoxification mechanism in mine waters (Lottermoser 2007).

3.4.2 *Assessing the Mobility of Metals in Solid Wastes and Contaminated Soils*

A rapid evaluation of the mobility of metals in contaminated wastes and soils can be done by so-called speciation analyses. Instrumental methods for the characterization of metals speciation were reviewed by D'Amore et al. (2005). Reeder and Schoonen (2006) discuss them in the context of bioavailability assessment. Modeling attempts of mobility at site scale also exist (Gandy and Younger 2007). Mineralogical techniques used for heavy-metal-bearing characterization are enumerated in Hudson-Edwards (2003). Especially interesting are quantitative applications of techniques such as X-ray powder diffraction (e.g., Hillier et al. 2001). The speciation methods are very diverse, but for reasons of technological accessibility since the seminal work of Tessier et al. (1979) sequential extraction methods had an exponential development (Usero et al. 1998), and are the most widely used in the evaluation of mobility, although the extractants do not reflect correctly the mineralogical and phase distributions of metals (McCarty et al. 1998). For instance, Cu and Ni in a smelting region were associated with soil forms that would not have been predicted by the sequential extraction alone (carbonaceous material, silicate spheres, and carbonate particles, Adamo et al. 1996). Iron oxide and sulfide particles were still visible in the residual soil after sequential extractions, indicating the uncompleted decomposition of these mineral phases after the stage supposed to lead to the extraction of metals associated with them (Adamo et al. 2002). The solids remaining after the acid digestion of the sequential extraction residue mainly consisted of white precipitates of aluminum fluoride salts, including traces of K, Mg, Ca, Fe, and Ni and by black carbon particles showing a porous texture (Adamo et al. 2002).

Sequential extraction schemes are considered an essential tool in establishing element fractionation in soils and sediments (Gleyzes et al. 2002). Thus, the determination of different ways of binding gives more information on trace metal mobility, as well as on their availability or toxicity, in comparison with the total element content (Tuzen 2003). Speciation of the metals can help to assess how strongly they are retained in soil and how easily they may be released into soil solution (Kaasalainen and Yli-Halla 2003), and finally how they can affect environmental and human health. These procedures present a series of different problems such as low reproducibility especially with large particles and encapsulated pollutants (Dahlin et al. 2002a), ambiguities over error propagations (Koeckritz et al. 2001), the strong influence of operative conditions (Koeckritz et al. 2001; Ngiam and Lim 2001), the effective selectivity of the extracting reagents (Nirel and Morel 1990), and the readsorption of metals during extraction (Rendell et al. 1980). As a consequence, sequential extraction procedures cannot be used as standalone evaluations to identify the actual form of metals in soils and should be accompanied by deeper experimental investigations and solid matrix characterization (Dahlin et al. 2002b). However, sequential extractions can be useful to have an operational classification of metals in different geochemical fractions.

Metallic oxides, hydroxides, amorphous aluminum silicates, and organic matter (Neagoe et al. 2012) possess reactive surfaces on which heavy metals tend to be selectively adsorbed. To quantify the concentration of heavy metals in this form, it is necessary to use an extractant able to form particularly stable bonds with these elements. Salomons (1995) relates metal speciation to potential relative mobility with the following distinctions: exchangeable cations – high mobility; metals associated with iron and manganese oxyhydroxides – medium mobility; metals bound/fixed inside organic substances – medium mobility; metals bound/fixed inside mineral particles – low mobility; metals associated with sulfidic phase – strongly dependent on environmental conditions; metals associated with silicate phase – unlikely to be release under the normal weathering conditions.

The fractions of elements determined by sequential extractions were operationally defined in function of the reagents used, the reaction times, temperatures, and solid-to-extraction solution ratio used in each extraction step. No single reagent, time, and temperature combination could be applied to all sample types to recover a given phase; extractions were matrix dependent. This extraction procedure also attempted to differentiate the amorphous versus crystalline iron-oxide and iron-hydroxide phases. There is a gradient from amorphous to cryptocrystalline to crystalline iron-oxides and hydroxides; Hall et al. (1996) discussed the subtleties in differentiating among the phases depending on reagent strength. Additional complicating factors included the possibility that occluded grains might persist past their designated dissolution step or factors such as grain size, mineralogy, or solid solution may affect the reactivity of phases.

In acidic soils, the content of metals in a ready mobile form (fractions one and two) and in the oxidizable fraction constitutes an important source of potentially available trace metals that could be remobilized and incorporated to the soil solution (Riba et al. 2002). The same pattern was shown in alkaline soils, where a big proportion of metals (about 80%) could be remobilized and are potentially available for leaching or plants.

For revealing the free and several of the stable chelated forms of metals present in soils and reflect the metal availability in both the short-term and relatively long-term, one can use single extraction with ethylenediamine tetraacetic acid (EDTA). Besides its relevance for assessing the rapid mobility in barren surface of mining areas, the EDTA-extractable fraction has been found to give the best correlation with the amounts of metals taken up by plants (Ure 1996; Lo and Yang 1999), so is a useful tool also for designing the bioremediation of the contaminated sites.

3.4.3 Information at Site Scale

Case-specific studies provide insights into the diversity of situations occurring and reveal that they are more complex than the basic (bio)geochemical processes. One can speak about the diversity of the types of sources in mining areas (especially in historical ones, Alvarez-Valero et al. 2008), and also about the diversity of each source type (e.g., bedrock, tailings, dumps, etc.). The first type of diversity reflects the structure the mining landscape, and the second one the hydrogeomorphological context of the contamination source and its structural and mineralogical characteristics.

Bedrock seldom functions as a source of acid mine drainage. In a study of a catchment draining, a bedrock rich in sulfide (gold deposit) the surface water was found to be neutral and slightly basic (up to pH 8.13). The speciation and mineralogy of stream sediments was not coupled directly to the mineralogy of the bedrock, but probably by dissolved metals in the hyporheic interception of groundwater (Drahota et al. 2010).

Each tailing has its own identity in terms of mineralogy and no general conclusions can be drowning concerning the potential hazard without a case-specific comprehensive characterization. For instance, in a case the primary mineral phases are quartz (SiO_2), calcite ($CaCO_3$), pyrite (FeS_2), pyrrhotite ($Fe_{(1-x)}S$), sphalerite ((Zn,Fe)S), and arsenopyrite (FeAsS), and the secondary minerals in oxidized part included gypsum ($CaSO_4{\cdot}2H_2O$), K-jarosite ($KFe_3^{3+}(OH)_6(SO_4)_2$), lepidocrocite (γ-FeO (OH)), goethite (FeO(OH)), beudantite ($PbFe_3(OH)_6$-SO_4AsO_4), and kaolinite ($Al_2Si_2O_5(OH)_4$) (Romero et al. 2006), with As released by oxidation incorporated in beudantatite and K-jarosite or immobilized on the surface of hydrous ferric oxides. In other (gold) tailings, two As–Pb-bearing minerals were scorodite ($FeAsO_4{\cdot}2H_2O$) and beudantite $PbFe_3(AsO_4)(SO_4)(OH)_6$ (Roussel et al. 2000). The immobilization of As in scorodite was not relevant in this case because of its solubility whatever the pH, but beudantite efficiently maintained low Pb concentration in waters (Roussel et al. 2000). The study of the secondary mineralogy and microtextures in mine waste rocks derived from sphalerite ore in quartz veins containing manganoan carbonates (Jeong and Lee 2003) revealed complicated patterns with resistance to oxidation from pyrite > galena > arsenopyrite > sphalerite > pyrrhotite, Rhodochrosite dissolved to form hydrohetaerolite pseudomorphs ($ZnMn_2O_4H_2O$), manganoan calcite ((Ca,Mn) CO_3) with an outer alteration rim of hydrohetaerolite and an inner zone of smithsonite ($ZnCO_3$), and rock and mineral fragments cemented by fine aggregates of plumbojarosite ($PbFe_6^{3+}(SO_4)_4(OH)_{12}$), Fe oxyhydroxides/sulfates, and manganates (Jeong and Lee 2003). A 50-year-old pyrite bearing waste dump included a large diversity of minerals: pyrite, gypsum, jarosite ($KFe_3^{3+}(OH)_6(SO_4)_2$), hydronium-jarosite ($(H_3O)Fe_3^{3+}(SO_4)_2(OH)_6$), plumbojarosite ($PbFe_6^{3+}(SO_4)_4(OH)_{12}$), anglesite($Pb(SO_4)$)), quartz, feldspars, goethite, hematite, pickeringite ($MgAl_2(SO_4)_4{\cdot}22(H_2O)$), alunogen ($Al_2(SO_4)_3{\cdot}17(H_2O)$), epsomite ($Mg(SO_4){\cdot}7(H_2O)$), hexahydrite ($Mg(SO_4){\cdot}6(H_2O)$), a smectite-group phase, galena, sphalerite, and trace calcite, rhomboclase, and copiapite ($Fe^{2+}Fe_4{}^{3+}(SO_4)_6(OH)_2{\cdot}20(H_2O)$) (Farkas et al. 2009). A comprehensive list of 43 sulfate minerals found in four mining waste sites is reviewed by Hammarstrom et al. (2005). In a silver tailing, the most abundant primary mineral phases in the mine wastes are hematite, hydrohematite, barite, quartz, muscovite, anorthite, calcite, and phillipsite ($(Ca,Na_2,K_2)_3Al_6Si_{10}O_{32}{\cdot}12H_2O$) (Navarro and Cardellach 2009), and the minor phase consisted of primary minerals including ankerite ($Ca(Fe, Mg, Mn)(CO_3)_2$), cinnabar (HgS), digenite (Cu_9S_5), magnesite ($MgCO_3$), stannite (Cu_2FeSnS_4), siderite ($FeCO_3$), and jamesonite ($Pb_4FeSb_6S_{14}$), and secondary minerals such as glauberite ($Na_2Ca(SO_4)_2$), szomolnokite ($Fe^{2+}(SO_4){\cdot}(H_2O)$), thenardite ($Na_2SO_4$), and uklonscovite. ($NaMg(SO_4)(OH)$) Dold and Fontbote (2001) provide a comprehensive case study for the influence of climate, primary mineralogy, and mineral processing on the secondary mineralogy of several mine tailings. The trade-off of evaporation and infiltration controls the mobility of minerals in the vertical profile and their secondary mineralization. A multiphase mineralogical model for precipitation-controlled and evaporation-controlled climate is provided.

The behavior and potential export of metals from a mining waste is metal dependent. Schuwirth et al. (2007) and Hofmann and Schuwirth (2008) describe the different behavior of Zn and Pb in a flotation dump as strongly dependent on pH and redox potential. While Zn was an easily mobilizable species transferred to water phase, Pb was enriched in top soil at pHs >4.

Variability of waste materials creates an internal heterogeneity structure characterized by wide variations in grain size, density, mineralogy (Loredo et al. 2008), and implicitly in hydrological properties. Contrasting past and present hydrology between different parts of a tailing induce differential weathering, with the well-aerated core showing the most intense near-surface weathering (Kovacs et al. 2006, for a discussion of Zn and Pb), with occurrence of metal attenuation zones at different depths.

The estimation of tailing properties can be dependent on the scale of the measurements and experiments. For instance, the weathering rate of mining waste minerals was found to be scale dependent (Banwart et al. 1998) in the sense that the rate was different between batch-experiments, column experiments, and field studies. Banwart et al. (1998) consider that scaling effect reflects differences in environmental temperature, pore water pH, particle size distribution, different weathering rates for some minerals, the heterogeneity of mineral content in field not reflected at smaller scale, and water flow patterns. The particular balance between acid and base generation depends on the particular type of generating minerals (Paktunc 1999), with consequences on the need for correction when extrapolating between wastes in terms of acid–base generating properties of classes of minerals.

The release of toxic metals by underground fluxes can be strongly reduced when the tailings are placed on a reactive permeable bed (7 wt%) of porous, alkaline pellets of transformed red mud (TRM) (Zijlstra et al. 2010). The mineralogy of permeable barriers for subsurface inorganic contaminants (acid mine drainage included) is reviewed by Jambor et al. (2005). The absence of contaminants from a tailing in the groundwater may also be caused by the carbonate-rich environment of "host-rocks" that limits their mobility (Navarro and Cardellach 2009).

Quantitative analyses of the mineral speciation forms in smelter-contaminated soil were performed for Zn (Manceau et al. 2000). Primary minerals franklinite ($ZnFe_2O_4$), willemite (Zn_2SiO_4), hemimorphite ($Zn_4Si_2O_7(OH)_2{\cdot}H_2O$), and Zn-containing magnetite ($(Fe,Zn)Fe_2O_4$) were identified as originating from atmospheric fallout of Zn dusts emitted during the pyrometallurgical smelting process, and they act as the main source of Zn in contaminated soils (Manceau et al. 2000). The Zn released from these minerals was taken up partly by phyllosilicates and, to a lesser extent, by Mn and Fe (oxyhydroxides) (Manceau et al. 2000). A detailed speciation of Zn in contaminated soil (Zn-rich phyllosilicate, Zn-layered double hydroxide, and hydrozincite) is reported by Jacquat et al. (2008) for calcareous soils. Romero et al. (2008) investigated the solid phase control on lead (Pb) bioaccessibility in soils (near neutral) impacted by smelter activities. They identified common Pb phases reported in similar contaminated environments (galena (PbS) and anglesite ($PbSO_4$)), solid lead arsenate phase (Romero et al. 2008), which contributed to the very low Pb availability in these soils. The sources of pollution in contaminated soils can be traced also by isotopic ratios in the case of elements such as Pb (Ettler et al. 2004), and this technique can be coupled with mineralogical analyses to reconstruct the pas emission of smelters (Sonke et al. 2002). One of the few studies comparing the mineralogy of smelter-contaminated soils and tailing is Morin et al. (1999). They also found in addition to the usual larger concentrations of Pb in tailings than contaminated soil clear pH and organic matter dependent mineralogical and phase differences with direct relevance for the mobility of the metal. Technological aspects like temperature controlled the mineralogy of air-borne residues near secondary Pb smelter (Ettler et al. 2005), anglesite ($PbSO_4$) and laurionite (Pb(OH)Cl), cotunnite ($PbCl_2$), $(Zn,Cd)_2SnO_4$, and $(Sb,As)_2O_3$ associated with high temperature, $KCl{\cdot}2PbCl_2$ and caracolite ($Na_3Pb_2(SO_4)_3Cl$) with low temperature, and metallic elements homogeneously distributed. Regardless of the temperature of production, the chlorides had a larger solubility and as a result larger mobility following the dispersal form the smelter (Ettler et al. 2005). The solubility of the phases in smelter dust is crucial for the toxicity of the associated elements (Davis et al. 1996). For instance in the case of As-bearing particles, calcium-iron arsenate indicates high bioaccessibility, amorphous iron arsenates, and arsenic-bearing iron(oxy)hydroxides are associated with intermediate bioaccessibility, and arsenopyrite or scorodite have the lowest bioaccessibility (Meunier et al. 2010).

In cold climates, particles reaching the soil around industrial sources can be traced by analyzing the snow covering the soil. Such depositions reflect both technogenic (sulfides, oxides, metallic phases and alloys, slag particles, coke) and geogenic sources (Gregurek et al. 1999), allowing the discrimination between industrial sources with different technologies in an area.

As in the case of mining wastes, treatment of contaminated soils provides opportunities for assessing the role of mineralogy in metals mobility. For instance, in both treated and untreated smelter contaminated soils (with cyclonic ash), 30–50% of Zn was present in smelter-related minerals (willemite (Zn_2SiO_4), hemimorphite ($Zn_4Si_2O_7(OH)_2 \cdot H_2O$) or gahnite ($ZnAl_2O_4$)), while 50–70% of Zn was incorporated into newly formed Zn precipitates (Nachtegaal et al. 2005). The treated soil did not contain gahnite or sphalerite, probably due to higher pH at the time of the treatment (Nachtegaal et al. 2005). A review of the stabilization of As, Cr, Cu, Pb, and Zn in soil using amendments (covering the role of primary and secondary minerals) can be found in Kumpiene et al. (2008).

A particular situation occurs when mine-soils originating with the origin in remediated waste materials are enriched in certain metals as a result of fertilizers and animal manure application (Vega et al. 2004). Still another case of metals' fluxes and contaminated site location is by commercial and domestic transfer of mine wastes. This process was found to be responsible for the dispersion of mine wastes 15–30 km around the mine sites (Hamilton 2000), and the distance was controlled by economic costs of removal and transport.

Smelting slags are another case with specific mineralogical aspects. They were found to be heterogeneous materials dominated by Ca and Ba compounds and Pb phases (Gee et al. 1997) and weathering products such as cerussite and Pb–iron oxide complexes. Primary phases in the slags of an abandoned mine site included olivine-group minerals, glass, spinels, sulfide minerals native metals, clinopyroxenes, and other unidentified metallic compounds (Piatak et al. 2004). Olivine-group minerals and pyroxenes were dominantly fayalitic (Fe_2SiO_4) and hedenbergitic ($CaFeSi_20_6$) in composition and contained Zn, spinel minerals ranged between magnetite and hercynite ($Fe^{2+}Al_2O_4$) in composition and contained Zn, Ti, and Cr, while Co, Ni, Cu, As, Ag, Sb, and Pb occurred in the glass phase, sulfides, metallic phases, and unidentified metallic compounds (Piatak et al. 2004). The leaching from slags was controlled by the dissolution of silicate and oxide phases, the oxidation of sulfide phases, and the precipitation of secondary phases (Piatak et al. 2004). Slags of a Cu–Co smelter included as primary phases Ca–Fe silicates (clinopyroxene, olivine) and leucite, oxides (spinel-series phases), ubiquitous silicate glass and sulfide/metallic droplets of various sizes (Vitkova et al. 2010), with Cu and Co found in all phases, but main carriers Cu sulfides (digenite, chalcocite, bornite, chalcopyrite), Co–Fe sulfides (cobaltpentlandite), Co-bearing intermetallic phases ($(Fe,Co)_2As$) and alloys. The weathering of these phases led to the formation of secondary metal-bearing phases malachite ($Cu_2(CO_3)(OH)_2$), brochantite ($Cu_4SO_4(OH)_6$), and sphaerocobaltite ($CoCO_3$) (Vitkova et al. 2010). Smelter slags contribute together with atmospheric sources (via deposition on soil) to the contamination of rivers sediments (Yang et al. 2010).

There are very few studies comparing the hazard associated with different types of sources in a mining area. As a rare example, Alvarez-Valero et al. (2008) found a hazard of the primary sources (based on total quantity of metals) in a mining area in the following order: modern slag > industrial landfills > country rocks > leaching tank refuses > gossan wastes > Roman slag > pyrite-rich samples iron oxides > smelting ashes, but with no detailed reference to the types of metals fluxes potentially going out of the stock of metals.

In conclusion, the role of mineralogy in the contaminated sites of mining areas is crucial for their future evolution. The most common tailings are those associated with polymetallic ores and their composition, comprise, along with sulfides, gangue minerals, such as silicates, carbonates, and secondary minerals (salts, Fe hydroxides, etc.). Oxidation of sulfides determines acidic solution producing. The more sulfides presents, the more acidic solutions are generated. In the presence of carbonates, which are acid-buffering minerals, the acid solution generating is retarded, and the system can reach the equilibrium if the gangue minerals are present in sufficiently large quantities. Other types of contamination sources such as soils and smelter slags are equally important in determining the hazard potential of the mining area and have their own, often unique and requesting a case-specific study, environmentally and technologically determined mineralogical characteristics controlling the mobility of metals.

3.4.4 Microbiological Conditions

Neagoe et al. (2012) synthesized the role of microorganisms in the mobility of metals in contaminated zones. Of these roles those involving the interactions with minerals (with and without need of organic carbon) are of primary importance. In this part, we screen the existing literature with a biogeochemist eye, looking at functions and scales. The reader with microbiological interests can find the names of species and detailed examples in the cited reviews and articles.

A recent review of the mineral–microbe interactions was done by Dong (2010). Also recently, Bini (2010) reviewed the Archaeal transformations of metals in the environment. The relationships between the geomicrobiology of metals and their use for bioremediation were reviewed by Gadd (2010), and the same author had extensively screened the biogeochemical transformations of rocks, minerals, and metals by fungi (Gadd 2007). The ecological relevance of mineral weathering by bacteria in the larger context of soil ecology was done by Uroz et al. (2009). An extensive description of the coupled hydrobiogeochemical processes (where the “bio” refers strictly to microorganisms) in groundwater for several heavy metals and radionuclides was done recently by Jardine (2008). We refer the reader to these syntheses for details about these processes. In this part of the chapter, we want only to underline the small scale of the ecosystems supported by microorganisms and its consequences for the internal heterogeneity of the role of microorganisms in the mining areas.

Taking the perceptual perspective of the microorganisms is needed in order to understand their interactions with the minerals. Mineral weathering may be influenced by a mineral's nutritional potential, with microorganisms destroying only "beneficial" minerals (Bennett et al. 2001). The remnant mineralogy of the rock reflects early microbial weathering of valuable minerals, leaving a residuum of nonvaluable minerals, where "value" reflects the needs of the organism in a particular diagenetic environment (Bennett et al. 2001). Mineral preferences exist also for *Arthrobacter*, who preferred sand grains other than the dominant mineral quartz (Gommeaux et al. 2010). Relatively, larger heavy-metal concentrations on a granite outcrop had a strong influence on bacterial community structure developing of its surface, with strong relationships found between certain ribotypes and particular chemical heavy-metal elements (Gleeson et al. 2005a). Not only bacteria, but also fungi showed such correlations revealing eco-physiological preferences (Gleeson et al. 2005b). Altering the mineral composition of soil by adding mica, basal and rock phosphate determined a shift in microbial community structure (Carson et al. 2007) and minerals in soil selected distinct bacterial communities in their microhabitats (Carson et al. 2009). The silicates dissolution in the presence of acidophilic microorganisms depends on the mineral type and the presence of ferrous iron (Dopson et al. 2009). Thus, minerals and rock surface are entities with different values for microorganisms. Colonization of rocks by microorganisms is interpretable in terms of primary ecosystem succession (Borin et al. 2010). The micro-ecosystem productivity (the initial rate of microbial activity) is determined by the surface area of the mineral, but the crystallinity of the mineral and the particles' aggregation are important as well (Cutting et al. 2009). Diverse minerals on rock outcrops may even induce selection of microorganisms at meter scale, because of their relatively smaller dimension and short length of the life cycle (Hutchens et al. 2010).

The minute scale of bacterial developmental systems and their associated ecological systems lead to small-scale heterogeneities, steep physical–chemical and microbial gradients, as well as hot spots of contaminants and biodegradation in the supposedly homogeneous aquifers (Anneser 2010). Sampling only groundwater in order to understand processes supporting the attenuation of pollutants is not enough in this context. When the bacterial diversity at micro-scale did vary between samples in a shallow subsurface, the number of bacteria still varied one order of magnitude (Musslewhite et al. 2003).

Starting from these facts one can infer that also in mining wastes there is a large heterogeneity of microbial communities due not only to large-scale structure of the wastes, but also to small scale variability in mineral composition. Studying the spatial scaling of microbial diversity (Green et al. 2004; Green and Bohannan 2006) in mining areas may provide very interesting results if related to the distribution of contaminant and their mineral phases.

In tailings, the largest numbers of bacteria (10^9 g^{-1} dry weight of tailings) were found at the oxidation front (the junction between the oxidation and neutralization zones), where sulfide minerals and oxygen were both present (Diaby et al. 2006). The bacterial biomass in Cu tailings was found to be much smaller than in the adjacent soil (Iglesia et al. 2006) and depending at micro-scale on Cu concentration, pH and organic matter. Tailings with larger pH have a more diverse bacterial community than very acid tailings (Mendez et al. 2008), and the richness increase as a result of phytoremediation. Microbial weathering for nutrients acquisition by bacteria (Mailloux et al. 2009) and fungi (Lian et al. 2008) alone or in association with roots (Calvaruso et al. 2006) may release also toxic elements such as As (Mailloux 2009). Mobilization of metals at micro-scale in contaminated sites may not necessarily lead to larger scale mobilization because of coupling with adsorption and precipitation processes (Quantin et al. 2001).

In the rhizosphere soil, the activity of microorganisms is coupled with that of the plants and rather difficult to be separated from it. The relationships between minerals and plants are bidirectional, and not necessarily direct. For example, the flooding of crops with tailing wastes after failure of a tailing dam led to P deficiency in plants by several coupled micro-scale processes (Nikolic et al. 2010). Such physiological effects are coupled with hydrological processes influencing also the vertical mobility of metals (Zhao et al. 2007). There can be an induced weathering of minerals by plants (Bormann et al. 1998), and also a crystallization of metalliferous minerals (Cabala and Teper 2007). Pb and Cd carbonates and Fe and Zn sulfates, for instance, seem to be associated with root growth (Cabala and Teper 2007). The mineralogy of the rhyzosphere changes with waste site, soil type, and grain size (Cabala and Teper 2007). Biominerals can occur also as a result of the excretion of metals from plants (e.g., vaterite grains Isaure et al. 2010). Another way at looking at the effects of microorganisms and vegetation on mineralogy is to assess the effect of phytoremediation. For instance, Panfili et al. (2005) found that in the untreated and unvegetated sediment, Zn was distributed as 50% (mole ratio of total Zn) sphalerite, 40% Zn-ferrihydrite, and 10–20% (Zn–Al)-hydrotalcite plus Zn-phyllosilicate, in unvegetated but amended sediments, ZnS and Zn-ferrihydrite each decreased by 10–20% and were replaced by Zn-phosphate (30–40%), and in the presence of plants, ZnS was almost completely dissolved, and the released Zn bound to phosphate (40–60%) and to Zn-phyllosilicate plus (Zn,Al)-hydrotalcite (20–40%). The mineral species and the co-addition of mineral amendment did not influence the speciation of Zn in the vegetated substrate (Panfili et al. 2005). Understanding the coupled microbial-plant conditions for the mobilization of metals from minerals requires an integrated approach (Iordache et al. 2012) involving the role of the organic carbon (Neagoe et al. 2012) and the plant species specific characteristics (Farcasanu et al. 2012).

3.5 Mobilization of Metals by Surface Water Fluxes

Although the management of contaminated soils and sediments should be inseparable problems (Apitz et al. 2006), by now they were treated mostly separately probably because of the different regulations covering them. There is, however, a growing literature concerning the transport of metals from mining wastes and contaminated lands into river systems. The role of geomorphic processes in metals redistribution in floodplains was reviewed by Miller (1997).

According to our knowledge, there is a single, but excellent, relatively recent review in this area, from a mineralogist perspective, done by Hudson-Edwards (2003). Table 3.3 presents from this source the heavy-metal-bearing minerals and phases in river sediments. For a review-type discussion of the same problems but strictly from the perspective of the Water Framework Direction implementation, see Macklin et al. (2006).

The general conclusions of Hudson-Edwards (2003) are as follows: "The mineralogy and geochemistry of these particles (identified in mining affected river sediments) is dependent upon the original ore mineralogy, and on processes that have occurred in the source areas (e.g., mining or extraction, oxidation of sulfides,

Table 3.3 Mineral and phases associated with heavy metals in river sediments (literature compilation reproduced from Hudson-Edwards 2003 with permission)

Mineral	Formula	Mineral	Formula
Al hydroxide	$Al(OH)_3$	Hydronium jarosite	$Fe_3(SO_4)_2(OH)_5{\cdot}2H_2O$
Anglesite	$PbSO_4$	Jamesonite	$FePb_4Sb_6S_{14}$
Arsenopyrite	$FeAsS$	Jarosite	$KFe_3(SO_4)_2(OH)_6$
Azurite	$Cu_3(CO_3)_2(OH)_2$	Kaolinite	$Al_2Si_2O_3(OH)_4$
Beaverite	$Pb(Fe,Cu)_3(SO_4)_2(OH)_6$	Lepidocrocite	γ-FeOOH
Bornite	Cu_5FeS_4	Malachite	$Cu_2(CO_3)(OH)_2$
Burononite	$PbCuSbS_3$	Melanterite	$Fe^{2+}SO_4{\cdot}7H_2O$
Cassiterite	SnO	Mercury (elemental)	Hg^0
Cerussite	$PbCO_3$	Mn oxide	Amorphous
Chalcanthite	$CuSO_4{\cdot}5H_2O$	Natrojarosite	$NaFe_3(SO_4)_2(OH)_6$
Chalcocite	Cu_2S	Plumbogummite	$PbAl_3(PO_4)_2(OH)_5{\cdot}H_2O$
Chalcopyrite	$CuFeS_2$	Plumbojarosite	$Pb_{0.5}Fe_3(SO_4)_2(OH)_6$
Cinnabar	HgS	Pyrite	FeS_2
Copiapite	$Fe^{2+}Fe_4{}^{3}{}^{+}(SO_4)_6(OH)_2{\cdot}20H_2O$	Pyromorphite	$(Pb,Ca)(PO_4)_3Cl$
Copper	Cu	Rozenite	$Fe^{2+}SO_4{\cdot}4H_2O$
Coquimbite	$Fe_2{}^{3+}(SO_4)_3{\cdot}9H_2O$	Schwertmannite	$Fe_8O_8(OH)_6SO_4$
Covellite	CuS	Scorodite	$FeAsO_4{\cdot}2H_2O$
Fe oxide	amorphous	Smectite	$(NaCa)_{0.3}(Al,Fe,Mg)_2Si_4O_{10}(OH)_2{\cdot}nH_2O$
Ferrihydrite	$5Fe_2O_3{\cdot}9H_2O$	Smithsonite	$ZnCO_3$
Galena	PbS	Sphalerite	ZnS
Gahnite	$ZnAl_2O_4$	Stannite	Cu_2FeSnS_4
Goethite	ς-FeO(OH)	Symplesite	$Fe_3(AsO_4)_2{\cdot}8H_2O$
Goslarite	$ZnSO_4{\cdot}7H_2O$	Tenorite	CuO
Gratonite	$Pb_9As_4S_{16}$	Tennantite	$(Cu,Ag,Fe,Zn)_{12}As_4S_{13}$
Hematite	Fe_2O_3	Terahedrite	$Cu_{12}Sb_4S_{13}$
Hydrohetaerolite	$Zn_2Mn_4O_8{\cdot}H_2O$	Wolframite	$FeWO_4$

post-depositional changes in remobilized floodplain alluvium, etc.), during transport and deposition (sorting, abrasion, adsorption/desorption reactions, complexation by dissolved organic matter, etc.) and during post-depositional early diagenesis (redox and pH changes, reduction of organic matter, etc.)." The general weathering reaction paragenesis of Pb-, Zn-, Cd-, and Cu-bearing minerals in river systems (floodplain included) is sulfides to carbonate, silicate, phosphate, and sulfate weathering products to iron and manganese oxyhydroxides (Hudson-Edwards et al. 1996). What is not known is "the precise effects of these processes on the heavy-metal-bearing particles, and the timing and rate at which they occur" and "the influence of fluvial geomorphology on heavy-metal-bearing particle mineralogy" (Hudson-Edwards 2003).

In this part of the chapter, we screen the literature not covered by the mentioned review, and extend the perspective with the specificity of larger river floodplains, the presence of contamination hotspots at distance from the source, and with the interaction between floodplain organisms and minerals.

As shown in another chapter of this book, (Iordache et al. 2012) rivers and floodplains integrate many types of underground and surface hydrological flows carrying metals. The downstream quality of water, river sediments, and riparian sediments depends on interplay of biogeochemical processes with different scales, from micro to regional. Understanding the kinetics of dissolution and precipitation of metals in large-scale environment, entities such as river systems would involve understanding biogeochemical reactions networks (Zhu 2009). We are still far from being able to do this more than qualitatively in the real contaminated floodplains, but there is a body of literature describing phenomenological patterns, which can be a base for future research on modeling and predicting the involved processes.

The export of metal from the mine wastes and contaminated soils by surface runoff is an erosion process coupled eventually with increased groundwater flow. For instance, the load of Fe in the river increased 21-fold, while loads of Cu and Mn increased by eightfold and sevenfold, respectively, during the storm runoff, but the sources were multiple: erosion of particulates from the upland, resuspension of colloidal material from the streambed, and increased groundwater inflow to the stream (Kimball et al. 2007). The input of metals from groundwater may be important at a catchment scale also at low flows (Kimball and Runkel 2009), and diel metals' mass flow characterization over longer periods are needed in order to understand how to correctly support the remediation measure (Kimball et al. 2010). We notice also that such mass flow approaches can be useful also for the identification of retention and remobilization processes involving minerals and the vegetation cover (see last part of this chapter for a biogeochemical approach to the management of rivers contaminated by metal mining), and this seemed to be a way to link the optimization of mining area management with the climate change problems (especially in terms of changing hydrological events – intensity and frequency). In practice, the situation is even more complex. Besides the hydrological fluxes taken separately, their coupling with dust deposition leads to site-specific patterns. But beyond a certain distance from the contamination source down slope, the hydrological fluxes become dominant (Aslibekian and Moles 2000).

The mechanical dispersion of tailings material resulted from erosion can be identified by comparing the mineral assemblages in the floodplain and in the tailings (Avila et al. 2008), but, as these authors point out, the secondary dispersion of metals depends also on other processes such as (a) precipitation of hydroxide, oxyhydroxide, or hydroxysulfate phases from aqueous species as pH increases; and (b) adsorption of metals onto neoformed mineral surfaces (carbonates or iron and manganese coatings) (Avila et al. 2008).

After export from the terrestrial zone, the metals can be retained for different periods in the river bed or in the flooding area. The retention in river bed (stream sediments) reflects either the settling of particulate metals or the phase change of dissolved metal. Especially in the acid mine drainage impacted rivers, the second process is of large importance and couple the river mineralogy with the impacted groundwater problem. For comprehensive information related to this issue, see for instance Lottermoser (2007). For instance, by an interaction between the mineralogy of As and Pb leading to the formation of arsenatian plumbojarosite, the high concentrations of Pb in surface water in a mining catchment were located only in the upper part of the catchment (Frau et al. 2009). Barite and beudantite, when present in stream sediments, serve as the most important control in the mobility of As, Ba, and Pb under acidic conditions (Romero et al. 2010), in contrast to Fe-oxy-hydroxides. As still another example, Pb binding to hydrozincite ($Zn_5(CO_3)_2(OH)_6$) was strong enough to make this mineral a potential sink for this metal in contaminated waters (Lattanzi et al. 2010). Post-depositional redistribution of the deposited minerals plays an important role in the mobility of metals. Audry et al. (2010) estimate that during a flood event 870 tons of Zn, 18 tons of Cd, 25 tons of Pb, and 17 tons of Cu have been mobilized from the downstream sediments along a river by oxidation of sulfide phases induced by their resuspension.

Pb isotopic analyses of stream sediments allowed the discrimination between historic mining and primary smelting (important sources) and secondary smelting (car battery processing, negligible source in that case Ettler et al. 2006).

The sediment being transported by stream includes grains of resistant minerals (quartz, K-feldspar, muscovite, zircon, garnet) as well as weathering products (oxyhydroxides of Fe, Al, Mn, kaolinite) and particles of organic matter. In addition, cations of chemical elements in solution are sorbed by films or organic molecules that cover the surfaces of the sediment particles (Faure and Mensing 2005).

The isotopic compositions of certain elements (e.g., Sr, Nd and Pb) that are sorbed to colloidal particles (diameters less than 1 μm) are similar to the isotopic compositions of the elements in solution. Particles of resistant minerals with diameters larger than about 1 μm in diameter contain Sr, Nd, and Pb, whose isotope compositions differ from those of the sorbed fraction because they depend on the ages of the mineral grains and on the respective parent–daughter ratios.

The isotopic compositions of sorbed Sr, Nd, and Pb provide information about the minerals that are weathering in the drainage basin and about anthropogenic contaminants that are released into stream.

The dimensions of the river sediment particles are of key importance from the mobility of the associated metals. Nano-sized Fe and Ti oxides (most commonly goethite, ferrihydrite and brookite), and aggregates of such nano-partciles were found hosting a large fractions of trace metals, with extended transport capabilities (Plathe 2010). Hochella et al. (2005) predict that toxic metals nanophases are far more bioavailable than from larger crystals of the same mineral due to enhanced solubility of the former.

Although the institutional perspective on heavy metals concentrations and mobility in contaminated river sediments recently prevail as a result of pressure from the implementation of Water Framework Directive (WFD, e.g., Byrne et al. 2010, but for a general methodological approach, see Macklin et al. 2006, or Bird et al. 2010), we believe that this view is far too limiting for understanding the real (not necessarily already regulated) consequences of metals presence mining catchments. Actually even within the WFD context, a more integrated approach for water quality as depending on all sources of metallic substances at catchment levels (not limited to mining areas) is promoted (Chon et al. 2010). So in the ecological context set up by this directive, it is rather a matter of interpreting the WFD the approach one adopts, more limited or more integrated.

Overbank sediments deposited in the ecotone (interface with terrestrial systems) of small streams and rivers are also involved in the buffering and retention of metals. Irrespective of the location, pH, organic carbon content and "mobilisable" metal concentrations were the most important factors explaining "mobile" metal concentrations in the alluvial soils as evaluated by leaching tests (Cappuyns and Swennen 2007). Pyrite and sphalerite were metal-containing phases found in the overbank sediment and not in the streambed in a catchment impacted by Pb–Zn mining (Gonzalez-Fernandez et al. 2010). Overbank concentration of Cu, Zn, Sn, and Pb were twice larger than the concentration in the river channel sediments (Gonzalez-Fernandez et al. 2010) and had larger coefficients of variation. Silicates, sulfides, and metal slag breaking down in the oxidizing at low-pH zones of the floodplain sediments released silica, alumina, metals, and sulfate to solution and led to the formation of (1) metal oxides, (2) sulfates, and (3) amorphous silica (Hochella et al. 2005). In the anoxic streambeds, the formation of new sulfides (with Zn, As, and Pb) with microbial origin was also observed together with poorly crystalline (Fe,Mn) hydrous oxides formed in the floodplain (Hochella et al. 2005). Crystalline Fe-oxides (goethite, hematite, and wustite) were almost not present. The mobilization, immobilization, and remobilization of metals from overbank sediments depends on acidification, organic and inorganic complexation, redox potential (Cappuyns and Swennen 2004; Van Damme et al. 2010), and the (micro)organisms influencing them. In upper reaches, near the mining source the remobilization by such processes interfere with dissolution and flushing of efflorescent minerals, which may play the most important role in specific circumstances (Byrne et al. 2009).

In higher order rivers or in particular geomorphological conditions, the retention overbank sediments take place at a larger scale as retention in floodplains. Dennis et al. (2009) estimate a retention of 32,000 tons of Pb within the mined tributaries and of 123,000 tons within the main channel of a mining catchment, representing approximately 28% of the Pb produced in the mining area. At present export rate, it would take over 5,000 years for all metals to be removed from the catchment (Dennis et al. 2009). Iordache (2009) comprehensively characterize the retention of metals in the lower Danube river system (large stretch of several hundreds of km of the Danube river) and in different riparian and insular local landscapes. A recent review of trace metal behavior in riverine floodplain soils and sediments was done by Du Laing et al. (2009).

In contrast to soils, the metals in floodplains of mining catchments are not correlated with the fine fraction, because part of the coarse fraction originates in mine wastes (Loredo et al. 2008). Lower reaches of the river systems usually receive only secondary minerals from the upper mining catchments, but failure of tailing dams can lead to the unusual spread of the primary mineral in downstream floodplains (Gallart et al. 1999). At even larger distance, the original tailing material may become not observable (Langedal 1997), but the pollution with metals manifest by transfer in transformed phases. In a catchment with primary ores dominated by As sulfides (arsenopyrite, FeAsS; realger, AsS; orpiment, As_2S_3), no mineral phase of As could be detected in the contaminated floodplain, but mostly phases of poorly crystalline and redox sensitive Fe(hydr)oxides (Ackermann et al. 2010). The interplay of retention and remobilization of metals form the floodplains of rivers with different order leads, and their interaction with the biogeochemistry of other (major) elements (Saedeleer et al. 2010) leads to a lag between the increase or decline of metals emissions from the mining area and the corresponding increasing or decreasing trends in lower floodplains (Martin 2009).

Profiles in the floodplain sediments are a usual tool for inferring conclusions about historical conclusion (Ciszewski 2003; Day et al. 2008). They should be used with care, however, because of the potential redistribution of metals between phases and depths (Hudson-Edwards et al. 1998), a process of a similar biogeochemical type as in the case of vertical redistribution in solid mining wastes, but different in details by the nature of involved mineralogical, organic matter and biological constituents and of the different environmental controlling factors such as hydrological regime. To be more effective, sediment profiles can be coupled with other chronometric tools such as particular isotopes (Lokas et al. 2010), also the isotopes coupled with mineralogical investigations providing information for tracing the multiple sources of the metals (Vojtěch et al. 2006).

The study of the metals mineralogy in the context of floodplain biogeochemical characteristics is crucial for understanding the different risks associated different to metals in floodplains (Vaněk et al. 2008; Sivry et al. 2010), not to speak of the potential failure of tailing dams. Limiting such studies to a geomorphological approach (Macklin et al. 2006) and in practice to comparison with sediment concentrations with regulated thresholds does not allow a proper quantification of the human health and ecotoxicological risks associated with the spread of mining waste materials. A better approach is to use metal budgets and fluxes, for instance, in the case of a Zn mining catchment by Pavlowsky (1996) allowing interesting conclusions about the stocks of metals in the parts of the abandoned mining area by particle classes with different dimensions (e.g., 69% still in tailing piles, 11% in floodplain deposits, 4% exported by fluvial sediment and the rest unaccounted), and about the timescale of the future impact. The prediction of the fluxes in time has to consider all variables involved, whether mineralogical, biological, hydrological, etc. Already existing information concerning the scale-dependent buffering capacity of rivers systems developed from nutrients in agricultural catchments (Viaud et al. 2004) could be screened for its relevance by extrapolation and adaptation to the buffering of metals in large catchments including mining areas.

One generally assumes an exponential decay pattern of metals with the distance from the mining area (Leenaers and Rang 1989), but these authors and ourselves remarked the importance of local deviations from such patterns due to coupled geomorphological-vegetation processes leading eventually to the formation of large distance contamination hotspots (Iordache et al. 2012). Besides such complications, another source of complex pattern in space is the different dimensions of the particles bearing metals, and to different densities of same size particles of different minerals. Vandeberg et al. (2010) for instance point out such effect resulted from the different densities: "In the form of galena, Pb-rich particles would have a higher specific gravity (7.4–7.6) and be transported shorter distances downstream than other similar-sized ore minerals such as arsenopyrite (6.07); the copper minerals tetrahedrite (4.6–5.1), bornite (5.06–5.08), and chalcopyrite (4.1–4.3); Mn bearing mineral hodochrosite (3.5–3.7) and the Zn bearing mineral sphalerite (3.9–4.1)."

The effect of land plants on the weathering of different minerals is documented (e.g., silicates Drever et al. 1994), but their effects on the mineralogical processes occurring in floodplains contaminated by metals are not clarified, and especially their relative importance in the immobilization – remobilization of metals in the context of all types of biogeochemical process. In the larger context of the mining, catchment plants can have also indirect effects on the export of metals in rivers systems. Zak et al. (2009), for instance, the effect of slope area deforestation on the erosion of fine-grained material from mine wastes leading to the formation of entirely new (contaminated) floodplains.

To conclude this part, the role of mineralogy in the transport of metals in river systems is related to primary, secondary, and ternary phases with specific space–time location along the river continuously. Their dynamic is controlled by a set of parameters including besides mineralogical, geomorphological, and hydrological ones also biological ones (mainly microbial and related to vegetation).

3.6 The Hazard Associated with Tailing Dams in Romania

The existence of several failures of tailing dams with transboundary consequences in the recent past, and the intensive public discussions raised by large-scale mining projects made the problem of hazard assessment of mining areas and in particular of tailing dams to receive a special attention recently from the governmental authorities in Romania (Mara 2010).

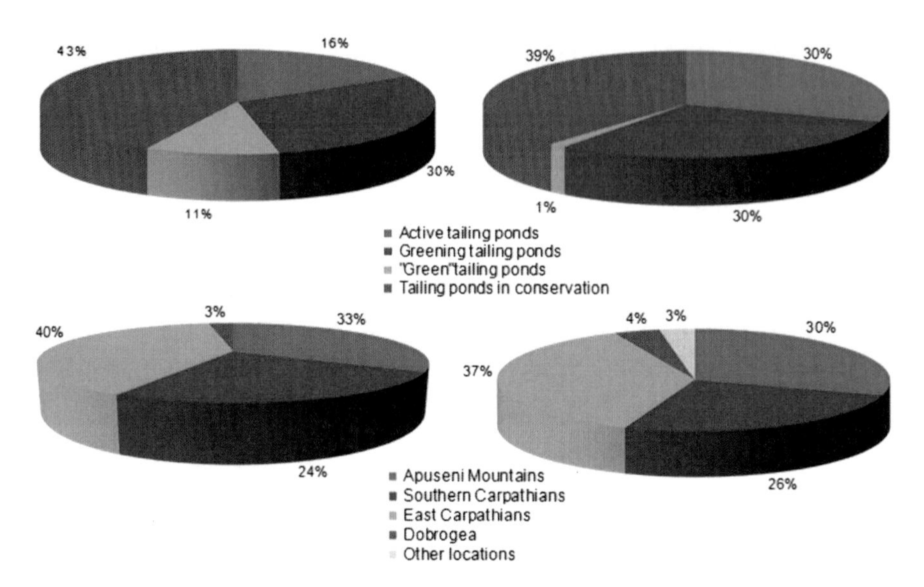

Fig. 3.6 Types of tailing dams and waste deposits (dumps, *up*) in Romania, and their geographical location (*down*)

The inventory of mining wastes allowed the identification of 1,749 tailing dumps: 1,661 mining waste deposits and 73 tailing ponds. From all tailing ponds and mining waste deposits, some are active, some are in conservation, while others are in "greening" state (Fig. 3.6). The largest number and density of tailings and dumps can be found in Eastern Carpathians and in Apuseni Mountains (Fig. 3.6). The Apuseni Mountains are drained by transboundary rivers, so in their case the management of mining areas has international relevance.

For the moment, there is no governmentally accepted procedure for the integrated risk assessment of mining areas at local and catchment scale. The recent national strategy for the management of contaminated sites focused only on the local risks associated with industrially contaminated areas (Iordache et al. 2010). Since the Baia-Mare spill that lead to cross-border contamination, several Romanian and foreign experts worked under the geomorphologic paradigm (Macklin et al. 2005) for the evaluation of current and future risks associated with mining areas, with results published in the international literature. The main problem, however, with the geomorphologic approach proposed by Macklin et al. (2005) is that its focus is on the concentrations of metals and regulated threshold values. The investigation of source-pathway-receptor relationship supposed by this approach as one of the final steps requires not only knowledge about contamination in concentration terms, but also knowledge about stocks and fluxes and their space–time distribution in the mining area, i.e., about hazard potential in the sense introduced in this chapter estimated at multiple scales, from local to regional, and even within local sources in order to devise appropriate remediation strategies considering the internal heterogeneity of the sources. It is worthy to look at the application of this approach in Romania.

Bird et al. (2003) identify, for instance, "exchangeable hotspots" at distance from the mining wastes in terms of extractable concentrations exceeding the guidelines. While this is correct from current regulatory point of view, from a science-based and real hazard estimation point of view, we prefer to define hotspots in terms potential hazard as defined here in the introduction, i.e., stocks (volume and concentrations), and outgoing fluxes of metals. Bird et al. (2008) discuss the system "recovery" after a tailing dam failure in Romania in terms of decrease of metals concentrations in sediment and are surprised by the still high concentrations found. Such an approach looking only for concentrations of metals and not for stocks in abiotic and biotic compartments and the corresponding retention times may also lead to risky conclusions concerning the opportunity for the development of new large mining and tailing facilities (Bird et al. 2005). In the framework introduced in this chapter, the potential hazard associated with the construction of huge tailing dams in existing mining areas is related not only to the eventual increase in the contamination of already contaminated floodplains, but also to the increase in the exportable stocks of metals, to the longer time of the potential export, and to the larger distances downstream reachable by the heavy metals. We have demonstrated (Iordache et al. 2012) that in a catchment with much lower mining and processing activity like Ampoi river large distance hotspots of contamination were formed in the vicinity of densely populated areas. Assessing the existence of such situation and of all other biogeochemical parameters in the Aries catchment is needed in order to have quantitative information on the hazard and risk associated with future mining developments like Rosia Montana. Simply from the dimensions of the extractions and processing envisaged, taking into consideration the surfacing of rocks with large concentration of metals when the current soil cover in the area is not with high geochemical background (and thus has low potential for metals export by erosion, Lacatusu et al. 2009) it is clear the available stock for potential export will be much higher and for a much longer time, and consequently the potential hazard (in the sense defined here) of Rosia Montana mining area would be much larger in case of the implementation of the envisaged project than it is in the current mine wastes situation.

In this context, a portfolio of national and international projects is currently implemented in order to link the local and regional aspects of risk assessment of mining areas into an integrated biogeochemical approach. The research strategy include an extensive research (characterization of the catchment contexts of all tailing dams and mining dumps coupled with screening of the main morphological, mineralogical, geochemical, and geophysical parameters of the waste body), and an intensive research in selected mine wastes.

The intensive research attempts to quantify the potential hazard as defined in the introduction part of this chapter. Tailings volume is estimated using geophysical methods. Geophysical profile characterizes the distribution of formations resistivity, both in the top of the pond and its slopes. By integrating seismic and resistivity surveys (apparent resistivity maps and inversion results), we identify and extrapolate the contact between tailings and natural environment and reconstruct the topographic profile of the valley, before the start of the tailings discharge

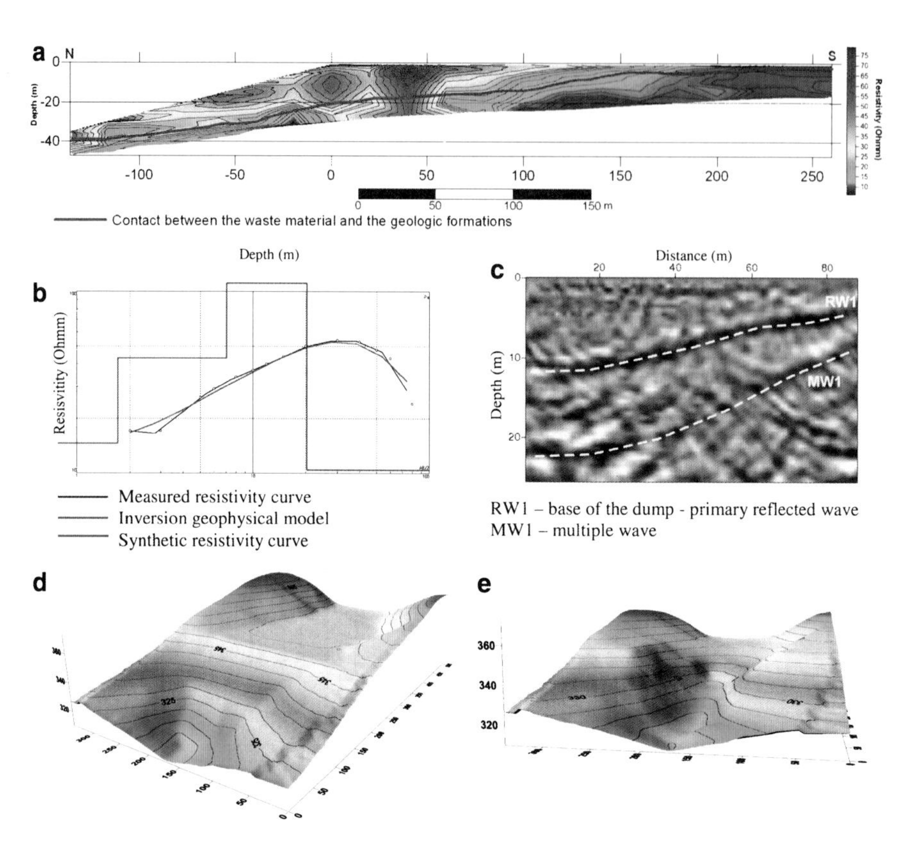

Fig. 3.7 Valea Mica tailing dam, Romania: apparent resistivity profile (**a**), inverse modeling on a vertical electric sounding (**b**), reflection seismic profile (**c**), present-day topographic surface (**d**), paleo-topography of the valley before the start of tailings discharge (**e**), (Orza et al. 2010)

(Orza et al. 2010) (Fig. 3.7). By making the difference between appropriate grids of the surface and paleo-valley topography, the volume in cubic meters is obtained (Radu and Iacob 2011).

Metal content is obtained using geochemical methods: tailings sampling at different levels within the pond and computation of an average metal concentration. An interesting fact observed was the large heterogeneity of the concentrations and pHs on the surface of the tailings (Fig. 3.8). This has important consequences for the design of the remediation strategy (control of the hazard).

After characterizing the stock of metals, we estimated the fluxes by dust, underground and surface water fluxes. Besides local estimations, investigations of the downstream floodplains are performed and tracing of sources of metals using mineralogical and for certain metals (Pb) isotopic proofs. The hazard potential scheme is applied also to the internal structure of the tailing dam in terms of layers and horizontal zones as revealed by geophysical and geochemical zones. For instance, in the case of the tailing dam presented in Fig. 3.2 by correlating mineralogical, geochemical, and geophysical information hazard areas can be

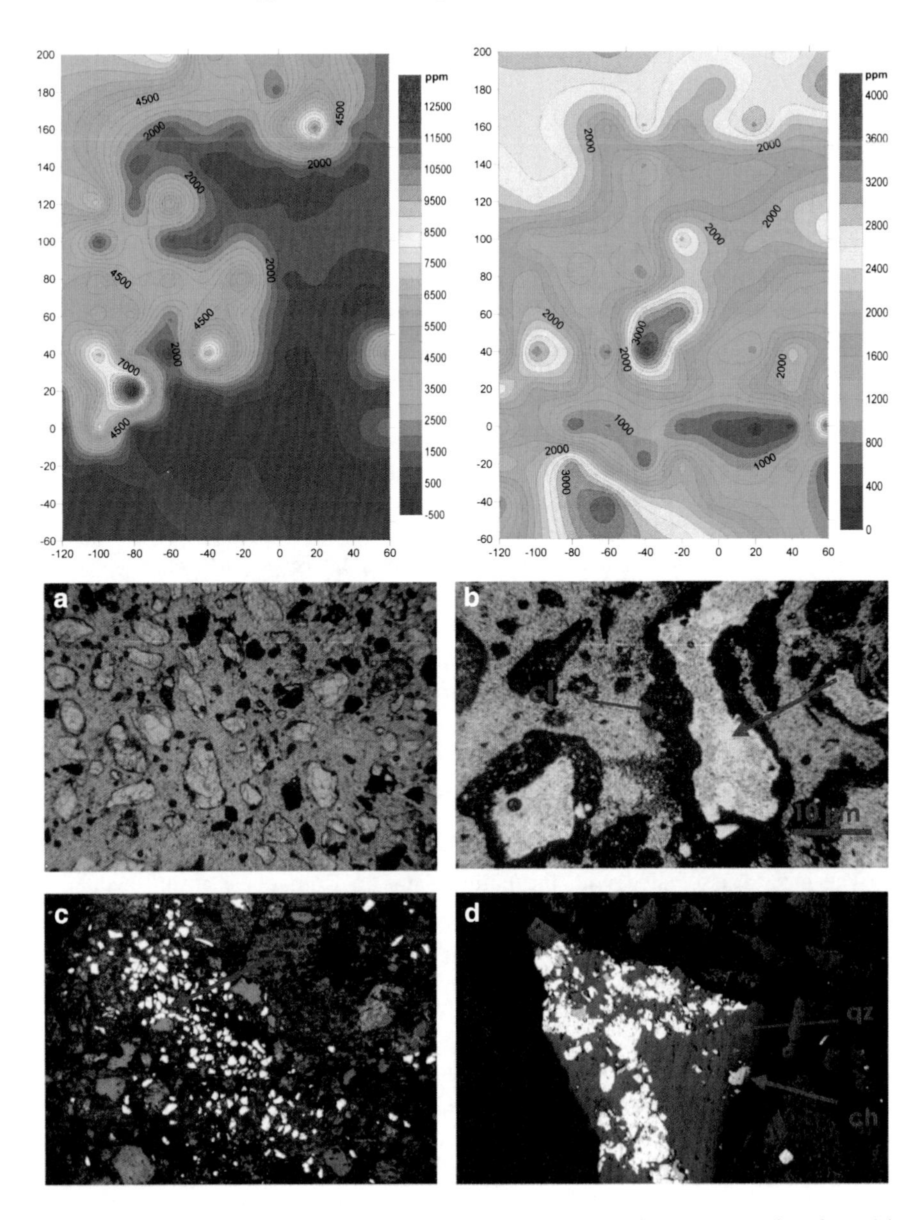

Fig. 3.8 Valea Mica tailing dam, Romania – Illustration of the surface heterogeneity of metals' concentrations (*up* – Cu and Pb from left to right, axes in meters, isolines in ppm d.w.) and of the mineralogy of the waste material (*down* – microscopic images in transmitted (**a**, **b**) and reflected (**c**, **d**) light)

shaped as follows: H1 – areas with metal-bearing sulfides, negative net neutralization potential – NNP and low pH (areas with sulfides partially oxidized, without carbonates; chemical reaction produces acid), H2a – areas with metal-bearing sulfides, positive NNP and high pH (areas with unoxidized sulfides but carbonates and/or secondary gypsum neutralize acid produced), H2b – areas where metal-bearing sulfides have been weathered and with negative NNP and low pH (area without metals possible to mobilize), H3 – areas without metal-bearing sulfides, negative NNP and high pH (mineralogical neutral areas; weathered minerals within these areas do not produce acid or metals possible to mobilize). The horizontal and vertical zonation of the hazard in the tailings is used finally for a structured evaluation of the overall hazard (in conjunction with the estimation of the outgoing fluxes) and for the design of a site-specific remediation plan (controlling the hazard).

Although the potential hazard is multiscale, controlling it when still not manifested is a matter of local measures. Although classical approaches involve the cover of tailings with a soil layer and planting vegetation, this seldom gave the intended results on the average and long term, because once the roots of the vegetation developed deep enough to rich the tailing material the not-adapted vegetation usually do not survive (the inventory of "green" tailing pond mentioned above revealed such situations). Our approach is to use cost-effective phytoremediation techniques with native species (adapted to the region and preferably also to the tailing) and a mixture of tailing material with clean soil and other amendments, Neagoe unpublished data. Case-specific challenges arise, for instance, in the tailings illustrated in Fig. 3.2 a problem was to find a phytoremediation solution able to cope with the large variability of metals and pH. Three solutions were identified: using a mixture of species with different preferences for pH, using one species widely tolerant, and using a widely tolerant species coupled with the seed bank species from a clean soil from the area. Experiments at three scales were performed with these solutions as mentioned in Iordache et al. (2012).

The hazard associated with the tailing dams in Romania is in an evaluation phase-based on the concept potential hazard introduced in this chapter.

3.7 Conclusions

In this chapter, we deduced a biogeochemical concept of potential hazard from a critical analysis of the risk assessment procedure and coupled with scale-specific processes of metals mobility as characterized in Iordache et al. (2012). This concept allows the differentiation between short-term and long-term hazard and local and regional hazard. The mineralogical aspects controlling the stocks of metals in contaminated areas and the outgoing fluxes of metals were then discussed analytically by type of source and type of flux. We will not reiterate here the partial conclusions formulated at the end of subchapters.

The following specific research questions are formulated Hudson-Edwards (2003), and we consider that they are still very actual: "(1) Where do the majority of the source (ore) mineral transformations occur: in mine tailings or waste piles? During transport? In the channel or floodplain after deposition? (2) What are the relative influences of physical processes (i.e., flow rate, abrasion, etc.) and chemical processes (sorption/desorption, dissolution, etc.) on the transformations of source metal-bearing minerals and dispersion of metal contaminants? (3) What are the mineralogical controls on the dispersal and storage of trace heavy metals such as Ag, Bi, Ni, Sb, and Tl within river systems? (4) What are the relative weathering rates of heavy-metal-bearing minerals, and magnitude of related metal release, in fluvial environments? (5) What are the heavy-metal-bearing minerals and mineral coatings present in the particles?"

The first question from above can be extended to all types of primary sources in mining area as referred in this chapter, and especially to contaminated soils. Question (2) can be extended in the light of our analyses to the full range of biogeochemical processes. The study of the influence of plants on the minerals in experimental ecosystems of dumps and contaminated floodplain is an important research direction. Questions (3) and (4) can be put in a large biogeochemical context as follows: which is the relative importance of minerals, organic carbon, microorganisms and plants in controlling the "spiral length" (in the sense introduced in Neagoe et al. 2012) of different metals at scales from within mining waste/contaminated soil to catchments of increasing order and in different conditions of the environmental variables (especially under climate changing scenarios)? Elucidating the environmentally context dependent effect of plants and their associated microorganisms on the retention time of metals in floodplains of mining catchments is a research priority for risk assessment of mining areas both at the hazard assessment and exposure assessment levels.

An applied research direction is the realistic hazard assessment of the mining areas at multiple scales based on a biogeochemical approach, and we have illustrated our strategy for Romanian tailing dams.

Acknowledgments This research was done in the Romanian Consortium for the Biogeochemistry of Trace Elements with financing from National Center for the Management of Projects (CNMP) by projects 31012/2007 FITORISC and 31043/2007 PECOTOX, from National University Research Council (CNCSIS) by project 291/2007 MECOTER, and in the international consortium of the project UMBRELLA, FP7-ENV-2008-1 no. 226870. We thank to the anonymous reviewers for their criticism, which improved the quality of the manuscript.

References

Ackermann J, Vetterlein D, Kuehn T, Aiser K, Jahn R (2010) Minerals controlling arsenic distribution in floodplain soils. Eur J Soil Sci 61:588–598

Adamo P, Dudka S, Wilson MJ, McHardy WJ (1996) Chemical and mineralogical forms of Cu and Ni in contaminated soils from the Sudbury mining and smelting region, Canada. Environ Pollut 91(1):11–19

Adamo P, Dudka S, Wilson MJ, Mchardy WJ (2002) Distribution of trace elements in soils from the Sudbury smelting area (Ontario, Canada). Water Air Soil Pollut 137:95–116

Almeida MP, Andrade JS Jr, Herrmann HJ (2007) Aeolian transport of sand. Eur Phys J 22:195–200

Alvarez-Valero AM, Perez Lopez R, Matos J, Capitan MA, Nieto JM, Saez R, Delgado J, Caraballo M (2008) Potential environmental impact at São Domingos mining district (Iberian Pyrite Belt, SW Iberian Peninsula): evidence from a chemical and mineralogical characterization. Environ Geol 55:1797–1809

Andreotti B (2004) A two species model of aeolian sand transport. J Fluid Mech 520:319

Anneser B, Pilloni G, Bayer A, Lueders T, Griebler C, Einsiedl F, Richters L (2010) High Resolution Analysis of Contaminated Aquifer Sediments and Groundwater—What Can be Learned in Terms of Natural Attenuation? Geomicrobiology Journal 27:130–142

Anterrieu O, Chouteau M, Aubertin M (2010) Geophysical characterization of the large-scale internal structure of waste rock pile from a hard rock mine. Bull Eng Geol Environ 69:533–548

Apitz SE, Brils J, Marcomini A, Critto A, Agostini P, Micheletti C, Pippa R, Scanferla P, Zuin S, Lanczos T, Dercova K, Kocan A, Petric J, Hucko P, Kusnir P (2006) Approaches and frameworks for managing contaminated sediments – a European perspective. In: Lanczos T, Reible D (eds) Assessment and remediation of contaminated sediments. Springer, Dordrecht, pp 5–82

Archibold OW (1985) The metal content of wind-blown dust from uranium tailings in northern Saskatchewan. Water Air Soil Pollut 24:63–76

Aslibekian O, Moles R (2000) Environmental contamination related to mine drainage distribution from old mine sites by waterways. In: Proceedings, 7th international mine water association congress Ustron, pp 49–58

Audry S, Grosbois C, Bril H, Schäfer J, Kierczak J, Blanc G (2010) Post-depositional redistribution of trace metals in reservoir sediments of mining/smelting-impacted watershed (the Lot River, SW France). Appl Geochem 25:778–794

Avila PF, Ferreira da Silva E, Salgueiro AR, Farinha JA (2008) Geochemistry and mineralogy of mill tailings impoundments from the Panasqueira Mine (Portugal): implications for the surrounding environment. Mine Water Environ 27:210–224

Banwart SA, Destouni G, Malmstrom M (1998) Assessing mine water pollution: from laboratory to field scale, groundwater quality: remediation and protection. In: Proceedings of the GQ'98 Conference held at Tubingen, Germany, IAHS Publ. No. 250, pp 307–311

Bennett PC, Rogers JR, Choi WJ (2001) Silicates, silicate weathering, and microbial ecology. Geomicrobiol J 18:3–19

Bini E (2010) Archaeal transformation of metals in the environment. Microbiol Ecol 73:1–16

Bird G, Brewer PA, Macklin MG, Balteanu D, Driga B, Serban M, Zaharia S (2003) The solid state partitioning of contaminant metals and As in river channel sediments of the mining affected Tisa drainage basin, northwestern Romania and eastern Hungary. Appl Geochem 18:1583–1595

Bird G, Brewer PA, Macklin MG, Serban M, Balteanu D, Driga B (2005) Heavy metal contamination in the Aries river catchment, western Romania: implications for development of the Rosia Montana gold deposit. Journal of Geochemical Exploration 86:26–48

Bird G, Brewer PA, Macklin MG, Balteanu D, Serban M, Driga B, Zaharia S (2008) River system recovery following the Novat-Rosu tailings dam failure. Maramures County, Romania

Bird G, Brewer PA, Macklin MG (2010) Management of the Danube drainage basin: implications of contaminant-metal dispersal for the implications of the EU Water Framework Directive. Int J River Basin Manage 8(1):63–78

Bonifacio E, Falsone G, Piazza S (2010) Linking Ni and Cr concentrations to soil mineralogy: does it help to assess metal contamination when the natural background is high? J Soils Sediments. doi:10.1007/s11368-010-0244-0

Borgegard SO, Rydin H (1989) Utilization of waste products and inorganic fertilizer in the restoration of iron-mine tailings. J Appl Ecol 26:1083–1088

Borin S, Ventura S, Tambone F, Mapelli F, Schubotz F, Brusetti L, Scaglia B, Acqui LPD, Solheim B, Turicchia S, Marasco R, Hinrichs KU, Baldi F, Adani F, Daffonchio D (2010) Rock weathering creates oases of life in a high arctic desert. Environ Microbiol 12(2):293–303

Bormann BT, Wang D, Bormann FH, Benoit G, April R, Snyder MC (1998) Rapid, plant induced weathering in an aggrading experimental ecosystem. Biogeochemistry 43:129–155

Bradshaw AD, Chadwick MJ (1980) The restoration of land: the ecology and reclamation of derelict and degraded land. University of California Press, Berkeley, Los Angeles
Broadhurst JK, Petrie JG, von Blottnitz H (2007) Understanding element distribution during primary metal production. Trans Inst Min Metall 116(1):1–16
Byrne P, Reid I, Wood PJ (2009) Short-term fluctuations in heavy metal concentrations during flood events through abandoned metal mines, with implications for aquatic ecology and mine water treatment. In: Abstracts of the international mine water conference, 19th–23rd October, Pretoria, South Africa, pp 124–129
Byrne P, Reid I, Wood PJ (2010) Sediment geochemistry of streams draining abandoned lead/zinc mines in central Wales: the Afon Twymyn. J Soil Sediments 10:683–697
Cabala J, Teper L (2007) Metalliferous constituents of rhizosphere soils contaminated by Zn-Pb mining in Southern Poland. Water Air Soil Pollut 178:351–362
Calvaruso C, Turpault MP, Klett P (2006) Root associated bacteria contribute to mineral weathering and to mineral nutrition in trees: a budgeting analysis. Appl Environ Microb 72(2):1258–1266
Campbell DL, Fitterman DV (2000) Geoelectrical methods for investigating mine dumps. ICARD 2000. In: Proceedings from the fifth international conference on acid rock drainage, pp 1513–1523
Campbell DL, Horton RJ, Bisdorf RJ, Fey DL, Powers MH, Fitterman DV (1999) Some geophysical methods for tailings/mine waste work, tailing and mine waste '99. In: Proceedings of the sixth international conference, Fort Collins, Colorado, January 24–27, 1999, Rotterdam, AA Balkema, pp 35–43
Cappuyns V, Swennen R (2004) Secondary mobilization of heavy metals in overbank sediments. J Environ Monit 6:434–440
Cappuyns V, Swennen R (2007) Classification of alluvial soils according to their potential environmental risk: a case study for Belgian cathments. J Environ Monit 9:319–328
Carlon C, Griove S, Agostini P, Critto A, Marcomini A (2004) The role of multi-criteria decision analysis in a decision support system for rehabilitation of contaminated sites (the DESYRE software). In: Pahl-Wostl C, Schmidt S, Rizzoli AE, Jakeman AJ (eds) Complexity of the integrated resources management, Transactions of the 2nd Biennial Metting of the International Modelling and Software Society. iEMSs, Manno, Switzeland. ISBN 88-900787-1-5
Carlon C, Pizzol L, Critto A, Marcomini A (2008) A spatial risk assessment methodology to support the remediation of contaminated land. Environ Int 34:397–411
Carson JK, Rooney D, Gleeson DB, Clipson N (2007) Altering the mineral composition of soil causes a shift in microbial community structure. Microbiol Ecol 61:414–423
Carson JK, Campbell L, Rooney D, Clipson N, Gleeson DB (2009) Minerals in soil selected distinct bacterial communities in their microhabitats. Microbiol Ecol 67:381–388
Chen W, Fryrear DW (2001) Aerodynamic and geometric diameters of airborne particles. J Sediment Res 71:365–371
Chon H-S, Ohandja D-G, Voulvoulis N (2010) Implementation of E.U. Water Framework Directive: source assessment of metallic substances at catchment levels. J Environ Monit 12:36–47
Chow JC, Watson JG (1998) Guideline on speciated particulate monitoring. Prepared for U.S. EPA. Desert Research Institute, Reno
Ciszewski D (2003) Heavy metals in vertical profiles on the middle Odra River overbank sediments: evidence for pollution changes. Water Air Soil Pollut 143:81–98
Coulthard TJ, Macklin MG (2003) Modeling long-term contamination in river systems from historical metal mining. Geology 31:451–454
Cutting RS, Coker VS, Fellowes JW, Lloyd HR, Vaughan DJ (2009) Mineralogial and morphological constraints on the reduction of Fe(III) minerals by *Geobacter sulfurreducens*. Geochim Cosmochim Acta 73:4004–4022
D'Amore JJ, Al-Abed SR, Scheckel KG, Ryan JA (2005) Methods for speciation of metals in soils: a review. J Environ Qual 34:1707–1745
Dahlin CL, Williamson CA, Collins WK, Dahlin DC (2002a) Part III-Heavy metals: can standard sequential extraction determinations effectively define heavy metal species in superfund site soils? Contam Soils 7:87–114

Dahlin CL, Williamson CA, Collins WK, Dahlin DC (2002b) Sequential extraction versus comprehensive characterization of heavy metal species in brownfield soils. Environ Forensic 3(2):191–201
Davis A, Ruby MV, Bloom M, Schoof R, Freeman G, Bergstrom PD (1996) Mineralogic constraints on the bioavailability of arsenic in smelter-impacted soils. Environ Sci Technol 30(2):392–299
Day G, Dietrich WE, Rowland JC, Marshall A (2008) The depositional web on the floodplain of the Fly River, Papua New Guinea. J Geophys Res 113(F01S02):1–19
DEFRA (2002) Contaminants in soil: collation of toxicological data and intake values for humans. CLR9. Department for the Environment, Food and Rural Affairs and the Environment Agency, Bristol, UK
Dennis IA, Coulthard TJ, Brewer P, Macklin MG (2009) The role of floodplains in attenuating contaminated sediment fluxes in formerly mined drainage basins. Earth Surf Process Landforms 34:453–466
Desenfant F, Petrovsky E, Rochette P (2004) Magnetic signature of industrial pollution of stream sediments and correlation with heavy metals: case study from South France. Water Air Soil Pollut 152:297–312
Diaby N, Dold B, Pfeifer GR, Hollinger C, Johnson DB, Hallberg KB (2006) Microbial communities in a porphyry copper tailings impoundment and their impact on the geochemical dynamics of the mine waste. Environ Microbiol. doi:10.1111/j.1462-2920.2006.01138.x
Dold B, Fontbote L (2001) Element cycling and secondary mineralogy in porphyry copper tailings as a function of climate, primary mineralogy, and mineral processing. J Geochem Explor 74:3–55
Dong H (2010) Mineral-microbe interactions: a review. Front Earth Sci China 4(2):127–147
Dopson M, Lövgren L, Boström D (2009) Silicate mineral dissolution in the presence of acidophilic microorganisms: implications for heap bioleachining. Hydrometallurgy 96:288–293
Drahota P, Filippi M (2009) Secondary arsenic minerals in the environment: a review. Environ Int 35:1243–1255
Drahota P, Mihaljevič M, Grygan T, Rohovec J, Pertold Z (2010) Seasonal variations of Zn, Cu, As and Mo in arsenic-rich stream at the Mokrsko gold deposit, Czech Republic. Environ Earth Sci. doi:10.1007/s12665-010-0538-y
Du Laing G, Rinklebe J, Vandecasteele B, Meers E, Tack FMG (2009) Trace metal behavior in estuarine and riverine floodplain soils and sediments: a review. Sci Total Environ 407:3972–3985
Ettler V, Johan Z, Baronnet A, Jankovsky F, Gilles C, Mihaljevi M, Sebek O, Strnad L, Bencika P (2005) Mineralogy of air-pollution-control residues from a secondary lead smelter: environmental implications. Environ Sci Technol 39(23):9309–9316
Ettler V, Mihaljevič M, Komarek M (2004) ICP-MS measurements of lead isotopic ratios in soils heavily contaminated by lead smelting: tracing the sources of pollution. Anal Bioanal Chem 378:311–317
Ettler V, Mihaljevic M, Sebek O, Molek M, Grygar T, Zeman J (2006) Geochemical and Pb isotopic evidence for sources and dispersal of metal contamination in stream sediments from the mining and smelting district of Prıbram, Czech Republic. Environ Pollut 142:409–417
Farcasanu I, Matache M, Neagoe A, Iordache V (2012) Hyperaccumulation: a key to heavy metal bioremediation. In: Kothe E, Varma A (eds) Bio-geo-interactions in contaminated soils. Springer, Berlin, Heidelberg
Farkas IM, Weiszburg TG, Pekker P, Kuzmann E (2009) A half-century of environmental mineral formation on a pyrite-bearing waste dump in the Mátra Mountains, Hungary. Can Mineral 47:509–524
Faure G, Mensing TM (2005) Isotopes – principles and applications, 3rd edn. Wiley, Hoboken, NJ
Frau F, Ardau C, Fanfari L (2009) Environmental geochemistry and mineralogy of lead at the old mine area of Caccu Locci (South-East Sardinia, Italy). J Geochem Explor 100:105–115
Gadd GM (2007) Geomycology: biogeochemical transformations of rocks, minerals, metals and radionuclides by fungi, bioweathering and bioremediation. Mycol Res 3:3–49

Gadd GM (2010) Metals, minerals and microbes: geomicrobiology and bioremediation. Microbiology 156:609–643
Gallart F, Benito G, Martin-Vide JP, Benito A, PRio JM, Regues D (1999) Fluvial geomorphology and hydrology in the dispersal and fate of pyrite mud particles released by the Aznalcollar mine tailings spill. Sci Total Environ 242:13–26
Gandy CJ, Younger PL (2007) An object-oriented particle tracking code for pyrite oxidation and pollutant transport in mine spoil heaps. J Hydroinform 9(4):293–304
Garrels RM, Christ CL (1965) Solution, minerals and equilibria. Harper & Row, New York
Gay JR, Korre A (2006) A spatially-evaluated methodology for assessing risk to a population from contaminated land. Environ Pollut 142:227–234
Gay RJ, Korre A (2009) Accounting for pH heterogeneity and variability in modeling human health risks from cadmium in contaminated land. Sci Total Environ 407:4231–4237
Gee C, Ramsey MH, Thornton I (1997) Mineralogy and weathering processes in historical smelting slags and their effect on the mobilization of lead. J Geochem Explor 58(2–3):249–257
Gleeson D, McDermott F, Clipson N (2005a) Structural diversity of bacterial communities in a heavy metal mineralized granite outcrop. Environ Microbiol. doi:doi:10.1111/j.1462-2920.2005.00903.x
Gleeson DB, Clipson N, Melville K, Gadd GM, McDermott FP (2005b) Characterization of fungal community structure on a weathered pegmatitic granite. Microb Ecol 0:1–9
Gleyzes C, Tellier S, Astruc M (2002) Fractionation studies of trace elements in contaminated soils and sediments: a review of sequential extraction procedures. Trends Anal Chem 21:451–467
Gommeaux M, Barakat M,Montagnac G,Christen R,Francois Guyot F,Heulin T (2010) Mineral and Bacterial Diversities of Desert Sand Grains from South-East Morocco, Geomicrobiology Journal 27:76–92
Gonzalez-Fernandez O, Jurado-Roldan AM, Queralt I (2010) Geochemical and mineralogical features of overbank and stream sediments of the Beal Wadi (Cartagena-La Union Mining District, SE Spain): relation to former lead-zinc mining activities and its environmental risk. Water Air Soil Pollut. doi:10.1007/s11270-010-0458-1
Green F, Bohannan BJM (2006) Spatial scaling of microbial biodiversity. Trends Ecol Evol 21(9):501–508
Green JL, Holmes AJ, Westoby M, Oliver I, Briscoe D, Dangerfield M, Gillings M, Beattie AJ (2004) Spatial scaling of microbial eukaryote diversity. Nature 432:747–751
Gregurek D, Melcher F, Pavlov VA, Reimann C, Stumpfl EF (1999) Mineralogy and mineral chemistry of snow filter residues in the vicinity of the nickel-copper processing industry, Kola Peninsula, NW Russia. Miner Petrol 65:87–111
Gromet LP, Haskin LA, Orotev RL, Dymek RF (1984) The "North American shale composite": its compilation, major and trace element characteristics. Geochim Cosmochim Acta 48(12):2469–2482
Hall GEM, Vaive JE, Beer R, Hoashi M (1996) Selective leaches revisited, with emphasis on the amorphous Fe oxyhydroxide phase extraction. J Geochem Explor 56:59–78
Hamilton EI (2000) Environmental variables in a holistic evaluation of land contaminated by historic mine wastes: a study of multi-element mine wastes in West Devon, England using arsenic as an element of potential concern to human health. Sci Total Environ 249:171–221
Hammarstrom JM, Seal RR II, Meier AL, Kornfeld JM (2005) Secondary sulfate minerals associated with acid drainage in the eastern US: recycling of metals and acidity in surficial environments. Chem Geol 215:407–431
Hanesch M, Scholger R (2002) Mapping of heavy metal loadings in soil by means of magnetic susceptibility measurements. Environ Geol 42:857–870
Hansen Y, Broadhurst JL, Petrie JG (2008) Modelling leachate generation and mobility from copper sulphide tailings – an integrated approach to impact assessment. Miner Eng 21:288–301
Hillier S, Suzuki K, Cotte-Howells J (2001) Quantitative determination of cerussite (lead carbonate) by X-ray powder diffraction and inferences for lead speciation and transport in stream sediments from a former lead mining area in Scotland. Appl Geochem 16(6):597–608

Hochella MF Jr, Moore JN, Putnis CV, Putnis A, Kasama T, Eberl DD (2005) Direct observation of heavy metal-mineral association from the Clark Fork River Superfund Complex: implications for metal transport and bioavailability. Geochim Cosmochim Acta 69(7):1651–1663

Hofmann T, Schuwirth N (2008) Zn and Pb release of sphalerite (ZnS)-bearing mine waste tailings. J Soils Sediments 8:433–441

Hudson-Edwards KA, Macklin MG, Curtin CH, Vaughan DJ (1996) Processes of formation and distribution of Pb-ZN-, Cd-, and Cu-bearing minerals in the Tyne Basin, Northeast England: implications for metal-contaminated river systems. Environ Sci Technol 30L:72–80

Hudson-Edwards KA, Macklin MG, Curtin CH, Vaughan DJ (1998) Chemical remobilization of contaminant metals within floodplain sediments in an incising river system: implications for dating and chemostratigraphy. Earth Surf Process Landforms 23:671–684

Hudson-Edwards KA (2003) Sources, mineralogy, chemistry and fate of heavy metal-bearing particles in mining-affected river systems. Mineral Mag 67(2):205–217

Hunt CP, Moskowitz BM, Banerjee SK (1995) Magnetic properties of rocks and minerals, rock physics and phase relation – a handbook of physical Constants. American Geophysical Union, ISBN 0-87590-853-5, pp 189–203

Hutchens E, Gleeson D, McDermott F, Caso-Luengo RM, Clipson N (2010) Meter-scale diversity of microbial communities on a wearhered pegmatite granite outcrop in the Wicklow Mountains, Ireland; Evidence for mineral induced selection? Geomicrobiol J 27(1):1–14

Iacob C, Orza R (2008) Integrated interpretation of geophysical data on metalliferous mining waste deposits. In: 14th european meeting of environmental and engineering geophysics, Krakow, Poland, Extended Abstracts, B02

Iacob C, Orza R, Jianu D (2009) Anomalous geo-magnetic effect of acid producing reactions in mine wastes. In: 71st EAGE conference & exhibition, Amsterdam, The Netherlands, Extended Abstracts, R011

Iacob C (2011) Geoelectric signatures of tailing ponds. In: 73rd EAGE conference and exhibition, Vienna, Austria, Extended Abstracts, P 187

Iglesia R, Castro D, Ginocchio R, van der Lelie D, González B (2006) Factors influencing the composition of bacterial communities found at abandoned copper-tailings dumps. J Appl Microbiol 100:537–544

Iordache V (2009) Ecotoxicologia metalelor grele in lunca Dunarii. Ars Docendi, Bucharest

Iordache V, Scradeanu D, Bodescu F, Jianu D, Petrescu L, Neagoe A (2010) Space-time scales of the risk associated to contaminated sites: scientific foundation vs. regulatory framework. In: International conference on environmental legislation, safety engineering and disaster management, Cluj-Napoca, Romania, Book of Abstracts: 19

Iordache V, Lăcătusu R, Scrădeanu D, Onete M, Jianu D, Bodescu F, Neagoe A, Purice D, Cobzaru I (2012) Contributions to the theoretical foundations of integrated modeling in biogeochemistry and their application in contaminated areas. In: Kothe E, Varma A (eds) Bio-geo interactions in contaminated soils. Springer, Berlin, Heidelberg

Isaure MP, Sarret G, Harada E, Choi YE, Marcus MA, Fakra SC, Geoffroy N, Pairis S, Susini J, Clemens S, Manceau A (2010) Calcium promotes cadmium elimination as variate grains by tobacco trichomes. Geochim Cosmichim Acta 74:5817–5834

Jacquat O, Voegelin A, Villard A, Ma M, Kretzschmar R (2008) Formation of Zn-rich phyllosilicate, Zn-layered double hydroxide and hydrozincite in contaminated calcareous soils. Geochim Cosmochim Acta 72:5037–5054

Jambor JL, Raudsepp M, Mountjoy K (2005) Mineralogy of permeable reactive barriers for the attenuation of subsarface contaminants, The Canadian Mineralogist 43: 2117–2140

Jardine PM (2008) Influence of coupled processes on contaminant fate and transport in sub-surface environments. Adv Agron 99:1–100

Jeong GY, Lee BY (2003) Secondary mineralogy and microtextures of weathered sulfide and manganoan carbonates in mine waste-rock dumps, with implications for heavy-metal fixation. Am Mineral 88(11–12):1933–1942

Jordanova D, Hoffmann V, Fehr KT (2004) Mineral magnetic characterization of anthropogenetic magnetic phases in the Danube river sediments (Bulgarian part). Earth Planet Sci Lett 221:71–89

Kaasalainen M, Yli-Halla M (2003) Use of sequential extraction to assess metal partitioning in soils. Environ Pollut 126:225–233

Kimball BA, Bianchi F, Walton-Day K, Runkel RL, Nannucci M, Salvadori A (2007) Quantification of changes in metal loading from storm runoff, Merse River (Tuscany, Italy). Mine Water Environ 26:209–216

Kimball BA, Runkel RL (2009) Spatially delailed quantification of metal loading for decision making: metal mass loading to Americam fork and Mary Ellen Gulch, Utah. Mine Water Environ 28:274–290

Kimball BA, Runkel RL, Walton-Day K (2010) An approach to quantify sources, seasonal change, and biogeochemical process affecting metal loading in streams: facilitating decisions for remediation of mine drainage. Appl Geochem 25:728–740

Koeckritz T, Thoming J, Gleyzes C, Odegard KE (2001) Simplification of sequential extraction scheme to determine mobilisable heavy metal pool in soils. Acta Hydrochimica et Hydrobiologica 29(4):197–205

Kovacs E, Dubbin WE, Tamas J (2006) Influence of hydrology on heavy metal speciation and mobility in a Pb-Zn mine tailing. Environ Pollut 141:310–320

Kukier U, Ishak CF, Sumner ME, Mille WR (2003) Composition and element solubility of magnetic and non-magnetic fly ash fractions. Environ Pollut 123(2):255–266

Kumpiene J, Lagerkvist A, Maurice C (2008) Stabilization of As, Cr, Cu, Pb and Zn in soil using amendments – a review. Waste Manage 28:215–225

Lacatusu R, Citu G, Aston J, Lungu M, Lacatusu AR (2009) Heavy metals soil pollution state in relation to potential future mining activities in the Rosia Montana Area. Carpathian J Earth Envioron Sci 4(2):39–50

Langedal M (1997) Dispersion of tailings in the Knabena–Kvina drainage basin, Norway, 2: mobility of Cu and Mo in tailings-derived fluvial sediments. J Geochem Explor 58 (2–3):173–183

Lattanzi P, Meneghini C, Giudici GD, Podda F (2010) Uptake of Pb by hydrozincite, $Zn_5(CO3)_2(OH)_6$ – implications for remediation. J Hazard Mater 177:1138–1144

Leenaers H, Rang MC (1989) Metal dispersal in the fluvial system of the River Geul: the role of discharge, distance to the source, and floodplain geometry. Sediment Environ 184:47–57

Lian B, Wang B, Pan M, Liu C, Teng HH (2008) Microbial release of potassium from K-bearing minerals by thermophilic fungus *Aspergillus fumigates*. Geochim Cosmochim Acta 71:87–98

Lo IMC, Yang XY (1999) EDTA extraction of heavy metals from different soil fractions and synthetic soils. Water Air Soil Pollut 109:219–236

Lokas E, Wachniew P, Ciszwski D, Owezarek P, Dinh Chau N (2010) Simultaneous use of trace metals, ^{210}Pb and ^{137}Cs in floodplain sediments of a lowland river as indicators of anthropogenic impacts. Water Air Soil Pollut 207:57–71

Loredo J, Álvarez R, Ordónez A, Bros T (2008) Mineralogy and geochemistry of the Texeo Cu-Co mine site (NW Spain): screening tools for environmental assessment. Environ Geol 55:1299–1310

Lottermoser BG (2007) Mine wastes: characterization, treatment and environmental impacts, 2nd edn. Springer, Berlin

Lu SG, Chen YY, Shana HD, Baia SQ (2009) Mineralogy and heavy metal leachability of magnetic fractions separated from some Chinese coal fly ashes, J. Hazardous Materials 169: 246–255

Macklin MG, Brewer PA, Hudson-Edwards KA, Bird G, Coulthard TJ, Dennis IA, Echler PJ, Miller JR, Turner JN (2006) A geomorphological approach to the management of rivers contaminated by metal mining. Geomorphology 79:423–447

Mailloux BJ, Alexandrova E, Keimowitz AR, Wovkulich K, Freyer GA, Herron M, Stolz JF, Kenna TC, Pichler T, Polizzotto ML, Dong H, Bishop M, Knappett PSK (2009) Microbial

mineral wathering for nutrient acquisition releases arsenic. Appl Environ Microbiol 75(8):2558–2565
Manceau A, Lanson B, Schlegel ML, Harge JC, Musso M, Eybert-Berard L, Hazzemann JL, Chateigner D, Lamble GM (2000) Quantitative Zn speciation in smelter-contaminated soils by exafs spectroscopy. Am J Sci 300:289–343
Mara S (2010) NATECH events related to tailings from mining industry in Romania. In: International conference on environmental legislation, safety engineering and disaster management, Cluj-Napoca, Romania, Book of Abstracts: 25
Martin CW (2009) Recent changes in heavy metal storage in flood-plain soils of the Lahn River, Central Germany. Environ Geol 58:803–814
McCarty DK, Moore JN, Marcus WA (1998) Mineralogy and trace element association in an acid mine drainage iron oxide precipitate; comparision of selective extractions. Appl Geochim 13(2):165–176
McDonald JC, Liddell FDK, Gibbs GW, Eyssen GE, McDonald AD (1980) Dust exposure and mortality in chrysotile mining, 1910-75. Br J Ind Med 37:11–24
McGrath D, Zhang C, Carton OT (2004) Geostatistical analyses and hazard assessment on soil lead in Silvermines area, Ireland. Environ Pollut 127:239–248
Mendez MO, Neilson JW, Maier RM (2008) Characterization of bacterial community in an abandoned semiarid lead-zinc mine tailing site. Appl Environ Microbiol 74(12):3899–3907
Meunier L, Walker SR, Wragg J, Parsons MB, Koch I, Jamieson HE, Reimer KJ (2010) Effects of soil composition and mineralogy on the bioaccessibility of arsenic from tailings and soil in gold mine districts of Nova Scotia. Environ Sci Technol 44(7):2667–2674
Mihalík J, Tlustoš P, Szaková J (2010) The impact of an abandoned uranium mining area on the contamination of agricultural land in its surroundings. Water Air Soil Pollut. doi:10.1007/s11270-010-0518-6
Miller JR (1997) The role of fluvial geomorphic processes in the dispersal of heavy metals from mine sites. J Geochem Explor 58(2):101–118
Modis K, Komnitsas K (2008) Dimensionality of heavy metal distribution in waste disposal sites using nonlinear dynamics. J Hazard Mater 156:285–291
Modis K, Papantonopoulos G, Komnitsas K, Papaodysseus K (2008) Mapping optimization based on sampling size in earth related and environmental phenomena. Stoch Environ Res Risk Assess 22:83–93
Monna F, Poujol M, Losno R, Dominik J, Annegarn H, Coetzee H (2006) Origin of atmospheric lead in Johannesburg, South Africa. Atmos Environ 40:6554–6566
Morin G, Jd O, Juillot F, Ildefonse P, Calas G, Brown GE Jr (1999) XAFS determination of the chemical from lead in smelter-contaminated soils and mine tailings: importance of adsorption processes. Am Mineral 84:420–434
Murgueytio AM, Evans RG, Robert D (1980) Relationship between soil and dust lead in a lead mining area and blood lead levels. University of California Press, Berkeley, Los Angeles, 302
Musslewhite CL, McInerney MJ, Dong H, Onstott TC, Green-Blum M, Swift D, Macnauughton WDC, Murray C, Chien YJ (2003) The factorial controlling microbial distribution and activity in the shallow subsurface. Geomicrobiol J 20(3):245–261
Nachtegaal M, Marcus MA, Sonke JE, Vangronsveld J, Livi KLT, van Der Leilie D, Sparls DL (2005) Effects of in situ remediation on the speciation and bioavailability of zinc in a smelter contaminated soil. Geochim Cosmochim Acta 69(19):4649–4664
Navarro A, Cardellach E (2009) Mobilization of Ag, heavy metals and Eu from the waste deposit of the Las Herrerias mine (Almeria, SE Spain). Environ Geol 56:1389–1404
Neagoe A, Iordache V, Farcasanu IC (2012) The role of organic matter in the mobility of metals in contaminated catchments. In: Kothe E, Varma A (eds) Bio-geo-interactions in contaminated soils. Springer, Berlin, Heidelberg
Neuman CM, Boulton JW, Sanderson S (2009) Wind tunnel simulation of environmental controls on fugitive dust emissions from mine tailings. Atmos Environ 43:520–529

Ngiam L, Lim P (2001) Speciation patterns of heavy metals in tropical estuarine anoxic and oxidized sediments by different sequential extraction schemes. Sci Total Environ 275:53–61

Nikolic N, Kostic L, Djordjevic A, Nikolic M (2010) Phosphorus deficiency is the major limiting factor for wheat on alluvium polluted by the copper mine pyrite tailings: a black box approach. Plant Soil. doi:10.1007/s11104-010-0605-x

Nirel PMV, Morel FMM (1990) Pitfalls of sequential extractions. Water Res 24:1055–1056

Orza R, Panea I, Iacob C (2010) Integrating seismic and resistivity surveys on mine wastes. In: 72nd EAGE conference and exhibition, Barcelona, Spain, Extended Abstracts, P514

Paktunc AD (1999) Mineralogical constraints on the determination of neutralization potential and prediction of acid mine drainage. Environ Geol 39(2):103–113

Panfili F, Manceau A, Sarret G, Spadini L, Kirpichtchikova T, Bert V, Laboudigue A, Marcus MA, Ahamdach N, Libert AF (2005) The effect of phytostabilization on Zn speciation in a dredged contaminated sediment using scanning electron microscopy, X-ray flourescente, EXAFS spectroscopy, and principal components analysis. Geochim Cosmichim Acta 69(9):2265–2284

Pavlowsky RT (1996) Fluvial transport and long term mobility of mining-related zinc. In: Tailing and mine waste, PP 395–404, ISBN 9054105941

Pearse MJ (2005) An overview of the use of chemical reagents in mineral processing. Miner Eng 18:139–149

Petrovsky E, Kapi A, Jordanova N, Boruvka L (2001) Magnetic properties of alluvial soils contaminated with lead, zinc and cadmium. J Appl Geophys 48(2):127–136

Piatak NM, Seal RR II, Hammarstrom JM (2004) Mineralogical and geochemical controls on the release of trece elements from slag produced by base- and precious-metal smelting at abandoned mine sites. Appl Geochem 19:1039–1064

Plathe KL (2010) Nanoparticle – heavy metal associations in river sediments. Ph.D. thesis, Polytechnic Institute and State University

Plumlee GS, Ziegler TL (2007) The medical geochemistry of dusts, soils, and other Earth materials, in Lollar BS (editor) Environmental Geochemistry, Elsevier, Amsterdam, pp. 263–310

Poisson J, Chouteau M, Aubertin M, Campos D (2009) Geophysical experiments to image the shallow internal structure and the moisture distribution of mine waste rock pile. J Appl Geophys 67:172–192

Quantin C, Becquer T, Rouiller JH, Berthelin J (2001) Oxide weathering and trace metal release by bacterial reduction in a New Caledonia Ferralsol. Biogeochemistry 53:323–340

Radu IB, Iacob C (2011) Optimizing the volume calculation for tailing ponds. In: 73rd EAGE conference and exhibition, Viena, Austria, Extended Abstracts, SP79

Reeder RJ, Schoonen MAA (2006) Metal speciation and its role in bioaccessibility and bioavailability. Rev Miner Geochem 64:59–113

Rendell PS, Batley GE, Camerun AJ (1980) Adsorption as a control of metal concentrations in sediment extracts. Environ Sci Technol 14:314–318

Riba I, DelValls TA, Forja JM, Gomez Parra A (2002) Influence of the Aznalcollar mining spill on the vertical distribution of heavy metals in sediments from Guadalquivir estuary (SW Spain). Mar Pollut Bull 44:39–47

Rodrigues SM, Pereira ME, Ferreira da Silva E, Hursthouse AS, Duarte AC (2009) A review of regulatory decisions for environmental protection: part I – challenges in the implementation of national soil policies. Environ Int 35:202–213

Romero FM, Armienta MA, Villasenor G, Gonzalez JL (2006) Mineralogical constraints on the mobility of arsenic in tailing from Zimapan, Hidalgo, Mexico. Int J Environ Pollut 26 (1–3):23–40

Romero FM, Prol-Ledesma RM, Canet C, Alvarez LN, Perez-Vazquez R (2010) Acid drainage at the inactive Santa Lucia mine, western Cuba: Natural attenuation of arsenic, barium and lead, and geochemical behavior of rare earth elements. Appl Geochem 25:716–727

Romero FM, Villalobos M, Aguirre R, Gutierrez ME (2008) Solid-phase control on lead bioaccessibility in smelter-impacted soils. Arch Environ Contam Toxicol 55:566–575

Roussel C, Neel C, Bril H (2000) Minerals controlling arsenic and lead solubility in an abandoned gold mine tailing. Sci Total Environ 263:209–219

Saedeleer VD, Cappuyns V, Cooman W, Swennen R (2010) Influence of major elements on heavy metal composition of river sediments. Geologica Belgica 13(3):257–268

Salomons W (1995) Environmental impact of metals derived from mining activities – processes, prediction, prevention. J Geochem Explor 52:5–23

Sangode SJ, Vhatkar K, Patil SK, Meshram DC, Pawar NJ, Gudadhe SS, Badekar AG, Kumaravel V (2010) Magnetic susceptibilitu distribution in the soils of Pune Metropolitan Region: implications to soil magnetometry of anthropogenic loading. Curr Sci 98(4):516–528

Schuwirth N, Voegelin A, Kretzschmar R, Hofmann T (2007) Vertical distribution and speciation of trace metals in weathering flotation residues of zinc/lead sulfide mine. J Environ Qual 36:61–69

Silva EF, Chaosheng Z, Serrano Pinto L, Patinha C, Reis P (2004) Hazard assessment on arsenic and lead in soils of Castromil gold mining area, Portugal. Appl Geochem 19:887–898

Sivry Y, Munoz M, Sappin-Didier V, Riotte J, Denaix L, de Parseval P, Destrigneville C, Dupre B (2010) Multimetallic contamination from Zn-ore smelter: solid speciation and potential mobility in riverine floodbank soils of the upper Lot River (SW France). Eur J Miner 22:679–691

Smith JL, Lee Kiyoung (2003) Soil as a source of dust and implications for human health, Academic, Adv Agron 80

Smith KS, Briggs PH, Campbell DL, Castle CJ, Desborough GA, Eppinger III RG, Fitterman DV, Hageman PL, Leinz RW, Meeker GP, Stanton MR, Sutley SJ, Swayze GA, Yager DB (2000) Tools for the rapid screening and characterization of historical metal-mining waste dumps, In Proceedings of the 2000 Billings Land Reclamation Symposium, Billings, Montana, March 20–24, 2000. Bozeman, Montana State University, Reclamation Research Unit Publication No. 00-01 (CD-ROM). p. 435–442

Smith KS, Campbell DL, Desborough GA, Hageman PL, Leinz RW, Stanton MR, Sutley SJ, Swayze GA, Yager DB (2002) Toolkit for the rapid screening and characterization of waste piles on abandoned mine lands. In: Searl RR II, Foley NK (eds) Geoenvironmental models of mineral deposits, U.S. Geological Survey Open-File Report 02-195, pp 55–64

Sonke Je, Hoogewerff JA, van der Laan SR, Vangronsveld J (2002) A chemical and mineralogical reconstruction of Zn-smelter emissions in the Kemper region (Belgium), based on onrganic pool sediment cores. Sci Total Environ 292(1–2):101–119

Stromberg B, Banwart SA (1999) Experimental study of acidity-consuming processes in mining waste rock: some influences of mineralogy and particle size. Appl Geochim 14(1):1–16

Taylor MP, Hudson-Edwards KA (2008) The dispersal and storage of sediment-associated metals in an arid river system: the Leichhardt River, Mount Isa, Queensland, Australia. Environ Pollut 152:193–204

Teper E (2009) Dust-particle migration around flotation tailings pounds: pine needles as passive samplers. Environ Monit Assess 154:383–391

Tessier A, Campbell PGX, Bisson M (1979) Sequential extraction procedure for the speciation of particulate trace metals. Anal Chem 51:844–851

Tolocka MP, Lake DA, Johnston MV, Wexler AS (2004) Number concentrations of fine and ultrafine particles containing metals. Atmos Environ 38:3262–3273

Tuzen M (2003) Determination of trace metals in the River Yesilirmak sediments in Tokat, Turkey using sequential extraction procedure. Microchem J 74:105–110

Ure AM (1996) Single extraction schemes for soils analysis and related applications. Sci Total Environ 178:3–10

Uroz S, Calvaruso C, Turpault MP, Frey-Klett P (2009) Mineral weathering by bacteria: ecology, actors and mechanisms. Trends Microbiol 17:378–387

Usero J, Gamero M, Morillo J, Gracia I (1998) Comparative study of three sequential extraction procedures for metals in marine sediments. Environ Int 24:487–497

Van Damme A, Degryse F, Smolder E, Sarret G, DEwit J, Swennen R, Manceau A (2010) Zinc speciation in mining and smelter contaminated overbank sediments by EXAFS spectroscopy. Geochim Comsochim Acta 74:3707–3720

Vandeberg GS, Martin CW, Pierzynski GM (2010) Spatial distribution of trace elements in floodplain alluvium of the upper Blackfood River, Montana. Environ Earth Sci. doi:10.1007/s12665-010-0637-9

Vaněk A, Ettler V, Grygar T, Boruvka L, Šebek O, Brabek O (2008) Combined chemical and mineralogical evidence for heavy metal binding in mining- and smelting-affected alluvial soils. Pedosphere 18(4):464–478

Vega FA, Covelo EF, Andrade ML, Marcel P (2004) Relationships between metals content and soil properties in minesoils. Anal Chim Acta 524:141–150

Viaud V, Merot P, Baurdty J (2004) Hydrochemical buffer assessment in agricultural landscapes: from local to catchment scale. Environ Manage 34(4):559–573

Vitkova M, Ettler V, Johan Z, Kribek B, Sebek O, Mihaljevic M (2010) Primary and secondary phases in copper-cobalt smelting slags from the Copperbelt Province, Zambia. Mineral Mag 74(4):581–600

Vojtěch E, Mihaljevič M, Šebek O, Molek M, Grygar T, Zemen J (2006) Geochemical and Pb isotopic evidence for sources and dispersal of metal contamination in stream sediments from the mining and smelting district of Příbram, Czech Republic. Environ Pollut 142:409–417

Yang Y, Li S, Bi X, Wu P, Liu T, Li F, Liu C (2010) Lead, Zn, and Cd in slags, stream sediments, and soils in an abandonated Zn smelting region, southwest of China, And Pb and S isotopes as source tracers. J Soils Sediments. doi:10.1007/s11368-010-0253-z

Zak K, Rohovec J, Navratil T (2009) Fluxes of heavy metals from a highly polluted watershed during flood events: a case study of the Litavka River, Czech Republic. Water Air Soil Pollut 203:343–358

Zhang J, Ngothai Y, Weng W, Zhang Y, Liang W, Zhao J, Mei Y, Xie M, Jia Y, Ma H, Liu P, Gao F, Wang H (2009) A literature survey on synthetic polymetic reagents used in sulfide minerals flotation. In: Proceedings: 8th world congress of chemical engineering, Montréal, Québec, Canada, August 23–27

Zhao LYL, Schulin R, Nowack B (2007) The effects of plants on the mobilization of Cu and Zn in soil columns. Environ Sci Technol 41:2770–2775

Zhu C (2009) Geochemical modeling of reaction paths and geochemical reaction networks. Rev Miner Geochem 70:533–569

Zijlstra JJP, Dessi R, Peretti R, Zucca A (2010) Treatment of percolate from metal sulfide mine tailings with a permeable reactive barrier of transformed red mud. Water Environ Res 82(4):319–327

Chapter 4
Rare Earth Elements in Acidic Systems – Biotic and Abiotic Impacts

Anja Grawunder and Dirk Merten

4.1 Introduction

Rare earth elements (REE) belong to a group of heavy metals with increasing technical application and importance in science. Taylor and McLennan (2003) stated that the term *"rare earth elements,"* or shortly REE, in geochemistry refers to the group La to Lu (atomic number 57–71). Nevertheless, in many investigations, Y is included to form REE or REY with Y because of its similar ionic radius being grouped to the series of REE between Dy and Ho.

Generally, REE are grouped into light rare earth elements (LREE; La to Nd or Pm), middle rare earth elements (MREE, Sm to Dy) and heavy rare earth elements (HREE, Ho to Lu). Promethium (Pm) is the only REE without stable isotopes. The chemical similarity of the REE is due to their electron configuration (Table 4.1). The outermost shell of La has the configuration [Xe] $5d^1$ $6s^2$, while Ce has [Xe] $4f^2$ $6s^2$. The following elements have the electrons entering the 4f sub-shell, until it is filled at Yb (Henderson 1984). These 4f electrons are well shielded by the electrons of the shells $5s^2$ and $5p^6$ and thus, not significantly involved in chemical interactions (Cotton 2006; Henderson 1984). For Gd and Lu, the additional electron enters the 5d and not the 4f shell.

Another characteristic is the decreasing ionic radius from La to Lu, which is also called *lanthanide contraction*. Due to different ionic radii, but also to different valences, REE can substitute elements like Ca^{2+} (most Ln^{3+}) or Sr^{2+} (Eu^{2+}) in a crystal's structure. In the nineteenth century, the interest in REE was rising because of their wide applicability in industry and science. At the same time, analytical methods and extraction methods became more effective. REE are mainly used in high-tech applications, with, e.g., Nd in the Nd:YAG (YAG = $Y_3Al_5O_{12}$) laser for laser ablation inductively coupled plasma mass spectrometry (LA-ICP-MS)

A. Grawunder (✉) • D. Merten
Institute of Geosciences, Friedrich Schiller University Jena, Burgweg 11, 07749 Jena, Germany
e-mail: anja.grawunder@uni-jena.de; dirk.merten@uni-jena.de

E. Kothe and A. Varma (eds.), *Bio-Geo Interactions in Metal-Contaminated Soils*, Soil Biology 31, DOI 10.1007/978-3-642-23327-2_4,

Table 4.1 Rare earth elements and their chemical properties after Wieser and Coplen (2011) (atomic weights) and Brookins (1989) (ionic radii)

Element	Symbol	Atomic number	Atomic weight	Ionic radius (Å)		Electron configuration
				CN6	CN8	
Lanthanum	La	57	138.91	1.032	1.160	[Xe] $5d^1$ $6s^2$
Cerium	Ce	58	140.12	1.011	1.143	[Xe] $4f^2$ $6s^2$
Praseodymium	Pr	59	140.91	0.990	1.126	[Xe] $4f^3$ $6s^2$
Neodymium	Nd	60	144.24	0.983	1.109	[Xe] $4f^4$ $6s^2$
Promethium	Pm	61	–	–	–	[Xe] $4f^5$ $6s^2$
Samarium	Sm	62	150.36	0.958	1.079	[Xe] $4f^6$ $6s^2$
Europium	Eu	63	151.96	0.947	1.066	[Xe] $4f^7$ $6s^2$
Gadolinium	Gd	64	157.25	0.938	1.053	[Xe] $4f^7$ 5d $6s^2$
Terbium	Tb	65	158.93	0.923	1.040	[Xe] $4f^9$ $6s^2$
Dysprosium	Dy	66	162.50	0.912	1.027	[Xe] $4f^{10}$ $6s^2$
Holmium	Ho	67	164.93	0.901	1.015	[Xe] $4f^{11}$ $6s^2$
Erbium	Er	68	167.26	0.890	1.004	[Xe] $4f^{12}$ $6s^2$
Thulium	Tm	69	168.93	0.880	0.994	[Xe] $4f^{13}$ $6s^2$
Ytterbium	Yb	70	173.05	0.868	0.985	[Xe] $4f^{14}$ $6s^2$
Lutetium	Lu	71	174.97	0.861	0.977	[Xe] $4f^{14}$ 5d $6s^2$

CN coordination number

(Cotton 2006). In medicine, Gd also plays an important role, since Gd compounds are used in magnetic resonance imaging (e.g., Thunus and Lejeune 1999).

At very low doses the lanthanides can enhance growth, the so-called *hormesis effect* (Wang et al. 1999). Thus, recently REE have been studied as growth promoters for pigs or other livestock as well as for agriculture (e.g., He et al. 2001). However, this is not a typical REE effect; rather it has been noted that stressors like toxic/radioactive elements in general may induce metabolic rates. A review on metabolism and toxicity of REE stated that many studies on the toxicity of REE have been carried out with rather high concentrations, whereas low-level administration of REE shows minor to nonacute toxic effects (Bulman 2003).

Additionally, REE are studied because of their chemical homology to trivalent actinides (Am^{3+}, Cm^{3+} and under reducing conditions, Np^{3+}), because of their similar behavior (e.g., Chapman and Smellie 1986; Choppin 1983). In geochemistry, Nd and Sm are applied for age determination, rock classification or in investigations of melting processes in Earth's crust and mantle. Fanton et al. (2002) used Nd isotopes for correlation of Ordovician carbonates within a sedimentary basin, while Lee et al. (2008) used $^{143}Nd/^{144}Nd$ ratio as a tool for provenance analysis of river sediments.

Another field of application is the use of REE in environmental studies as process indicators, relating specific fractionations to various processes or factors (Lee et al. 2003; Merten et al. 2005; Xu et al. 2007). Thus, they are applied as tracers, for example in hydrogeology (e.g., tracer in interaction of groundwater and lake Naivasha/Kenya, Ojiambo et al. 2003), in soil erosion studies (Zhu et al. 2010), in pedogenetic processes (Laveuf and Cornu 2009), or in the soil–water–plant system (e.g., Ding et al. 2007; Merten et al. 2005; Worrall and Pearson 2001).

High concentrations in water are especially known not only from acid mine drainage (AMD)-influenced areas, but also from naturally acidic systems. This work will give an introduction in the (bio)geochemistry of REE and their applicability as tracers in soil–water–biota cycling with special focus on low-pH areas. Such areas are often contaminated with a series of heavy metals including REE, which have upcoming importance in economy.

4.2 Abundance of REE in the Environment

4.2.1 REE in Solids

The most common REE minerals are monazite ([Ce,La,Nd,Th][PO_4,SiO_4]) and bastnaesite ([Ce,La]CO_3[F,OH]), which contain especially the LREE La and Ce and sometimes lower amounts of Pr and Nd. Quite often, one REE occurs together with other REE, which was the reason for the long time needed for discovery and separation in history. Primarily, REE appear not only in accessory minerals such as titanite, zircone, or apatite, but also in amphiboles, pyroxenes, feldspars, micas, or garnets (Gaspar et al. 2008; Henderson 1996; Pohl 2005). Further studies on REE as trace elements in minerals were carried out, e.g., for pyrite (Mao et al. 2009), fluorite (Schönenberger et al. 2008; Schwinn and Markl 2005), or chalcopyrite (Rimskaya-Korsakova et al. 2003).

In most cases, REE occur as trace metals, with igneous and sedimentary environments bearing some of the largest REE deposits. Kanazawa and Kamitani (2006) divided REE deposits into three main groups: (1) igneous deposits: hydrothermal deposits, carbonatites, alkaline rocks and alkaline granites; (2) sedimentary deposits: placers and conglomerates; and (3) secondary deposits: weathered residuals of granite. Furthermore, they described the major REE deposits in the world: the Bayan Obo deposit in China, the Mt. Pass carbonatite deposit in USA, the Mt. Weld carbonatite and placer deposits in Australia, and ion adsorption clays, e.g., in Nanling/China (Kanazawa and Kamitani 2006).

REE contents in soils are, like the contents of other heavy metals, the result of a variety of processes, including sorption, surface complexation, precipitation, coprecipitation or dissolution, depending on physico-chemical parameters like pH and Eh, on the parent material, fertilization schemes, or input via groundwater (Laveuf and Cornu 2009). For 30 Swedish top soil samples, REE concentrations in the range of 32–183 μg/g were measured (Tyler and Olsson 2002). Liang et al. (2005) reported higher concentrations for Chinese soils (85–523 μg/g). Åström et al. (2010) worked on acidic sulfate-rich soils of a boreal landscape and found a total REE content of 197 μg/g, whereas Grawunder et al. (2009) reported for

AMD-influenced Quaternary sediments of a former uranium mining area REE contents of 117–278 μg/g.

Sorption of REE, like for other metals, strongly depends on the composition of the sediment and the supply of sorbents. Clay minerals were found to be better sorbents for HREE (Aagaard 1974; Coppin et al. 2002). Laveuf and Cornu (2009) summarized the REE behavior toward different clay minerals from literature and described both: depletion of LREE compared to HREE (kaolinite, chlorite, smectite, montmorillonite) and LREE enrichment compared to HREE (illite, vermiculite).

Quinn et al. (2006a, b, 2007) investigated REE sorption to amorphous ferric hydroxides and found only weak dependence of REE sorption on ionic strength, while dependence on pH was strong and sorption increased with increasing temperature. Ferric (hydr)oxides are important and widespread minerals in AMD. Especially changes in pH and Eh values are influencing both, Fe precipitate formation and REE mobility and/or concentration in groundwater as well as soil solution (e.g., Welch et al. 2009). Zhao et al. (2007) analyzed Fe-bearing AMD precipitates from the Sitai coal mine in China and found REE contents of 50–202 μg/g, indicating an enrichment compared to the coal in their working area with only 18 μg REE/g. Regarding partitioning toward ferric precipitates, REE remain in solution below pH 5.1 (Verplanck et al. 2004). Verplanck et al. (1999) carried out field and laboratory experiments indicating that dissolved REE are affected by Fe- and Al-colloid formation which at pH >4.5 is stronger for Al than for Fe. Furthermore, they observed that during Fe-colloid formation below pH 4.5, REE are removed from solution without altering the REE pattern of the solution.

Coprecipitation mechanisms of REE are not well investigated until today. For seawater, Byrne and Kim (1993) assumed that the REE concentration may be limited by REE-phosphate coprecipitation. Welch et al. (2009) measured the REE enrichment in jarosite and suggested that this mineral is in equilibrium with REE in pore water and highly enriched in LREE. Protano and Riccobono (2002) stated that Al-rich flocs in their environment enrich HREE. Similar results were found for Al-flocs by Grawunder et al. (2010). However, under natural conditions like in groundwater or river water, it is rather impossible to distinguish between adsorption, surface precipitation and coprecipitation (Bozau et al. 2008).

4.2.2 Abundance in Water

The abundance of REE in water strongly depends on the processes discussed above. Thus, waters without AMD influence show lower REE concentrations. In the following, the range of REE in precipitation, river water, seawater, and groundwater without AMD influence and acidic waters will be reviewed.

REE distribution in rain has been studied, e.g., in Japan and the East China Sea (Zhang and Liu 2004) resulting in a quite high variation of REE in rain water ranging between 60 and 1,703 pmol/kg. Lower concentrations have been reported from the central part of South Korea (33–272 ng/L, Ryu et al. 2007). Aubert et al. (2002) found values in the same range for snow sampled in the Alps with about 45–66 ng/L, for rainwater sampled in the Vosges about 10 ng/L and for rainwater collected in Strasbourg about 261 ng/L.

In comparison, higher REE concentrations have been reported from non-acidic river and lake water. An average concentration of 0.16 μg/L for total REE was reported from 15 major rivers in Japan (Uchida et al. 2006). For big rivers like Amazon, Indus, Mississippi, and Ohio total REE concentrations of up to 2 μg/L were found (Goldstein and Jacobsen 1988), or in case of the Xijiang River/South China 22–354 ng REE/L (Xu and Han 2009).

In seawater, input by river water (e.g., Goldstein and Jacobsen 1988; Sholkovitz and Szymczak 2000) and hydrothermal alteration of the oceanic crust (Michard 1989; Michard and Albarède 1986) resulted in 3–146 pmol REE/kg (e.g., Zhang and Nozaki 1996; Kulaksız and Bau 2007). Local differences, especially from anthropogenic Gd release (Kulaksız and Bau 2007), but also a general trend of increase in concentration with depth of the oceanic water column (Alibo and Nozaki 1999) can be seen. Sholkovitz and Szymczak (2000) studied the geochemical behavior of REE in estuaries and their fractionation behavior at different salt concentrations. In detail, they found decreased LREE at low salinity, with a preferential release of HREE at middle to high salinity. In combination, water reaching the ocean shows a typical HREE-enriched pattern for seawater.

Water at circumneutral and slightly acidic pH generally shows REE concentrations in a wide range from few ng/L to some hundred μg/L (Table 4.2). At Carnmenellis (SW England), REE concentrations up to 229 μg/L were found for an aquifer in metasediments (Smedley 1991), whereas organic-rich groundwater in Kervidy/

Table 4.2 The maximum total REE concentrations in different selected waters ranging from AMD-influenced to naturally acidic and circumneutrals

Σ REE (mg/L)	pH	Location	References
AMD-influenced			
29	3.6	Osamu Utsumi mine, Brazil	Miekeley et al. (1992)
8.15	3.4	Gessenwiese, Germany	own data
1.94	3.5	Montevecchio, Sardinia/Italy	Cidu et al. (2011)
0.77	4.1	Lusatia, Germany	Bozau et al. (2004)
0.19	4.3	Guadiamar aquifer, Spain	Olías et al. (2005)
0.07	3.6	Shanxi, China	Zhao et al. (2007)
0.06	3.6	Montana, USA	Verplanck et al. (2004)
Naturally acidic			
5.1	0.3	Copahue volcano, Chile	Varekamp et al. (2009)
3.9	0.1	Ijen volcano, Indonesia	own data
0.23	5.4	Carnmenellis, UK	Smedley (1991)
Circumneutral			
0.023	6.5	Kervidy, France	Dia et al. (2000)
0.004	6.4	Kangwon, Korea	Choi et al. (2009)
0.0005	7.5	Alpine Aquifer, Alps	Biddau et al. (2009)

Coët-Dan (France) had only 23 μg REE/L (Dia et al. 2000) and Alpine carbonate aquifers were even lower with only 516 ng REE/L (Biddau et al. 2009). Another significant fraction of REE in natural waters is associated with organic as well as inorganic colloidal phases (Dia et al. 2000; Gammons et al. 2005; Welch et al. 2009).

Acidic water can reach low as well as very high concentrations depending on the source rock and physicochemical parameters. The highest total REE concentration of 29 mg/L at pH 3.6 has been reported from the Osamu Utsumi uranium mine, an AMD-influenced environment in Brazil with phonolites as bedrock (Table 4.2) (Miekeley et al. 1992). From a former uranium mining site in Thurinia/Germany total REE concentrations of up to 8.15 mg/L at pH 3.4 were determined for groundwater sampled in Quaternary sediments influenced by heap leaching of Paleozoic slates (own data). For Montevecchio/Sardinia, a former Zn–Pb mining site in silicate-dominated host rocks, REE concentrations of up to 1.94 mg/L were found (Cidu et al. 2011). For lignite mining areas, total REE concentrations of 0.07–0.77 mg/L were reported (Bozau et al. 2004; Zhao et al. 2007). Streams and aquifers affected by AMD have lower concentrations due to dilution ranging between 59 and 194 μg/L (Verplanck et al. 2004; Olías et al. 2005).

Naturally acidic systems like the Copahue volcano hot springs (Varekamp et al. 2009) and the Ijen volcano discharge (Indonesia, own data) can have high concentration in REE as well. At the Ijen volcano, 3.9 mg/L Σ REE were determined at a pH of 0.1. At the Copahue volcano, concentrations of up to 5.1 mg/L at pH 0.3 were measured (Varekamp et al. 2009).

4.3 Pathways and Fractionation of REE in AMD-Influenced Areas

4.3.1 Normalization and Anomalies

Besides the ionic radius, redox behavior and complexation behavior control the geochemical behavior of the REE (Taylor and McLennan 2003). When plotting REE abundances against the atomic numbers, the natural REE abundance shows a characteristic zig-zag pattern due to the higher stability of even masses, the Oddo-Harkins effect. To recognize slight variations in the behavior of REE in different samples, it is helpful to normalize the REE concentrations to a reference standard resulting in a characteristic graph called REE pattern, which reflects the normalized concentrations vs. the atomic numbers. Most common standards are the C1-chondrite (Anders and Grevesse 1989), the upper continental crust (UCC), the North American Shale Composite (NASC; Gromet et al. 1984), and the Post Archean Australian Shale (PAAS; McLennan 1989; Nance and Taylor 1976; Taylor and McLennan 1985). C1-chondrite is rather used for interpretations related to earth's estimated primitive mantle and shale standards for recent sedimentological processes.

For interpretation of the normalized values, fractionation ratios and anomalies are calculated. As fractionation, the process of enrichment or depletion of a defined group of REE compared to the other REE is understood. Especially, the ratios between normalized HREE to LREE [e.g. $(Lu/La)_{normalized}$], MREE to LREE [$(Sm/La)_{normalized}$], and HREE to MREE [$(Lu/Sm)_{normalized}$] are of special interest. Moreover, in the normalized patterns of samples taken in areas without AMD influence, but also in AMD-influenced areas often anomalies occur for Ce, Eu, and Gd (Fig. 4.1). These anomalies can be calculated using the normalized nearest neighbors (that do not cause anomalies) of an REE (formulas, e.g., summarized by Lawrence et al. 2006). Values less than 0.95 or higher than 1.05 indicate depletion and enrichment, respectively, as stated by Taylor and McLennan (1985) for the Eu anomaly.

Ce and Eu are two exceptions among the REE, normally existing in trivalent state as dominating valence. Ce appears as Ce^{4+} under oxidizing conditions and Eu as Eu^{2+} under reducing conditions (Brookins 1989; Henderson 1984). Thus, Eu^{2+} can substitute for Ca^{2+} or Sr^{2+} and be incorporated, e.g., in feldspars (Aubert et al. 2001; Compton et al. 2003). Eu anomalies are common in metamorphic-altered regions and under hydrothermal conditions (e.g., Kikawada et al. 2001; Whitney and Olmsted 1998). Bau (1991) found that the Eu^{3+}/Eu^{2+} redox potential depends strongly on temperature, slightly on pH and is almost unaffected by pressure.

Ce anomalies can have various origins, abiotic as well as biotic ones. Abiotic Ce oxidation was described as scavenging process by Fe-/Mn-(hydr)oxides (e.g., Bau 1999;

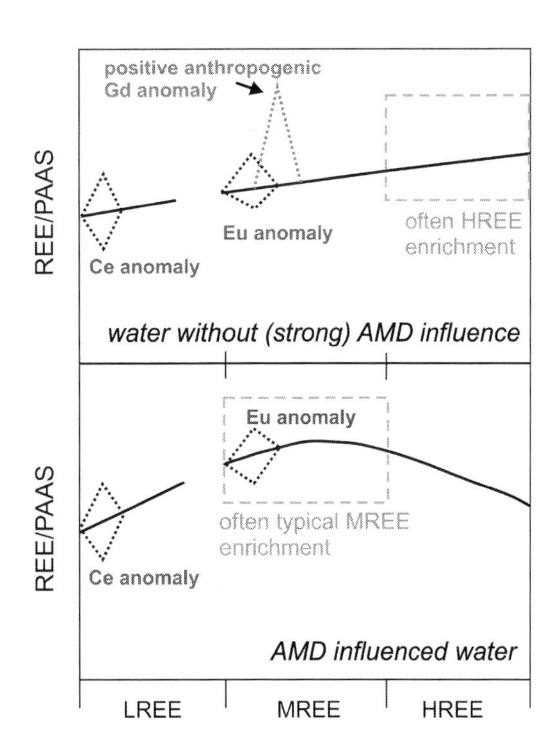

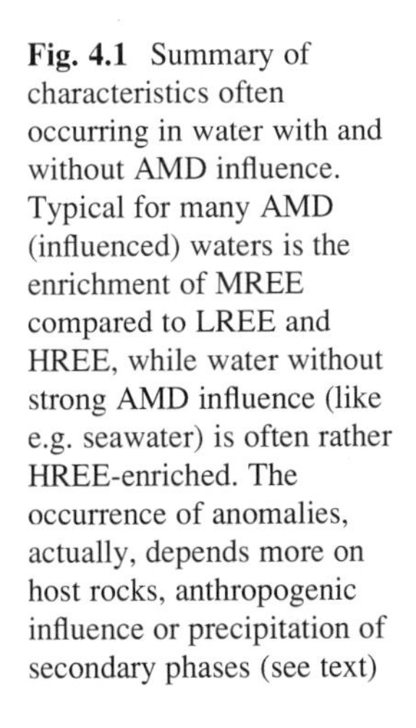
Fig. 4.1 Summary of characteristics often occurring in water with and without AMD influence. Typical for many AMD (influenced) waters is the enrichment of MREE compared to LREE and HREE, while water without strong AMD influence (like e.g. seawater) is often rather HREE-enriched. The occurrence of anomalies, actually, depends more on host rocks, anthropogenic influence or precipitation of secondary phases (see text)

Ohta and Kawabe 2001). Bau (1999) investigated the REE scavenging of precipitating Fe-oxyhydroxides and found an oxidative scavenging of Ce for pH ≤4.6. He explained the Ce enrichment for ferric (hydr)oxides in three steps. After sorption of trivalent REE to Fe-(hydr)oxides surface, a partial oxidation of Ce^{3+} to Ce^{4+} follows. Trivalent REE desorb preferentially compared to Ce^{4+}, resulting in an enrichment of Ce on Fe-(hydr)oxides (Bau 1999). This is most likely transferrable to Mn-(hydr)oxides that were found to be enriched in Ce, as are Mn nodules or crusts (Cui et al. 2009; Nagender Nath et al. 1992; Neaman et al. 2004). Microorganisms are able to oxidize Ce^{3+} (Moffett 1990, 1994) to the immobile Ce^{4+} causing negative Ce anomalies in the aqueous and positive in the solid phase as well (De Baar et al. 1988). Generally, negative Ce anomalies are a characteristic of oceanic water (De Baar et al. 1988; Elderfield and Greaves 1982), but also common in groundwater and river water (Dia et al. 2000; Smedley 1991). In some studies, the impact of humic acids on the development of Ce anomalies was investigated resulting in negative Ce anomalies in alkaline organic-rich waters and positive Ce anomalies in alkaline, waters rather poor in organics (Pourret et al. 2008). Another REE showing anomalies is Gd. Most publications dealing with Gd anomalies, especially in river water of densely settled areas describe them as anthropogenic anomalies (e.g., Bau and Dulski 1996; Knappe et al. 2005; Möller et al. 2000). Bau and Dulski (1996) presented strongly positive Gd anomalies for Berlin and attributed them to the use of Gd compounds in magnetic resonance tomography (MRT) because of the high magnetic moment of Gd. In the following years, positive Gd anomalies also have been reported for rivers, e.g., from Japan (Nozaki et al. 2000) and France (Elbaz-Poulichet et al. 2002; Rabiet et al. 2009). Knappe et al. (2005) stated that the half-filled 4f electron shell of Gd leads to a higher solution complexation compared to the other REE. Consequently, Gd would remain in solution in higher amounts than the neighboring elements.

Apart from possible anomalies, AMD influenced sediments or soils, in many cases show flat shale-normalized patterns (Fernández-Caliani et al. 2009; Grawunder et al. 2009). Mine soils with a slight HREE depletion have also been described (Pérez-López et al. 2010). Depending on structure and formation, even single minerals can carry different signatures and thus, their formation or dissolution influence the REE patterns of water. For example, uraninite and apatite were found to be MREE-enriched (e.g., Grandjean-Lécuyer et al. 1993; Lécuyer et al. 2004; McLennan 1994). A detailed review on the use of REE to trace pedogenetic processes was given by Laveuf and Cornu (2009). Amongst others, they focused on the release of REE in soils during weathering, their fixation in secondary minerals, and also on the impact of argilluviation or redox potential and their recycling in the system soil–plant.

REE patterns in water are a combination of release at a source due to mineral weathering or desorption and solution complexation. In contrast to seawater or river water, which is rather HREE-enriched (e.g., Tricca et al. 1999, Fig. 4.1), acidic water often shows MREE enrichment after normalization to a shale standard (e.g., Johannesson and Lyons 1995; Johannesson and Zhou 1999, Fig. 4.1). In the case of the water samples of a slate-dominated area, MREE and at the same time HREE were enriched (Merten et al. 2005), while from an acidic coal mining environment in Lusatia/Germany, MREE as well as LREE enrichment was reported (Bozau et al. 2004). The source of the widely spread MREE enrichment is not well-understood until now and still under discussion. Besides dissolution of MREE-enriched minerals as, e.g., secondary formed sulfate efflorescent salts in the São Domingos mine area/Spain (Pérez-López et al. 2010), also leaching of Fe–Mn-(oxyhydr)oxides (Johannesson and Zhou 1999) or phosphate dissolution were taken into account (Hannigan and Sholkovitz 2001).

Further fractionation can be caused by sorption and coprecipitation to sediments or secondary precipitates formed on site (e.g., Al-flocs enriched in HREE, Protano and Riccobono 2002; Jarosite enriched in LREE, Welch et al. 2009) leading, e.g., to a change in the REE pattern from a source area along the flow path.

So far, only a basic understanding of sources, fractionations and anomalies has been reached, which should be studied more in detail since application of REE as tracers in the environment is a promising tool.

4.3.2 REE Speciation in Natural and AMD-Influenced Water

In aqueous solution, the REE speciation strongly depends on pH, since REE also form strong complexes with carbonate under circumneutral conditions (Johannesson et al. 1996b; Tang and Johannesson 2005). A detailed review summarizing different complexation distributions for different pH conditions was presented by Johannesson and Zhou (1997). For circumneutral conditions, they stated that carbonate complexation is the dominating process with increasing proportions of REE–carbonate complexes formed from LREE to HREE. Under acidic conditions, like in AMD-influenced environments, the dominating ligand for REE is SO_4^{2-} (Gimeno Serrano et al. 2000; Johannesson and Lyons 1995; Verplanck et al. 2004). Besides sulfate complexes, REE in acidic environments exist as free metal ions. With increasing ionic strength, cation competition (e.g., with Ca^{2+} or Mg^{2+}) becomes more important and thus, the proportion of free, uncomplexed REE increases (Johannesson et al. 1996a). In acidic water, also F^- was found to be an important complexing agent for REE, but high concentrations of Al can lead to competitive effects, because Al–F complexes are much stronger than REE–F complexes (Gimeno Serrano et al. 2000).

In organic-rich water, REE species including organic compounds are known to be important. However, research is not yet distributed widely. Complexation of REE with dissolved humic substances can prevent Ce anomaly development during partitioning with Fe- or Mn-oxyhydroxides, respectively (Davranche et al. 2004, 2005; Pourret et al. 2007a). Complexation experiments with REE and humic acids at various pH conditions could show that the higher the pH, the higher the amount of REE bound to humic acids (Pourret et al. 2007b).

4.4 REE and the Biosphere

4.4.1 Microbial Influences

In AMD-influenced areas, consortia of microorganisms adapted to these conditions colonize the soils. These organisms can tolerate high metal concentrations, low pH conditions, and low nutrient supply. In bioremediation, such microorganisms are applied to improve the metal uptake from soil into plants since they support the element solubilization and transfer (e.g., Aouad et al. 2006). However, bacteria and fungi are also able to stabilize metals in the soil by intracellular or extracellular complexation, thus minimizing the translocation to plants, depending on species and environmental conditions.

In AMD-influenced areas, *Acidithiobacillus ferrooxidans* can catalyze pyrite oxidation (Wakao et al. 1982) being the main source of AMD. Thus, pH and concentration of metals in the solid and aqueous phase are directly influenced microbially. Bacteria adapted to high concentrations of metals such as Ni (e.g., Haferburg et al. 2008) or Cd (e.g., Sinha and Mukherjee 2008; Siñeriz et al. 2009) as well as fungi tolerant to REE at 50 mM (D'Aquino et al. 2009) have been described. Tsuruta (2005, 2007) investigated accumulation and separation of REE in solution by microorganisms and found preferential removal of single REE, a potential cause of fractionation in natural samples, especially with gram-positive bacteria such as *Bacillus subtilis* or *Rhodococcus elythropolis* (Tsuruta 2006, 2007). Binding of metals to cell walls has been studied widely for biosorption. Sorption of Gd depending on bacterial physiological stage was found for, e.g., living *Mycobacterium smegmatis* (Andrès et al. 2000). Haferburg et al. (2007a) used microorganisms isolated from an acid mine drainage area to carry out sorption experiments. Normalized to the AMD stock solution, they found preferential removal of HREE during the first week for all used microorganisms. However, after 4 weeks and presumably related to death of microbes, the REE concentrations in the solution again increased indicating that the sorption process is at least partially reversible due to lysis of cells after a longer incubation time. A fungal strain isolated from a former uranium mining site induced a fractionation among the REE by increasing the ratio of light (La) to heavy (Lu) REE (Haferburg et al. 2007b). Incubation with *Escherichia coli* and *Schizophyllum commune* in water sampled in the same AMD-influenced area revealed no significant fractionation for *E. coli*, while *S. commune* enriched especially HREE (Merten et al. 2004). This is in contrast to Takahashi et al. (2005), who performed adsorption experiments using *B. subtilis* and *E. coli* and found HREE to have a higher affinity for sorption. In this case, especially the different behavior of *E. coli* is interesting, which might be the result of different experimental conditions such as pH or competing ions in solution.

In future studies on this topic, a focus should be laid on the boundary conditions influencing the sorption to bacterial cells and the differentiation of fractionation due to sorption or uptake under these conditions.

4.4.2 Behavior of REE in plants

Uptake of REE into plants has been studied on a series of different plant species. REE are part of many fertilizers and their uptake into plants is governed by factors like plant species, growing season, environmental factors and geochemistry/availability of REE in soil – in turn influenced by soil microbes (Aidid 1994; Wang et al. 1997). Zhang et al. (2002) suggested that there is no significant fractionation of REE from soil to root, when studying various plants in a granite weathering zone in China. REE are connected to the water-soluble fraction of soil, however, fractionation during transport through the plant occurs as a result of active uptake and delivery of anions into xylem vessels (Zhang et al. 2002). Aidid (1994) studied the abundance and distribution of REE in *Pelthophorum pterocarpum* and *Impatiens balsamina*. For *P. pterocarpum*, they found increasing accumulation of REE with age controlled by the available amount of REE in soil. For *I. balsamina*, a stronger enrichment of REE in the roots as compared to the leaves was found (Aidid 1994). In ferns, enrichment for REE in leaves (*Dicranopteris dichotoma*, 3,358 μg Σ REE/g leaves, 38.6 μg REE/g root; Wang et al. 1997) was observed. Fractionation of REE in plants including characteristics and mechanisms has been reviewed by Liang et al. (2008), who stated that fractionation can be used to trace REE's pathways from soil to plant. They also concluded that concentrations in plants under natural conditions are very low and suggested the necessity of finding suitable study areas and plant species for studies on REE fractionation in plants. AMD-influenced areas can provide the necessarily high available concentrations of REE, but, of course, also stress factors for plants like acidic pH or low nutrients. Merten et al. (2005) investigated plants (*Geum urbanum* and *Geranium robertianum*) from a former AMD-influenced waste rock dump. They found that for a single plant species the REE distribution was quite similar, and that REE were more strongly accumulated in the leaves than in the shoots. Furthermore, they investigated the REE patterns of trees (*Populus balsamifera-Hybr.*, *Betula pendula*, and *Robinia pseudoacacia*) and found higher total concentrations for *P. balsamifera-Hybr.* than for *B. pendula* and *R. pseudoacacia* (Merten et al. 2005) with MREE enrichment and negative Ce anomalies for *P. balsamifera-Hybr.* and *R. pseudoacacia*.

4.5 Conclusion

REE are of high and increasing importance in high-tech applications and also in agriculture. They occur especially in acidic environments with higher mobility, leading to higher concentrations in water, secondary formed minerals and biota. This higher mobility allows to study REE fractionation in the system soil–water–biota and to delineate processes relevant for the cycling of heavy metals in general such as coprecipitation, sorption, uptake by biota and translocation in plants. Since many microorganisms can act as growth promoters for plants, future studies of REE related to microbial consortia will gain more and more interest in future.

References

Aagaard P (1974) Rare earth elements adsorption on clay minerals. Bull Groupe Fr Argiles 26(2):193–199

Aidid SB (1994) Rare earth element abundance and distribution patterns in plant materials. J Radioanal Nucl Chem 183(2):351–358

Alibo DS, Nozaki Y (1999) Rare earth elements in seawater: particle association, shale-normalization, and Ce oxidation. Geochim Cosmochim Acta 63(3/4):363–372

Anders E, Grevesse N (1989) Abundances of the elements – meteoritic and solar. Geochim Cosmochim Acta 53(1):197–214

Andrès Y, Thouand G, Boualam M, Mergeay M (2000) Factors influencing the biosorption of gadolinium by micro-organisms and its mobilisation from sand. Appl Microbiol Biotechnol 54:262–267

Aouad G, Stille P, Crovisier JL, Geoffroy VA, Meyer JM, Lahd-Geagea M (2006) Influence of bacteria on lanthanide and actinide transfer from specific soil components (humus, soil minerals and vitrified municipal solid waste incinerator bottom ash) to corn plants: Sr-Nd isotope evidence. Sci Total Environ 370:545–551

Åström ME, Nystrand M, Gustafsson JP, Österholm P, Nordmyr L, Reynolds JK, Peltola P (2010) Lanthanoid behaviour in an acidic landscape. Geochim Cosmochim Acta 74:829–845

Aubert D, Stille P, Probst A (2001) REE fractionation during granite weathering and removal by waters and suspended loads: Sr and Nd isotopic evidence. Geochim Cosmochim Acta 65(3): 387–406

Aubert D, Stille P, Probst A, Gauthier-Lafaye F, Pourcelot L, Del Nero M (2002) Characterization and migration of atmospheric REE in soils and surface waters. Geochim Cosmochim Acta 66(19):3339–3350

Bau M (1991) Rare-earth element mobility during hydrothermal and metamorphic fluid-rock interaction and the significance of the oxidation state of europium. Chem Geol 93(3–4): 219–230

Bau M (1999) Scavenging of dissolved yttrium and rare earths by precipitating iron oxyhydroxide: experimental evidence for Ce oxidation, Y-Ho fractionation, and lanthanide tetrad effect. Geochim Cosmochim Acta 63(1):67–77

Bau M, Dulski P (1996) Anthropogenic origin of positive gadolinium anomalies in river waters. Earth Planet Sci Lett 143(1–4):245–255

Biddau R, Bensimon M, Cidu R, Parriaux A (2009) Rare earth elements in groundwater from different Alpine aquifers. Chem Erde 69(4):327–339

Bozau E, Leblanc M, Seidel JL, Stärk HJ (2004) Light rare earth elements enrichment in an acidic mine lake (Lusatia, Germany). Appl Geochem 19:261–271

Bozau E, Göttlicher J, Stärk HJ (2008) Rare earth element fractionation during the precipitation and crystallisation of hydrous ferric oxides from anoxic lake water. Appl Geochem 23(12): 3473–3486

Brookins DG (1989) Aqueous geochemistry of rare earth elements. Rev Mineral Geochem 21(1):201–225

Bulman RA (2003) Metabolism and toxicity of the lanthanides. In: Sigel A, Sigel H (eds) The lanthanides and their interrelations with biosystems. Metal ions in biological systems. CRC, New York, pp 683–706

Byrne RH, Kim KH (1993) Rare-earth precipitation and coprecipitation behavior – the limiting role of PO_4^{3-} on dissolved rare-earth concentrations in seawater. Geochim Cosmochim Acta 57(3):519–526

Chapman NA, Smellie JAT (1986) Natural analogues to the conditions around a final repository for high-level radioactive waste – introduction and summary of the workshop. Chem Geol 55(3–4):167–173

Choi HS, Yun ST, Koh YK, Mayer B, Park SS, Hutcheon I (2009) Geochemical behavior of rare earth elements during the evolution of CO_2-rich groundwater: a study from the Kangwon district, South Korea. Chem Geol 262:318–327

Choppin GR (1983) Comparison of the solution chemistry of the actinides and lanthanides. J Less-Common Met 93(2):323–330
Cidu R, Frau F, Da Pelo S (2011) Drainage at abandoned mine sites: natural attenuation of contaminants in different seasons. Mine Water Environ 30:113–126
Compton JS, White RA, Smith M (2003) Rare earth element behavior in soils and salt pan sediments of a semi-arid granitic terrain in the Western Cape, South Africa. Chem Geol 201 (3–4):239–255
Coppin F, Berger G, Bauer A, Castet S, Loubet M (2002) Sorption of lanthanides on smectite and kaolinite. Chem Geol 182:57–68
Cotton S (2006) Lanthanide and actinide chemistry. Wiley, Chichester
Cui Y, Liu J, Ren X, Shi X (2009) Geochemistry of rare earth elements in cobalt-rich crusts from the Mid-Pacific M seamount. J Rare Earths 27(1):169–176
D'Aquino L, Morgana M, Carboni MA, Staiano M, Antisari MV, Re M, Lorito M, Vinale F, Abadi KM, Woo SL (2009) Effect of some rare earth elements on the growth and lanthanide accumulation in different Trichoderma strains. Soil Biol Biochem 41:2406–2413
Davranche M, Pourret O, Gruau G, Dia A (2004) Impact of humate complexation on the adsorption of REE onto Fe oxyhydroxide. J Colloid Interface Sci 277(2):271–279
Davranche M, Pourret O, Gruau G, Dia A, Le Coz-Bouhnik M (2005) Adsorption of REE(III)-humate complexes onto MnO_2: experimental evidence for cerium anomaly and lanthanide tetrad effect suppression. Geochim Cosmochim Acta 69(20):4825–4835
De Baar HJW, German CR, Elderfield H, Vangaans P (1988) Rare-earth element distributions in anoxic waters of the Cariaco trench. Geochim Cosmochim Acta 52(5):1203–1219
Dia A, Gruau G, Olivie-Lauquet G, Riou C, Molenat J, Curmi P (2000) The distribution of rare earth elements in groundwaters: assessing the role of source-rock composition, redox changes and colloidal particles. Geochim Cosmochim Acta 64(24):4131–4151
Ding SM, Liang T, Yan JC, Zhang ZL, Huang ZC, Xie YN (2007) Fractionations of rare earth elements in plants and their conceptive model. Sci China Ser C Life Sci 50(1):47–55
Elbaz-Poulichet F, Seidel JL, Othoniel C (2002) Occurrence of an anthropogenic gadolinium anomaly in river and coastal waters of Southern France. Water Res 36(4):1102–1105
Elderfield H, Greaves MJ (1982) The rare earth elements in seawater. Nature 296:214–219
Fanton KC, Holmden C, Nowlan GS, Haidl FM (2002) Nd-143/Nd-144 and Sm/Nd stratigraphy of upper Ordovician epeiric sea carbonates. Geochim Cosmochim Acta 66(2):241–255
Fernández-Caliani JC, Barba-Brioso C, De la Rosa JD (2009) Mobility and speciation of rare earth elements in acidic minesoils and geochemical implications for river waters in the southwestern Iberian margin. Geoderma 149:393–401
Gammons CH, Wood SA, Nimick DA (2005) Diel behaviour of rare earth elements in a mountain stream with acidic to neutral pH. Geochim Cosmochim Acta 69(15):3747–3758
Gaspar M, Knaack C, Meinert LD, Moretti R (2008) REE in skarn systems: a LA-ICP-MS study of garnets from the Crown Jewel gold deposit. Geochim Cosmochim Acta 72(1):185–205
Gimeno Serrano MJ, Sanz LFA, Nordstrom DK (2000) REE speciation in low-temperature acidic waters and the competitive effects of aluminum. Chem Geol 165(3–4):167–180
Goldstein SJ, Jacobsen SB (1988) Rare-earth elements in river waters. Earth Planet Sci Lett 89(1):35–47
Grandjean-Lécuyer P, Feist R, Albarede F (1993) Rare-earth elements in old biogenic apatites. Geochim Cosmochim Acta 57(11):2507–2514
Grawunder A, Lonschinski M, Merten D, Büchel G (2009) Distribution and bonding of residual contamination in glacial sediments at the former uranium mining leaching heap of Gessen/Thuringia, Germany. Chem Erde 69(S2):5–19
Grawunder A, Lonschinski M, Boisselet T, Merten D, Büchel G (2010) Hydrogeochemistry of rare earth elements in an AMD-influenced area. In: Proceedings of IMWA 2010: "Mine Water and Innovative Thinking", Sydney/Canada, September 05.-09, 2010, pp. 347–350
Gromet LP, Dymek RF, Haskin LA, Korotev RL (1984) The "North American shale composite": its compilation, major and trace element characteristics. Geochim Cosmochim Acta 48:2469–2482

Haferburg G, Merten D, Büchel G, Kothe E (2007a) Biosorption of metal and salt tolerant microbial isolates from a former uranium mining area. Their impact on changes in rare earth element patterns in acid mine drainage. J Basic Microbiol 47(6):474–484

Haferburg G, Reinicke M, Merten D, Büchel G, Kothe E (2007b) Microbes adapted to acid mine drainage as source for strains active in retention of aluminium or uranium. J Geochem Expl 92:196–204

Haferburg G, Kloess G, Schmitz W, Kothe E (2008) "Ni-struvite" – a new biomineral formed by a nickel resistant *Streptomyces acidiscabies*. Chemosphere 72(3):517–523

Hannigan RE, Sholkovitz ER (2001) The development of middle rare earth element enrichments in freshwaters: weathering of phosphate minerals. Chem Geol 175:495–508

He ML, Ranz D, Rambeck WA (2001) Study on the performance enhancing effect of rare earth elements in growing and fattening pigs. J Anim Physiol Anim Nutr 85:263–270

Henderson P (1984) General geochemical properties and abundance of the rare earth elements. In: Henderson P (ed) Rare earth element geochemistry. Elsevier, Amsterdam, pp 1–32

Henderson P (1996) The rare earth elements: introduction and review. In: Jones AP, Wall F, Williams CT (eds) Rare earth minerals: chemistry, origin and ore deposits. Chapman & Hall, London, pp 1–17

Johannesson KH, Lyons WB (1995) Rare-earth element geochemistry of Color Lake, an acidic fresh-water lake on Axel-Heiberg-Island, Northwest-Territories, Canada. Chem Geol 119 (1–4):209–223

Johannesson KH, Lyons WB, Yelken MA, Gaudette HE, Stetzenbach KJ (1996a) Geochemistry of the rare-earth elements in hypersaline and dilute acidic natural terrestrial waters: complexation behavior and middle rare-earth element enrichments. Chem Geol 133(1–4):125–144

Johannesson KH, Stetzenbach KJ, Hodge VF, Lyons WB (1996b) Rare earth element complexation behavior in circumneutral pH groundwaters: assessing the role of carbonate and phosphate ions. Earth Planet Sci Lett 139:305–319

Johannesson KH, Zhou X (1997) Geochemistry of the rare earth elements in natural terrestrial waters: a review of what is currently known. Chin J Geochem 16(1):20–42

Johannesson KH, Zhou X (1999) Origin of middle rare earth element enrichments in acidic waters of a Canadian High Arctic lake. Geochim Cosmochim Acta 63(1):153–165

Kanazawa Y, Kamitani M (2006) Rare earth minerals and resources in the world. J Alloys Compd 408:1339–1343

Kikawada Y, Ossaka T, Oi T, Honda T (2001) Experimental studies on the mobility of lanthanides accompanying alteration of andesite by acidic hot spring water. Chem Geol 176(1–4):137–149

Knappe A, Möller P, Dulski P, Pekdeger A (2005) Positive gadolinium anomaly in surface water and ground water of the urban area Berlin, Germany. Chem Erde 65(2):167–189

Kulaksız S, Bau M (2007) Contrasting behaviour of anthropogenic gadolinium and natural rare earth elements in estuaries and the gadolinium input into the North Sea. Earth Planet Sci Lett 260:361–371

Lawrence MG, Greig A, Collerson KS, Kamber BS (2006) Rare earth element and yttrium variability in South East Queensland waterways. Aquat Geochem 12:39–72

Laveuf C, Cornu S (2009) A review on the potentiality of rare earth elements to trace pedogenetic processes. Geoderma 154:1–12

Lécuyer C, Reynard B, Grandjean P (2004) Rare earth element evolution of Phanerozoic seawater recorded in biogenic apatites. Chem Geol 204(1–2):63–102

Lee SG, Kim JK, Yang DY, Kim JY (2008) Rare earth element geochemistry and Nd isotope composition of stream sediments, south Han River drainage basin, Korea. Quat Int 176–177:121–134

Lee SG, Lee DH, Kim Y, Chae BG, Kim WY, Woo NC (2003) Rare earth elements as indicators of groundwater environment changes in a fractured rock system: evidence from fracture-filling calcite. Appl Geochem 18(1):135–143

Liang T, Zhang S, Wang L, Kung HT, Wang Y, Hu A, Ding S (2005) Environmental biogeochemical behaviors of rare earth elements in soil–plant systems. Environ Geochem Health 27:301–311

Liang T, Ding S, Song W, Chong Z, Zhang C, Li H (2008) A review of fractionations of rare earth elements in plants. J Rare Earths 26:7–15

Mao G, Hua R, Gao J, Li W, Zhao K, Long G, Lu H (2009) Existing forms of REE in gold-bearing pyrite of the Jinshan gold deposit, Jiangxi Province, China. J Rare Earths 27(6):1079–1087

McLennan SM (1989) Rare earth elements in sedimentary rocks; influence of provenance and sedimentary processes. Rev Min Geochem 21(1):169–200

McLennan SM (1994) Rare-earth element geochemistry and the tetrad effect. Geochim Cosmochim Acta 58(9):2025–2033

Merten D, Kothe E, Büchel G (2004) Studies in microbial heavy metal retention from uranium mine drainage water with special emphasis on rare earth elements. Mine Water Environ 23:34–43

Merten D, Geletneky J, Bergmann H, Haferburg G, Kothe E, Büchel G (2005) Rare earth element patterns: a tool for understanding processes in remediation of acid mine drainage. Chem Erde 65(S1):97–114

Michard A (1989) Rare earth element systematics in hydrothermal fluids. Geochim Cosmochim Acta 53(3):745–750

Michard A, Albarède F (1986) The REE content of some hydrothermal fluids. Chem Geol 55(1–2):51–60

Miekeley N, Coutinho de Jesus H, Porto da Silveira CL, Linsalata P, Morse R (1992) Rare-earth elements in groundwaters from the Osamu Utsumi mine and Morri do Ferro analogue study sites, Poços de Caldas, Brazil. J Geochem Explor 45:365–387

Moffett JW (1990) Microbially mediated cerium oxidation in sea-water. Nature 345(6274):421–423

Moffett JW (1994) A radiotracer study of cerium and manganese uptake onto suspended particles in Chesapeake Bay. Geochim Cosmochim Acta 58(2):695–703

Möller P, Dulski P, Bau M, Knappe A, Pekdeger A, Sommer-von Jarmersted C (2000) Anthropogenic gadolinium as a conservative tracer in hydrology. J Geochem Explor 69:409–414

Nagender Nath B, Balaram V, Sudhakar M, Plüger WL (1992) Rare earth element geochemistry of ferromanganese deposits from the Indian Ocean. Mar Chem 38(3–4):185–208

Nance WB, Taylor SR (1976) Rare earth element patterns and crustal evolution – I. Australian post-Archean sedimentary rocks. Geochim Cosmochim Acta 40:1539–1551

Neaman A, Mouele F, Trolard F, Bourrie G (2004) Improved methods for selective dissolution of Mn oxides: applications for studying trace element associations. Appl Geochem 19(6):973–979

Nozaki Y, Lerche D, Alibo DS, Tsutsumi M (2000) Dissolved indium and rare earth elements in three Japanese rivers and Tokyo Bay: evidence for anthropogenic Gd and In. Geochim Cosmochim Acta 64(23):3975–3982

Ohta A, Kawabe I (2001) REE(III) adsorption onto Mn dioxide (δ-MnO_2) and Fe oxyhydroxide: Ce(III) oxidation by δ-MnO_2. Geochim Cosmochim Acta 65(5):695–703

Ojiambo SB, Lyons WB, Welch KA, Poreda RJ, Johannesson KH (2003) Strontium isotopes and rare earth elements as tracers of groundwater-lake water interactions, Lake Naivasha, Kenya. Appl Geochem 18:1789–1805

Olías M, Cerón JC, Fernández I, De la Rosa J (2005) Distribution of rare earth elements in an alluvial aquifer affected by acid mine drainage: the Guadiamar aquifer (SW Spain). Environ Pollut 135:53–64

Pérez-López R, Gelgado J, Nieto JM, Márquez-García B (2010) Rare earth element geochemistry of sulphide weathering in the São Domingos mine area (Iberian Pyrite Belt): a proxy for fluid-rock interaction and acient mining pollution. Chem Geol 276:29–40

Pohl WL (2005) Mineralische und Energie-Rohstoffe. Eine Einführung zur Entstehung und nachhaltigen Nutzung von Lagerstätten. Schweizerbart'sche Verlagsbuchhandlung, Stuttgart

Pourret O, Davranche M, Gruau G, Dia A (2007a) Competition between humic acid and carbonates for rare earth elements complexation. J Colloid Interface Sci 305(1):25–31

Pourret O, Davranche M, Gruau G, Dia A (2007b) Rare earth elements complexation with humic acid. Chem Geol 243(1–2):128–141

Pourret O, Davranche M, Gruau G, Dia A (2008) New insights into cerium anomalies in organic-rich alkaline waters. Chem Geol 251(1–4):120–127

Protano G, Riccobono F (2002) High contents of rare earth elements (REEs) in stream waters of a Cu-Pb-Zn mining area. Environ Pollut 177:499–514

Quinn KA, Byrne RH, Schijf J (2006a) Sorption of yttrium and rare earth elements by amorphous ferric hydroxide: influence of pH and ionic strength. Mar Chem 99(1–4):128–150

Quinn KA, Byrne RH, Schijf J (2006b) Sorption of yttrium and rare earth elements by amorphous ferric hydroxide: influence of solution complexation with carbonate. Geochim Cosmochim Acta 70(16):4151–4165

Quinn KA, Byrne RH, Schijf J (2007) Sorption of yttrium and rare earth elements by amorphous ferric hydroxide: influence of temperature. Environ Sci Technol 41(2):541–546

Rabiet M, Brissaud F, Seidel JL, Pistre S, Elbaz-Poulichet F (2009) Positive gadolinium anomalies in wastewater treatment plant effluents and aquatic environment in the Herault watershed (South France). Chemosphere 75(8):1057–1064

Rimskaya-Korsakova MN, Dubinin AV, Ivanov VM (2003) Determination of rare-earth elements in sulfide minerals by inductively coupled plasma mass spectrometry with ion-exchange preconcentration. J Anal Chem 58(9):870–874

Ryu JS, Lee KS, Lee SG, Lee D, Chang HW (2007) Seasonal and spatial variations of rare earth elements in rainwaters, river waters and total suspended particles in air in South Korea. J Alloys Compd 437:344–350

Schönenberger J, Köhler J, Markl G (2008) REE systematics of fluorides, calcite and siderite in peralkaline plutonic rocks from the Gardar Province, South Greenland. Chem Geol 247 (1–2):16–35

Schwinn G, Markl G (2005) REE systematics in hydrothermal fluorite. Chem Geol 216 (3–4):225–248

Sholkovitz E, Szymczak R (2000) The estuarine chemistry of rare earth elements: comparison of the Amazon, Fly, Sepik and the Gulf of Papua systems. Earth Planet Sci Lett 179(2):299–309

Sinha S, Mukherjee SK (2008) Cadmium-induced siderophore production by a high Cd-resistant bacterial strain relieved Cd toxicity in plants through root colonization. Curr Microbiol 56:55–60

Siñeriz ML, Kothe E, Abate CM (2009) Cadmium biosorption by *Streptomyces* sp. F4 isolated from former uranium mine. J Basic Microbiol 49:55–62

Smedley PL (1991) The geochemistry of rare-earth elements in groundwater from the Carnmenellis Area, Southwest England. Geochim Cosmochim Acta 55(10):2767–2779

Takahashi Y, Chatellier X, Hattori KH, Kato K, Fortin D (2005) Adsorption of rare earth elements onto bacterial cell walls and its implication for REE sorption onto natural microbial mats. Chem Geol 219(1–4):53–67

Tang J, Johannesson KH (2005) Rare earth element concentrations, speciation and fractionation along groundwater flow paths: the Carrizo Sand (Texas) and Upper Floridan aquifers. In: Johannesson KH (ed) Rare earth elements in groundwater flow systems. Springer, Dordrecht, pp 223–251

Taylor SR, McLennan SM (1985) The continental crust: its composition and evolution – an examination of the geochemical record preserved in sedimentary rocks. Blackwell Scientific, Oxford

Taylor SR, McLennan SM (2003) Distribution of the lanthanides in the earth's crust. In: Sigel A, Sigel H (eds) The lanthanides and their interrelations with biosystems. Metal ions in biological systems. CRC, New York

Thunus L, Lejeune R (1999) Overview of transition metal and lanthanide complexes as diagnostic tools. Coord Chem Rev 184:125–155

Tricca A, Stille P, Steinmann M, Kiefel B, Samuel J, Eikenberg J (1999) Rare earth elements and Sr and Nd isotopic composition of dissolved and suspended loads from small river systems in the Vosges mountains (France), the river Rhine and groundwater. Chem Geol 160:139–158

Tsuruta T (2005) Separation of rare earth elements by microorganisms. J Nucl Radiochem Sci 6(1):81–84
Tsuruta T (2006) Selective accumulation of light or heavy rare earth elements using gram-positive bacteria. Colloid Surf B Biointerfaces 52:117–122
Tsuruta T (2007) Accumulation of rare earth elements in various microorganisms. J Rare Earths 25:526–532
Tyler G, Olsson T (2002) Conditions related to solubility of rare and minor elements in forest soils. J Plant Nutr Soil Sci 165:594–601
Uchida S, Tagami K, Tabei K, Hirai I (2006) Concentrations of REEs, Th and U in river waters collected in Japan. J Alloys Compd 408:525–528
Varekamp JC, Ouimette AP, Herman SW, Flynn KS, Bermudez A, Delpino D (2009) Naturally acidic waters from Copahue volcano, Argentina. Appl Geochem 24:208–220
Verplanck PL, Nordstrom DK, Taylor HE (1999) Overview of rare earth element investigations in acid waters of US Geological Survey abandoned mine lands watersheds. US Geol Surv Water-Resour Invest Rep 4018A:83–92
Verplanck PL, Nordstrom DK, Taylor HE, Kimball BA (2004) Rare earth element partitioning between hydrous ferric oxides and acid mine water during iron oxidation. Appl Geochem 19(8):1339–1354
Wakao N, Mishina M, Sakurai Y, Shiota H (1982) Bacterial pyrite oxidation I. The effect of pure and mixed cultures of *Thiobacillus ferrooxidans* and *Thiobacillus thiooxidans* on release of iron. J Gen Appl Microbiol 28:331–343
Wang YQ, Sun JX, Chen HM, Guo FQ (1997) Determination of the contents and distribution characteristics of REE in natural plants by NAA. J Radioanal Nucl Chem 219(1):99–103
Wang K, Li RC, Cheng Y, Zhu B (1999) Lanthanides – the future drugs? Coord Chem Rev 192:297–308
Welch SA, Christy AG, Isaacson L, Kirste D (2009) Mineralogical control of rare earth elements in acid sulfate soils. Geochim Cosmochim Acta 73(1):44–64
Whitney PR, Olmsted JF (1998) Rare earth element metasomatism in hydrothermal systems: the Willsboro-Lewis wollastonite ores, New York, USA. Geochim Cosmochim Acta 62(17):2965–2977
Wieser ME, Coplen TB (2011) Atomic weights of the elements 2009 (IUPAC Technical Report). Pure Appl Chem 83(2):359–396
Worrall F, Pearson DG (2001) Water-rock interaction in an acidic mine discharge as indicated by rare earth element patterns. Geochim Cosmochim Acta 65(18):3027–3040
Xu C, Campbell IH, Allen CM, Huang Z, Qi L, Zhang H, Zhang G (2007) Flat rare earth element patterns as an indicator of cumulate processes in the Lesser Qinling carbonatites, China. Lithos 95(3–4):267–278
Xu Z, Han G (2009) Rare earth elements (REE) of dissolved and suspended loads in the Xijiang river, South China. Appl Geochem 24:1803–1816
Zhang J, Nozaki Y (1996) Rare earth elements and yttrium in seawater: ICP-MS determinations in the East Caroline, Coral Sea, and South Fiji basins of the western South Pacific Ocean. Geochim Cosmochim Acta 60(23):4631–4644
Zhang ZY, Wang YQ, Li FL, Xiao HQ, Chai ZF (2002) Distribution characteristics of rare earth elements in plants from a rare earth ore area. J Radioanal Nucl Chem 252(3):461–465
Zhang J, Liu CQ (2004) Major and rare earth elements in rainwater from Japan and East China Sea: natural and anthropogenic sources. Chem Geol 209:315–326
Zhao FH, Cong ZY, Sun HF, Ren DY (2007) The geochemistry of rare earth elements (REE) in acid mine drainage from the Sitai coal mine, Shanxi Province, North China. Int J Coal Geol 70 (1–3):184–192
Zhu MY, Tan SD, Liu WZ, Zhang QF (2010) A review of REE tracer method used in soil erosion studies. Agric Sci China 9(8):1167–1174

Chapter 5
Geomicrobial Manganese Redox Reactions in Metal-Contaminated Soil Substrates

Christian Lorenz, Dirk Merten, Götz Haferburg, Erika Kothe, and Georg Büchel

5.1 Introduction

Soils contaminated with heavy metals including manganese are found as the result of mining activities, specifically where processes of acid mine drainage dominate. The former uranium mining activities in Eastern Thuringia are one such example where large open pit mines, waste rock piles, and production dumps (Jakubick et al. 1997; Rüger and Dietel 1998) were left after closure of the mine. One former leaching heap site has been studied for transfer processes of heavy metals including radionuclides in the resulting soil substrate after remediation efforts (Grawunder et al. 2009). It could be shown that reactive transport processes in water and soil are directly linked to microbial metabolism, or are at least moderated by microbial activities. Previous investigations showed several effects of microbial strains inoculated in the field (Kothe et al. 2005; Merten et al. 2004, 2005). During these field tests and additional laboratory investigations, it became evident that microbial activities can lead to major changes in heavy metal mobilities.

Besides their toxic potential, a number of heavy metals are also essential micronutrients, which lead to the development of specific detoxification and transport mechanisms. These include the active uptake of metals into the cell using efflux transporters and the transformation of these metals via specific redox

C. Lorenz (✉)
Chair of Environmental Geology, Brandenburg Technical University, Erich-Weinert-Str. 1, 03046 Cottbus, Germany
e-mail: christian.lorenz@tu-cottbus.de

D. Merten • G. Büchel
Institute of Earth Sciences, Friedrich-Schiller-University Jena, Burgweg 11, 07749 Jena, Germany

G. Haferburg • E. Kothe
Institute of Microbiology, Friedrich-Schiller-University Jena, Neugasse 25, 07743 Jena, Germany

E. Kothe and A. Varma (eds.), *Bio-Geo Interactions in Metal-Contaminated Soils*, Soil Biology 31, DOI 10.1007/978-3-642-23327-2_5,

reactions leading to changes in speciation and mobility (Haferburg et al. 2007; Kothe et al. 2005; Merten et al. 2005).

Manganese is an essential cofactor for enzymes like superoxide dismutases (SOD) or for the photosystem II of phototrophic organisms (Atlas and Bartha 1992; Gounot 1994; McKenzie 1989; Lovley 1991). Organisms able to use Fe(III) as electron acceptor under anaerobic conditions often can also utilize Mn(IV) (Lovley 1991; Sposito 1989) for respiratory electron transport under anaerobic conditions (Lovley 1991; Tack et al. 2006). Oxidized Fe and Mn species are common in soils and act as electron acceptors in many microbially mediated redox systems, if accessible oxygen and nitrate have been consumed (Kashem and Singh 2001). Reduction of Mn oxides results in mobile Mn(II), and was observed at redox conditions lower than those found for Fe(III) reduction (Sposito 1989; Nealson and Myers 1992). Thus, Mn shows a much more mobile behavior under comparable Eh–pH conditions than Fe.

A number of microbially mediated processes affecting metal mobility are known including biomineralization or mobilization through excretion of chelators (Kothe et al. 2005). Plant root-associated intracellular and extracellular fungi also produce chelating compounds leading to changes in pH conditions and redox potential in the rhizosphere. These habitat modifications influence the bioavailability and thus metal mobility due to phytoextraction and phytostabilization (Kothe et al. 2005). Additionally, microbial activities provide the conditions for metal mobilization and immobilization due to precipitation and co-precipitation reactions as well as sorption onto the cell surface and bioextraction (Merten et al. 2004, 2005; Kothe et al. 2005; Haferburg et al. 2007).

Often, contaminants including hydrolyzed cations such as nickel and zinc are specifically bound to immobile, metal–organic complexes of Mn and Fe compounds like pedogenic oxides and (hydr)oxides (McKenzie 1989). Due to changing Eh–pH conditions, mobilization of these compounds can result in the release of contaminants including copper, zinc, or cadmium (Lovley 1991; Sposito 1989). Mobilization thus can ensue through the formation of hydrous Mn and Fe oxides without changes in the redox state, while microbial changes in oxidation states of formerly stable Mn and Fe (hydr)oxides would provide a mobilization potential (Tack et al. 2006; Kashem and Singh 2001).

Here, previously isolated metal-resistant *Streptomyces tendae* F4 and *Streptomyces acidsicabies* E13 (Amoroso et al. 2000; Haferburg et al. 2007; Schmidt et al. 2005) were used to spike columns. The high resistance toward nickel of *S. acidiscabies* E13 was accompanied by tolerance toward Cd, Co, Cu, and Mn (Schmidt et al. 2005).

Preliminary batch experiments had demonstrated noticeable Mn concentrations in the water-extractable fraction of the test site material. Thus, potential microbial influences on mobilization and immobilization of heavy metals with special emphasis on Mn were investigated using spiked vs. nonspiked and nonpoisoned vs. poisoned column elution experiments.

5.2 Soil Characterization

About 100 kg of subsurface material (30 cm below ground) was taken from the south-western part of the test field site at Ronneburg, Germany, and stored at 8°C. Elution experiments were carried out using sieved material with a grain size below 2 mm. Grain density of this material ($n = 5$) was determined with a Gay-Lussac pycnometer (BRAND, Germany). The current proton activity pH_{H2O} and effective proton activity pH_{CaCl2} of the substrate were measured in the supernatant of a substrate–water and substrate–$CaCl_2$ suspension according to DIN 19684:1977-02 ($n = 3$; 10 g soil per 25 ml water) applying a pH meter (WTW pH/Oxi 340i, WTW, Germany) with a pH electrode (SenTix 41, WTW, Germany).

The effective cation exchange capacity CEC_{eff} was determined according to DIN ISO 11260:1997-05 using 2.5 g of air-dried material (<2 mm). A control sample (blank) without soil was prepared accordingly. The physicochemical characteristics of the silty sand (<2 mm) showed near-neutral pH_{H2O} value of 6.41 ± 0.04 and a slightly acidic pH_{CaCl_2} of 5.49 ± 0.01. The grain density (2.58 ± 0.01 g cm^{-3}) and CEC_{eff} (11 $cmol_c$ kg^{-1}) are comparable with clayish sands or clayish silts.

Total digestions were performed by using a pressure digestion system (DAS, PicoTrace, Germany) using 40% HF (Suprapur, Merck, Germany) and 4 ml 70% $HClO_4$ (Suprapur, Merck, Germany) at 180°C for 12 h. After acids evaporation for 12 h, the solids were incubated with 2 ml HNO_3 (65%, subboiled), 0.6 ml HCl (37%, subboiled) and 7 ml of pure water (Pure Lab Plus, USF, Germany), and heated to 150°C for 10 h. The cooled samples were transferred to calibrated 25 ml PMP flasks (Vitlab, Germany) and the solution was adjusted to 25 ml with ultrapure water. The concentrations of As, Cd, Cr, Ni, and Pb were still well below under prescriptive German Threshold Values for different land uses (Table 5.1). For Cu, Ni and Zn, measured values were above the German Precautionary Value indicating a contamination. The concentrations of Cd, Cr, and Pb were close to these indicated values. It is thus assumed that the measured amount of metal content is a possible source of future contamination.

5.3 Column Experiments

The column equipment consisted of four identical acrylic columns (50 cm × 8 cm, 2,513 cm^3) filled with nonautoclaved soil substrate (<2 mm) including autochthonous microorganisms. The substrate filled into the columns had a density of 2.58 g cm^{-3}, an available porosity of 0.32 (derived from pore water content of approximately 800 ml). Porous plates (100 μm pore width) in the column top and bottom units and upstream filter units (0.45 and 0.22 μm cellulose acetate, Sartorius, Germany) were placed to avoid the loss of fine-grained material and biomass.

Table 5.1 Element content (μg g^{-1}) of the nonleached substrate from the testsite "Gessenwiese"

	Mean	Standard deviation	TV[a]	TV[b]	PV[c]		Mean	Standard deviation	TV[a]	TV[b]	PV[c]
Al	43,938	543	–[d]	–[d]	–[d]	Mg	2,543	221	–[d]	–[d]	–[d]
As	7.2	1.9	125	140	–[d]	Mn	233	43	–[d]	–[d]	–[d]
Ba	512	2	–[d]	–[d]	–[d]	Na	3,022	119	–[d]	–[d]	–[d]
Ca	1,205	11	–[d]	–[d]	–[d]	Ni	28	4	350	900	15–70
Cd	0.25	0.05	50	60	0.4–1.5	Pb	17	2	1,000	2,000	40–100
Ce	96	56	–[d]	–[d]	–[d]	Sc	5.3	1.1	–[d]	–[d]	–[d]
Co	11	3	–[d]	–[d]	–[d]	Sr	76	3	–[d]	–[d]	–[d]
Cr	18	5	1,000	1,000	30–100	Th	11	2	–[d]	–[d]	–[d]
Cs	4.8	0.1	–[d]	–[d]	–[d]	Ti	4,996	639	–[d]	–[d]	–[d]
Cu	29	5	–[d]	–[d]	20–60	U	1.9	0.5	–[d]	–[d]	–[d]
Fe	32,477	3,152	–[d]	–[d]	–[d]	V	63	14	–[d]	–[d]	–[d]
K	19,672	196	–[d]	–[d]	–[d]	Y	14	2	–[d]	–[d]	–[d]
La	40	18	–[d]	–[d]	–[d]	Zn	65	0	–[d]	–[d]	60–200
Li	30	6	–[d]	–[d]	–[d]						

[a]German trigger values for soils in parks and recreational facilities (http://www.gesetze-im- internet.de/bbodschv/anhang_2_27.html, Accessed 17 June 2011)
[b]German trigger values for soils used for industrial and commercial purposes (http://www.gesetze-im-internet.de/bbodschv/anhang_2_27.html, Accessed 17 June 2011)
[c]German precautionary values for different soil types (range from sand to clay) (http://www.gesetze-im-internet.de/bbodschv/anhang_2_27.html, Accessed 17 June 2011)
[d]No trigger or precautionary values defined

After column passage, sampling ports and downstream flow-through cells allowed to measure electrical conductivity (LTG 1/24L, Sensortechnik Meinsberg, Germany), redox potential (EMC 30L, Sensortechnik Meinsberg, Germany), pH (EGA 151L, Sensortechnik Meinsberg, Germany), and temperature (Pt 100 element) on-line. A pumping rate of 0.5 ml min^{-1} allowed 0.9 pore volumes (PV) to be exchanged per day. Beside on-line measurements, electrical conductivity (EC) of 10 ml of each eluate sample was determined using a conductivity meter (WTW pH/Cond 340i, WTW, Germany) with electrical conductivity sensor (TetraCon 325, WTW, Germany).

Cations were analyzed by ICP-OES (Spectroflame, Spectro, Germany) and total organic carbon (TOC) was measured by a simultaneous TOC/TN analyzer (multi N/C 2100, Analytik Jena, Germany).

Each elution consisted of four columns, two nonpoisoned and two poisoned ones. Both nonpoisoned columns (A-columns) were eluted with ultrapure water (Elix 5 & A10, Millipore, France) of lower KCl concentration (0.5 mM) during the first column trial and higher KCl concentration (1 mM) during the second one. The remaining two columns (P-columns) were poisoned (NaN_3, 15 mM) and also leached by KCl (15 mM) again first column trial: 0.5 mM, second trial: 1 mM containing ultrapure water. The KCl was added to increase the ionic strength of the eluent in order to minimize the electrical double layer of formed colloids and the associated electrostatic repulsion between them (Marre 2003).

The first elution test was used to leach the substrate without additional microbial inoculation. The second column trial used all four columns microbially inoculated with a mixed culture of *S. tendae* F4 and *S. acidiscabies* E13 (Amoroso et al. 2000). Between 1.1 and 1.4 g wet weight biomass was used for inoculation of the columns with 125 ml spent cultivation media (originally composed of D(+)glucose, soluble starch, casein peptone, cornsteep, yeast extract, $(NH_4)_2SO_4$, KH_2PO_4, $CaCO_3$ and R2 medium ($ZnCl_2$, $FeCl_3{\cdot}6H_2O$, $CuCl_3{\cdot}6H_2O$, $MnCl_2{\cdot}4H_2O$, $Na_2B_4O_7$, $(NH_4)_2Mo_7O_{24}{\cdot}6H_2O$)).

5.3.1 *Inoculation Effects During Nonpoisoned Geosubstrate Elution*

The elution yielded detectable amounts of Ni, Sr and Zn, which were present at low contents in the solid phase. On-line measured differences of electrical conductivity resulted from the varying ionic strength of the eluent. The noninoculated columns A1 and A2 were eluted by a 0.5 mM KCl solution with an inlet EC of about 75 $\mu S\ cm^{-1}$, whereas the inoculated columns A3 and A4 were leached by a 1 mM KCl eluent solution demonstrating an EC value of 150 $\mu S\ cm^{-1}$. During column breakthrough, all four columns showed the maximum values for EC (A1/2: 570–700 $\mu S\ cm^{-1}$; A3/4: 620–690 $\mu S\ cm^{-1}$). This decreased to the initial eluent value during the following three exchanged pore volumes (EPV) for columns A1 to A3 and after seven pore volumes for column A4. Noninoculated column A1 showed slightly increasing pH values from 5.7 to 6.2 up to 1.1 EPV which remained stable at pH 6.1 from that point on. In comparison, the inoculated column A3 demonstrated stepwise increase of pH values from 6.0 to 7.0. Up to 1.6 EPV, pH values increased from pH 6 to pH 6.5. Another increase was observed from 1.9 to 5.1 EPV from pH 6.4 to pH 6.7. Four further pH increases peaking at pH 7.0 followed, which were only partially associated with eluent bottle change. Most remarkable changes could be demonstrated by the redox potential (Fig. 5.1).

Both nonpoisoned, noninoculated columns showed quite stable, slightly increasing Eh values of +210 to +300 mV during the entire leaching time. Inoculation resulted in a strong reduction of Eh in both nonpoisoned columns (A3 to approximately −200 mV and A4 to approximately −260 mV). Changing the eluent bottle led to short-term aerification at 1.8, 5.5, 6 and 10 EPV, which again decreased sharply within 0.5 EPV.

Only low-level metal concentrations with the exception of Mn were observed during the noninoculated and inoculated elution (Fig. 5.2). The detectable cations showed the "first flush" effect at the beginning of the elution. Elevated element concentrations for the untreated approach were measured for Mn (144 $\mu g\ l^{-1}$) and for Ni, Zn and Sr (not shown). Mn concentrations decreased after this first release to values around 30 $\mu g\ l^{-1}$, which were about twice as high as the Mn release observed in previous batch elution experiments (14 $mg\ l^{-1}$).

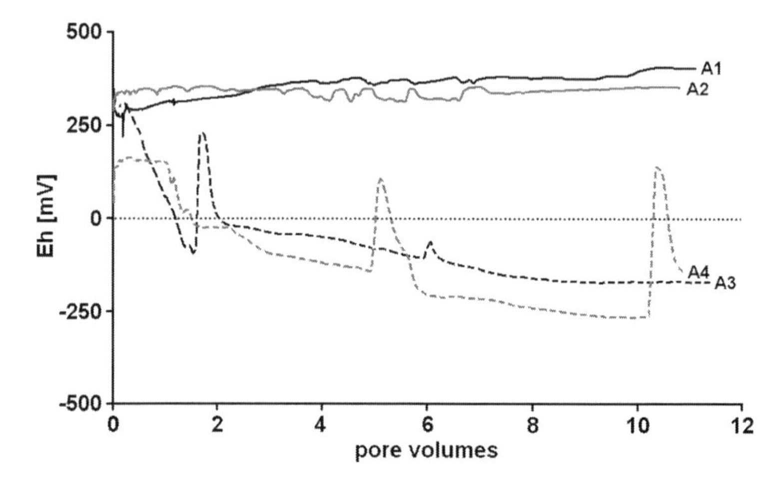

Fig. 5.1 Eh values of nonpoisoned columns A1 and A2 (noninoculated) and A3 and A4 (inoculated)

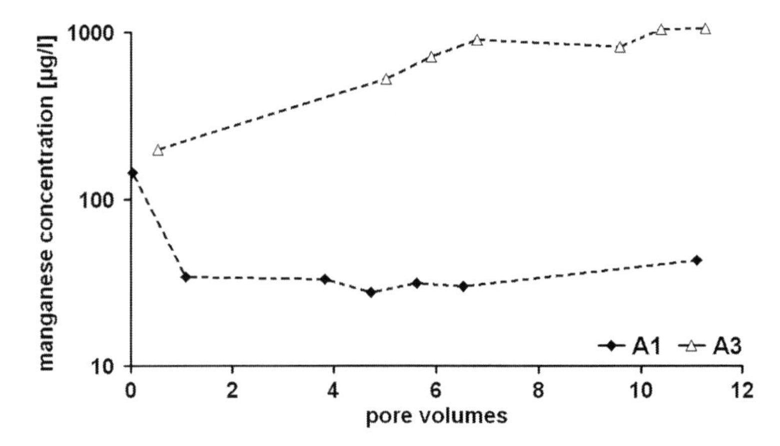

Fig. 5.2 Manganese concentrations within the eluate of nonpoisoned columns A1 (noninoculated) and A3 (inoculated)

A substantial difference, however, was observed for Mn in the inoculated columns; starting with approximately 200 $\mu g\ l^{-1}$ comparable to the noninoculated columns. After 11 EPV approx. 1,100 $\mu g\ l^{-1}$ Mn were released from the *Streptomyces*-inoculated column A3. These results were confirmed by the corresponding columns A2 and A4 (data not shown). Previous noninoculated batch investigations using 1 mM KCl eluent showed Mn concentrations of up to 28 $\mu g\ l^{-1}$.

5.3.2 *Inoculation Effects During Poisoned Geosubstrate Elution*

To inhibit microbial respiration processes during leaching, two noninoculated (P1/P2) and two inoculated columns (P3/P4) were poisoned by adding sodium azide. The online measured physicochemical data for electrical conductivity (EC) demonstrated higher values for all poisoned columns due to the addition of 15 mM NaN_3. The average EC value over the entire time is similar between all four poisoned columns with 1,580–1,610 $\mu S\ cm^{-1}$.

Both noninoculated columns showed higher maximum EC values with 1,860 $\mu S\ cm^{-1}$ (P1) and 1,940 $\mu S\ cm^{-1}$ (P2) compared to the inoculated pair of columns with 1,750 $\mu S\ cm^{-1}$ (P3 and P4). Both noninoculated columns reached stable EC values after 0.6 EPV (P1) and 1.3 EPV, while the inoculated pair of columns reached this status earlier (after 0.3 EPV). Thus, all columns demonstrated a quite rapid increase of EC at the beginning of elution.

The poisoned columns revealed increasing EC values during column saturation. Both noninoculated poisoned columns showed slightly increasing pH values from pH 5.5 to 6.4. The inoculation caused a general increase from pH 6.3 at the beginning to 6.8 at the end and much more variable conditions with maximum values of up to pH 8.8 in-between.

The redox potential (Fig. 5.3) for the poisoned, inoculated columns showed constant Eh values (+300 mV) from the beginning to nine to ten EPV followed by a significant redox potential increase to +370 mV (P3) and +320 mV (P4). Noninoculated columns demonstrated stable Eh values after one EPV in the range between +300 and +400 mV, which were comparable to the nonpoisoned experiments. The poisoned columns (Fig. 5.4) showed a higher element release, but comparable elution characteristics to the nonpoisoned, noninoculated columns. However, the previously observed Mn release from the nonpoisoned, inoculated column had been even higher.

The element contents measured showed constantly decreasing concentrations in the eluate, with Mn concentrations decreasing from 540 $\mu g\ l^{-1}$ (after column breakthrough) to 120 $\mu g\ l^{-1}$ at the end of the experiment. Previous noninoculated batch experiments had shown maximum Mn concentration of up to 31 $\mu g\ l^{-1}$. Zn demonstrated the same behavior over time, while Ni showed a massive loss during the first EPV and could not be detected at the end of the poisoned elution test.

5.3.3 *Total Organic Carbon Release*

Total Organic Carbon (TOC) measurements were performed for the inoculated columns A3 and P3. Nonpoisoned column A3 showed higher initial TOC concentrations (1,240 mg l^{-1}) as compared to the poisoned approach (380 mg l^{-1}). However, TOC elution characteristics until up to seven EPV were quite different comparing both column types. The nonpoisoned columns showed a

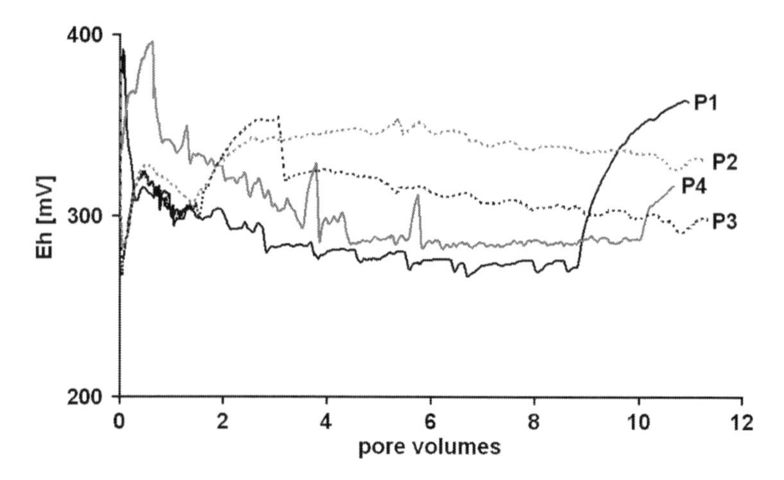

Fig. 5.3 Eh values of poisoned columns P1 and P2 (noninoculated) and P3 and P4 (inoculated)

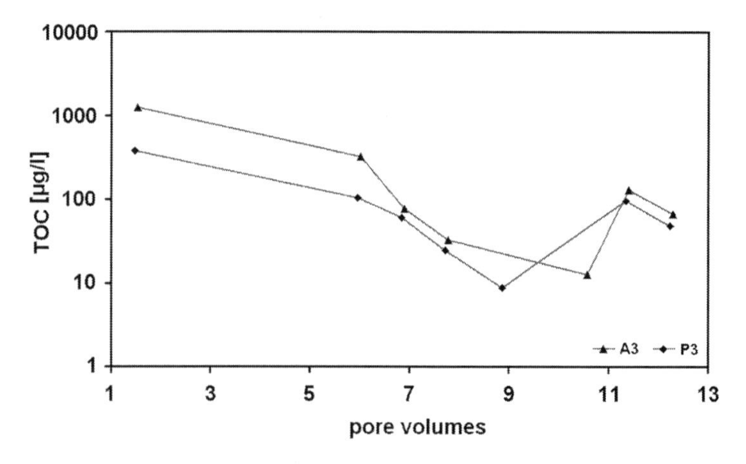

Fig. 5.4 Manganese concentrations within the eluate of poisoned columns P1 (noninoculated) and P3 (inoculated)

delayed elution of TOC, while the poisoned ones released TOC much faster. After seven EPV, both approaches yielded comparable TOC values (A3: 24 mg l^{-1}; P3: 32 mg l^{-1}), which increased up to 96 and 130 mg l^{-1}, respectively, when ten pore volumes were exchanged. Afterwards, TOC values decreased in a comparable way to 66 mg l^{-1} (A3) and 48 mg l^{-1} (P3).

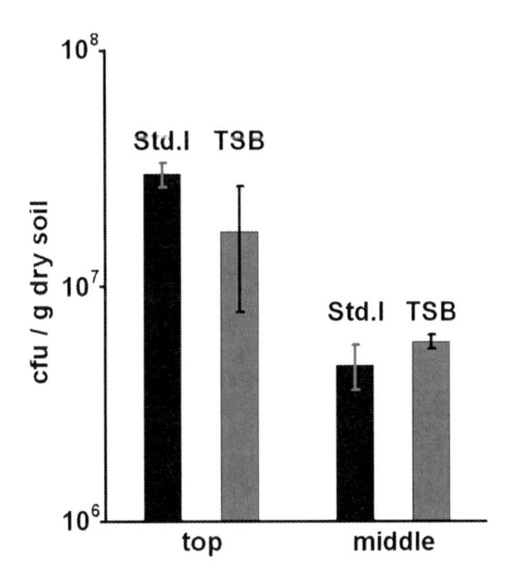

Fig. 5.5 Colony-forming units (cfu) obtained from eluted substrate of inoculated column A3 using soil samples from top and middle part of the column plated on two different complex media (Std I – standard I, TSB – tryptic soy broth)

5.3.4 Microbial Population

Since the *Streptomyces*-inoculated columns indicated elevated Mn release compared to the noninoculated ones, an isolation of obligate and facultative aerobic Mn-tolerant microorganisms from nonpoisoned, inoculated column A3 was performed. Samples from the top of the column and at 30 cm depth were taken at the end of the experiment. The colony-forming units (cfu) were determined using two different complex media (Std.I, Merck, Germany, and TSB, Bacto, France) under aerobic growth conditions, where 3.0×10^6– 4.6×10^7 cfu could be detected (Fig. 5.5).

No reisolation of the inoculated *Streptomyces* strains could be detected after the 2-week experiment. However, autochthonous streptomycetes were found when investigating noneluted, heat-treated substrate using soil extract agar (modified after Thiemann et al. 1968) with multiple different colonies. To select heat-resistant spores of actinobacteria, including streptomycetes, the samples were heated to 80°C for 2 h prior to isolation. The streptomycetes identified could be grown on casein medium agar (0.8 g nonhydrolyzed casein in 1,000 ml sterilized medium containing 0.5 g l^{-1} KCl, 1.0 g l^{-1} K_2HPO_4, 0.5 g l^{-1} $MgSO_4 \cdot 7H_2O$, 0.01 g l^{-1} $FeSO_4 \cdot 7H_2O$ and 20 g l^{-1} agar; modified after Altenburger et al. 1996). Their macroscopical characterization confirmed the existence of autochthonous *Streptomyces* within the natural substrate.

Furthermore, the capacity of single cultures to tolerate enhanced Mn concentrations was studied. Isolates tolerating 10, 20, or 30 mM $MnCl_2$ in minimal medium (10 g l^{-1} glucose, 0.5 g l^{-1} asparagine, 0.5 g l^{-1} K_2HPO_4, 0.2 g l^{-1} $MgSO_4 \cdot 7H_2O$, 0.01 g l^{-1} $FeSO_4 \cdot 7H_2O$, 18 g l^{-1} agar) were obtained by plating. Five prominent colony types were selected for further characterization. The pure cultures were used for DNA extraction and sequencing of a 16S rDNA resulting in the identification of these strains. Three strains, *Pseudomonas fluorescens*, *Leifsonia shinshuensis* and *Cupriavidus metallidurans* tolerated up to 30 mM Mn (II). These strains were further used for investigation of Mn reduction capabilities. Investigations using AMR medium (Bratina et al. 1998) under aerobic conditions showed no Mn reduction indicators, which means the formation of a bright halo around colonies capable to reduce the dark brown Mn(IV) in the form of MnO_2 to Mn(II). Even though the strains presented no discoloration, they survived and grew under the comparatively high Mn conditions (200 mM) of the medium overlay (17 g l^{-1} MnO_2, 7 g l^{-1} agar; Di-Ruggiero and Gounot 1990), while on $KMnO_4$ medium (1.0 g l^{-1} K_2HPO_4, 1 g l^{-1} $MgSO_4$, 0.12 g l^{-1} $FeCl_2{\cdot}4H_2O$, 0.01 g l^{-1} $Na_2MoO_4{\cdot}2H_2O$, 0.02 g l^{-1} $ZnSO_4{\cdot}7H_2O$, 0.001 g l^{-1} $CuSO_4{\cdot}5H_2O$, 10 g l^{-1} agar, with $KMnO_4$ added after autoclaving to give a final concentration of 3.5 mM Mn; modified after Ridge and Rovira 1971) no growth occurred.

5.4 Discussion

A high release of most elements was observed for the poisoned columns with a "first flush" effect due to weakly bound element fractions. Due to geochemical reactions such as dissociation of the added sodium azide and possible formation of various nitrogen compounds, the eluted element concentrations were higher in the poisoned columns where, additionally, higher ionic strength led to increased ion exchange reactions resulting in higher element release.

For Ni, Sr, Zn and Mn, a distinctive leaching effect was seen. Higher redox potentials and low Mn release of the noninoculated columns might be due to readsorption to column material of released Mn as hydrated Mn(hydr)oxides or reduced Mn(II).

The most significant change in this experiment was seen with the redox potential. The Eh values of the inoculated poisoned columns were comparable to the noninoculated columns without poisoning. But the inoculated nonpoisoned column showed constantly decreasing Eh values while temporarily increasing Eh values could be contributed to eluent change. The oxygen input into the previously closed system and possible reoxidation of formerly reduced elements potentially may lead to immobilization effects (Gounot 1994). However, this was not the case in our experiment.

Total organic carbon washout was quite similar for both inoculated columns. While inoculation led to higher starting TOC concentrations due to the growth medium used for inoculation, both types of columns had reached an equilibrium after approximately seven EPV with minimal and identical release thereafter (data not shown). This evidence suggests that the inoculated microorganisms, as well as the autochthonous microbial consortia, have not been lysed and washed out. This is important since at approximately six EPV, the redox potential in the inoculated columns had dropped below levels at which anaerobic growth can be expected of viable biomass still present, which might well result in direct microbial Mn reduction or in moderating the geochemical processes under anaerobic conditions within the column.

Decreasing Mn concentrations from both inoculated and poisoned columns during highest pH values were probably due to dissolution and readsorption effects within the substrate matrix (Khattack and Page 1992; Xiang and Banin 1996). In the course of decreasing pH values, Mn(IV) and the intermediate Mn(III) could have been transferred through geochemical redox reactions like ion exchange (Vandenabeele et al. 1995) into soluble Mn(II), or through geochemical hydration effects into more mobile hydrous Mn(hydr)oxides, and higher Mn concentrations were measured.

The addition of two streptomycetes to the substrate led to decreasing Eh values due to microbial respiration. The conditions within the inoculated, nonpoisoned columns changed from aerobic to anaerobic. Most likely due to an insufficient nutrition situation and anaerobic conditions within the column, the inoculated *Streptomyces* could not be reisolated. However, survival of *Streptomyces* strains experiencing anaerobic stress has been observed (Van Keulen et al. 2003, 2007).

Autochthonous strains, which were able to reduce Mn under anaerobic conditions in addition to possible geochemical reduction processes, led to elevated Mn release during the first seven EPV and after 9.6 EPV. Microbial dissimilatory Mn release through reduction of higher valent, accessible Mn compounds such as more crystalline or hydrated Mn(hydr)oxides to free soluble Mn, was observed.

Sequential extractions of comparable subsurface material demonstrated a large amount of mobile and easily mobilized Mn (20% of total Mn content) available for biogeochemical processes (Schindler 2007). The measured Eh values in the range of +300 to -175 mV led to the formation of hydrous Mn(hydr)oxides Mn(IV) and reduced Mn(II). Oxygen was not detected under these conditions and nitrate was already consumed. Thus, Mn as part of stable or already hydrated (hydr)oxides is one possible electron acceptor for anaerobic respiratory chains in microbial metabolism. The identified *C. metallidurans*, a bacterium known for its potential resistance to a variety of metals (Zn, Cr, Co) and its ability to live under anaerobic conditions (Mergeay et al. 1985, 2003; Nies 1992), might be capable of active Mn(IV) reduction. Microbial Mn reduction processes are known to take place under anaerobic and aerobic conditions (Mergeay et al. 2003; Nies 1992). Biogeochemical redox reactions led to the reduction of bound and hydrated Mn compounds to soluble Mn(II). Organic compounds excreted by the microorganisms or inorganic substances like nitrite or sulfide may serve for very efficient Mn and Fe reductants especially in acidic habitats (Atlas and Bartha 1992; Gounot 1994; Lovley 1991; Tack et al. 2006; Kashem and Singh 2001; Di-Ruggiero and Gounot 1990; Burdige and Nealson 1985). Reduced Fe, as well as Al species are often additionally bound to the organic matter and displace Mn(II) from the available exchanging sites (Trimble and Ehrlich 1970; Nealson and Saffarini 1994; Wieder and Lang 1986; Wieder and Linton 1990). Thus, reduced Mn is preferentially released from biomass.

The investigations under aerobic conditions on AMR medium as well as nonselective medium containing Mn(IV) did not indicate Mn reduction. The anaerobic conditions within the nonpoisoned columns might have induced microbial respiration processes using alternative electron acceptors (Lovley 1991). Facultative anaerobic Mn reducers would have preferred oxygen over Mn(IV) as electron acceptor during the plating test. Growth of the colonies on the AMR plates demonstrated that the tested strains were tolerant toward the high Mn concentrations supplied. A potential for Mn reduction can only be inferred from one of the identified strains, *P. fluorescens*, which is reported to be able to solubilize Mn under aerobic conditions (Burdige and Nealson 1985), and is also capable of oxidizing Mn(II). The respiratory poison sodium azide tripped the balance to even lower Mn(IV) reduction (Di-Ruggiero and Gounot 1990).

As the column with high Mn output presented strictly anaerobic conditions, the possibility of anaerobic reduction processes was given. Thus, during nonpoisoned elution, the detected members of the column's bacterial consortia might have been able to reduce Mn directly via microbial respiration. Potentially, Mn release was driven through the moderation of geochemical reactions, like redox potential change from oxic to anoxic, or the stimulation of Mn reduction through co-metabolic processes and their products.

5.5 Conclusions

Column elution experiments were carried out to improve the understanding of metal mobilization processes with special emphasis on Mn using various microbial treatments. The effect of microbial consortia on metal transfer in the soil–water system in metal-contaminated soils could be shown for Mn. The column trial without microbial treatment showed decreasing metal concentrations during elution according to the dominating Eh–pH conditions. All noninoculated columns, as well as the inoculated poisoned ones, showed comparable elution characteristics including "first flush" effect. Inoculation with streptomycetes resulted in an increased release of soluble Mn from nonpoisoned columns and a missing "first flush" release of accessible soluble elements. This indicates the high potential of microbial biomass to adsorb (heavy) metals to their biomass. The observed Mn release was associated with decreasing redox potential values, which was not detected in the eluates of all noninoculated or poisoned approaches. The Mn release can be attributed to either of two processes. First, direct microbial metabolisms might lead to Mn reduction through the transformation of higher valent solid Mn species to reduced and mobile Mn(II) and mobilization of Mn through hydration of Mn (hydr)oxides. Alternatively, the change of physicochemical conditions such as redox potential or excretion of chemical compounds might lead to abiotic reduction processes. This would indicate that microbial activity moderated the geochemical release. In both cases, an active role of microorganisms on metal release with leachate is seen.

Acknowledgements The authors thank Ulrike Buhler, Ines Kamp, Gundula Rudolph and Gerit Weinzierl for technical assistance. This work was supported by the German Federal Ministry of Education & Research grant no. 02S8294 KOBIOGEO and JSMC through the DFG Research Training Group 1257.

References

Altenburger P, Kämpfer P, Makristathis A, Lubitz W, Busse H-J (1996) Classification of bacteria islolated from a medieval wall painting. J Biotechnol 47:39–52

Atlas RM, Bartha R (1992) Microbial ecology: fundamentals and applications. The Benjamin Cummings, Redwood City, CA

Amoroso MJ, Schubert D, Mitscherlich P, Schumann P, Kothe E (2000) Evidence for high affinity nickel transporter genes in heavy metal resistant *Streptomyces* sp. J Basic Microbiol 40:295–301

Bratina BJ, Stevenson BS, Green WJ, Schmidt TM (1998) Manganese reduction by microbes from oxic regions of the Lake Vanda (Antarctica) Water Column. Appl Environ Microbiol 64:3791–3797

Burdige DJ, Nealson KH (1985) Microbial manganese reduction by enrichment cultures from coastal marine sediments. Appl Environ Microbiol 50:491–497

DIN ISO 11260:1997–05: Determination of effective cation exchange capacity and base saturation level using barium chloride solution. In: Deutsches Institut für Normung (DIN) (ed) Handbuch der Bodenuntersuchung, Beuth Verlag, Berlin

DIN 19684-1:1977–02 Methods of soil investigations for agricultural engineering; chemical laboratory tests; determination of pH-value of the soil and lime requirement. In: Deutsches Institut für Normung (DIN) (ed) Handbuch der Bodenuntersuchung, Beuth Verlag, Berlin

Di-Ruggiero J, Gounot A-M (1990) Microbial manganese reduction mediated by bacterial strains isolated from aquifer sediments. Microb Ecol 20:53–63

Gounot A-M (1994) Microbial oxidation and reduction of manganese: consequences in groundwater and applications. FEMS Microbiol Rev 14:339–349

Grawunder A, Lonschinski M, Merten D, Büchel G (2009) Distribution and bonding of the residual contamination in the glacial sediments at the former uranium mining leaching heap of Gessen/Thuringia. Chem Erde 69:5–19

Haferburg G, Reinicke M, Merten D, Büchel G, Kothe E (2007) Microbes adapted to acid mine drainage as source for strains active in retention of aluminum or uranium. J Geochem Explor 92:196–204

Jakubick AT, Gatzweiler R, Mager D, Robertson AM (1997) The Wismut waste rock pile remediation program of the Ronneburg mining district, Germany. In: Fourth international conference acid rock drainage, Vancouver, BC, Canada, May 31–June 6, 1997, vol III. pp 1285–1301

Kashem MA, Singh BR (2001) Metal availability in contaminated soils: I. Effects of flooding and organic matter on changes in Eh, pH and solubility of Cd, Ni and Zn. Nutr Cycl Agroecosyst 61:247–255

Khattack RA, Page AL (1992) Mechanism of manganese adsorption on soil constituents. In: Adriano DC (ed) Biogeochemistry of trace metals. Lewis, Boca Raton, FL, pp 383–400

Kothe E, Bergmann H, Büchel G (2005) Molecular mechanisms in bio-geo-interactions: from a case study to general mechanisms. Chem Erde 65:7–27

Lovley DR (1991) Dissimilatory Fe (III) and Mn (IV) reduction. Microbiol Rev 55:259–287

Marre D (2003) Untersuchungen zum Vorkommen und Transportverhalten von Partikeln in Grundwässern und Abschätzung ihrer Relevanz für den Schadstofftransport. Ph.D. thesis, Technical University Dresden

McKenzie RM (1989) Manganese oxides and hydroxides. In: Dixon J, Weed S (eds) Minerals in soil environments. Soil Science Society of America, Madison, WI, pp 439–465

Mergeay M, Nies D, Schlegel HG, Gerits J, Charles P, Van Gijsegem F (1985) Alcaligenes eutrophus CH34 is a facultative chemolithotroph with plasmid-bound resistance to heavy metals. J Bacteriol 162:328–334

Mergeay M, Monchy S, Vallaeys T, Auquier V, Benotmane A, Bertin P, Taghavi S, Dunn J, van der Lelie D, Wattiez R (2003) Ralstonia metallidurans, a bacterium specially adapted to toxic metals: towards a tentative catalogue of metal-responsive genes. FEMS Microbiol Rev 27:385–410

Merten D, Büchel G, Kothe E (2004) Studies on microbial heavy metal retention from uranium mining drainage water with special emphasis on rare earth elements. Mine Water Environ 23:34–43

Merten D, Geletneky J, Bergmann H, Haferburg G, Kothe E, Büchel G (2005) Rare earth element patterns: a tool for understanding processes in remediation of acid mine drainage. Chem Erde 65:97–114

Nealson KH, Myers CR (1992) Microbial reduction of manganese and iron: new approaches to carbon cycling. Appl Environ Microbiol 58:439–443

Nealson KH, Saffarini D (1994) Iron and manganese in anaerobic respiration. Annu Rev Microbiol 48:311–343

Nies DH (1992) Resistance to cadmium, cobalt, zinc, nickel in microbes. Plasmid 27:17–28

Ridge EH, Rovira AD (1971) Phosphatase activity of intact young wheat roots under sterile and nonsterile conditions. New Phytol 70:1017–1026

Rüger F, Dietel W (1998) Vier Jahrzehnte Uranerzbau um Ronneburg. Lapis 98:65–74

Schindler F (2007) Untersuchungen zur mikrobiellen Aktivität und Diversität im Bereich des schwermetallbelasteten Testfelds "Gessenwiese". M.Sc. thesis, Friedrich Schiller University Jena

Schmidt A, Haferburg G, Sineriz M, Merten D, Büchel G, Kothe E (2005) Heavy metal resistance mechanisms in actinobacteria for survival in AMD contaminated soils. Chem Erde 65:131–144

Sposito G (1989) The Chemistry of Soils. Oxford University Press, New York

Tack FMG, van Ranst E, Lievens C, Vandenberghe RE (2006) Soil solution Cd, Cu and Zn concentrations as affected by short-time drying or wetting: The role of hydrous oxides of Fe and Mn. Geoderma 137:83–89

Thiemann JE, Pagani H, Beretta G (1968) A new genus of the Actinomycetales: Microtetraspora gen. nov. J Gen Microbiol 50:295–303

Trimble RB, Ehrlich HL (1970) Bacteriology of manganese nodules: IV. Induction of an MnO_2-reductase system in a marine *Bacillus*. Appl Environ Microbiol 19:966–972

Van Keulen G, Jonkers HM, Claessen D, Dijkhuizen L, Wosten HAB (2003) Differentiation and anaerobiosis in standing liquid cultures of *Streptomyces coelicolor*. J Bacteriol 185:1455–1458

Van Keulen G, Alderson J, White J, Sawers RG (2007) The obligate aerobic actinomycete *Streptomyces coelicolor* A3(2) survives extended periods of anaerobic stress. Environ Microbiol 9:3143–3149

Vandenabeele J, de Beer D, Germonpre R, Van de Sande R, Verstraete W (1995) Influence of nitrate on manganese removing microbial consortia from sand filters. Water Res 29:579–587

Wieder RK, Lang GE (1986) Fe, Al, Mn, and S chemistry of Sphagnum peat in four peatlands with different metal and sulfur input. Water Air Soil Pollut 29:309–320

Wieder RK, Linton MN (1990) Laboratory mesocosm studies of Fe, Al, Mn, Ca, and Mg dynamics in wetlands exposed to synthetic acid coal mine drainage. Water Air Soil Pollut 51:181–196

Xiang HF, Banin A (1996) Solid-phase manganese fractionation changes in saturated arid-zone soils: pathways and kinetics. Soil Sci Soc Am J 60:1072–1080

Chapter 6
Natural Biomineralization in the Contaminated Sediment-Water System at the Ingurtosu Abandoned Mine

D. Medas, R. Cidu, P. Lattanzi, F. Podda, and G. De Giudici

6.1 Introduction

The Ingurtosu mine belongs to the Arburese mining district, located in SW Sardinia (Fig. 6.1). It was a prime source of Pb and Zn for more than a century. This area is now part of the Sardinian Geomining Park (Parco Geominerario Storico e Ambientale della Sardegna) sponsored by UNESCO. The results of mining activities caused significant environmental degradation, particularly due to heavy metal pollution. The Ingurtosu mine is included in the list of abandoned mine sites to be reclaimed in Sardinia (RAS 2003, see also the web page http://www.regione.sardegna.it/j/v/25?s=9020&v=2&c=9&t=1). A peculiar biomineralization phenomenon, leading to natural polishing of heavy-metal contaminated waters through precipitation of hydrozincite [$Zn_5(CO_3)_2(OH)_6$], and other Zn-rich phase(s) is observed in this area, and is described in this chapter.

6.2 Geological Setting and Hydrological Data

Figure 6.1 shows the schematic geological setting of the Arburese district. The main unit is represented by an allochthonous low-grade metamorphic complex made up of sedimentary and volcano-sedimentary successions, dating to Cambrian-Ordovician time (Carmignani et al. 1996). This complex (Arburese tectonic unit) is part of the external nappe zone of the Variscan chain in south-western Sardinia, and was thrusted over the autochthonous successions of the Iglesiente complex during the Variscan orogeny. At the end of the orogeny (ca. 300 Ma), the Arburese igneous complex was emplaced. It is characterized by a roughly concentric structure, with dominant

D. Medas • R. Cidu • P. Lattanzi • F. Podda • G. De Giudici (✉)
Department of Earth Sciences, University of Cagliari, Via Trentino 51, 09127 Cagliari, Italy
e-mail: dmedas@unica.it; cidur@unica.it; lattanzp@unica.it; fpodda@unica.it; gbgiudic@unica.it

E. Kothe and A. Varma (eds.), *Bio-Geo Interactions in Metal-Contaminated Soils*, Soil Biology 31, DOI 10.1007/978-3-642-23327-2_6,

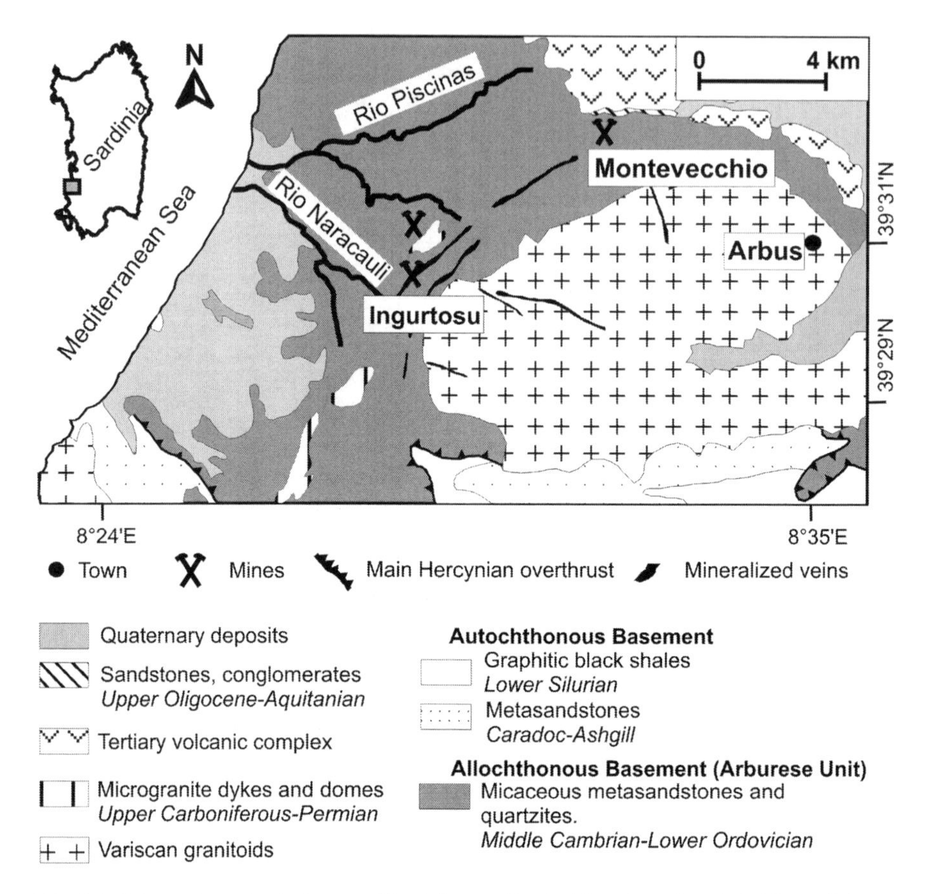

Fig. 6.1 Schematic geological map of the Arburese district (Carmignani et al. 1996, modified)

granodiorite in the border zone, and leucogranite in the core (Secchi et al. 1991). The igneous complex is characterized by radial fractures filled by acid and basic magmatic dykes, and by quartz and metalliferous hydrothermal veins. The mineralized vein system of Montevecchio-Ingurtosu occurs at the northern side of the Arburese igneous complex; minor deposits occur at the southern border.

Post-Paleozoic formations include a Cenozoic sedimentary complex of sandy, conglomeratic, marly and carbonate facies (Carmignani et al. 1996), and the volcanic complex of Monte Arcuentu. The latter is made up of basaltic flows overlapped by breccias and pillow lavas (Assorgia et al. 1984), dating to Upper Oligocene–Lower Miocene. Aeolian dunes along the coast (Piscinas dunal complex) and fluvial deposits in the internal area constitute the Quaternary covers.

Based on records collected in 1979–2009 (RAS, see the web page http://www.regione.sardegna.it/j/v/25?s=131338&v=2&c=5650&t=1), total annual rainfall is derived using Thiessen polygons. Rainfall ranges from 483 mm (1994–1995 hydrological year) to 1,040 mm (2008–2009 hydrological year) with a mean of 723 mm. Maximum rainfall occurs in autumn, especially in November. During spring the

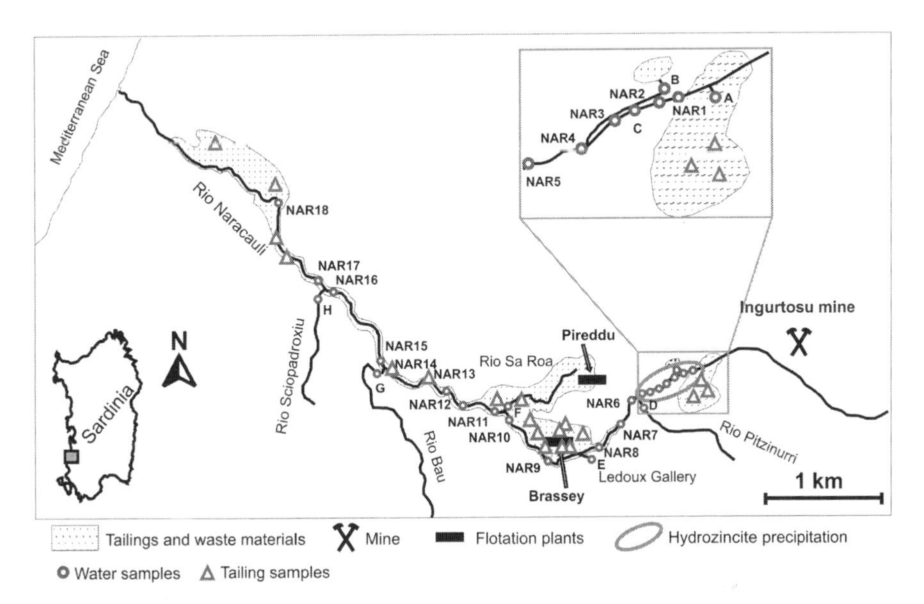

Fig. 6.2 Map showing the location of water and solid samples, the distribution of mining-related residues and the stations where hydrozincite bioprecipitation occurs

highest rainfall is in April, then it decreases reaching a minimum in summer (July). The mean annual temperature is 17°C (RAS, see the web page http://www.regione.sardegna.it/j/v/25?s=131338&v=2&c=5650&t=1).

The main water stream in the Ingurtosu area is Rio Naracauli (Fig. 6.2). It flows from Punta Tintillonis at 552 m.a.s.l. down to the west and into the Mediterranean Sea. Its course, which is 8.2 km long, develops prevalently across a landscape characterized by a moderate relief and hilly morphology. Its pattern is structurally controlled. It receives drainage from three tributaries: Rio Pitzinurri, Rio Bau, and Rio Sciopadroxiu. Furthermore, Rio Naracauli receives drainage from soil seepage and adits. The flow regime is typically torrential, with large flow variations between the wet and dry seasons.

6.3 Mining History

6.3.1 The Arburese Mining District

Because of the abundance of several metal minerals and rocks used in industry, Sardinia has for centuries been a prominent mining area in Europe. Specifically, the Iglesiente and Arburese districts were a world class resource of Pb and Zn. Mining activity started during pre-Roman age and peaked in the second half of the nineteenth century and first half of twentieth century. The decline began in the late 1950s; the last base metal mine closed in 1997.

At Montevecchio-Ingurtosu, mineralization consisted of a vein system extending in a NNE-SSW direction for at least 12 km. The vein bodies were mostly emplaced within the thermometamorphic aureole of the Variscan intrusion, but partly cut the intrusion itself (so-called radial veins). There are no recent descriptions of the mines; the most complete report remains that of Cavinato and Zuffardi (1948). The main ore minerals were galena and sphalerite, in a gangue of quartz and Fe-bearing carbonates (ankerite and siderite). Total tonnage is estimated at about 50–60 $\times$ 10^6 tons of crude ore, grading 10–11% combined Pb + Zn, 500–1,000 g Ag per ton of Pb, and 1,000 g of Cd per ton of Zn (Marcello et al. 2004). Metal production from Montevecchio ores between 1848 and 1973 was 1.6 $\times$ 10^6 tons of lead, 1.1 $\times$ 10^6 tons of Zn, and about 1 $\times$ 10^3 tons of Ag. By-products were Bi, Cd, Cu, Sb, and Ge (Salvadori and Zuffardi 1973). Currently, the estimated volume of mining-related wastes abandoned on the ground in the Iglesiente and Arburese districts is roughly 30 $\times$ 10^6 m^3. They occupy a surface of about 10^7 m^2 (RAS 2008).

6.3.2 *The Ingurtosu Mine*

The Ingurtosu Pb–Zn mine started operations in the 1870s and closed in 1968. Similar to other mines of the district, galena and sphalerite were the targets of exploitation. Chalcopyrite and pyrite were present in small amounts. Gangue minerals were quartz and siderite. Other accessory minerals were greenockite, arsenopyrite, Ni and Co sulfides, sulfosalts, goethite, hematite, and cuprite (Stara et al. 1996). Calcite and dolomite were locally found in the gangue. Secondary minerals include smithsonite, cerussite, azurite, malachite, barite, anglesite, pyromorphite, and mimetite. A treatment plant (Brassey) began operation in 1900 (Concas and Caroli 1994) (Fig. 6.2). Because of the increased production after the First World War, the Pireddu plant was established in the 1920s. Mining-related waste materials and flotation tailings were banked near to flotation plants or along the valleys; their distribution is shown in Fig. 6.2. The total volume of mining residues in this area is about 75 $\times$ 10^4 m^3 (Loi 1992). To avoid wind erosion and transport, flotation muds were covered with thin layers of gravel. Until 1968, runoff and creek waters were diverted by a system of galleries and channels. Tailing dams were built by simply using gravel and old train cars. Since the galleries and channel system were abandoned, the dams and the tailings have been deeply eroded by the action of water.

6.4 Materials and Analytical Methods

6.4.1 *Water Samples*

Water samples in the Rio Naracauli catchment were collected from 1997 to 2009, under different seasonal conditions. Early data (1997) were reported by Podda et al. (2000) and Zuddas and Podda (2005).

Water sampling points are shown in Fig. 6.2. Water samples A, B, C, and F drain mine tailings and wastes. At stations A and B, water flows first through an artificial channel, then into the Rio Naracauli. Station C is located at the base of the tailings on the left side of the Rio Naracauli. Water flow at station C is perennial, unlike stations A and B, which are dry in summer. Station F is at Rio Sa Roa, which receives contribution from tailings located upstream (Fig. 6.2). Sample E is the water flowing out of the Ledoux Gallery. Samples D, G, and H were collected from the tributaries Rio Pitzinurri, Rio Bau, and Rio Sciopadroxiu. Samples from NAR1 to NAR18 were collected in the Rio Naracauli stream.

Water temperature, conductivity, pH, redox potential, and alkalinity were measured in situ. For metal analysis filtered (0.4 μm), samples were collected and acidified with 1% HNO_3. Ca, Mg, Na, K, Sr, Ba, Si, Zn, and Cd were analyzed by inductively coupled plasma optical emission spectrometry (ICP-OES) and Pb, Cu, Ni, and Co by inductively coupled plasma mass spectrometry (ICP-MS). The anions were analyzed in nonacidified aliquots by ion chromatography.

Speciation and equilibrium calculations were performed using the PHREEQC 2.17.4137 computer code (Parkhurst and Appelo 1999) and the Lawrence Livermore National Laboratory (LLNL) thermodynamic database (LLNL database has the id: llnl.dat 4023 2010-02-09). The saturation indexes (SI) with respect to hydrozincite were calculated using the formula $SI = \log(IAP/K_s)$, where IAP is the ionic activity product in the water, and K_s is the solubility equilibrium constant at the specific water temperature. The solubility equilibrium constant of hydrozincite used to calculate SI is $10^{-14.85}$ (Mercy et al. 1998).

6.4.2 Solid Samples

During different sampling campaigns, about 30 hydrozincite samples were collected at several sites in the Rio Naracauli, in different years and under different seasonal conditions. The location of solid samples is shown in Fig. 6.2. Hydrozincite collected at station A occurs either as white material in suspension or white crusts covering the channel. In station B, hydrozincite precipitates as white crusts on stones. From point NAR1 to point NAR5, hydrozincite consists of crusts with variable thickness and color (white, ochre, green) depending on the amount of organic matter and iron oxides. Hydrozincite precipitates either on inorganic substrates like stones or on organic substrates such as roots, leaves, and twigs. At some sampling points, polycarbonate sheets were nailed on the riverbed to facilitate hydrozincite collection. During the rainy season, this material is largely washed away by flowing waters. In some years, a residual white precipitate has been observed in the sediment at stations NAR7 and NAR8.

In stations NAR12 and NAR13, a white colloidal precipitate often occurs; it is well distinguished from hydrozincite crusts observed upstream. Bulk chemical analysis indicates a rough composition of amorphous Zn-silicate (Zuddas et al. 1998).

Locations of tailing samples collected in the Ingurtosu area are shown in Fig. 6.2. Based on grain size analysis, three fractions were considered: the first fraction is made up by particles larger than 2 mm, the second fraction by sand + silt (2 mm–3.9 μm), and the third fraction by silt + clay (<3.9 μm). Mineralogical and chemical analyses on these samples were carried out by Caboi et al. (1993). Also, sequential extraction procedure tests were applied on tailings from the nearby Rio Piscinas catchment (Fanfani et al. 1997).

Conventional source X-ray powder diffraction (XRPD) patterns of precipitates from the Rio Naracauli waters have been performed since the beginning of environmental studies in this area and are now part of the analytical routine on new samples. Currently, we are using a Panalytical X'pert system, with Cu Kα radiation. More recently, high precision XRPD patterns were collected with a synchrotron source at ESRF, Grenoble (BM08 and ID31 beamlines; Lattanzi et al. 2007, 2010c).

Microscopic surface features of hydrozincite were investigated using an environmental scanning electron microscope (ESEM QUANTA 200, FEI). Specifically, the morphological analysis on hydrozincite samples was performed from SEM photos using the ImageJ software. The autofluorescence of the microbial community due to photosynthetic pigments was observed with scanning confocal laser microscopy (SCLM; Bio-Rad Microscience Division, model MRC-500, see Podda et al. 2000, for more details). HRTEM images were collected using a JEM 2010UHR (Jeol) microscope with a LaB_6 thermoionic source operating at 200 kV, and equipped with a Gatan imaging filter (GIF). Finally, X-ray absorption spectroscopy (XAS) measurements were carried out at ESRF (BM08 beamline), see Lattanzi et al. (2010a, b) for more details.

6.5 Results and Discussion

6.5.1 Mine Tailings Characterization

Table 6.1 reports the minimum, maximum, mean (X), and standard deviation (σ) values of metals in the tailings samples collected in the Ingurtosu district (see Fig. 6.2). The most abundant metals are Pb and Zn; their average concentrations far

Table 6.1 Minimum, maximum, mean (X), and standard deviation (σ) values of metals in the Ingurtosu tailings (Caboi et al. 1993, modified). *N* indicates number of the samples

	Sand + silt			Silt + clay			Gravel		
	N = 10			N = 5			N = 5		
Element	Min	Max	$X \pm \sigma$	Min	Max	$X \pm \sigma$	Min	Max	$X \pm \sigma$
Pb (g/kg)	1.7	71.1	13.3 ± 21.2	3.25	59	32.3 ± 25.1	0.32	3.6	1.76 ± 1.28
Zn (g/kg)	1.1	29.3	12.1 ± 7.9	1.88	27.5	19.7 ± 10.2	3.85	15.2	8.21 ± 4.23
Cd (mg/kg)	20	140	66 ± 35	100	190	145 ± 36	30	95	57 ± 24
Cu (mg/kg)	38	115	78 ± 25	45	120	86 ± 30	135	230	191 ± 36

exceed the limits (1,000 mg/kg for Pb, and 1,500 mg/kg for Zn) established by Italian regulations for soils in industrial/commercial sites. Cd also exceeds the 15 mg/kg limit, whereas Cu concentration is well below the 600 mg/kg limit. The highest Pb content was measured in samples taken in tailings (near Ingurtosu village), due most likely to the presence of anglesite ($PbSO_4$), because the treatment plants were designed to separate Pb and Zn sulfides, thus leaving the oxidized minerals in the residues (Caboi et al. 1993).

Three main fractions can be distinguished in these tailings (see previous paragraph). The gravel fraction shows the lowest content of metals (Table 6.1). The size of particles in the other two fractions varies depending on the way in which the flotation tailings were deposited. The sand + silt fraction corresponds to material discharged close to the pipes; tailings reworked by the river also belong to this fraction. This fraction consists of scarce sphalerite, galena, and pyrite; some cerussite and anglesite; abundant siderite, ankerite, quartz, barite, and dolomite; and minor zircon, ilmenite, monazite, and corundum. Also in this fraction, newly formed minerals, comprising Fe and Zn-sulfate and silicate minerals derive from interaction between water and tailings. The silt + clay fraction corresponds to materials located away from the discharge pipes. This fraction consists of Zn-rich illite; Zn, Ca, and Mg sulfates; and iron oxides (Caboi et al. 1993).

Sequential extraction techniques (according to Tessier et al. 1979, see more details in Fanfani et al. 1997) applied on similar tailings in a nearby area (Rio Piscinas catchment) showed that Pb, Zn, and Cd were easily extracted in the soluble + exchangeable fraction. Therefore, they can be considered potentially bioavailable elements (Chessa 1994; Fanfani et al. 1997; Biddau et al. 2001).

6.5.2 *Water Geochemistry*

Waters in the Rio Naracauli catchment have pH values (6.2–8.5) near neutral to slightly alkaline, suggesting a buffering effect of carbonate minerals. Indeed, most waters are at equilibrium or supersaturated with respect to calcite ($SI_{calcite}$: 0–1). The lowest pH values (pH < 7) were measured in waters from stations A and B that drain the tailings. Redox potential values (mainly between 400 and 520 mV) indicate oxidizing conditions. Conductivity varies significantly in the Rio Naracauli catchment (460–2,760 μS/cm). The drainages from the tailings (stations A and B) have the highest values (1,180–2,760 μS/cm); waters from tributaries (460–1,000 μS/cm) and Ledoux Gallery (630–860 μS/cm) show the lowest values. Accordingly, the highest values of conductivity in the Rio Naracauli were observed upstream of the Rio Pitzinurri (see Fig. 6.2), while conductivity decreases downstream due to dilution by less saline tributaries.

The Piper diagram shown in Fig. 6.3 was specifically adapted to take into account of Zn as major component. The Rio Naracauli waters have a dominant Ca–Mg-sulfate composition, reflecting the composition of tributaries, the interaction with tailings banked close to this stream, and inputs from stations A, B, and C

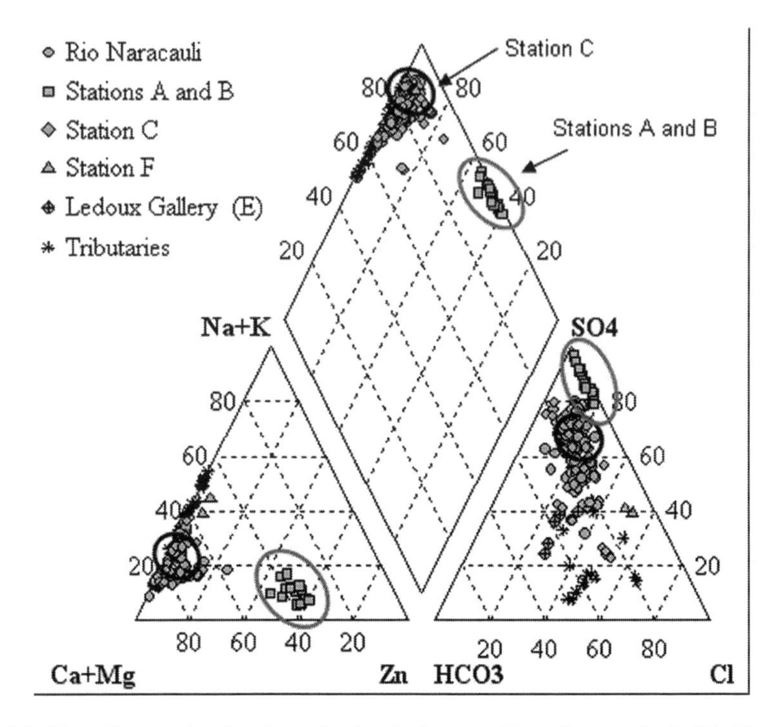

Fig. 6.3 Piper diagram showing the main chemical composition of waters in the Rio Naracauli catchment; Zn is included in the main cations because of its high concentration in some samples

(see Fig. 6.2). Tributaries have variable composition. Water in Rio Bau has Ca–Mg-sulfate composition. Waters in Rio Pitzinurri and Rio Sciopadroxiu show Na–Ca-chloride and Ca–Na-chloride composition, respectively. Waters draining tailings of the Rio Sa Roa valley (station F) have a Na–Mg-chloride(sulfate) composition. Sulfate and Zn concentrations reach maximum values in waters at station A, B, and C, where the ratio between volume of mine wastes and volume of interacting water is very high. Specifically, waters at stations A and B show a marked Zn-sulfate composition, while Zn is less abundant at station C, where waters have a Ca–Mg-sulfate character.

Table 6.2 reports the minimum, maximum, mean (X), and standard deviation (σ) values of metals in waters of the Rio Naracauli catchment. Many waters have high metal contents. As previously said, Zn is the most abundant metal, with concentrations from 0.7 to 200 mg/L in the Rio Naracauli waters. The highest values are observed in stations NAR1–NAR3. Cd (6.6–2,200 μg/L) is the second most abundant metal, with 240 μg/L on average. Also, Pb and Ni have significant concentrations in the range of 1.1–700 and 6.7–180 μg/L, respectively. Waters draining tailings show the highest concentration of metals, especially Cd and Zn in

Table 6.2 Main heavy metals content in the Rio Naracauli catchment waters: minimum, maximum, mean (*X*), and standard deviation (σ). *N* indicates number of samples

Rio Naracauli				Tributaries			
$N = 297$	Min	Max	$X \pm \sigma$	$N = 20$	Min	Max	$X \pm \sigma$
Zn (mg/L)	0.7	200	20 ± 22	Zn (mg/L)	0.02	0.3	0.93 ± 1.3
Cd (μg/L)	6.6	2200	240 ± 210	Cd (μg/L)	0.2	39	7.9 ± 12
Pb (μg/L)	1.1	700	50 ± 65	Pb (μg/L)	0.46	22	6 ± 5.6
Ni (μg/L)	6.7	180	75 ± 40	Ni (μg/L)	0.45	16	4.4 ± 4.2
Cu (μg/L)	1	142	4.6 ± 9.6	Cu (μg/L)	0.9	6.1	2.5 ± 1.7
Station F				Station A			
$N = 2$	Min	Max	$X \pm \sigma$	$N = 6$	Min	Max	$X \pm \sigma$
Zn (mg/L)	20	26	23 ± 4.4	Zn (mg/L)	214	800	600 ± 200
Cd (μg/L)	204	305	254 ± 71	Cd (μg/L)	2000	7600	5500 ± 1900
Pb (μg/L)	21	22	22 ± 0.70	Pb (μg/L)	99	1250	720 ± 400
Ni (μg/L)	16	26	21 ± 7.1	Ni (μg/L)	100	375	270 ± 92
Cu (μg/L)	3.4	5.7	4.6 ± 1.6	Cu (μg/L)	12	29	19 ± 6.6
Station B				Station C			
$N = 8$	Min	Max	$X \pm \sigma$	$N = 30$	Min	Max	$X \pm \sigma$
Zn (mg/L)	300	530	460 ± 84	Zn (mg/L)	13	41	20 ± 4.8
Cd (μg/L)	2600	5040	4000 ± 830	Cd (μg/L)	121	321	166 ± 35
Pb (μg/L)	35	250	120 ± 70	Pb (μg/L)	73	1050	232 ± 172
Ni (μg/L)	76	134	110 ± 18	Ni (μg/L)	70	151	97 ± 21
Cu (μg/L)	4.2	7.5	5.5 ± 1.1	Cu (μg/L)	4.3	18	8 ± 2.8

waters from stations A and B. Tributaries have the lowest content of Zn, Cd, Pb, Ni, and Cu, listed in decreasing order of abundance.

Results of chemical speciation modeling of dissolved Zn, Cd, Pb, and Cu indicate that Zn^{2+}, $ZnSO_4{}^0$, Cd^{2+}, $CdCl^+$, $PbCO_3{}^0$, Pb^{2+}, and $CuCO_3{}^0$ are the dominant species in Rio Naracauli waters. Pb^{2+} and Cu^{2+} species are more abundant than $PbCO_3{}^0$ and $CuCO_3{}^0$ in waters at stations A and B, reflecting their lower pH values.

Taking into consideration stations where bioprecipitation occurs (from NAR1 to NAR5), concentrations of major and trace elements change under different seasonal conditions. Specifically, Ca, Mg, B, Li, and Sr show higher concentrations in water samples collected under low-flow conditions (May–September), probably due to evaporation processes and/or a higher contribution of groundwater to the streams during low rain periods. On the contrary, concentrations of Pb, Ni, Zn, and Cd are higher under high-flow condition (October–April), probably due to the high runoff through the tailings. Similar seasonal variations have been already observed in previous studies of Sardinian river waters sampled under different flow conditions. Increasing concentrations of Pb, Ni, Zn, and Cd in winter under high flow conditions were attributed to solid transport of these metals as very fine particles (Cidu and Biddau 2007; Cidu and Frau 2009).

6.5.3 Biomineralization

The precipitation of secondary minerals causes metal attenuation in the Rio Naracauli. This process especially affects the dispersion of zinc, cadmium, and lead through hydrozincite ($Zn_5(CO_3)_2(OH)_6$) formation (Podda et al. 2000; Zuddas and Podda 2005). This natural process mainly occurs in late spring, in the Rio Naracauli before the inflow of Rio Pitzinurri. Hydrozincite precipitation is mediated by a photosynthetic microbial community constituted of a filamentous cyanobacterium (*Scytonema* sp. strain ING-1) and a microalga (*Chlorella* sp. strain SA1; see Podda et al. 2000 for more details). Figure 6.4 shows hydrozincite biomineralization observed by Confocal Laser Microscopy. Cyanobacteria can be recognized as a series of attached cells, often fluorescent, that result in filaments tens of micrometers long. The biomineralization consists of globules, external to the bacterial cells which attach themselves one to another to form sheaths. On the basis of the observed morphologies, De Giudici et al. (2007) proposed that the process is biologically controlled. Morphological details of hydrozincite biomineralization can be observed by scanning electron microscopy. Figure 6.5a shows how hydrozincite sheaths are made up of micrometric globules which attach themselves one to another (De Giudici et al. 2009). In turn, sheaths several tens of micrometers long are interlaced to form a rough superstructure (Fig. 6.5b). The globules are made of small platelets flattened onto the (100) crystal face (De Giudici et al. 2009).

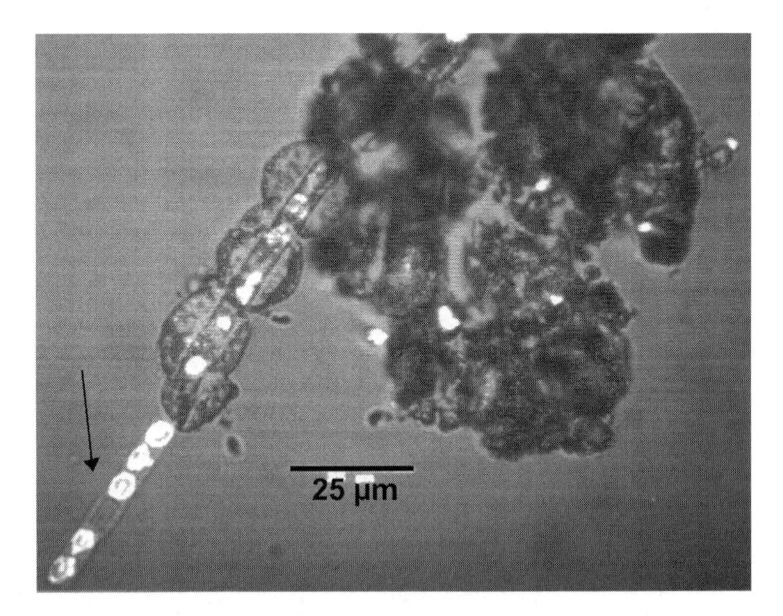

Fig. 6.4 Image collected by confocal microscopy in the transmission mode. The *Scytonema* sp. is visible as a series of attached cells that result in filaments that are tens of micrometers long, while biomineralization forms globules external to the bacterial cells that attach one to another to form sheaths. A thick-wall heterocyst (*short arrow*) is clearly visible

Fig. 6.5 SEM images of Rio Naracauli biomineral. The morphology is characterized by sheaths and organic matter filaments. Hydrozincite sheaths are made of micrometric globules that attach one to another (**a**). Sheaths several tens of micrometer long are interlaced to form a rough superstructure (**b**)

Analysis of Rio Naracauli hydrozincite samples by High Resolution-Transmission Electron Microscopy (HR-TEM) indicates that the biomineral is made of nanocrystals, 3–4 nm long, that are misaligned and aggregated following an imperfectly oriented aggregation mechanism (De Giudici et al. 2009).

In addition, morphological analysis of Rio Naracauli hydrozincite has shown that the shape of globules and sheaths can change depending on time of sampling. Taking into account the diameter distribution in some hydrozincite samples collected in 2009, 90% of measurements performed on hydrozincite collected in May (Fig. 6.6a) is between 20 and 40 µm, with a mean diameter of 30 µm; while

Fig. 6.6 SEM images of Rio Naracauli biomineral. Hydrozincite globules and sheaths in samples collected in May (**a**) and in August (**b**)

hydrozincite collected in August (Fig. 6.6b) shows a greater variability, with 90% of measurements between 30 and 80 μm, and a mean diameter of 50 μm. Furthermore, hydrozincite collected in August is characterized by short-length sheaths and by irregular globules. The difference in diameter may be explained taking into account that bioprecipitation mainly occurs in late spring and continues during the summer. At the end of the summer, the amount of hydrozincite precipitated on the bacterium is greater than that precipitated in spring, leading to an increase in diameter.

At stations where bioprecipitation occurs, the variations in Zn concentration, alkalinity, pH, and SI values with respect to hydrozincite were examined in 2009 under periods of contrasting rainfall. Soon after heavy rainfalls dissolved Zn increased up to 150% with respect to the mean value, due to high runoff interacting with tailings and mine wastes, whereas alkalinity, pH, and SI values decreased. Under such conditions the biomineralization process was not observed. Precipitation of hydrozincite resumed only after the stream flow; dissolved Zn concentration, alkalinity, pH, and SI returned to prestorm values. On the contrary, under scarce rain conditions significant changes in Zn concentration, alkalinity, and pH were not observed. Thus, nearly constant SI values were observed. These features match field observations: bioprecipitation was scanty and even absent after heavy rain events, while hydrozincite precipitation was not significantly affected by slight rainfall events.

Biomineralization process is more efficient in late spring, a period usually characterized by low rain and stationary hydraulic conditions. In 2009, peaks in bioprecipitation were observed; the highest hydrozincite SI values were recorded between late May and early June, when pH reached the highest values. This is due to the higher stability of hydrozincite in contact with slightly alkaline waters (see also Zuddas and Podda 2005).

As previously mentioned, in addition to hydrozincite biomineralization, the precipitation of amorphous minerals with a bulk composition of Zn-silicate is effective in the abatement of zinc and other heavy metals in the Rio Naracauli waters. This process occurs downstream of hydrozincite precipitation and is comparatively less well known. Also this process is seasonal, but generally occurs in summer, i.e., later than hydrozincite biomineralization. During surveys from 1997 to 2010, this precipitate was sampled periodically. In contrast with hydrozincite, analysis of SEM images does not show any distinctive morphological feature; the grains have a dust-like appearance and a composition made of Zn, Si, and O. These grains have a micrometric size and do not seem to aggregate. Very rarely (Fig. 6.7), some isolated sheaths can be recognized. Focusing the electron beam on the sheath shown in Fig. 6.7b, microanalysis indicates a chemical composition compatible with hydrozincite.

In the course of a study on heavy metal resistance of bacteria from Ingurtosu, Sprocati et al. (2006) obtained samples of white precipitates between NAR11 and NAR14 (see Fig. 6.2), and found several bacteria consortia, including some filamentous bacteria. In a biofilm sampled at station NAR13 on July 2010, cyanobacteria were found together with the Zn-silicate precipitates (Fig. 6.8, courtesy of K. Turnau). *Scytonema* sp. however was not positively identified. This indicates that the microbiological composition changes to some extent along the riverbed of Rio Naracauli stream.

6.5.4 X-ray Diffraction and Absorption Studies

XRPD clearly showed the mineralogical identity (hydrozincite) of the precipitate in the upper part of Rio Naracauli, and the amorphous nature of the white precipitate

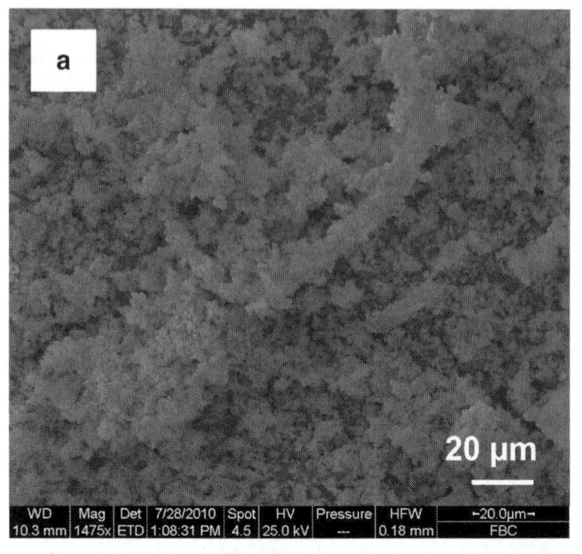

Fig. 6.7 SEM images of white precipitate sampled on July 2010 at NAR13. (**a**) shows the presence of nonaggregated grains and sheaths. Some isolated sheaths can be recognized (**b**); focusing the electron beam on the sheath shown in (**b**), microanalysis indicates a chemical composition compatible with hydrozincite

downstream of the Brassey plant (see Fig. 6.2). While identification of hydrozincite is unequivocal, other features do not always match data from past studies (e.g., PDF card 19-1458, see De Giudici et al. 2009). Specifically, peak shapes suggest a lower crystallinity of the Rio Naracauli material compared to reference samples (typically obtained from supergene Zn deposits). Aging effects were never detected (i.e., XRPD patterns of the studied samples remain essentially unchanged after several years from collection). Recent studies include a systematic comparison of XRPD (from both conventional and synchrotron sources) of Rio Naracauli hydrozincite with synthetic material and "typical" hydrozincite from Zn oxide deposits (De Giudici

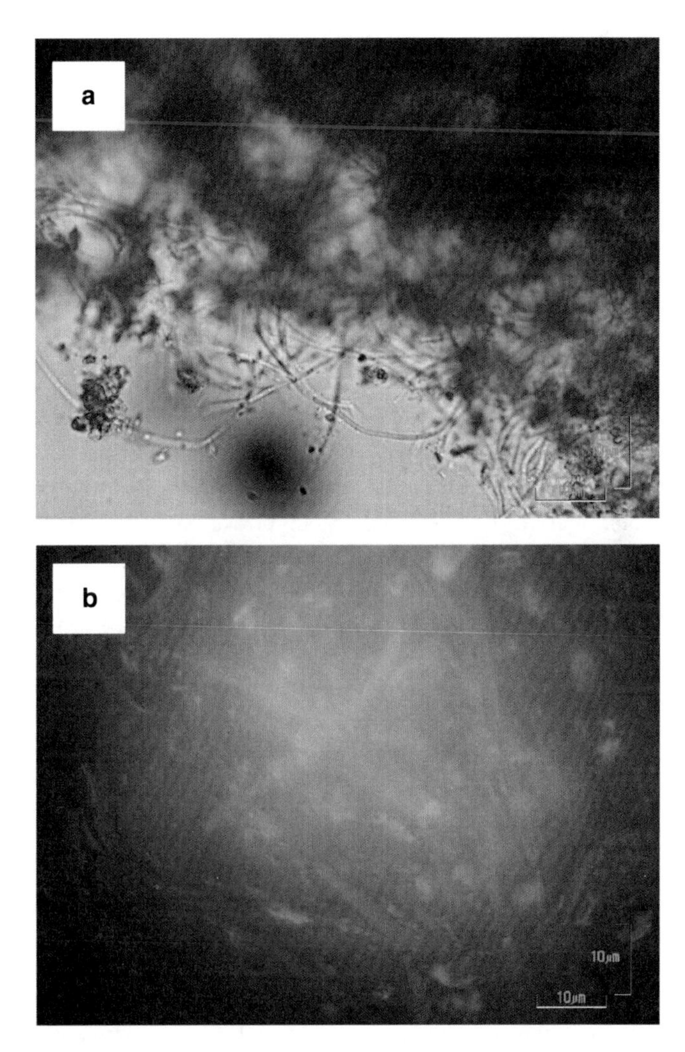

Fig. 6.8 White precipitate sampled at NAR13. Optical microscope imaging indicates the occurrence of a biofilm (**a**). Using fluorescent microprobes (**b**) cyanobacteria appear colored in *red*, the zinc-bearing mineral appears colored in *blue*. *Scytonema* sp. was not recognized while diatoms are abundant (personal communication of K. Turnau). These images are courtesy of K. Turnau

et al. 2009; Lattanzi et al. 2007, 2010a, b, c). The results confirm the differences between the Rio Naracauli samples and "typical" hydrozincites. Specifically, high precision synchrotron-based XRPD patterns of Rio Naracauli hydrozincites cannot be fitted to known structures of this mineral, which were obtained from "typical" hydrozincites. The largest difference is recorded for the a_0 cell parameter ("typical" hydrozincite: 13.59 ± 0.03 Å; Rio Naracauli hydrozincite 13.832 ± 0.006 Å), and suggests a difference in the stacking sequence of ZnO_6 octahedral sheets present in the hydrozincite structure. Synthetic hydrozincites, including Cd-doped and Pb-doped samples, show wide variations of the cell parameters; however, these are not correlated with the bulk content of either Cd or Pb. The binding nature of these metals to hydrozincite was investigated by X-ray absorption spectroscopy (XAS). For Cd, the results of extended X-ray absorption fine structure (EXAFS) analysis, backed by anomalous X-ray diffraction, suggest a disordered mode of occurrence, presumably as an amorphous surface precipitate (Lattanzi et al. 2010b). For Pb a more complex model is suggested (Lattanzi et al. 2010a): this metal is supposed to occur partly as an amorphous surface carbonate, partly either as a substituting ion in the tetrahedral Zn site of the hydrozincite structure or as an inner-sphere surface complex.

6.6 Summary and Conclusions

A survey of biomineralization processes at Ingurtosu has been carried out since 1997. Two distinct processes were recognized: formation of hydrozincite and formation of an amorphous Zn silicate. Because of its distinct morphology and crystallinity, the precipitation of hydrozincite in association with cyanobacteria and microalgae is better understood. This process occurs in spring and varies in intensity from year to year. In a first attempt to find out relationships between seasonal variability and the intensity of biomineralization, this work clearly indicates that the latter is favored by high values of annual rainfall intensity. The process appears, however, inhibited when heavy rain events occur in late spring, due to scouring of the streambed and undersaturation of rain-diluted stream water with respect to hydrozincite.

SEM morphological analysis of the shape of sheaths produced by cyanobacteria indicates some seasonal evolution in sheath shape during the biomineralization process. In fact, while in spring sheaths clearly look narrow and elongated along the direction of the globule attachment, the few samples of bacteria sheaths surviving until late August look broader.

TEM analysis has shown that nanocrystals which are approximately 3 nm long aggregate by an imperfectly oriented aggregation mechanism. This results in the formation of mesocrystals. These are the observed hydrozincite platelets flattened onto the (100) crystal face. Hydrozincite globules are made by a further aggregation of these micrometric platelets. As mentioned earlier, both the shape and size of globules are apparently influenced by environmental conditions (namely, sunlight, rainfall, runoff, and stationary hydraulic conditions). The cell parameters of Rio Naracauli hydrozincite are all similar, through slight differences with respect to the published structures, possibly suggest a different stacking order along a_o.

An important implication of hydrozincite biomineralization is its effectiveness in the abatement of bioavailable heavy metals. In addition to Zn removal, the behavior of Cd and Pb was investigated. Data by XAS suggest that Cd is weakly bound to hydrozincite surfaces. In contrast, Pb is coprecipitated with hydrozincite and/or retained onto its surfaces as an inner sphere complex. Hydrozincite can thus be regarded as an effective trap for Pb, while any sequestration of Cd is only ephemeral.

Acknowledgment This work was funded by the EU UMBRELLA project (grant number 226870). Rainfall data were kindly provided by hydrological service of Regione Autonoma della Sardegna. Katarzyna Turnau (Krakow University) is acknowledged for microscope work and scientific discussion.

References

Assorgia A, Brotzu P, Morbidelli L, Nicoletti M, Traversa G (1984) Successione e cronologia (K-Ar) degli eventi vulcanici del complesso calco-alcalino oligo-miocenico dell'Arcuentu (Sardegna centro-occidentale). Period Mineral 53:89–102

Biddau R, Da Pelo S, Dadea C (2001) The abandoned mining area of Montevecchio-Ingurtosu. Rend Sem Fac Sc Univ Cagliari 71(2):109–123

Caboi R, Cidu R, Cristini A, Fanfani L, Massoli-Novelli R, Zuddas P (1993) The abandoned Pb-Zn mine of Ingurtosu, Sardinia (Italy). Eng Geol 34:211–218

Carmignani L, Barca S, Oggiano G, Pertusati I, Conti P, Eltrudis A, Funedda A, Pasci S (1996) Carta Geologica della Sardegna 1:200.000. Serv Geol d'Italia

Cavinato A, Zuffardi P (1948) Geologia della miniera di Montevecchio. In: Notizie sull'industria del Piombo e dello Zinco in Italia. vol 1, pp 427–464. Montevecchio Società Italiana del Piombo e dello Zinco

Chessa A (1994) Studio geochimico-ambientale degli sterili nel bacino idrografico del Rio Piscinas (Sardegna SW). Degree thesis, University of Cagliari

Cidu R, Biddau R (2007) Transport of trace elements at different seasonal conditions: effects on the quality of river water in a Mediterranean area. Appl Geochem 22:2777–2794

Cidu R, Frau R (2009) Distribution of trace elements in filtered and non filtered aqueous fractions: insights from rivers and streams of Sardinia (Italy). Appl Geochem 24:611–623

Concas E, Caroli S (1994) Le Miniere di Gennamari e Ingurtosu. Pezzini, Viareggio

De Giudici G, Podda F, Sanna R, Musu E, Tombolini R, Cannas C, Musinu A, Casu M (2009) Structural properties of biologically controlled hydrozincite: an HRTEM and NMR spectroscopic study. Am Mineral 94:1698–1706

De Giudici G, Podda F, Caredda A, Tombolini R, Casu M, Ricci C (2007) In vitro investigation of hydrozincite biomineralization. In: Bullen TD, Wang Y (eds) Water rock interaction 12, vol 2. Taylor & Francis, London, pp 415–419

Fanfani L, Zuddas P, Chessa A (1997) Heavy metals speciation analysis as a tool for studying mine tailings weathering. J Geochem Explor 58:241–248

Lattanzi P, Meneghini C, De Giudici G, Podda F (2007) Report of experiment 08-02-636. ESRF, Grenoble

Lattanzi P, Meneghini C, De Giudici G, Podda F (2010a) Uptake of Pb by hydrozincite, $Zn_5(CO_3)_2(OH)_6$ – implications for remediation. J Hazard Mater 177:1138–1144

Lattanzi P, Maurizio C, Meneghini C, De Giudici G, Podda F (2010b) Uptake of Cd in hydrozincite, $Zn_5(CO_3)_2(OH)_6$: evidence from X-ray absorption spectroscopy and anomalous X-ray diffraction. Eur J Mineral 22:557–564

Lattanzi P, Meneghini C, De Giudici G, Medas D, Podda F (2010c) Report of experiment CH-2838. ESRF, Grenoble

Loi M (1992) Studio degli sterili della miniera di Ingurtosu e loro interazione con le acque del Rio Naracauli. Degree thesis, University of Cagliari

Marcello A, Pretti S, Valera P, Agus M, Boni M, Fiori M (2004) Metallogeny in Sardinia (Italy): from the Cambrian to the Tertiary, 32nd international geological congress, APAT 4:14–36, Firenze

Mercy MA, Rock PA, Casey WH, Mokarram MM (1998) Gibbs energies of formation for hydrocerussite [$Pb(OH)_2$. $(PbCO_3)_2(S)$] and hydrozincite [$Zn(OH)_2]_3$. $(ZnCO_3)_2(S)$] at 298 K and 1 bar from electrochemical cell measurements. Am Mineral 83:739–745

Parkhurst DL, Appelo CAJ (1999) User's guide to PHREEQC (version 2) – a computer program for speciation, batch-reaction, one-dimensional transport, and inverse geochemical calculations. US Geol Surv Water-Resour Invest Rep, 99–4259

Podda F, Zuddas P, Minacci A, Pepi M, Baldi F (2000) Heavy metal coprecipitation with hydrozincite [Zn_5 $(CO_3)_2(OH)_6$] from mine waters caused by photosynthetic microorganisms. Appl Environ Microbiol 66:5092–5098

Regione Autonoma della Sardegna (RAS) (2003) Piano Regionale di gestione dei rifiuti – Piano di bonifica siti inquinati

Regione Autonoma della Sardegna (RAS) (2008) Piano di bonifica delle aree minerarie dismesse del Sulcis-Iglesiente-Guspinese

Salvadori I, Zuffardi P (1973) Guida per l'escursione a Montevecchio e all'Arcuentu. Itinerari Geologici, Mineralogici e Giacimentologici in Sardegna 1:29–46

Secchi FAG, Brotzu P, Callegari E (1991) The Arburese igneous complex (SW Sardinia). An example of dominant igneous fractionation leading to peraluminous cordierite-bearing leucogranites as residual melts. Chem Geol 92:213–249

Sprocati AR, Alisi C, Segre L, Tasso F, Galletti M, Cremisini C (2006) Investigating heavy metal resistance, bioaccumulation and metabolic profile of a metallophile microbial consortium native to an abandoned mine. Sci Total Environ 366:649–658

Stara P, Rizzo R, Tanca GA (1996) Iglesiente e Arburese. Miniere e Minerali, vol 2. EMSA, Cagliari

Tessier A, Campbell PG, Bisson M (1979) Sequential extraction procedure for speciation of particulate trace metals. Anal Chem 51:844–850

Zuddas P, Podda F, Lay A (1998) Flocculation of metal rich-colloids in a stream affected by mine drainage. In: Arehart GB, Hulston JR (eds) Water rock interaction 9. A.A. Balkema, Rotterdam, pp 1009–1013

Zuddas P, Podda F (2005) Variations in physico-chemical properties of water associated with bio-precipitation of hydrozincite [$Zn_5(CO_3)_2(OH)_6$] in the waters of Rio Naracauli, Sardinia (Italy). Appl Geochem 20:507–517

Chapter 7
Speciation of Uranium in Seepage and Pore Waters of Heavy Metal-Contaminated Soil

Nils Baumann, Thuro Arnold, and Martin Lonschinski

7.1 Introduction

Plants take up nutrients and toxins from the surrounding pore water. If this water contains increased uranium concentrations, uranium may enter the plants and accumulate in plant compartments. Different plant species incorporate different amounts of uranium. This is a decisive criterion for selecting plants for remediation of heavy metal-contaminated soils by bio/phytoremediation.

The aim of this chapter is to identify the speciation of uranium in seepage and pore waters of the "Gessenwiese" environment, which provides dissolved nutrients for plants growing on the test site which is used in a phytoremediation study. The Gessenwiese is a recultivated former uranium mining heap close to Ronneburg in Eastern Thuringia. This test field was installed as a part of a research program of the Friedrich Schiller University Jena for investigations of acid mining drainage (AMD) and heavy metals retention, especially uranium (Grawunder et al. 2009), as well as heavy metal uptake and accumulation into plants. AMD is a severe environmental problem characterized by very metal-rich waters, high sulfate concentrations, and low pH. Mining of sulfidic ore bodies or lignite deposits containing significant amounts of pyrite is the result of the formation of AMD. In combination with uranium mining, AMD is either generated by technical processing or by in situ leaching of uranium ores with sulfuric acid. Despite the hostile environment in the Gessenbach heap, many microorganisms thrive in this low pH environment and have been identified (Haferburg and Kothe 2007; Haferburg et al. 2007).

N. Baumann (✉) • T. Arnold
Helmholtz-Zentrum Dresden-Rossendorf, Institut für Radiochemie, PF 51 01 19 01314 Dresden, Germany
e-mail: n.baumann@hzdr.de

M. Lonschinski
Friedrich-Schiller University, Institute of Geosciences, Jena 07743, Germany

E. Kothe and A. Varma (eds.), *Bio-Geo Interactions in Metal-Contaminated Soils*, Soil Biology 31, DOI 10.1007/978-3-642-23327-2_7,

Uranium is a potentially dangerous substance, which in enriched concentrations represents a major health hazard. Uranium has various oxidation states (i.e., III–VI) and it is well established that its transport behavior strongly depends on its oxidation state (Arnold et al. 2011). In contrast to tetravalent uranium, U(VI) is much more soluble and may migrate in the environment in dissolved form via the water path. Geochemical calculations indicate that aqueous uranium in low pH acid mine water environments generally occur either as free uranyl ion or as uranyl sulfate species (Espana et al. 2005).

Sulfate is also able to form strong complexes with uranium and thereby strongly affects its speciation and migration (Hennig et al. 2007). Uranium sulfate complexation has been studied by time-resolved laser-induced fluorescence spectroscopy (TRLFS) and X-ray absorption spectroscopy (EXAFS) for simple two-component solutions prepared in the laboratory (Hennig et al. 2007, 2008; Geipel et al. 1996; Vercouter et al. 2008) and thermodynamic data are available for three sulfato species, $UO_2SO_{4(aq)}$, $UO_2(SO_4)_2^{2-}$, and $UO_2(SO_4)_3^{4-}$ (Geipel et al. 1996; Bernhard et al. 1998; Guillaumont et al. 2003).

TRLFS combines very low detection limits and a high sensitivity toward U(VI) complex formation in aqueous solutions (Moulin et al. 1990, 1995). Currently, the detection limit of TRLFS for uranium(VI) species in solution is 0.2 μg/l (Bernhard and Geipel 2007). TRLFS is sensitive to the electron states induced by the ligand arrangements (Hennig et al. 2008) and thus provides direct evidence about uranium speciation.

Elemental uranium does not emit fluorescence (Baumann et al. 2006), whereas U(IV) (Kirishima et al. 2003, 2004), U(V) (Großmann et al. 2007, 2009; Steudtner et al. 2006), and, in particular, U(VI) emit characteristic fluorescence signals and are thus detectable with TRLFS. Aqueous complexes of hexavalent uranium have been well investigated (Billard et al. 2003; Brendler et al. 1996; Wang et al. 2004) as also have solid uranium(VI) phases (Amayri et al. 2004a, b; Arnold and Baumann 2009) and uranium(VI) species adsorbed on mineral surfaces (Baumann et al. 2005; Arnold et al. 2006; Křepelová et al. 2007).

The uranium speciation and hence its migration behavior in natural waters might be different in comparison with laboratory samples, since these waters are no pure two-component systems but do contain additional complexing agents, e.g., inorganic and organic components as well as microbial metabolites. In this study, TRLFS was used to identify the uranium speciation in seepage and soil waters of the test field Gessenwiese. Such investigations are important, because speciation determines chemical reactivity and thus its transport behavior and bioavailability (Carrière et al. 2005) in the biosphere.

7.2 Sampling Campaign

The surface water samples SrfWtr1 and SrfWtr2 were collected in and close to the test site Gessenwiese (see Fig. 7.1) from a little creek in the north of the test site on 29th of March in 2009.

Three pore water samples (SoilWtr1–3) were extracted from Gessenwiese soil (30–100 cm depth) of the site section MF3 (Fig. 7.1) on the 19th of November 2009.

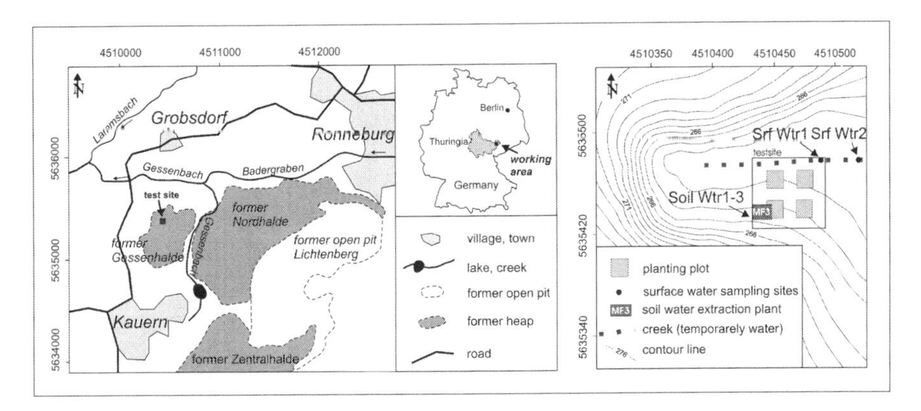

Fig. 7.1 Test site Gessenwiese near Ronneburg in eastern Thuringia (*left side*) with sampling sites (*right side*). Coordinates are in the German grid (Gauß-Krüger zone 4)

The test field "Gessenwiese" is located on a former leaching heap of the former uranium mining area Ronneburg, between the villages Kauern and Grobsdorf (eastern Thuringia), which was covered up during remediation by a soil layer. For more details see Grawunder et al. (2009).

7.3 Water Analyses

All water samples investigated in this study show a low pH and contain high concentration of heavy metals, including uranium, and a high concentration of sulfate. These results are in agreement with data published in Merten et al. (2004).

The chemical composition of seepage (SrfWtr1 and -2) and soil water samples (SoilWtr1, -2, and -3) were analyzed for cations by inductively coupled plasma mass spectrometry (ICP-MS), atomic absorption spectrometry (AAS) for Na, K, Mg, and Ca, and graphite furnace atomic absorption spectrometry (GFAAS) for Fe. An ELAN 9000 type ICP-MS spectrometer (Perkin Elmer SCIEX, Waltham, Massachusetts, USA), a Perkin Elmer 4100 AAS, and an AAS-6F ZEEnit 600s Graphite furnace AAS with Zeeman background correction (Analytik Jena, Jena, Germany) were used for these analyses. The error of the chemical analysis for AAS analyses is only 1–2% and 5–10% for ICP-MS measurements. Anions, i.e., chloride, nitrate, phosphate, and sulfate were determined by ion chromatography (IC-system 732/733, Metrohm, Filderstadt, Germany). The analytical error for the anion analyses is smaller than 5%. Total organic carbon (TOC) and total nitrogen (TN_b) were obtained by Multi-N/C 2100 (Analytik Jena, Jena, Germany). The concentration of selected anions from the two investigated surface water samples and the three soil water samples, together with the measured pH values and concentrations of cations are listed in Tables 7.1 and 7.2, respectively.

Table 7.1 Concentrations of anions, total organic carbon (TOC), and total nitrogen (TN_b) in mg/l; and pH of the water samples

	SrfWtr1	SrfWtr2	SoilWtr1	SoilWtr2	SoilWtr3
Chloride	48.8	82.6	3.75	1.85	11.4
Nitrate	<2	<2	<2	<2	<2
Phosphate	<5	<5	<5	<5	<5
Sulfate	3,520	4,480	966	1,010	2,530
TOC	7.54	4.90	6.38	2.36	2.04
TN_b	0.50	0.46	<0.2	0.60	0.25
pH	3.84	3.76	4.02	3.40	3.27

Table 7.2 Concentrations of cations in μg/l of the water samples

	SrfWtr1	SrfWtr2	SoilWtr1	SoilWtr2	SoilWtr3
Na	7,040	8,570	1,920	1,540	2,520
Mg	622,000	849,000	52,500	11,000	133,000
Al	30,100	43,100	21,500	31,700	100,000
Si	18,600	22,300	28,900	32,800	49,500
K	4,890	5,350	830	1,240	2,010
Ca	344,000	340,000	209,000	244,000	389,000
Mn	86,600	97,400	12,000	1,870	23,700
Fe	8,070	5,390	2,420	15,500	18,300
Co	2,050	2,910	331	102	1,000
Ni	11,400	14,200	1,920	751	6,370
Cu	54.3	293	1,120	1,490	3,600
Zn	1,760	2,870	1,130	1,490	2,840
As	<0.1	<0.1	<0.1	<0.1	<0.1
Sr	518	612	136	81.2	251
Cd	14.1	43.3	24.4	9.76	107
Ba	17.2	16.1	10.6	7.02	7.36
Pb	64.4	31.1	1.87	0.508	8.63
U	75.1	291	322	156	890

The reason for the scatter of the uranium concentrations in the different water samples could be related to the heterogeneity in the metal contents of the soil (Grawunder et al. 2009) or to the precipitation history prior to sampling.

7.4 Description of the TRLFS Analyses

TRLFS exhibits some outstanding features, above all a very low detection limit in case of fluorescent heavy metal ions. The predominance of TRLFS for analyzing uranium(VI) phosphate species, compared to other spectroscopic techniques, e.g., XRD and IR, was shown in Baumann et al. (2008).

TRLFS measurements of the water samples presented here were carried out at room temperature. The TRLFS system consists of a Nd:YAG diode laser (Inlite Continuum), where the actual laser power was monitored with an optical power meter (model LabMaster, Ultima Coherent, USA) to allow for corrections due to fluctuations in the laser power. The excitation wavelength was 266 nm. The resulting generated fluorescence signal was collected perpendicular to the excitation beam and focused into a fiber optic cable that was coupled to the slit of a triple-grating spectrograph (0.5 m spectrograph, model 1236 OMA, Princeton Applied Research, USA). The fluorescence spectra were measured by a charge-coupled device (CCD) camera (model 7467-0008, Princeton Instruments, Inc., USA) cooled to 18°C.

The spectra were recorded in a wavelength ranging from 430 to 600 nm. Exposure time from the CCD camera was set to 2 μs. The recording of the fluorescence signal occur in certain time intervals after the laser excitation pulse, in successive steps of 100 ns, in a range from 30 to 8030 ns. The average laser power was approximately 3 mJ. For every delay time, every fluorescence spectrum was measured three times, and for each spectrum, 100 laser shots were averaged. All functions (time controlling, device settings, recording of the spectra, data storage) of the spectrometer were computer-controlled. The computer software WinSpec/32 version 2.5.19.0 (Galactic Industries Corporation, USA) was used for the deconvolution of the spectra. The time dependencies of the spectra were calculated with the Origin 7.5G (OriginLab Corporation, Northampton, MA, USA) program. Another more detailed data processing procedure, described in Baumann et al. (2008), was used in the data evaluation process. Both methods delivered similar numerical results. The acquired TRLFS spectra displayed a high signal-to-noise ratio and sharp emission bands.

7.5 Results

Results obtained by TRLFS on U(VI) species on the investigated natural surface water samples and naturally occurring soil water samples – not on samples synthesized in the laboratory – were presented. Generally, TRLFS provides two kinds of spectroscopic information: the position of the emission maxima and the fluorescence lifetime. In a fingerprinting procedure, the TRLFS spectra obtained are then identified with the help of previously recorded reference substances. At an excitation wavelength of 266 nm, the fluorescence quantum yield for uranium(VI) is very high (Billard et al. 2003), and the respective uranium(VI) fluorescence is detected in the range 470–600 nm (Billard et al. 2003).

As shown in previous investigation, TRLFS is a useful tool for identifying the speciation of uranium in water, which are associated with plants.

7.5.1 Surface Water Samples

Both samples, i.e., SrfWtr1 and SrfWtr2, provided an evaluable fluorescence signal for TRLFS; one example of a time-resolved fluorescence signal is shown in Fig. 7.2. The positions of the six peak maxima from these signals in both water samples are shown in Table 7.3 and are in agreement with data for uranium sulfate species published in the literature (Geipel et al. 1996; Vercouter et al. 2008; Vetešník et al. 2009; Arnold et al. 2011). The intensity of the fluorescence signal from sample SrfWtr1 is about half of the intensity from sample SrfWtr2 in analogy to the different uranium content of these samples (Table 7.2).

The TRLFS signals of both water samples possess a mono-exponential decay, indicating the presence of only one main species. These two characteristics, i.e., positions of peak maxima and lifetimes revealed without doubt that the uranium speciation in the seepage water is dominated by the uranium(VI) sulfate species

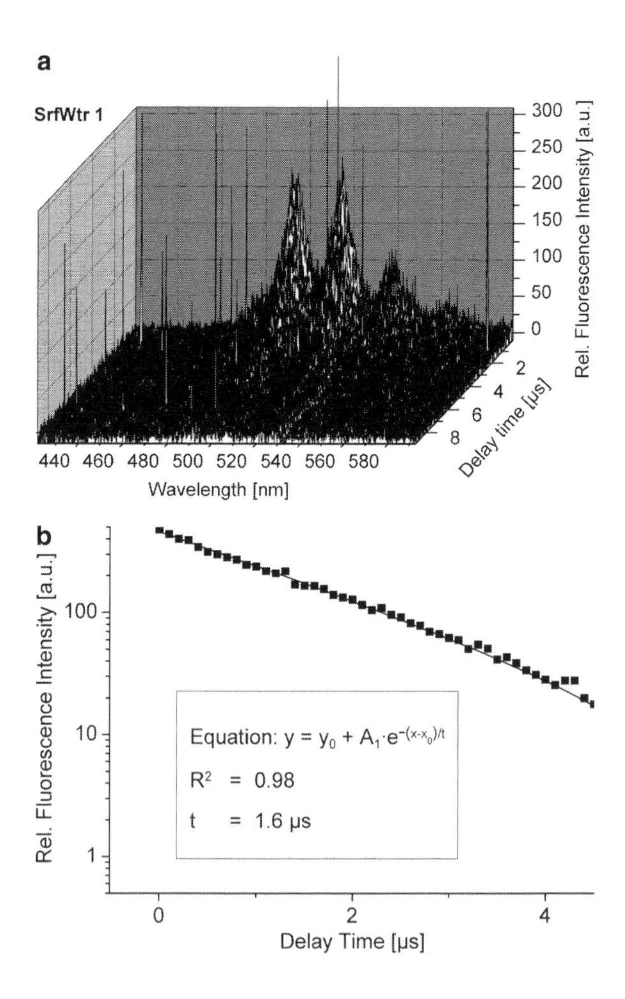

Fig. 7.2 Time-resolved laser-induced fluorescence signal obtained from sample SrfWtr1 (*top*) and lifetime curve from the fluorescence signal sample SrfWtr1 (*bottom*)

Table 7.3 Positions of the peak maxima of the fluorescence signals SrfWtr1 and SrfWtr2 in nm, rounded in whole nm, and lifetime of the signals *t* in μs, compared with data published in Vercouter et al. (2008)

	First peak	Second peak	Third peak	Fourth peak	Fifth peak	Sixth peak	*t*
SrfWtr1	477	491	513	537	562	591	1.6
SrfWtr2	478	492	513	538	562	590	1.0
Vercouter et al. (2008)	477	493	515	538	565	–	–

$UO_2SO_{4(aq)}$. A mono-exponential decay curve from sample SrfWtr1 is shown in Fig. 7.2 (right).

The observed lifetimes of the two uranium fluorescence signals were 1.6 and 1.0 μs, respectively, and thus (Table 7.3) shorter than reported lifetime (4.3 ± 0.5 μs) for uranium sulfate species found in the literature (Geipel et al. 1996). These shortened lifetimes of the uranium sulfate fluorescence signal from the investigated natural samples were attributed to the presence of quenchers in the samples, e.g., iron, manganese, and organic substances.

A reason for the shorter lifetime of the fluorescence signal from SrfWtr2 compared to the fluorescence signal from SrfWtr1 is attributed to the higher concentration of manganese in SrfWtr2, because manganese quenches the fluorescence signal of uranium(VI).

7.5.2 Pore Water Samples

All three pore water samples from the test field showed an evaluable fluorescence signal for TRLFS. The positions of the peak maxima from these signals in all pore water samples are also in concordance with data for uranium sulfate species published in the literature (Geipel et al. 1996; Vercouter et al. 2008; Vetešník et al. 2009; Arnold et al. 2011).

A time-resolved fluorescence measurement is a row of single spectra, in which every single fluorescence measurement (every single spectrum) is occurs temporal offset a certain time interval after the excitation laser pulse (Fig. 7.2). In measurements presented here, every single time interval between two consecutive single spectra is 100 ns wide. Because the fluorescence signal decays immediately after laser excitation, the first fluorescence spectrum directly acquired after excitation shows the highest fluorescence intensity.

The three spectra with the highest fluorescence intensity of the three soil water samples are shown in Fig. 7.3 and the respective emission maxima and fluorescence lifetimes are shown in Table 7.4.

Figure 7.3 shows that the fluorescence intensity contains information on the uranium(VI) concentration. The higher the uranium concentration (Table 7.2), the more intensive is the fluorescence intensity of the first spectrum (Fig. 7.3). A comparison with literature data, e.g., with Vercouter et al. (2008), indicate that the TRLFS results, based

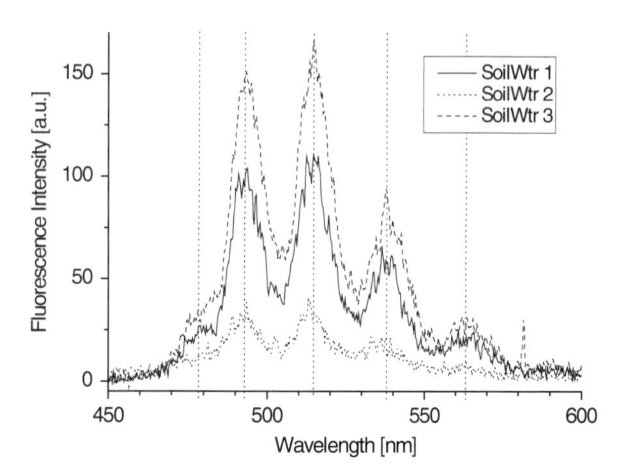

Fig. 7.3 Intensity from the fluorescence signals of the three pore water samples, in comparison with positions of peak maxima published in Vercouter et al. (2008) for $UO_2SO_{4(aq)}$ (*dashed vertical lines*)

Table 7.4 Positions of the peak maxima from the fluorescence signals of three pore water samples in nm, rounded in whole nm, compared with data published in the literature and lifetime of the signals *t* in µs

	First peak	Second peak	Third peak	Fourth peak	Fifth peak	Sixth peak	*t*
SoilWtr1	480	492	514	537	562	590	2.6
SoilWtr2	473	492	513	535	562	–	1.7
SoilWtr3	477	493	514	538	562	590	1.0
Vercouter et al. (2008)	477	493	515	538	565	–	–

on the positions of the peak maxima and the mono-exponential decay of the fluorescence signal, are interpreted in such a way that the observed uranium speciation in the soil water from the "Gessenwiese" is dominated by the presence of the $UO_2SO_{4(aq)}$ species. This shows that the speciation in the seepage waters as well as in the soil waters are both dominated by the presence of the $UO_2SO_{4(aq)}$ species.

The different lifetimes of the fluorescence signals in the three soil water samples investigated here were explained with the different concentration of quenching substances in these samples. In case of the soil water samples, there seems to be a dependency between lifetime of the uranium fluorescence signal and the iron concentration in the sample: Fe behaves as a quencher for the uranium(VI) fluorescence signal. The higher the Fe concentration in the soil water sample, the shorter is the uranium fluorescence signal (Tables 7.2 and 7.4). The differences in the positions of the first peaks in Table 7.4 are an effect of the comparatively small size of these peaks, which are rather shoulders than real peaks.

7.6 Conclusions

It was shown that TRFLS is a useful tool for identifying the speciation of uranium in natural waters characterized by low pH and high metal concentration.

Uranium speciation in two naturally occurring seepage water samples, SrfWtr1 and SrfWtr2, and three soil water samples, SoilWtr1, -2, and -3, all samples from test site Gessenwiese, was analyzed by TRLFS. All five samples showed an evaluable fluorescence signal for TRLFS. The positions of the six peak maxima from these signals are in all five water samples in agreement with data for uranium sulfate species $UO_2SO_{4(aq)}$ (Geipel et al. 1996; Vercouter et al. 2008). Moreover, the TRLFS signals of all five water samples possess a mono-exponential decay, indicating the presence of one main uranium species. These two characteristics, i.e., positions of peak maxima and the observed mono-exponential decay of the fluorescence signal revealed that the uranium speciation in the seepage water is dominated by the uranium(VI) sulfate species $UO_2SO_{4(aq)}$. The fluorescence lifetime of this $UO_2SO_{4(aq)}$ species decreases in the presence of iron ions. The higher the iron concentration, the shorter becomes the uranium fluorescence lifetime. This indicates that iron plays an important role in quenching the uranium(VI) fluorescence signal lifetime in naturally occurring water samples, an additional quenching effect is assumed as a result of the high manganese concentrations in these waters.

The analyses were performed to compare the results on the identified uranium speciation in a later stage with the uranium speciation in plants, which grow on the grassland test site Gessenwiese and take up the same uranium-contaminated water during their growth.

TRLFS measurements with plant compartments (e.g., roots, leaves, shoots), which grow in association with the seepage water, will be carried out in future investigations and will be compared to the identified uranium speciation in surface water and soil water of the Gessenheap. Samples could be obtained from the plant compartments by centrifugation as cell sap, or as solid milled plant compartment sample and subsequently analyzed by TRLFS. The reactivity and toxicity of uranium depends on the speciation of heavy metals, and has to be considered as important possible risk factor as uranium may enter economic plants and eventually arrives in the food chain.

Acknowledgments The authors thank the EU for funding UMBRELLA, GA No 226870, project within FP7 topic "Recovery of degraded soil resources," and the Bundesministerium für Bildung und Forschung (BMBF), Project No 02NUK015F also for support, and E. Kothe and G. Büchel (both Friedrich-Schiller Universität Jena) for getting access to test site "Gessenwiese."

Supplemental Information

Original sample name	Sample name in that report
GB 3	SrfWtr1
GB 6	SrfWtr2
MF3/60/1	SoilWtr1
MF3/30/2	SoilWtr2
MF3/100/3	SoilWtr3

References

Amayri S, Arnold T, Foerstendorf H, Geipel G, Bernhard G (2004a) Spectroscopic characterization of synthetic becquerelite, $Ca[(UO_2)_6O_4(OH)_6]\cdot 8H_2O$, and swartzite, $CaMg[UO_2(CO_3)_3]\cdot 12H_2O$. Can Mineral 42(4):953–962

Amayri S, Arnold T, Reich T, Foerstendorf H, Geipel G, Bernhard G, Massanek A (2004b) Spectroscopic characterization of the uranium carbonate andersonite $Na_2Ca[UO_2(CO_3)_3]\ 6H_2O$. Environ Sci Technol 38:6032–6036

Arnold T, Baumann N (2009) Boltwoodite $[K(UO_2)(SiO_3OH)(H_2O)_{1.5}]$ and compreignacite $K_2[(UO_2)_3O_2(OH)_3]_2\ 7H_2O$ characterized by laser fluorescence spectroscopy. Spectrochim Acta A Mol Biomol Spectrosc 71:1964–2968

Arnold T, Utsunomiya S, Geipel G, Ewing RC, Baumann N, Brendler V (2006) Adsorbed U(VI) surface species on muscovite identified by laser fluorescence spectroscopy and transmission electron microscopy. Environ Sci Technol 40:4646–4652

Arnold T, Baumann N, Krawczyk-Bärsch E, Brockmann S, Zimmermann U, Jenk U, Weiß S (2011) Identification of the uranium speciation in an underground acid mine drainage environment. Geochim Cosmochim Acta. 75(8):2200–2212

Baumann N, Brendler V, Arnold T, Geipel G, Bernhard G (2005) Uranyl sorption onto gibbsite studied by time-resolved laser-induced fluorescence spectroscopy (TRLFS). J Colloid Interface Sci 290:318–324

Baumann N, Arnold T, Geipel G, Trueman E, Black S, Read D (2006) Detection of U(VI) on the surface of altered depleted uranium by time-resolved laser-induced fluorescence spectroscopy (TRLFS). Sci Total Environ 366:905–909

Baumann N, Arnold T, Foerstendorf H, Read D (2008) Spectroscopic verification of the mineralogy of an ultra-thin mineral film on depleted uranium. Environ Sci Technol 42:8266–8269

Bernhard G, Geipel G (2007) Bestimmung der Bindungsform des Urans in Mineralwässern. Vom Wasser 105(3):7–10

Bernhard G, Geipel G, Brendler V, Nitsche H (1998) Uranium speciation in waters of different uranium mining areas. J Alloys Compd 271–273:201–205

Billard I, Ansoborlo E, Apperson K, Arpigny S, Azenha ME, Birch D, Bros P, Burrows HD, Choppin G, Couston L, Dubois V, Fanghänel T, Geipel G, Hubert S, Kim JI, Kimura T, Klenze R, Kronenberg A, Kumke M, Lagarde G, Lamarque G, Lis S, Madic C, Meinrath G, Nagaishi R, Parker D, Plancque G, Scherbaum F, Simoni E, Sinkov S, Viallesoubranne C (2003) Aqueous solutions of uranium(VI) as studied by time-resolved emission spectroscopy: a round-robin test. Appl Spectrosc 57(8):1027–1038

Brendler V, Geipel G, Bernhard G, Nitsche H (1996) Complexation in the system $UO_2^{2+}/PO_4^{3-}/OH_{(aq)}^-$: potentiometric and spectroscopic investigations at very low ionic strengths. Radiochimica Acta 74:75–80

Carrière M, Gouget B, Gallien JP, Avoscan L, Gobin R, Verbavatz JM, Khodja H (2005) Cellular distribution of uranium after acute exposure of renal epithelial cells: SEM, TEM and nuclear microscopy analysis. Nucl Instrum Methods Phys Res Sect B-Beam Interact Mater Atoms 231:268–273

Espana JS, Pamo EL, Santofimia E, Aduvire O, Reyes J, Barettino D (2005) Acid mine drainage in the Iberian Pyrite Belt (Odiel river watershed, Huelva, SW Spain): geochemistry, mineralogy and environmental implications. Appl Geochem 20(7):1320–1356

Geipel G, Brachmann A, Brendler V, Bernhard G, Nitsche H (1996) Uranium(VI) sulfate complexation studied by time-resolved laser-induced fluorescence spectroscopy (TRLFS). Radiochimica Acta 75:199–204

Grawunder A, Lonschinski M, Merten D, Büchel G (2009) Distribution and bonding of residual contamination in glacial sediments at the former uranium mining leaching heap of Gessen/Thuringia, Germany. Chemie der Erde – Geochemistry 69:5–19

Großmann K, Arnold T, Krawczyk-Bärsch E, Diessner S, Wobus A, Bernhard G, Krawietz R (2007) Identification of fluorescent U(V) and U(VI) microparticles in a multispecies biofilm by confocal laser scanning microscopy and fluorescence spectroscopy. Environ Sci Technol 41:6498–6504

Großmann K, Arnold T, Ikeda-Ohno A, Steudtner R, Geipel G, Bernhard G (2009) Fluorescence properties of a uranyl(V)-carbonate species $[U(V)O_2(CO_3)_3]^{5-}$ at low temperature. Spectrochim Acta A Mol Biomol Spectrosc 72:449–453

Guillaumont R, Fanghänel T, Fuger J, Grenthe I, Neck V, Palmer DA, Rand MH (2003) Update on the chemical thermodynamics of uranium, neptunium, plutonium, americium and technetium. In: OECD Nuclear Energy Agency (ed) Chemical thermodynamics, vol 1. Elsevier, Amsterdam, pp 230–233

Günther A, Bernhard G, Geipel G, Reich T, Roßberg A, Nitsche H (2003) Uranium speciation in plants. Radiochimica Acta 91:319–328

Haferburg G, Kothe E (2007) Microbes and metals: interactions in the environment. J Basic Microbiol 47:453–467

Haferburg G, Merten D, Büchel G, Kothe E (2007) Biosorption of metal and salt tolerant microbial isolates from a former uranium mining area. Their impact on changes in rate earth element patterns in acid mine drainage. J Basic Microbiol 47:474–484

Hennig C, Schmeide K, Brendler V, Moll H, Tsushima S, Scheinost AC (2007) EXAFS investigations of U(VI), U(IV), and Th(IV) sulfato complexes in aqueous solution. Inorg Chem 46:5882–5892

Hennig C, Ikeda A, Schmeide K, Brendler V, Moll H, Tsushima S, Scheinost AC, Skanthakumar S, Wilson R, Soderholm L, Servaes K, Görrler-Walrand C, van Deun R (2008) The relationship of monodentate and bidentate coordinated uranium(VI) sulphate in aqueous solution. Radiochimica Acta 96:607–611

Kirishima A, Kimura T, Tochiyama O, Yoshida Z (2003) Luminescence study of tetravalent uranium in aqueous solution. Chem Commun 7:910–911

Kirishima A, Kimura T, Nagaishi R, Tochiyama O (2004) Luminescence properties of tetravalent uranium in aqueous solution. Radiochimica Acta 92:705–710

Křepelová A, Brendler V, Sachs S, Baumann N, Bernhard G (2007) U(VI)-kaolinite surface complexation in absence and presence of humic acid studied by TRLFS. Environ Sci Technol 41:6142–6147

Merten D, Kothe E, Büchel G (2004) Studies on microbial heavy metal retention from uranium mine drainage water with special emphasis on rare earth elements. Mine Water Environ 23:34–43

Moulin C, Beaucaire C, Decambox P, Mauchien P (1990) Determination of uranium in solution at the ng L^{-1} level by time-resolved laser-induced spectrofluorimetry – application to geological samples. Anal Chim Acta 238(2):291–296

Moulin C, Decambox P, Moulin V, Decaillon JG (1995) Uranium speciation in solution by time-resolved laser-induced fluorescence. Anal Chem 67(2):348–353

Steudtner R, Arnold T, Großmann K, Geipel G, Brendler V (2006) Luminescence spectrum of uranyl(V) in 2-propanol perchlorate solution. Inorg Chem Commun 9:939–941

Vercouter T, Vitorge P, Amekraz B, Moulin C (2008) Stoichiometries and thermodynamic stabilities for aqueous sulfate complexes of U(VI). Inorg Chem 47(6):2180–2189

Vetešník A, Semelová M, Štamberg K, Vopálka D (2009) Uranium(VI) sulfate complexation as a function of temperature and ionic strength studied by TRLFS. In: Merkel BJ, Hasche-Berger A (eds) Uranium mining and hydrogeology. Springer, Berlin, pp 623–630

Wang Z, Zachara JM, Yantasee W, Gassman PL, Liu C, Joly AG (2004) Cryogenic laser induced fluorescence characterization of U(VI) in Hanford vadose zone pore waters. Environ Sci Technol 38:5591–5597

Chapter 8
Plant–Microbe Interaction in Heavy-Metal-Contaminated Soils

Neeru Narula, Martin Reinicke, Götz Haferburg, Erika Kothe, and Rishi Kumar Behl

8.1 Introduction

Heavy metals, as soil and water pollutants, cause severe threat to the environment. Heavy metal pollution affects the production and quality of crops, the quality of atmosphere and water bodies, and thus threatens the human and animal health. The metal species commonly found in the soils as a result of human activities include copper (Cu), zinc (Zn), nickel (Ni), lead (Pb), cadmium (Cd), cobalt (Co), mercury (Hg), chromium (Cr), arsenic (As), etc. Some of these act as micronutrients at small concentrations for living organisms for their normal physiological activities, but their accumulation is toxic to most life forms. The most common human activities resulting in entry of heavy metal into land are disposal of industrial effluents, disposal of waste such as sewage sludge, atmospheric deposition from industrial activities, mining activities, domestic and industrial wastes, land fill operations, and use of agrochemicals. Release of heavy metals from various industrial sources, agrochemicals, and sewage sludge presents a major threat to the soil environment. Generally, heavy metals are not degraded biologically and persist in the environment indefinitely.

The toxic heavy metals, upon accumulation in the soils, inversely affect the microbial compositions, including plant growth promoting rhizobacteria (PGPR) in the rhizosphere and their metabolic activities. In addition, the elevated concentration of metals in soils and their uptake by plants adversely affect the growth, symbiosis, and consequently the yields of crops (Wani et al. 2008a) by

N. Narula (✉)
Department of Microbiology, CCS Haryana Agricultural University Hisar, Haryana, India
e-mail: neeru_narula@yahoo.com

M. Reinicke • G. Haferburg • E. Kothe
Institut für Mikrobiologie, Friedrich-Schiller Universität, Neugasse 25, D-07743 Jena

R.K. Behl
Department of Plant Breeding, CCS Haryana Agricultural University Hisar, Haryana, India

E. Kothe and A. Varma (eds.), *Bio-Geo Interactions in Metal-Contaminated Soils*, Soil Biology 31, DOI 10.1007/978-3-642-23327-2_8,

disintegrating cell organelles, and disrupting the membranes (Stresty and Madhava Rao 1999), acting as genotoxic substance disrupting the physiological process, such as photosynthesis (Wani et al. 2007b) or by inactivating the respiration, protein synthesis, and carbohydrate metabolism (Shakolnik 1984). The remediation of metal-contaminated soils thus becomes important, as these are rendered unsuitable for sustainable agriculture. This chapter provides an overview on plant–microbe interactions toward phyto- and bio-remediation.

8.2 Phytoremediation

Phytoremediation strategy involves the cultivation of metal accumulating higher plants to remove, transfer, or stabilize the contaminants from metal polluted soils (Brooks 1998). In this approach, plants capable of accumulating high levels of metals are grown in contaminated soils. At maturity metal enriched above ground biomass is harvested and soil metal contamination is removed. Successful plant-based decontamination of even moderately contaminated soils would require crops able to concentrate metals in excess of 1–2%. Accumulation of such high levels of heavy metals is highly toxic and would certainly kill the common nonaccumulator plants. However, in hyperaccumulator species, such concentrations are attainable. Nevertheless, the extent of metal removal is ultimately limited by the plants ability to extract and tolerate only a finite amount of metals. On a dry weight basis, this threshold is around 3% for Zn and Ni, and considerably less for more toxic metals, such as Cd and Pb. The other biological parameter which limits the potential for metal phytoextraction is biomass production. With highly productive species, the potential for biomass production is about 100 tons fresh weight/ha. The values of these parameters limit the annual removal potential to a maximum of 400 kg metal/ha/yr. It should be mentioned, however, that most metal hyperaccumulators are slow growing and produce little biomass. These characteristics severely limit the use of hyperaccumulator plants for environment cleanup. Practices have been developed to increase the potential of common nonaccumulator plants for Pb phytoextraction. Particularly, the uptake-inducing properties of synthetic chelates open the possibility of using high biomass-producing crops for Pb phytoextraction. Under chelate-induced conditions, maize (Huang and Cunningham 1996) and Indian mustard (Blaylock et al. 1997) have been successfully used to remove Pb from solution culture and contaminated soil, respectively. Physical characteristics of soil contamination are also important for the selection of remediating plants. For example, for the remediation of surface-contaminated soils, shallow rooted species would be appropriate to use, whereas deep-rooted plants would be the choice for more profound contamination.

The identification of metal hyperaccumulators, plants capable of accumulating high metal levels, demonstrates that some plants have genetic potential to clean up metal-contaminated soils. In general, the concentration of metals in hyperaccumulators is about 10- to 100-fold higher than most other plants growing on metal-

Table 8.1 Plants capable of hyperaccumulating metals

Plant species	Metal	Accumulated concentration, mg/kg dry matter
Thlaspi caerulescens (Brassicaceae)	Zn	10,000
Sebertia acuminate (Sapotaceae)	Ni	10,000
Alyssum lesbiacum (Brassicaceae)	Ni	20,000
Arabidopsis halleri (Brassicaceae)	Cd	1,000
Thlaspi rotundifolium (Brassicaceae)	Pb	8,200
Astralagus sp. (Leguminosae)	Se	1,000
Pteris vittata (Fern)	As	22,630

contaminated soils. It has been possible through bioengineering to develop plants (Raskin 1996) capable of removing methyl mercury from the contaminated soil. To detoxify this compound, such bioengineered plants express modified bacterial genes *merB* and *merA* which convert methyl mercury to elemental mercury. About 400 plant species have been identified as hyperaccumulator. The Indian mustard plant (*Brassica juncea*) can extract both heavy metals and radionuclides from soil. Panwar et al. (2002) reported that *B. juncea* has the potential to be hyperaccumulator of Ni. Survey of literature reveals that the rate of metal removal depends upon the plant species, soil contaminating heavy metal(s), and biomass harvested and metal concentration in harvested biomass as summarized in Table 8.1.

8.3 Rhizosphere

Soil is the living environment that supports the extremely diverse communities of macro- and microorganisms and often considered as "black box." Rhizosphere is the zone of soil surrounding a plant root where in biology and chemistry of the soil is influenced by the roots, rather it is an area of intense activity (biological, chemical, and physical) influenced by compounds exuded by roots and by microorganisms feeding on these compounds (Bais et al. 2006; Kumar et al. 2007). Generally, soil contains bacterial numbers in the range of 10^7–10^{10} cells per gm dry soil. But microbiological activity in the rhizosphere is much greater (10^8) than in soil away from plant roots (10^5) and also microorganisms provide or make available nutrients for the plants (Walker et al. 2003). Many of these microbes live there as a part of a distinct community surrounding plant roots. Heterotrophic bacteria are able to use organic compounds excreted in root exudates, whereas their metabolites can be used by other microbes, which in the end creates a network of closely connected microorganisms. This phenomenon of highly active microorganisms in root-associated soil is known as the "rhizosphere effect." Thus, microbial population is one of the essential parts of the rhizosphere that affect the rhizosphere soil by its various activities such as water and nutrient uptake, exudation, and all the biological transformations. Among the fast growing and early colonizing bacteria attracted by the plant exudates are members of genera *Bacillus* and *Pseudomonas* besides N fixing bacteria *Azospirillum* and *Rhizobium*.

Cultivation-based methods show that the *Pseudomonas* spp. are generally more abundant in rhizosphere than in the bulk soil. But number of related clones to *Pseudomonas* is higher in rhizosphere of ryegrass and white clover. On the basis of 16s rRNA gene clones, plant roots have further been shown to have a selective effect toward γ-proteobacteria leading to predominance of *Pseudomonas* spp. in rhizosphere as compared to bulk soil. Free living and associative diazotrophs are more numerous in the rhizosphere than in bulk soils, indicating their dependence on organic compounds exuded by roots in the rhizosphere. Various nitrogen-fixing microorganisms have been found to be present in the rhizosphere of agricultural plants, but the contribution of fixed nitrogen to plant nutrition is controversial (Lima et al. 2006). Diazotrophs found in the soil or associated with roots include *Azotobacter chroococcum, Azospirillum brasilense,* and *Gluconacetobacter diazotrophicus* (formerly *Acetobacter diazotrophicus*) and the positive responses of plants to inoculation with these bacteria are attributed to nitrogen (N_2) fixation besides several other factors like phytohormone and/or ammonium production, etc. (Okon 1985). Plant-associated rhizobacteria and mycorrhizae may significantly increase the bioavailability of various heavy metal ions for their uptake by plants. Also, they are known to catalyze redox transformations leading to changes in heavy-metal bioavailability (Yang et al. 2005).

Plant association with diazotrophs or any colonizer indicates a high degree of adaptation between the host plant and the most abundant diazotrophs. The micro-habitat provided by the host plant seems to generate a selection pressure in favor of the microorganisms, which in turn best benefit the host. Nutrients and metals are typically present in the soil solution at low concentrations and tend to form sparingly soluble minerals (except nitrogen, sulfur, and boron) or may be adsorbed to a solid phase through ion exchange, hydrogen bonding, or complexation (White 2003). The extent to which they are transferred from the soil to the biota (i.e., microbes or plants) is dependent on the biogeochemical interactions (N, P, S) and/or processes among the soil, plant roots, and microorganisms in the rhizosphere (Abbott and Murphy 2003). At this interface, the presence of root exudates may influence chemical reaction kinetics within the soil environment and subsequently affect biological activities. The exudates act as messengers that stimulate biological and physical interactions between root and soil organisms. In addition to the adsorption and conduction, roots also produce hormones and other substances that help to regulate the plants development and structure, help in modifying the biochemical and physical properties of the rhizosphere (Abbott and Murphy 2003), and contribute to root growth and plant survival.

8.4 Plant Growth Promoting Rhizobacteria

Free-living soil bacteria beneficial to plant growth are usually referred to as PGPR. These are capable of promoting plant growth by colonizing and establishing around the plant root (Narula et al. 2006). PGPRs have been found to play a potential role in developing sustainable systems in crop production (Shoebitz et al. 2009). Soil bacteria have been used as biofertilizer for ameliorating the soil fertility and enhancing crop production for decades. The main functions of these bacteria are (1) to supply nutrients to crops; (2) to stimulate plant growth, e.g., through the production of plant hormones; (3) to control or inhibit the activity of plant pathogens; (4) to improve soil structure; and (5) to act as bioaccumulator in microbial leaching of inorganics (Ehrlich 1990). More recently, bacteria have also been used in soil for the mineralization of organic pollutants, i.e., bioremediation of polluted soils (Zaidi et al. 2008). Generally, PGPR functions in three different ways (Glick 1995, 2001): (a) synthesizing particular compounds for the plants (Dobbelaere et al. 2003), (b) facilitating the uptake of certain nutrients from the soil (Çakmakçi et al. 2006), and (c) lessening or preventing the plants from diseases. Besides their role in protecting the plants from metal toxicity, the PGPR are also known for their role in enhancing the soil fertility and promoting crop productivity by providing essential nutrients (Zaidi and Khan 2006) and plant growth regulators (Kumar et al. 2007). Glick et al. (2002) reported that these PGPRs also promote the growth of plants by alleviating the stress induced by ethylene-mediated impact on plants by synthesizing 1-aminocyclopropane-1-carboxylate (ACC) deaminase (Belimov et al. 2005). According to their relationship with the plants, PGPRs can be divided into two groups (Khan 2005a). First group "Symbiotic bacteria" with species of *Rhizobium* (*Rhizobium*, *Mesorhizobium*, *Bradyrhizobium*, *Azorhizobium*, *Allorhizobium*, and *Sinorhizobium*) have been successfully used worldwide to permit an effective establishment of the nitrogen-fixing symbiosis with leguminous crop plants (Bottomley and Dughri 1989; Bottomley and Maggard 1990; Lynch 1990). Second group "Free-living rhizobacteria" with nonsymbiotic nitrogen-fixing bacteria such as *Azotobacter, Azospirillum, Bacillus,* and *Klebsiella* sp. are also used to inoculate a large area of arable land in the world with the aim of enhancing plant productivity (Lynch 1983). In addition, phosphate solubilizing bacteria such as species of *Bacillus* and *Paenibacillus* (formerly *Bacillus*) have been applied to soils to specifically enhance the phosphorus status of plants.

8.5 Microbial Adaptation to Heavy Metals

To overcome the metal stress, numbers of mechanisms have been evolved by microorganisms of agronomic importance by which they tolerate and promote the uptake of heavy metal ions. Such mechanisms include (1) the pumping of metal ions exterior to the cell, (2) accumulation and sequestration of the metal ions inside the cell, (3) transformation of toxic metal to less toxic forms (Wani et al. 2008b) and adsorption/desorption of metals (Mamaril et al. 1997). One or more of these defense mechanisms allows these microorganisms to function metabolically in environment polluted by metals. These mechanisms could be constitutive or inducible. The bacterial resistance mechanisms are encoded generally on plasmids and transposons, and it is probably by gene transfer or spontaneous mutation that bacteria acquire their resistance to heavy metals. For example, in gram-negative bacteria (e.g., *Ralstonia eutropha*), the *czc* system is responsible for the resistance to cadmium, zinc, and cobalt. The *czc*-genes encode for a cation-proton antiporter (CzcABC), which exports these metals (Nies 1999). A similar mechanism, called *ncc* system, has been found in *Alcaligenes xylosoxidans*, which provides resistant against nickel, cadmium, and cobalt. On the contrary, the cadmium resistance mechanism in gram-positive bacteria (e.g., *Staphylococcus*, *Bacillus*, or *Listeria*) is through Cd-efflux ATPase. Plasmid-encoded energy-dependent metal efflux systems involving ATPases and chemiosmotic ion/proton pumps are also reported for arsenic, chromium, and cadmium resistance in other PGPR (Roane and Pepper 2000).

Due to these properties, when PGPR including nitrogen fixers, used as seed inoculants (Narula et al. 2005a, b), were applied to soil, either treated/amended intentionally with metals or already contaminated, have shown a substantial increase in accumulation of heavy metals and concomitantly improved the overall growth and yield of chickpea (*Cicer arietinum*) (Gupta et al. 2004), green gram (*Vigna radiata* L. wilczek), and pea (*Pisum sativum*) (Wani et al. 2007a, c). An experiment on effect of rhizobial inoculation on metal accumulation potential of the chickpea plant was done by Gupta et al. (2004). The chickpea plants were harvested and accumulation of metal was studied in root and shoot parts of the plant. Accumulation of Cu, Zn, Cr, and Cd was more in shoot whereas Fe was more in root part of the plant. The chickpea variety CSG 8962 showed about 50% increase in Fe accumulation followed by 46% increase in Cu and Zn and 25% increase in Cr in roots when inoculated with *Rhizobium*. These studies are significant while using the plant for metal decontamination of fly ash land fills. The use of such microbes possessing multiple properties like plasmid borne (Saitia et al. 1989) metal resistance/reduction (Saitia and Narula 1989; Narula et al. 2011b) and their ability to promote plant growth through different mechanisms in metal-contaminated soils is chosen as a suitable choice for bioremediation studies.

The efficiency of phytoremediation technique is, however, influenced by the activity of rhizosphere microbes (Khan et al. 2009) and the speciation and concentration of metals deposited into soil (Wang et al. 1989; Khan 2005a, b). For instance, Lippmann et al. (1995) used PGPR *Pseudomonas* and *Acinetobacter* which enhanced phytoremediation abilities (Fe and Zn) of nonhyperaccumulating maize (*Zea mays* L.) plants by increasing their growth and biomass. Many metals such as Zn, Mn, Ni, and Cu are essential micronutrients. In common nonaccumulator plants, accumulation of these micronutrients does not exceed their metabolic needs (<10 ppm). In contrast, metal hyperaccumulator plants can accumulate exceptionally high amounts of metals (in the thousands of ppm). Corticeiro et al. (2006) reported that plants growing in metal-stressed soils can protect themselves from metal toxicity by synthesizing antioxidant enzymes, scavenging the toxicity of reactive oxygen species generated by plants (Cardoso et al. 2005), and by associative bacteria under metal stress.

8.6 Plant–Microbe Interactions in the Rhizosphere

In the era of sustainable crop production, the plant–microbe interactions in the rhizosphere play an important role in transformation, mobilization, solubilization, etc., of nutrients from a limited nutrient pool, and subsequently uptake of essential nutrients by plants to realize their full genetic potential. An understanding of the mechanisms, which is important for the initiation and establishment of the association between host and bacterium, can be reached from the analyses of influences exerted by each interaction partner on the other. In addition, there is a need to know how diazotrophs may benefit the plant. Investigations on the production of phytohormones and the action of siderophores produced by *Azotobacter* strains might help to understand this aspect of interactions (Narula et al. 2005a, b, 2006). Plant–microbe interactions are important for both the partners, i.e., macro as higher plants and micropartners as the plant-associated bacteria (Somers et al. 2004). Microbial partners can induce antagonistic (in case of phytopathogens) or symbiotic interactions. Different types of interactions involving plants roots in the rhizosphere have been reviewed by Bais et al. (2004, 2006). These include root–root, root–insect, and root–microbe interactions. The rhizosphere represents a highly dynamic front to study interaction between roots and pathogenic as well as beneficial soil microbes, invertebrates, and root systems of competitors (Bais et al. 2004). In recent years, several plant scientists have recognized the importance of root exudates in mediating these biological interactions. However, because plant roots are always hidden below ground, many of the interesting phenomena, their attractions, love and hate relationship in which they are involved have remained largely unnoticed. Especially the role of chemical signals (Peters et al. 1986) in mediating below ground interactions is only beginning to be understood. Chemical signaling between plant roots and other soil organisms, including the roots of neighboring plants, is often based on chemicals exuded from the roots. The same chemical signals may elicit dissimilar responses from different recipients. Most importantly, chemical components of root exudates may be the one to deter one organism while attracting another, or two very different organisms may be attracted with differing consequences to the plant. Therefore, the mechanisms used by roots to communicate and interpret these signals which they receive from other roots and soil microbes in the rhizosphere are largely unknown. The rhizosphere has some positive or negative and neutral associations. Much has to be done still to determine and elucidate whether the chemical signature of a plant root exudates will be perceived as a negative or a positive signal. However, evidences over the years suggest that root exudates are the determining factor to identify interactions in the rhizosphere and, ultimately, plant and soil community dynamics (Bais et al. 2006; Narula et al. 2009).

8.6.1 Biotic and Chemical Interferences of Heavy Metals in Plant–Microbe Interactions

In the last decades, the essential role of soil microbiota has been emphasized for many processes that are directly related to soil bioremediation and, especially, phytoremediation, in which plant–microbe interactions have been documented to be of vital importance (Khan 2005a). Phytoremediation is a process that uses actively growing plant roots to stimulate a diverse population of soil microorganisms, some of which have the capability to metabolize contaminants. In case of phytoextraction, inoculation of plants with PGPR may be beneficial mainly due to plant growth stimulation and a corresponding increase in the above-ground biomass, rather than to an increase in metal concentrations in plant tissues, which may actually be weakly influenced (Wu et al. 2006b). Such highly persistent and hardly avoidable pollutants as heavy metals, in contrast to organic contaminations (Karthikeyan and Kulakow 2003), cannot be biologically or chemically degraded, but can only be deactivated (by detoxification and/or immobilization) within the soil or removed from the contaminated site (Ruml and Kotrba 2003). On the other hand, they can directly or indirectly affect plant–microbe interactions in the rhizosphere. Their direct influence involves possible toxic or other detrimental effects of particular chemical metal species on the biochemistry and physiology of plant-associated microorganisms (Renella et al. 2005) and on the growth and development of plants and these effects comprise the biotic interferences. The chemical reactivity of particular metal species or their catalytic effects, which result in chemical binding or degradation of biogenic organics involved in plant–microbe interactions (i.e., remote exchange of molecular signals, quorum sensing within bacterial communities, release and exchange of nutrients, etc., Gunatilaka 2006), may thus be classified as indirect effects comprising chemical interferences. Biotic processes, induced in the rhizosphere by biogeochemical interactions or formation of chemically active metal species, result in chemical depletion, inactivation or degradation of any biomolecules directly involved in plant–microbe interactions via their binding, and/or by redox transformation, invariably affect these biologically specific interactions as described by various researchers (Lasat 2000; Kamnev 2008).

8.7 Microorganism-Mediated Metal Bioavailability for Uptake into Roots

A major factor limiting metal uptake into roots is slow transport from soil particles to root surfaces. There are two mechanisms (Barber 1984) responsible for metal transport from the bulk soil to plant roots: (1) convection or mass flow, and (2) diffusion. Due to convection, soluble metal ions move from soil solids to root surface. From the rhizosphere, water is absorbed by roots to replace water transpired by leaves.

Water uptake from rhizosphere creates a hydraulic gradient directed from the bulk soil to the root surface. Some ions are absorbed by roots faster than the rate of supply via mass flow. Thus, a depleted zone is created in soil immediately adjacent to the root. This generates a concentration gradient directed from the bulk soil solution and soil particles holding the adsorbed elements to the solution in contact with the root surface. This concentration gradient drives the diffusion of ions toward the depleted layer surrounding the roots. Plants have evolved specialized mechanisms to increase the concentration of metal ions in soil solution. For example, at low ion supply, plants may alter the chemical environment of the rhizosphere to stimulate the desorption of ions from soil solids into solution. Such a mechanism comprises rhizosphere acidification due to H^+ extrusion from roots (Crowley et al. 1991). Protons compete with and replace metal ions from binding sites, stimulating their desorption from soil solids into solution. In addition, some plants can regulate metal solubility in the rhizosphere by exuding a variety of organic compounds from roots. Root exudates form a complex with metal ions keeping them in solution available for uptake into roots (Römheld and Marschner 1986).

8.7.1 Role of PGPR in Biotic Interferences of Heavy Metals

Both microorganisms and plants have developed diverse strategies helping them to overcome or to adapt to unfavorable environmental conditions. A number of PGPR and other microorganisms have been reported to be resistant to relatively high concentrations of heavy metals (Ruggiero et al. 2005) and remain active in moderately acidic soils (Yasmin et al. 2004) which are widespread, comprising over 30% of only arable territories (Von Uexkull and Mutert 1995). Such phytostimulating rhizobacteria could contribute to plant growth promotion in metal-contaminated sites both indirectly, by increasing the overall fertility of the contaminated soil, as well as providing nutrients, phytohormones, and exhibiting other PGPR-related traits (Rosenblueth and Martinez-Romero 2006) but also catabolizing certain organics and/or partly oxidized intermediate biodegradation products.

As for PGPR of the genus *Azospirillum*, studies have shown that these phytostimulating microaerophilic rhizobacteria are relatively tolerant to submillimolar concentrations of heavy metals at which, in many cases, growth of the bacterial culture is not significantly suppressed (Kamnev et al. 2006). Moderate concentrations of heavy metals in soil can often be found as a result of contamination and/or bioleaching of minerals, and assessing their impact on the bacterial metabolism is of importance for a deeper insight into their biology and presents an obvious biotechnological and agricultural interest, considering the plant-growth-promoting abilities of azospirilla (Somers et al. 2004). In particular, *A. brasilense* represents a relatively rare chance to compare two strains of the same species which essentially differ in their mode of plant-root colonization. For example, *A. brasilense* strain Sp245 is known to be a facultative endophyte, whereas nonendophytic strain Sp7 colonizes the root surface only (Kirchhof et al. 1997). The levels of the auxin phytohormone [indole-3-acetic acid (IAA)] production for each of the two strains were analyzed in a standard medium (control) and in the presence of 0.2 mM $CuSO_4$ or $CdCl_2$ under microaerobic conditions that are favorable for these microaerophilic bacteria. In parallel, the bacterial growth rate [in terms of colony-forming units (CFU)] was also assessed. Using strain Sp7, both copper(II) and cadmium ions were found to result in a significant decrease in the level of IAA production, whereas the bacterial growth rate was virtually not affected. In Sp245, the overall level of IAA production in the control was approximately three times higher than for Sp7. Nevertheless, in this respect strain Sp245 appeared to be more sensitive to heavy metals: a noticeable decrease in IAA production was observed under the effect of both the metals, especially with Cu(II); note also that for strain Sp245 copper(II) was somewhat more toxic decreasing also the bacterial counts. Thus, the decrease in IAA production by the bacteria in presence of copper(II) and cadmium(II) may directly affect the efficiency of associative plant-bacterial symbioses in heavy-metal-polluted soils. To compare heavy-metal resistance of some plant-associated soil bacteria, the values of minimum inhibitory concentrations (MIC) and minimal lethal concentrations (MLC) for several PGPR grown on the agar medium (aerobic conditions) have been reported by Tugarova et al. (2006). MIC of cadmium(II) found for both the two *A. brasilense* strains (5 mM) was much higher than those reported for the other associative rhizobacteria (0.35–0.6 mM), i.e., by an order of magnitude, while for the Cd-sensitive rhizobacterium *Flavobacterium* sp. L30 the MLC was 0.01 mM (Belimov et al. 2005). By their cadmium resistance, the following soil bacteria isolated from the rhizoplane of Indian mustard (*Brassica juncea* L. Czern.) plants (Belimov et al. 2005) were closer to *A. brasilense*: *Variovorax paradoxus* (MLC = 0.6–3.5 mM for different strains), a *Rhodococcus* sp., a *Ralstonia* sp., a *Flavobacterium* sp., a *Pseudomonas* sp., and some other strains (MLC = 2.0–4.0 mM).

Soil microorganisms can also directly influence metal solubility by altering their chemical properties. For example, a strain of *Pseudomonas maltophilia* was shown to reduce the mobile and toxic Cr(VI) to nontoxic and immobile Cr(III), and also to minimize environmental mobility of other toxic ions such as Hg(II), Pb(II), and Cd(II) (Blake et al. 1993; Park et al. 1999). In addition, it has been estimated that microbial reduction of Hg(II) generates a significant fraction of global atmospheric Hg 0 emissions (Keating et al. 1997).

Simon et al. (2008) reported that pretreatment of Indian mustard with *Pseudomonas fluorescens* before cadmium and nickel application slightly enhanced the Cd or Ni accumulation in roots. This enhancement was not observed when pseudomonads were applied simultaneously with metals. More cadmium was detected in the roots of Indian mustard when plants were pretreated with Cd-tolerant than Cd-sensitive *Pseudomonas cepacia*. Cell number of pseudomonads in the surface of Cd-treated Indian mustard roots was significantly lower when plants were pretreated with Cd-sensitive *P. cepacia*. Only for *Bacillus* and *Pseudomonas* species present in Se-contaminated soils can be supposed to stimulate the Se phytoextraction and/or phytovolatization in the rhizosphere of higher plants (Simon et al. 2007). Pal et al. (2010) reported influence of Pb and Hg on *Azotobacter* spp.

and *Pseudomonas* spp. multiplication using Indian Mustard as a test crop on an artificially spiked sandy loam soil. The results showed that the minimum bacterial counts were observed in Pb_{200} and Hg_{40} treatments at 54 days after sowing of crop.

8.7.2 Effect of Root Exudates on Metal Uptake

Root growth affects the properties of the rhizospheric soil and stimulates the growth of the microbial consortium. In turn, rhizospheric microorganisms may interact symbiotically with roots to enhance the potential for metal uptake. In addition, some microorganisms may excrete organic compounds (Narula et al. 2009) which increase bioavailability and facilitate root absorption of essential metals, such as Fe (Crowley et al. 1991) and Mn (Barber and Lee 1974), as well as nonessential metals, such as Cd (Salt et al. 1995). Rhizosphere effect is often expressed quantitatively as the ratio of number of microorganisms in rhizosphere soil to the number of microorganisms in nonrhizosphere soils (R/S ratio). This type of increased microbial activity in the rhizosphere may be responsible for the increased metabolic degradation rate of certain xenobiotic compounds in the rhizosphere. The actual composition of the microbial community in the rhizosphere is dependent on root type, plant species, plant age, and soil type, as well as other factors. The bacteria clearly showed preference for one exudate over the other. This might well be related to the substances found in each exudate. It might be argued that the energy yield from C6 compounds is better, thus resulting in observed preferences (Kumar et al. 2007). Generally, rhizosphere is colonized by gram-negative bacteria. The interaction between plants and microbial communities in the rhizosphere is complex and has evolved to the mutual benefit of both organisms. Plants sustain large microbial population in the rhizosphere by secreting substances such as carbohydrates and amino acids through root cells and by sloughing root epidermal cells. The magnitude of rhizodeposition can be quite large. In addition, root cells secrete mucigel, gelatinous substances that are lubricant for root penetration through the soil during growth. This mucigel along with other cell secretions consists of root exudates. Beneficial effects of the root exudates are stimulated by adding organic acids to the soil. This enhances the microbial population and evolution of greater amounts of carbon dioxide. Microbial counts are generally more in the soil from vegetation and plants; this explains the help given by root exudation. Microbial activity, biomass, and degradation are greater than in the nonvegetated soil.

Root exudates have an important role in the acquisition of several essential metals. For example, some grass species can exude from roots a class of organic acids called siderophores (mugineic and avenic acids), which were shown to significantly enhance the bioavailability of soil-bound iron (Kanazawa et al. 1994), and possibly zinc (Cakmak et al. 1996a, b). In addition, root exudates have been shown to be involved in plant tolerance. In support of this, it has been demonstrated that some plant species tolerate Al in the rhizosphere, by a mechanism involving exudation of citric and malic acids (Pellet et al. 1995; Larsen et al. 1998). These organic acids chelate rhizospheric Al(III) which is highly phototoxic to form a significantly less toxic complex. Microbial transformations occur more in the rhizosphere (Rovira et al. 1979) than in the soils or in the soil external to the plant roots.

8.7.3 Chemical Interferences

Microorganisms have been shown to alter metal bioavailability. For example, microbes have been documented to catalyze redox reactions leading to changes in metal mobility in soil. A strain of *Xanthomonas maltophyla* was shown to catalyze the reduction and precipitation of highly mobile Cr(VI) to Cr(III), a significantly less mobile and environmentally less hazardous component. Ionic Hg(II) which is susceptible to methylation and subsequent volatilization generally of toxic methyl mercury was shown to be reduced by same bacteria to elemental Hg which possess less environmental risk. A heavy-metal-resistant *Bacillus circulans* have been isolated from soil which is capable of biosorption of 25, 22, 20, 13, 12 mg/l of Mn, Zn, Cu, Ni, and Co, respectively.

Besides the biotic reduction of iron(III) mediated by specific soil bacteria – dissimilatory reducers, a range of secondary metabolites produced both by soil microorganisms and by plant roots can abiotically reduce Fe(III) (Kovacs et al. 2006). Another rapidly developing field within the "underground hemisphere" of plant–microbe interactions, that can be affected by chemical interferences in soil, includes biosynthesis and exudation of intercellular bacterial signaling (quorum-sensing molecules), as well as a variety of already known, undisclosed biomolecules by which plant root–microbe signaling and interactions are implemented (Sugawara et al. 2006) or by which plants mimic, quench, or otherwise interfere with bacterial signals (Teplitski et al. 2004; Adonizio et al. 2006). In particular, noteworthy is the active role of PGPR cell-surface polysaccharide-containing macromolecular complexes (Somers et al. 2004) and cell-surface lectins, in bacterial interactions with host-plant roots involving also plant lectins. As an example, wheat lectin (wheat germ agglutinin, WGA), which is exposed on the root surface and can also dissolve there from in the aqueous medium, has been documented to induce a number of metabolic responses in azospirilla at low concentrations and therefore is proposed to act as a molecular signal for these wheat-associated bacteria (Narula et al. 2011a). Integrated approach for phyto-bioremediation of heavy-metal-contaminated soils using *Brassica* as model plant is illustrated in Fig. 8.1.

8.7.4 Bioleaching and Chemical Transformations of Heavy Metals by Soil Rhizosphere Microorganisms

Soil microflora is an important and active participant of biogeochemical processes, leading to diverse chemical transformations of soil minerals. In this respect, the rhizospheric microbial communities supported by host plants (along with a

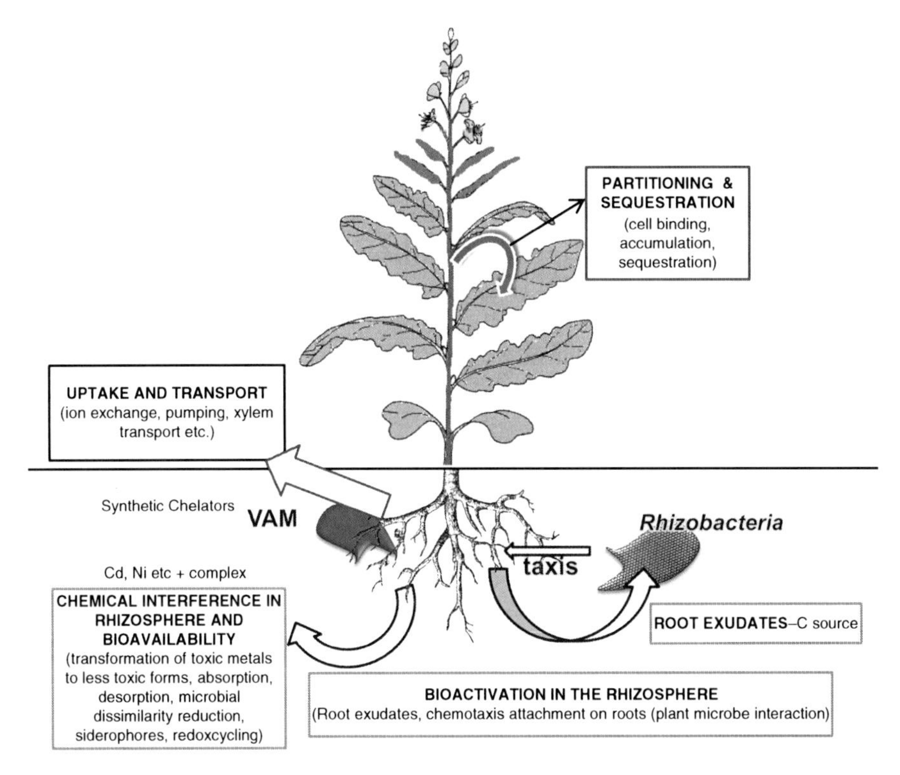

Fig. 8.1 Integrated approach for phyto-bioremediation of heavy metal contaminated soils using *Brassica* as model plant

contribution from plant-root exudates) can demonstrate even higher activities (11%) (Chander et al. 1997; Mukerji et al. 2006). The rhizosphere, as compared to the bulk soil, is highly populated by various microorganisms mainly comprising bacteria (predominating gram-negative bacteria) and mycorrhizal fungi (Karthikeyan and Kulakow 2003; Lucy et al. 2004) showing higher metabolic activity even in polluted soils (Biró et al. 2006). Some of diverse routes leading to microbially mediated bioleaching and biotransformation of heavy metals and radionuclide entrapped within soil minerals (Zachara et al. 2001; Kamnev et al. 2006) are (1) microbial dissimilatory reduction and concomitant solubilization of metal ions (e.g., Fe, Mn, and many other less abundant redox-active metals) via different mechanisms as reviewed by Lovley and Coates (2000), siderophores, carboxylic acids, acidic exopolysaccharides, and other organics production and excretion capable of binding and/or reducing metal cations (Faraldo-Gomez and Sansom 2003). (2) Contact interactions of microbial cells with metal complexes and solid compounds via microbial cell-wall biomacromolecules, sometimes with further involvement of metal ions in cellular metabolism or their inactivation via special intracellular mechanisms, etc. (Langley and Beveridge 1999). The bioavailability of iron, one of the few microelements vitally important for virtually all organisms (Sigel and Sigel 1998), which is commonly abundant in most soils, is yet well known to be extremely low within a wide range of environmental conditions. This is due to the very low solubility of ferric oxide or hydrated oxide species that are most stable under aerobic and microaerobic conditions. Thus, iron(III) largely represented by oxides or oxy hydroxides of variable crystallinity is an essential component of the majority of soil minerals. (3) Shuttle-assisted reduction involving specific compounds capable of redox cycling that serve as an electron shuttle between the bacterial cells and iron(III)-bearing mineral.

All the three mechanisms are clearly fundamentally different. Nevertheless, as for common features and distinctions, the first two mechanisms involve iron(III) reduction at the bacterial cell surface while the third one does not, which represents its main difference in the mode of electron transfer to iron(III).

In addition to biotic (enzymatic) dissimilatory reduction by bacteria reducers, humic substances, as well as their lower molecular weight precursors, are able to abiotically (chemically) reduce, e.g., highly soluble Cr(VI). It is interesting to note that some dissimilatory metal-reducing bacteria were found to be capable of sensing the depletion of soluble electron acceptors, with the subsequent synthesis of pili and flagella when growing on insoluble iron(III) or Mn(IV) minerals (but not on their soluble forms) (Lloyd 2003). Thereby, the bacteria can adapt to the conditions by "switching on" their active chemotaxis mechanism to move toward the insoluble sources of electron acceptors (solid Fe(III)- or Mn(IV)-containing mineral particles), with subsequent attachment, thus passing from shuttle-assisted reduction [mechanism (3), in view of shuttle deficiency, to contact reduction by mechanism (1)].

8.8 Engineering Plant–Microbe Symbiosis for Rhizoremediation of Heavy Metals

Bioaugmentation is the practice of inoculating specialized microbes into polluted matrices to accelerate transformation of specific pollutants. It is a strategy that possesses a great potential for overcoming contaminant persistence at sites that lack competent indigenous microbes for the degradation or adoption of particular contaminant (Shann 1995). Genetic engineered plants and microbes to enhance biodegradation of contaminants or to improve growth characteristics seem to be cost-effective process (Wu et al. 2006a). Kingsley et al. (1993) reported the beneficial effects of bioaugmentation on soil by adding competent bacteria capable of degrading, in the rhizosphere. Despite availability of a number of heavy metal hyperaccumulating plant species, the strategy is not applicable for remediating sites with multiple contaminants. It is advisable solution to combine the advantages of plant–microbe symbiosis within the plant rhizosphere into an effective cleanup technology. Researchers have exploited this symbiotic relationship for rhizoremediation. Cindy et al. (2006) demonstrated that expression of a metal-binding peptide (EC20) in a rhizobacterium, *Pseudomonas putida* 06909, not only improved cadmium binding but also alleviated the cellular toxicity of cadmium. More importantly, inoculation of sunflower roots with the engineered rhizobacterium resulted in a marked decrease in cadmium phytotoxicity and a 40% increase in cadmium accumulation in the plant root. Owing to the significantly improved growth characteristics of both the rhizobacterium and plant, the use of EC20-expressing *P. putida* endowed with organic compound-degrading capabilities seems to be a promising strategy to remediate mixed organic-metal-contaminated sites.

Biosorption using microbially produced phytochelatins ECs has been shown to be a promising technique for ameliorating heavy-metal contamination. Bacteria such as *Escherichia coli* and *Moraxella* sp. expressing EC20 (with 20 cysteines) on the cell surface or intracellularly have been shown to accumulate up to 25-fold more cadmium (Barea et al. 2005) or mercury (Bashan et al. 2004) than the wild-type strain. However, one major obstacle for utilizing these engineered microbes is sustaining the recombinant bacterial population in soil, with various environmental conditions and competition from native bacterial populations.

8.9 Conclusions and Outlook

The bacterial mobilization of heavy metals and radionuclide are poorly understood processes (Ruggiero et al. 2005). Microorganisms may, in principle, be advantageous and biotechnologically beneficial when occurring in the rhizosphere of metal-tolerant plants (special hyperaccumulator plants), thereby facilitating phytoremediation processes (Vassilev et al. 2004; Yang et al. 2005). Interaction of metal ions with biological matter is essential as well as important for various biological processes for all organisms (Sigel and Sigel 1996), and in related fields (biogeochemistry, bioremediation and phytoremediation, biomining, biotechnology of metal extraction, sorption and recovery, etc.). It can hardly be overemphasized that lack of understanding of molecular mechanisms underlying the effects of soil microorganisms and plant-root exudates on the state of metal compounds in the rhizosphere is a serious impediment in use of phyto-bioremediation technology for cleaning heavy soils contaminated with metals. One can overcome by widely using a variety of modern powerful physicochemical techniques in environmental and life sciences. On the other hand, knowing the mechanisms and routes of metal transformations may open ways for a variety of practical applications (Lucy et al. 2004). It would be most imperative to genetically and/or biotechnologically tailor plants for being hyperaccumulators, microbes to possess greater potential to convert toxic form of heavy metals to less or nontoxic types, and to harness the synergy of both partners to achieve better efficiencies for phyto-bioremediation of polluted soils through favorable plant–microbe interactions. Plant and microbial geneticists / bioengineers, microbiologists, soil chemists, and physiologists should work in close cooperation to develop a coherent soil pollution cleansing technology.

References

Abbott L, Murphy D (2003) Soil biology fertility: a key to sustainable land use in agriculture. Kluwer, Dordrecht, pp 187–203

Adonizio AL, Downum K, Bennett BC, Mathee K (2006) Anti-quorum sensing activity of medicinal plants in southern Florida. J Ethnopharmacol 105(3):427–435

Bais HP, Fall R, Vivanco JM (2004) Biocontrol of *Bacillus subtilis* against infection of *Arabidopsis* roots by *Pseudomonas syringae* is facilitated by biofilm formation and surfactin production. Plant Physiol 134:307–319

Bais HP, Tiffony L, Weir LT, Perry LG, Gilroy S, Vivanco JM (2006) The role of root exudates in rhizosphere interactions with plant and other organisms. Annu Rev Plant Biol 57:233–266

Barber SA (1984) Soil nutrient bioavailability. Wiley, New York

Barber SA, Lee RB (1974) The effect of microorganisms on the absorption of manganese by plants. New Phytol 73:97–106

Barea JM, Pozo MJ, Azcon R, Azcon-Aguilar C (2005) Microbial co-operation in the rhizosphere. J Exp Bot 56:1761–1778

Bashan Y, Holguin G, de Bashan LE (2004) *Azospirillum* plant relationship: physiological, molecular, agricultural and environmental advances (1997–2003). Can J Microbiol 50:521–577

Belimov AA, Hontzeas N, Safronova VI, Demchinskaya SV, Piluzza G, Bullitta S, Glick BR (2005) Cadmium-tolerant plant growth promoting rhizobacteria associated with the roots of Indian mustard (*Brassica juncea* L. Czern.). Soil Biol Biochem 37:241–250

Biró B, Köves-Péchy K, Tsimilli-Michael M, Strasser RJ (2006) Role of beneficial microsymbionts on the plant performance and plant fitness. In: Mukerji KG, Manoharachary C, Singh J (eds) Microbial activity in the rhizosphere, vol 7, Soil biology. Springer, Berlin, pp 265–296

Blake RC, Choate DM, Bardhan S, Revis N, Barton LL, Zocco TG (1993) Chemical transformation of toxic metals by a *Pseudomonas* strain from a toxic waste site. Environ Toxicol Chem 12:1365–1376

Blaylock MJ, Salt DE, Dushenkov S, Zakharova O, Gussman C (1997) Enhanced accumulation of Pb in Indian mustard by soil-applied chelating agents. Environ Sci Technol 31:860–865

Bottomley PJ, Dughri MH (1989) Population size and distribution of *Rhizobium leguminosarum* biovar *trifolii* in relation to total soil bacteria and soil depth. Appl Environ Microbiol 55:959–964

Bottomley PJ, Maggard SP (1990) Determination of viability within serotypes of a soil population of *Rhizobium leguminosarum* biovar *trifolii*. Appl Environ Microbiol 56:533–540

Brooks RR (1998) Geobotany and hyperaccumulators. In: Brook RR (ed) Plants that hyperaccumulate heavy metals. CAB International, Wallingford, CT, pp 55–94

Cakmak I, Ozturk L, Karanlik S, Marschner H, Ekiz H (1996a) Zinc-efficient wild grasses enhance release of phytosiderophores under Zn deficiency. J Plant Nutr 19:551–563

Cakmak I, Sari N, Marschner H, Ekiz H, Kalayci M (1996b) Phytosiderophore release in bread and durum what genotypes differing in zinc efficiency. Plant Soil 180:183–189

Çakmakçi R, Dönmez F, Aydın A, Şahin F (2006) Growth promotion of plants by plant growth-promoting rhizobacteria under greenhouse and two different field soil conditions. Soil Biol Biochem 38:482–1487

Cardoso PF, Priscila LG, Rui AG, Leonardo OM, Ricardo AA (2005) Response of *Crotalaria junceae* to nickel exposure. Braz J Plant Physiol 17:267–272

Chander K, Goyal S, Mundra MC, Kapoor KK (1997) Organic matter, microbial biomass and enzyme activity of soils under different crop rotations in the tropics. Biol Fertil Soils 24:306–310

Cindy HWu, Wood TK, Mulchandani A, Chen W (2006) Engineering plant-microbe symbiosis for rhizoremediation of heavy metals. Appl Environ Microbiol 72(2):1129–1134

Corticeiro CS, Lima AIG, Figueira EMAP (2006) The importance of glutathione in oxidative status of *Rhizobium leguminosarum* biovar *viciae* under cadmium stress. Environ Microbiol Technol 40:132–137

Crowley DE, Wang YC, Reid CPP, Szansiszlo PJ (1991) Mechanism of iron acquisition from siderophores by microorganisms and plants. Plant Soil 130:179–198

Dobbelaere S, Vanderleyden J, Okon Y (2003) Plant growth promoting effects of diazotrophs in the rhizosphere. Crit Rev Plant Sci 22:107–149

Ehrlich HL (1990) Geomicrobiology, 2nd edn. Dekker, New York, p 646

Faraldo-Gomez JD, Sansom MSP (2003) Acquisition of siderophores in gram-negative bacteria. Nat Rev Mol Cell Biol 4:105–116

Glick BR (1995) The enhancement of plant growth by free-living bacteria. Can J Microbiol 41:109–117

Glick BR (2001) Phytoremediation: synergistic use of plants and bacteria to cleanup the environment. Biotechnol Adv 21(3):83–93

Glick BR, Penrose DM, Li JA (2002) Model for the lowering of plant ethylene concentrations by plant growth promoting bacteria. J Theor Biol 190:63–68

Gunatilaka AAL (2006) Natural products for plant associated microorganisms: distribution, structural diversity, bioactivity and implication of their occurrence. J Nat Prod 69:509–526

Gupta DK, Rai UN, Sinha S, Tripathi RD, Nautiyal BD, Rai P, Inouhe M (2004) Role of *Rhizobium* (CA-1) inoculation in increasing growth and metal accumulation in *Cicer arietinum* L. growing under fly-ash stress condition. Bull Environ Contam Toxicol 73:424–431

Huang JW, Cunningham SD (1996) Lead phytoextraction: species variation in lead uptake and translocation. New Phytol 134:75–84

Kamnev AA (2008) Metals in soil verses plant-microbe interactions: biotic and chemical interferences. In: Barka EA, Clément C (eds) Plant-microbe interactions. Research Signpost, Trivandrum, Kerala, pp 291–318

Kamnev AA, Kovács K, Shchelochkov AG, Kulikov LA, Perfiliev YuD, Kuzmann E, Vértes A (2006) Bioleaching and chemical transformations of heavy metals and radionuclides mediated by soil microorganisms. In: Alpoim MC, Morais PV, Santos MA, Cristóvão AJ, Centeno JA, Collery Ph (eds) Metal ions in biology and medicine, vol. 9. John Libbey Eurotext, Paris, p 220–225

Kanazawa K, Higuchi K, Nishizawa NK, Fushiya S, Chino M, Mori S (1994) Nicotianamine aminotransferase activities are correlated to the phytosiderophore secretion under Fe-deficient conditions in Gramineae. J Exp Bot 45:1903–1906

Karthikeyan R, Kulakow PA (2003) Soil plant microbe interactions in phytoremediation. Adv Biochem Eng Biotechnol 78:51–74

Keating MH, Mahaffey KR, Schoney R, Rice GE, Bullock OR, Ambrose RB, Swartout J, Nichols JW (1997) Mercury study report to congress. EPA-452/R-97-003, Vol I, Sec 3, Washington, DC, pp 6–7

Khan AG (2005a) Role of soil microbes in the rhizospheres of plant grown in trace metals contaminated soils in phytoremediation. J Trace Elem Med Biol 18:355–364

Khan AG (2005b) Mycorrhizas and phytoremediation. In: Willey N (ed) Method in biotechnology-phytoremediation: methods and reviews. Humana, Totowa, NJ

Khan MS, Zaidi A, Zaidi PA, Mohammad O (2009) Role of plant growth promoting rhizobacteria in the remediation of metal contaminated soils. Environ Chem Lett 7:1–19

Kingsley MT, Metting FB Jr, Fredrickson JK, Seidler RJ (1993) *In situ* stimulation vs. bioaugmentation: can plant inoculation enhance biodegradation of organic compounds. Paper 93-WA-89.04, Proceedings of the 86th annual meeting and exhibition, air and waste management association, Pittsburgh, PA

Kirchhof G, Schloter M, Asmus B, Hartmann A (1997) Molecular microbial ecology approaches applied to diazotrophs associated with non-legumes. Soil Biol Biochem 29:853–860

Kovacs K, Kamnev AA, Mink J, Nemeth CS, Kuzmann E, Megyes T, Grosz T, Medzihradszky-Schweiger H, Vertes A (2006) Mössbauer, vibrational spectroscopic and solution X-ray

diffraction studies of the structure of iron (III) complexes formed with indole-3-alkanoic acids in acidic aqueous solutions. Struct Chem 17:105–120
Kumar R, Bhatia R, Kukreja K, Behl RK, Dudeja SS, Narula N (2007) Establishment of *Azotobacter* on plant roots: chemotactic response, development and analysis of root exudates of cotton (*G. hirsutum* L) and wheat (*T. aestivum* L). J Basic Microbiol 47:436–439
Langley S, Beveridge TJ (1999) Effect of O-side-chain-lipopolysaccharide chemistry on metal binding. Appl Environ Microbiol 65:489–498
Larsen PB, Degenhardt J, Tai CY, Stenzler LM, Howell SH, Kochian LV (1998) Aluminum-resistant *Arabidopsis* mutants that exhibit altered patterns of aluminum accumulation and organic acid release from roots. Plant Physiol 117:19–27
Lasat MM (2000) Phytoextraction of metals from contaminated soil: a review of plant-soil-metal interaction and assessment of pertinent agronomic issues. J Hazard Sub Res 2:5–25
Lima AIG, Pereira SAI, Figueira EMAP, Caldeira GCN, Caldeira HDQM (2006) Cadmium detoxification in roots of *Pisum sativum* seedlings: relationship between toxicity levels, thiol pool alterations and growth. Environ Exp Bot 55:149–162
Lippmann B, Leinhos V, Bergmann H (1995) Influence of auxin producing rhizobacteria on root morphology and nutrient accumulation of crops. 1. Changes in root morphology and nutrient accumulation in maize (*Zea mays* L.) caused by inoculation with indole-3-acetic acid (IAA) producing *Pseudomonas* and *Acinetobacter* strains or IAA applied exogenously. Angew Bot 69:31–36
Lloyd JR (2003) Microbial reduction of metals and radionuclides. FEMS Microb Rev 27:411–425
Lovley DR, Coates JD (2000) Novel forms of an aerobic respiration of environmental relevance. Curr Opin Microbiol 3:252–256
Lucy M, Reed E, Glick B (2004) Applications of free living plant growth promoting rhizobacteria. Antonie Van Leeuwenhoek 86:1–25
Lynch JM (1983) Soil biotechnology. Blackwell, Oxford
Lynch JM (1990) Beneficial interactions between microorganisms and roots. Biotechnol Adv 8:335–346
Mamaril JC, Paner ET, Alpante BM (1997) Biosorption and desorption studies of chromium (iii) by free and immobilized *Rhizobium* (BJVr 12) cell biomass. Biodegradation 8:275–285
Mukerji KG, Manoharachary C, Singh J (eds) (2006) Soil biology, vol 7, Microbial activity in the rhizosphere. Springer, Berlin
Narula N, Kumar V, Saharan BS, Bhatia R, Lakshminarayana K (2005a) Impact of the use of biofertilizers on grain yield of wheat under varying soil fertility conditions and wheat-cotton rotation. Arch Agron Soil Sci (Germany) 51(1):79–89
Narula N, Saharan BS, Kumar V, Bhatia R, Bishnoi LK, Lather BPS, Lakshminarayana K (2005b) Impact of the use of biofertilizers on cotton (*Gossypium hirusetum*) crop under irrigated agro-ecosystem. Arch Agron Soil Sci (Germany) 51(2):69–77
Narula N, Deubel A, Gans W, Behl RK, Merbach W (2006) Colonization and induction of Para nodules of wheat roots by phytohormone producing soil bacteria. Plant Soil Environ (Czech Republic) 52(3):119–129
Narula N, Kothe K, Behl RK (2009) Role of root exudates in plant-microbe interactions. J Appl Bot Food Qual 82:122–130
Narula N, Bhatia R, Anand RC, Gera R, Behl RK (2011a) Interaction of *A. chroococcum* with wheat germ agglutinin (WGA). Icfai Univ J Life Sci (India) IV(1):15–20
Narula N, Kothe E, Behl RK (2011b) Heavy metal resistance among *Azotobacter* spp.: survival in garden and heavy metal contaminated soil using Indian Mustard. Icfai Univ J Life Sci (India) IV (2):55–62
Nies DH (1999) Microbial heavy metal resistance. Appl Microbiol Biotechnol 51:730–750
Okon Y (1985) *Azospirillum* as a potential inoculants for agriculture. Trends Biotechnol 3:223–228
Pal MKI, Panwar BS, Goyal SI, Attila A, Laszlo M (2010) Effect of lead (Pb) and mercury (Hg) on plant growth promoting rhizobacterias (PGPB) as *Azotobacter chroococcum* and *Pseudomonas*

spp. using Indian Mustard (*Brassica juncea* L.) test crop. Proceedings of the international seminar on crop science for sustain and food security, Szeged, Hungary, 1–3 June

Panwar BS, Ahmed KS, Mittal SB (2002) Phytoremediation of nickel contaminated soils by *Brassica* species. J Environ Dev Sustain 1:1–6

Park CH, Keyhan M, Matin A (1999) Purification and characterization of chromate reductase in *Pseudomonas putida*. Abs Gen Meet Am Soc Microbiol 99:536–548

Pellet MD, Grunes DL, Kochian LV (1995) Organic acid exudation as an aluminum tolerance mechanism in maize (*Zea mays* L.). Planta 196:788–795

Peters NK, Frost JW, Long SR (1986) A plant flavones, luteolin, induces expression of *Rhizobium meliloti* nodulation genes. Science 233:977–980

Raskin I (1996) Plant genetic engineering may help with environmental cleanup. Proc Natl Acad Sci USA 93:3164–3166

Renella G, Mench M, Landi L, Nannipieri P (2005) Microbial activity and hydrolase synthesis in long-term Cd-contaminated soils. Soil Biol Biochem 37:133–137

Roane TM, Pepper IL (2000) Microbial responses to environmentally toxic cadmium. Microb Ecol 38:358–364

Römheld V, Marschner H (1986) Mobilization of iron in the rhizosphere of different plant species. Adv Plant Nutr 2:155–204

Rosenblueth M, Martinez-Romero E (2006) Bacterial endophytes and their interactions with hosts. Mol Plant Microbe Interact 19:827–837

Rovira AD, Foster R, Martyin JK (1979) Origin, nature and nomenclature of the organic materials in the rhizosphere. In: Harley JL, Russel RS (eds) The soil interface. Academic, London, pp 1–4

Ruggiero CE, Boukhalfa H, Forsythe JH, Lack JG, Hersman LE, Neu MP (2005) Actinide and metal toxicity to prospective bioremediation bacteria. Environ Microbiol 7(1):88–97

Ruml T, Kotrba P (2003) Recent advances in marine biotechnology. In: Fingerman M, Nagabhushanam R (eds) Bioremediation, vol 8. Science, Enfield, NH, pp 81–155

Saitia S, Narula N (1989) Heavy metal resistance and hydrocarbon utilization in *Azotobacter chroococcum*. Indian J Microbiol 29(3):213–215

Saitia S, Narula N, Lakshminarayana K (1989) Nature and role of plasmids in *Azotobacter chroococcum*. Biotechnol Lett 11:713–716

Salt DE, Blaylock M, Kumar PBAN, Dushenkov V, Ensley BD, Chet I, Raskin I (1995) Phytoremediation: a novel strategy for the removal of toxic metals from the environment using plants. Biotechnology 13:468–475

Shakolnik MY (1984) Trace elements in plants. Elsevier, New York, pp 140–171

Shann JR (1995) The role of plants and plant/microbial systems in the reduction of exposure. Environ Health Perspect 103(5):13–15

Shoebitz M, Ribaudo CM, Pardo MA, Cantore ML, Ciampi L, Curá JA (2009) Plant growth promoting properties of a strain of *Enterobacter ludwigii* isolated from *Lolium perenne* rhizosphere. Soil Biol Biochem 41(9):1768–1774

Sigel H, Sigel A (eds) (1996) Metal ions in biological systems, vol 32, Interaction of metal ions with nucleotides nucleic acids, and their constituents. Dekker, New York

Sigel A, Sigel H (eds) (1998) Metal ions in biological systems, vol 35, Iron transport and storage in microorganisms, plants, and animals. Dekker, New York

Simon L, Széles É, Balázsy S, Biró B (2007) Selenium phytoextraction, speciation and microbial groups in Se contaminated soils. Proceedings of the ICOBTE, 9–2007

Simon L, Biró B, Balázsy S (2008) Impact of pseudomonads and ethylene on the cadmium and nickel rhizofiltration of sunflower, squash and Indian mustard. Comm Soil Sci Plant Anal 36(15–16):2440–2455

Somers E, Vanderleyden J, Srinivasan M (2004) Rhizosphere bacterial signalling: a love parade beneath our feet. Crit Rev Microbiol 30:205–240

Stresty EV, Madhava Rao KV (1999) Ultrastructural alterations in response to zinc and nickel stress in the root cells of pigeonpea. Environ Exp Bot 41:3–13

Sugawara M, Okazaki S, Nukui N, Ezura H, Mitsui H, Minamisawa K (2006) Rhizobitoxine modulates plant–microbe interactions by ethylene inhibition. Biotechnol Adv 24:382–388

Teplitski M, Chen H, Rajamani S, Gao M, Merighi M, Sayre RT, Robinson JB, Rolfe BG, Bauer WD (2004) *Chlamydomonas reinhardtii* secretes compounds that mimic bacterial signals and interfere with quorum sensing regulation in bacteria. Plant Physiol 134:137–146

Tugarova AV, Kamnev AA, Antonyuk LP, Gardiner PHE (2006) Metal ions in biology and medicine, vol 9. John Libbey Eurotext, Paris, 242

Vassilev A, Schwitzguébel J-P, Thewys T, van der Lelie D, Vangronsveld J (2004) The use of plants for remediation of metal contaminated soils. Sci World J 4:9–34

Von Uexkull HR, Mutert E (1995) Global extent, development and economic impact of acid soils. Plant Soil 171:1–15

Walker TS, Bais HP, Grotewold E, Vivanco JM (2003) Root exudation and rhizosphere biology. Plant Physiol 132:44–51

Wang PC, Mori T, Komori K, Sasatsu M, Toda K, Ohtake H (1989) Isolation and characterization of an *Enterobacter cloacae* strain that reduces hexavalent chromium under anaerobic conditions. Appl Environ Microbiol 55:1665–1669

Wani PA, Khan MS, Zaidi A (2007a) Cadmium, chromium and copper in green gram plants. Agron Sustain Dev 27:145–153

Wani PA, Khan MS, Zaidi A (2007b) Impact of heavy metal toxicity on plant growth, symbiosis, seed yield and nitrogen and metal uptake in chickpea. Aust J Exp Agr 47:712–720

Wani PA, Khan MS, Zaidi A (2007c) Effect of metal tolerant plant growth promoting *Rhizobium* on the performance of pea grown in metal amended soil. Arch Environ Contam Toxicol. doi:10.1007/00244-9097-y

Wani PA, Khan MS, Zaidi A (2008a) Effect of heavy metal toxicity on growth, symbiosis, seed yield and metal uptake in pea grown in metal amended soil. Bull Environ Contam Toxicol. doi:10.1007/s00128-008-9383-z

Wani PA, Khan MS, Zaidi A (2008b) Chromium reducing and plant growth promoting *Mesorhizobium* improves chickpea growth in chromium amended soil. Biotechnol Lett 30:159–163

White PJ (2003) Ion transport. In: Thomas B, Murphy DJ, Murray DJ (eds) Encyclopedia of plant sciences. Academic, London, pp 625–634

Wu CH, Wood TK, Mulchandani A, Chen W (2006a) Engineering plant-microbe symbiosis for rhizoremediation of heavy metals. Appl Environ Microbiol 72:1129–1134

Wu SC, Cheung KC, Luo YM, Wong MH (2006b) Effects of an inoculum of plant growth promoting rhizobacteria on metal uptake by *Brassica juncea*. Environ Pollut 140:124–135

Yang X, Feng Y, He Z, Stoffella PJ (2005) Molecular mechanisms of heavy metal hyperaccumulation and phytoremediation. J Trace Elem Med Biol 18(4):339–353

Yasmin S, Bakar MAR, Malik KA, Hafeez FY (2004) Isolation, characterization and beneficial effects of rice associated PGPB from Zanzibar soils. J Basic Microbiol 44:241–252

Zachara JM, Fredrickson JK, Smith SC, Gassman PL (2001) Solubilization of Fe(III) oxide-bound trace metals by a dissimilatory Fe(III) reducing bacterium. Geochim Cosmochim Acta 65:75–93

Zaidi A, Khan MS (2006) Co-inoculation effects of phosphate solubilizing microorganisms and *Glomus fasciculatum* on green gram-*Bradyrhizobium* symbiosis. Turk J Agr Forest 30:223–230

Zaidi S, Usmani S, Singh BR, Musarrat J (2008) Significance of *Bacillus subtilis* strains SJ-101 as a bioinoculant for concurrent plant growth promotion and nickel accumulation in *Brassica juncea*. Chemosphere 64:991–997

Chapter 9
Heavy Metal-Resistant Streptomycetes in Soil

Eileen Schütze and Erika Kothe

9.1 Introduction

The microbial population of soils has been shown to contain different taxa of gram-positive and gram-negative bacteria with total cell numbers of 10^6–10^9 cells per cm^3 soil. A large group within this population generally is made up by the streptomycetes, which is especially true for metal-contaminated soils (Haferburg and Kothe 2007). They are gram-positive, aerobic, and heterotrophic bacteria with a high genomic G + C content belonging to the taxon Actinobacteria. Similar to fungi, they develop branching hyphae and form a mycelium, albeit of smaller cell diameter. After spore germination, polarized growth occurs in a germ tube that develops into a vegetative substrate mycelium by tip extension and branching (Fig. 9.1). Production of multiple secondary metabolites including antibiotics – in addition to the volatile geosmins responsible for the odor of fresh soil – is connected to morphological differentiation and is initiated by nutrient depletion and quorum sensing entering stationary phase (Flärdh and Buttner 2009; Scherr and Nguyen 2009). Quorum sensing, the detection of the density of cells of the same species, has been shown to be dependent on the A-factor, which has been termed pheromone for *Streptomyces coelicolor*. Upon detection of this pheromone, formation of aerial mycelium and subsequently sporogenesis occurs (Onaka et al. 1998).

The aerial hyphae, after breaking surface tension and growing into the air, divide into a long chain of prespore compartments, which then develop into mature spores with thick, lysozyme-resistant spore walls. The dormant spores are resistant to adverse conditions like heating to 80°C and dryness, and they start germination

E. Schütze • E. Kothe (✉)
Institute of Microbiology, Microbial Phytopathology, Friedrich Schiller University Jena, Neugasse 25, Jena 07743, Germany
e-mail: erika.kothe@uni-jena.de

E. Kothe and A. Varma (eds.), *Bio-Geo Interactions in Metal-Contaminated Soils*, Soil Biology 31, DOI 10.1007/978-3-642-23327-2_9,

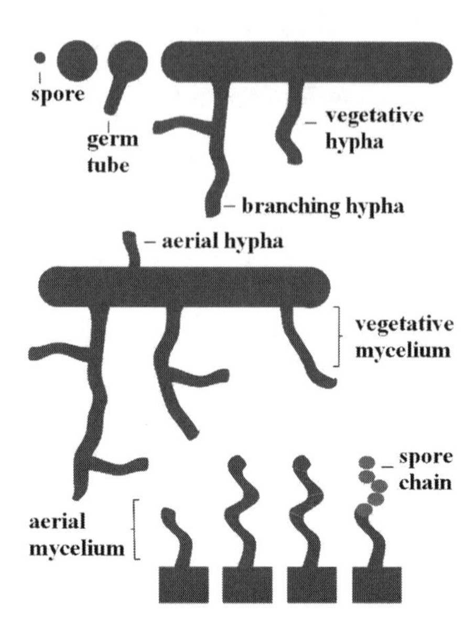

Fig. 9.1 Spore germination and development of aerial hyphae in streptomycetes

when favorable conditions and nutrient supply are re-established (Flärdh and Buttner 2009; Scherr and Nguyen 2009). In contrast to gram-positive firmicutes, these spores are not formed intracellularly, thus they are not homologous to endospores, and they are also less resistant to adverse conditions (Kothe et al. 2010).

9.2 Streptomycetes in Soil

During their complex, vegetative life cycle streptomycetes are able to excrete numerous extracellular molecules, e.g., hydrolytic enzymes like proteases, cellulases, lignocellulase, xylanases and chitinases, and a high number of secondary metabolites, like antibiotics, fungicides, siderophores, signaling molecules, modulators of immune response, effectors of plant growth like indol acetic acid, amphipathic proteins and geosmins (Chater et al. 2010). This versatility in secondary metabolite production makes them important tools for pharmaceutical, medical and biotechnological applications as well as for bioremediation approaches. The genome of *S. coelicolor* encodes 60 proteases, 13 chitinases/chitosanases of two different families, eight cellulases/endoglucanases, many high- and low-molecular-weight compounds and enzymes for the production of secondary metabolites, 7,000 of which have been described for different *Streptomyces* species and isolates (Chater et al. 2010), with more than 80% of all antibiotics used originate from actinomycetes. Their production usually is coupled to morphological differentiation and is regulated by pathway-specific regulatory genes, like *Streptomyces* Antibiotic Regulatory Proteins (SARPs), and by small signaling molecules, such as the γ-butyrolactone A-factor (Bibb and Hesketh 2009).

The genes encoding biosynthetic enzymes, regulatory genes, as well as genes involved into antibiotic export and self-resistance are often grouped in clusters on the chromosome or on a plasmid (Bibb and Hesketh 2009). The *S. coelicolor* chromosome is linear, with a G + C content of more than 70% and a size of 8–9 Mb. It consists of a core region, containing essential genes controlling replication, transcription, and translation, as well as central metabolism, while the chromosome arms mostly carry conditionally adaptive genes, coding for, e.g., cellulases, chitinases or secondary metabolism. The linear chromosomal DNA replicates bidirectionally by a special mode from a central origin. While the leading strand is synthesized to the chromosome ends, the lagging strand is completed by end-patching DNA synthesis primed from a terminally attached protein. Genome instability can be seen, which is thought to be dependent on loss of large terminal segments of chromosomal ends during unequal crossing over, which often results in colony sectors with different phenotypes, like chloramphenicol resistance or arginine auxotrophy. This phenomenon is often found in laboratory strains and the relevance under natural conditions is not clear yet (Hopwood 2006).

Because orthology is seen between the *S. coelicolor* chromosomal core region and parts of the genomes of *S. avermitilis*, as well as *Mycobacterium tuberculosis* while the arms carry more nonorthologous genes, the idea of a common actinomycete ancestor with a 4 Mb circular genome containing essential, housekeeping genes arose. Different sets of new genes then presumably were acquired via horizontal gene transfer, by integration of transposons, transposon relics, unequal crossing over, plasmid-mediated DNA transfer with subsequent integration into the bacterial chromosome and competence for uptake of extracellular DNA-fragments. This resulted in pattern offering the opportunity of adaptation to changing conditions in soil and ultimately in an advantage for specialization toward particular habitats (Hopwood 2006).

9.3 Heavy Metal-Contaminated Soil

Soils can be viewed as the basis and main organizer of the terrestrial ecosystem. They consist of mineral components, organic matter and microorganisms, interacting with each other. Due to these interactions, biochemical reactions like formation of metal oxides, mineral formation, formation of humic substances, enzyme stabilization, aggregate formation and turnover, cycling of C, N, P and S, as well as the fate and transformation of inorganic and organic pollutants, are controlled (Huang et al. 2005). The formation of soil structures is highly dynamic and affected by energy input, moisture, solid composition and composition of organic matter and clay interactions (de Jonge et al. 2009). However, these soil structures, wettability and moisture content are scale-dependent (Lamparter et al. 2009) and therefore difficult to investigate under natural conditions. It could be shown that in mesoscale, laboratory conditions, carbon sequestration was moisture-dependent, whereas carbon mineralization was reduced with decreasing pH and increasing C/N ratio (Lamparter et al. 2009). For aggregated soil samples, a higher respiration rate could be measured compared to homogenized samples, indicating that soil structure had significant influence. Thus, it could be concluded that chemical as well as physical properties are important parameters for carbon mineralization in soil (Lamparter et al. 2009).

Since heterogeneity in carbon content and microbial communities at field scale are higher in subsoil than in topsoil, it could be suggested that more attention should be paid to subsoil in order to get a better understanding of the global carbon cycle dynamics (Salomé et al. 2010). Nonliving organic material, defined as soil organic matter, is one of the most important factors of soil quality and fertility and the largest pool in terrestrial C cycle. It could be shown that detection of labeled genes from *Escherichia coli* was still possible after 224 days in soil experiments, indicating stabilization of DNA with the soil organic matter (Kindler et al. 2006). The number of added, viable cells declined very fast and half of their carbon pool was mineralized, 25% incorporated into microbial biomass and 25% was stabilized as compounds associated with soil organic matter (Kindler et al. 2006). The fate of fatty acids revealed similar results of rapid metabolization after cell death, and incorporation of 75% of the initially added fatty acids into microbial biomass was seen, as well as association of 14% with soil organic matter (Kindler et al. 2009). In contrast, amino acids were found to be very stable. More than 95% of the initial, labeled amino acids remained in soil and only about 6% were bound to living microbial biomass after 224 days. These results demonstrated that even peptides can be very stable in soil, which is due to stabilization to dead cells, aggregation to cell fragments or other general stabilization mechanism in soil, contributing to the formation of soil organic matter. Since amino acid pattern did not change during incubation, it could be suggested that peptides remained intact. Thus, microbial biomass is an important source for soil organic matter formation (Miltner et al. 2009) playing a central role in soil quality.

Heavy metal-contaminated soils are often found as a result of oxygen introduction through mining activities. The mobilization of metals is due to sulfuric acid generation caused by oxidation of pyrite and other sulfidic minerals (e.g., ZnS) during mining operations, a process accelerated microbiologically (Banks et al. 1997; Kothe et al. 2005). The complex cycle of reactions during pyrite oxidation is described in a simplified manner with the following reactions (Banks et al. 1997):

1. $2FeS_2 + 2H_2O + 7O_2 \rightarrow 4Fe^{2+} + 4SO_4^{2-} + 4H^+_{(aq)}$ [pyrite oxidation]

   pyrite + water + oxygen ferrous iron + sulphate + acid

2. $4Fe^{2+} + 4H^+_{(aq)} + O_2 \rightarrow 4Fe^{3+} + 2H_2O$

   ferrous iron + protons(acid) + oxygen ferric iron + water

   [oxidation of ferrous iron]

3. $$\underset{\text{pyrite + ferric iron + water ferrous iron + sulphate}}{FeS_2 + 14Fe^{3+} + 8H_2O \rightarrow 15Fe^{2+} + 2SO_4^{2-} + 16H^+_{(aq)}}$$

[further pyrite oxidation]

and/or

4. $$Fe^{3+} + 3H_2O \rightarrow Fe(OH)_3 + 3H^+_{(aq)} \quad \text{[hydrolysis]}$$

Σ. $$\underset{\text{pyrite + water + oxygen ferric iron hydroxide + sulphate + acid}}{4FeS_2 + 14H_2O + 15O_2 \rightarrow 4Fe(OH)_3 + 8SO_4^{2-} + 16H^+_{(aq)}}$$

The oxidation of pyrite is accelerated by orders of magnitude due to the action of iron or sulfide oxidizing bacteria, e.g., *Acidithiobacillus ferrooxidans*, *Leptospirillium ferrooxidans* and *Ferroplasma acidiphilum* or *F. acidarmanus* (Banks et al. 1997; Rawlings 2002). The resulting acid and sulfate production, as well as the release of rare earth elements, heavy metals and arsenic, are characterizing acid mine drainage (Fig. 9.2) and acid rock drainage (Banks et al. 1997; Kothe et al. 2005; Liang and Thomson 2009).

The treatment of contaminated areas is a matter of intense research due to the complexicity of acid mine drainage and environmental pollution. Many studies reported, at least temporary, neutralization effect with limestone application, or addition of sewage sludge, slurries of bauxol and brucite, calcite, gypsum and combinations of thereof. Incorporation of organic carbon by adding pig or sheep manure, or enhancing growth of sulfate-reducing bacteria with lignocellulose or cellulose materials also has been shown to have a positive effect (Liang and Thomson 2009). Other strategies to treat heavy metal-contaminated areas are phytoremediation (Table 9.1), which can be subdivided into phytoextraction, rhizofiltration, phytostabilization, phytovolatilization or biodegradation (Jabeen et al. 2009). Microbially enhanced phytoremediation, using heavy metal-resistant

Fig. 9.2 Acid mine drainage waters (**a**) and tailing pond for heavy metal-contaminated drainage waters (**b**)

Table 9.1 Advantages and disadvantages of bioremediation (modified from Shao et al. 2010)

Advantages	Disadvantages
Environmentally friendly, cost-effective, and esthetically pleasant	Relies on natural cycles and therefore takes time
Metals absorbed in plant shoots may be extracted from harvested plant biomass and then be sustainably recycled	Phytoremediation only works if contaminant is within the reach of the plant roots and dissolved in water
Phytoremediation can be used to clean up a large variety of contaminants simultaneously	Plants that absorb high loads of metals can be a potential risk for the food chain
Control of hydraulic conditions by transpiration offers a way to reduce entry of contaminants into ground and surface water systems	Depending on climatic conditions

soil bacteria and fungi, rhizosphere bacteria or mycorrhizal fungi, has been shown to have particularly positive effects on heavy metal-contaminated soils (Kothe et al. 2005; Nogueira et al. 2007; Kamaludeen and Ramasamy 2008; Kuffner et al. 2008; Shao et al. 2010).

9.4 Heavy Metal-Resistant Streptomycetes in Soil

It could be shown by cultivation and cultivation-independent, DNA-based methods, that streptomycetes are a dominant group of bacteria in heavy metal-contaminated soil (Haferburg et al. 2007b; Kothe et al. 2010). In contrast, in pristine soils, actinobacteria formed only about 20% of the population, indicating that streptomycetes developed special strategies to cope with different metal contaminations. Additionally, their specific features like filamentous growth, the formation of hyphae, the highly active secondary metabolism and the production of spores can be seen as adaption to living in soil, especially to extreme environmental conditions such as scant nutrients, intense salt load and low pH, as well as contamination with high metal loads (Kothe et al. 2010). Because of the complexity of soil, Schmidt et al. (2009a) used heavy metal-contaminated soil with, e.g., high nickel, zinc manganese, cobalt and copper loads in mobile fractions for preparing soil agar. In plating experiments, it could be shown that the heavy metal-sensitive control strain *S. coelicolor* A3(2) only was able to grow on plates with up to 12.5% heavy metal-containing soil, whereas strains isolated from a former uranium mining site could cope with higher concentrations of soil in the medium. Further experiments with one of these strains, *S. mirabilis* P16B-1, revealed its ability to grow in highly contaminated soil without the addition of any media (Fig. 9.3).

Microorganisms developed both intracellular and extracellular defense mechanisms to cope with heavy metals in the environment (Fig. 9.4). One strategy is the adsorption of metals to extracellular surfaces to reduce the available metal concentration in the surroundings. This process of biosorption is due to binding of metals to cell wall components, and this process is specifically predominant in gram-positive bacteria, while in gram-negative bacteria an outer membrane is protecting the cell wall surface

Fig. 9.3 Scanning electron micrograph of *S. mirabilis* P16B-1, grown in a microcosm consisting of contaminated soil, without addition of media components. The strain showed growth, as well as spore production, on the soil surface and throughout the soil matrix

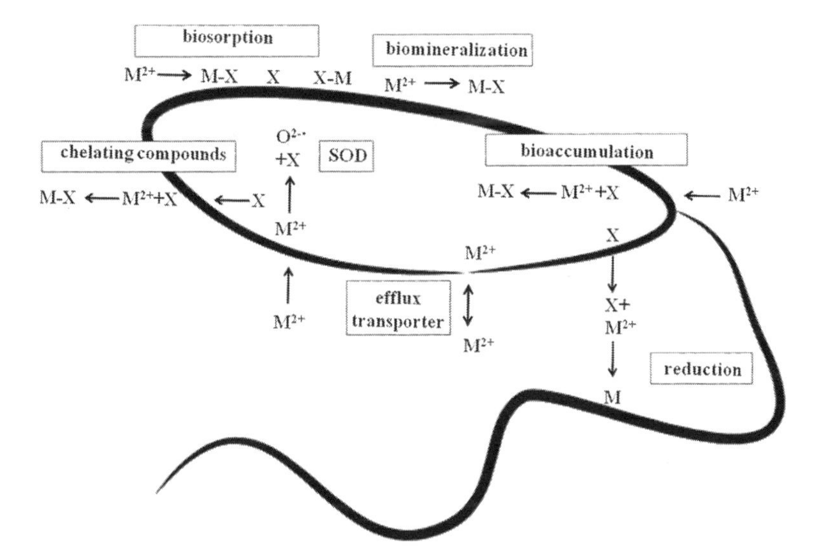

Fig. 9.4 Microbial resistance mechanisms: cells interacting with cations use different strategies to cope with heavy metal stress or heavy metal-mediated toxicity

(Haferburg and Kothe 2007; Schmidt et al. 2005). Additionally, microbes are able to excrete low-molecular weight compounds with high affinity to iron, the siderophores, which are involved in acquisition of the essential iron. However, these molecules show also (lower) affinity to other metals. Thus, siderophores can, aside from enhancing uptake of iron, complex harmful metals thereby protecting the cells (Dimkpa et al.

2009a). The production of, e.g., melanin-like pigments can be involved in metal resistance and protection from resulting oxidative stress in cells encountering heavy metals. Intracellular storage by binding to chaperone molecules, storage proteins or small peptides like metallothioneins and metallohistins can sequester metals and protect cells from their toxic effects even after uptake of heavy metals into the cytosol (Schmidt et al. 2010). After entering the cell by unspecific uptake, another important strategy is an enhanced efflux mediated by high affinity transporters leading to reduction of intracellular toxic metal concentrations. For some metals, mechanisms like reduction to form less toxic oxidation states, or biomineralization can be found to cope with metal stress, whereas the production of superoxide dismutase is a general defense mechanism to eliminate toxic radicals produced during Fenton reactions in the presence of heavy metal ions (Schmidt et al. 2007; Haferburg and Kothe 2007; Kothe et al. 2010).

Schmidt et al. (2005) and Haferburg et al. (2007b) could show that streptomycetes isolated from a former uranium mining site in Thuringia, Germany, cope with high concentrations of heavy metals, even 100 mM concentration. In comparison to control soils, a significantly higher amount of resistant strains could be isolated from contaminated soils, indicating that adaptation can occur within very short time periods of 40 years of pollution in this mining area. For two strains, an extracellular, diffusible compound could be shown, allowing other strains to grow in their vicinity (Schmidt et al. 2005). Shifts in the production of secondary metabolites and antibiotic compounds could be shown for strains isolated from long- and short-term adaptation sites, as well as from uncontaminated areas under heavy metal stress in minimal and nutrient-rich media. This suggests the induction of sleeping genes by metals like nickel or cadmium (Haferburg et al. 2009).

9.4.1 Biosorption

Sorption of metals by microbial cells is due to ion exchange, complexation, electrostatic interaction and precipitation to the cell surface and cell wall components. Bacterial biosorbents are often gram-positive cells, and include specifically firmicutes of the genera *Bacillus*, but also *Streptomyces*. However, it has been reported that viable biomass may exhibit lower affinity to metal ions, owing to competing protons excreted during metabolisms. Biosorption with regeneration of the sorbents can be an important industrial tool to recover metal ions from the liquid phase (Acheampong et al. 2009). Haferburg et al. (2007a) used minimal media containing AMD water to test 40 different strains isolated from soil samples for their biosorptive capacities. The strain *S. acidiscabies* E13 reduced the concentration of uranium 66%, while aluminum, manganese, cobalt, nickel, copper and strontium were also adsorbed. Highly chromium-resistant *Streptomyces* spp. isolates from Salí river sediments in Argentina were able to accumulate 10.0 g and 5.6 Cr per g of dry weight from starch-casein media containing 50 mg/l (mg l^{-1}) chromate (Amoroso et al. 2001). However, conditions and pretreatments differ in most publications, rendering a direct comparison of sorption capacities hardly possible.

However, while streptomycetes could be shown to possess great potential as biosorbents for removal of high amounts of metal ions from liquids, not much is known about biosorption in the environment, e.g., in soils or sediments. Preliminary results with *S. mirabilis* P16B-1 isolated from a former uranium mining site in Thuringia, Germany, grown in highly contaminated soil without the addition of any media, could show reduction of the bioavailable fraction of metals present and thus is an interesting basis for further research.

9.4.2 Metal Reduction

Redox transformation by bacterial reduction results in an insoluble and unavailable form of heavy metals and has been shown to occur for U(VI), Se(VI), Cr(VI), Cr(IV), Mo(VI), Se(IV), Hg(II), Ag(I), and other metals (Merroun and Selenka-Pobell 2008; Ramírez-Díaz et al. 2008). Uranium reduction is described for more than 25 species of phylogenetic diverse prokaryotes, e.g., mesophilic sulfate-reducing, Fe(III)-reducing, thermophillic, fermentative, acidotolerant bacteria, myxobacteria, as well as hyperthermophillic archae (Merroun and Selenska-Pobell 2008). Mercuric reductase, a flavin-containing disulfide oxidoreductase, is able to transform Hg(II) into Hg which is volatile and thus leaves the microorganisms' environment. Organic mercury can also be degraded by organomercurial lyase and the resulting Hg(II) again is reduced to Hg (Barkay et al. 1992).

Extracellular cupric reductase activity, which catalyses the reduction of Cu(II) to Cu(I), was shown by Albarracín et al. (2008) for three *Streptomyces* strains isolated from copper-polluted sediments. Enzyme activity was detected in both copper-adapted cells and nonadapted cells, but enzyme activity was up to 100-fold higher in preadapted cells of *Streptomyces* sp. AB2A. The authors suggested a direct relation between cupric reductase activity and copper uptake processes implying a role for the enzyme in copper resistance. *Streptomyces* strain CG252, tolerant to 500 μg/ml (μg ml^{-1}) copper in a glucose medium, is capable to reduce Cr(VI) to less toxic Cr(III), especially while growing in biofilms on glass beads. The enzyme was able to reduce 100 μg/ml (μg ml^{-1}) of $K_2Cr_2O_7$ to Cr(III) and its activity was cell-associated and not found to be secreted. A pretreatment of the cells with low concentrations of chromium was not necessary to express the enzyme (Morales et al. 2007). *Streptomyces* sp. MC1, isolated from sugarcane and highly resistant to Cr(IV) (890 mg/l (mg l^{-1}) was able to reduce 98% of Cr(IV) and accumulate 3.54 mg of Cr(III) from liquid minimal medium with 50 mg/l (mg l^{-1}) $K_2Cr_2O_7$. Reduction of Cr(IV) started during exponential growth and continued throughout stationary phase. No chromate re-oxidation could be observed after 70 days of incubation (Polti et al. 2010). Under aerobic conditions, most chromate reductase enzymes use NADH or NADPH as cofactors catalyzing the reduction of Cr(VI) or Cr(IV) to Cr(III), which cannot cross cellular membranes and is, if generated extracellularly, an effective resistance mechanism (Ramírez-Díaz et al. 2008). Reducing enzymes producing insoluble forms of the metals thus can be used for bioremediation approaches in contaminated soil, but still there is scarce information on streptomycetes employing such resistance mechanisms.

9.4.3 Extracellular Chelators

Extracellular metal binding, e.g., polysaccharides or proteins, can immobilize metals and thereby prevent the cells from metal toxicity by complex formation. Microorganisms, fungi, and plants developed different strategies to scavenge and adsorb essential metals from the environment. Since many metals are essential, these include both accumulation and protection mechanisms.

The most common and effective way to accumulate iron is the production of siderophores. These low-molecular weight compounds possess a high affinity and selectivity for iron (Hider and Kong 2010; Rajkumar et al. 2010), but also can bind other metals (Dimkpa et al. 2009a). In contrast to gram-negative bacteria, less information on iron chelation and transport is available for gram-positives (Hider and Kong 2010). It could be shown that it is common in streptomycetes to produce different siderophores at the same time, which may provide an advantage in competition with other microorganisms in soil (Chater et al. 2010). Actinomycetes isolated from medicinal plant rhizosphere soils mainly, with 89%, belonged to the genus *Streptomyces*, a high number of which were able to produce siderophores (Khamna et al. 2009).

Siderophores produced by heavy metal-resistant streptomycetes lowered the formation of free radicals and stimulated plant growth via enhancement of auxin production, as well as by protecting auxins from degradation. Additionally, microbial siderophores could be shown to supply plants with iron even in the presence of elevated metal concentrations of aluminum, nickel, copper, manganese and uranium. Besides, it could be shown that cadmium, copper, aluminum and nickel stimulated siderophore production. Thus, a higher efficiency of phytoextraction in siderophore-treated plants was observed (Dimkpa et al. 2008, 2009a, b).

Another strategy for extracellular binding is the excretion of melanins and other pigments, as well as antibiotics which are able to reduce the bioavailable metal fraction (Haferburg and Kothe 2007). For two extremely metal-resistant *S. mirabilis* strains isolated from a former uranium mining site, production of a brownish pigment on nutrient-rich, nickel-supplemented media was observed. *S. mirabilis* P16B-1, which probably encodes this resistance factor on a mega-plasmid, was not able to regulate pigment production with increasing metal content and was not able to grow on similar high nickel concentration in nutrient-rich media compared to minimal media (Schmidt et al. 2009a). However, *S. mirabilis* P10A-3, without plasmid and therefore encoding resistance factors on the chromosome, was able to regulate pigment production. Thus, it can be speculated that energy consuming pigment production is effective only for low concentration of metals, whereas elevated concentrations of metals necessitate other resistance mechanisms like, e.g., efflux transporters (Schmidt et al. 2009a) making regulation necessary for a concentration-dependent, suitable answer to heavy metal stress.

Other redox-active pigments and antibiotics, e.g., actinorhodin and prodigines of *S. coelicolor*, have been reported to serve as sensors of redox stress and to chelate metals, which also could increase heavy metal resistance (Kothe et al. 2007a; Chater et al. 2010). Thus, the microorganisms can benefit from excretion of extracellular chelators by preventing metals from entering the cells.

9.4.4 *Efflux Transport Systems*

Efflux proteins are widespread among microorganisms. Several transporter systems may cause insensitivity to antibiotics and antibacterial drugs, but also metal resistance has been found to be due to efflux transporters (Saidijam et al. 2006; Eitinger and Mandrand-Berthelot 2000). The majority of transporters belong to either the ATP-binding cassette (ABC) superfamily or to the major facilitator superfamily (MFS), using energy from ATP hydrolysis or from the electrochemical proton gradient for antiport mechanisms, respectively (Fig. 9.5). Other efflux systems are the drug metabolite transporter family (DMT) and its subclass of the small multidrug resistance family (SMR), resistance/nodulation/division family (RND), multidrug/oligosaccharidyl-lipid/polysaccharide flippase superfamily (MOP), and its subclass multidrug and toxic compound extrusion family (MATE). However, their distribution varies between different microorganisms and substrates of many efflux systems have not yet been identified (Saidijam et al. 2006).

Most metal efflux pumps, like high affinity nickel transporters, belong to the class of ABC superfamily (Nik system in *E. coli*), P-type ATPases (*Helicobacter pylori*) and transition metal permeases (HoxN in *Ralstonia eutropha*) which were identified in gram-positive and gram-negative bacteria, as well as in yeast (*Schizosaccharomyces pombe*) (Eitinger and Mandrand-Berthelot 2000). Genome analysis of *S. coelicolor* revealed the presence of 137 ABC transporters, probably necessary for adaptation to reducing conditions in soil and to the occurrence of potentially toxic compounds in this environment (Bentley et al. 2002). By transcriptome analysis, it could be shown that 56 genes of *S. coelicolor* encoding efflux pumps were up-regulated upon pH shock (from pH 7.2 to pH 4.0). Since the substrates for most of the efflux transporters are unknown and some of the proteins show homologies to cation efflux systems of other organisms (Kim et al. 2008), these results may indicate heavy metal efflux pumps involved in resistance, especially in relation to acidic mine drainage (AMD) and high metal bioavailability in such contaminated environment. For two *Streptomyces* strains, *S. tendae* F4 and *Streptomyces* sp. E8 isolated from a former uranium mining site, evidence for P-type ATPase transporter genes was found. The deduced amino acid sequence revealed similarity to the high affinity nickel transporter genes of *R. eutropha* and *H. pylori*. Sequences of both strains show high divergence, probably resulting from independent evolution (Amoroso et al. 2000). For several *S. mirabilis* strains

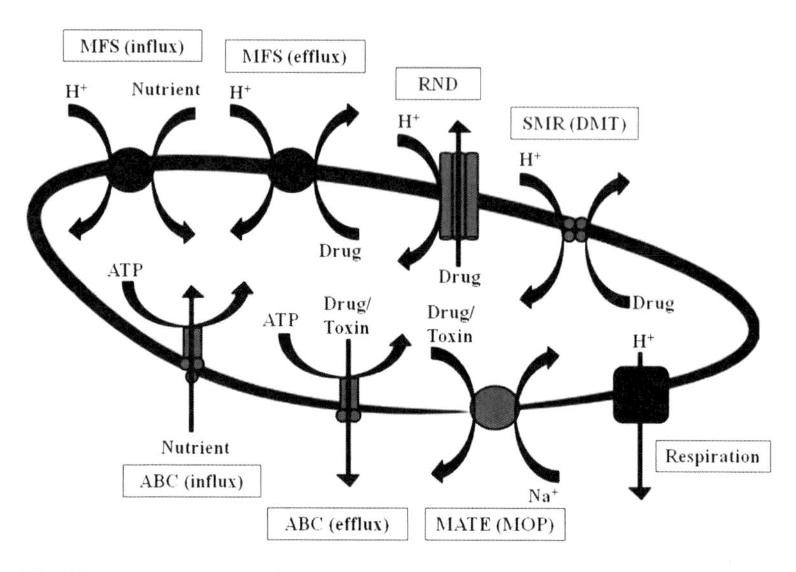

Fig. 9.5 Efflux systems and their respective source of energy to transport. Primary active transport is dependent on ATP hydrolysis, while secondary active transport depends on the use of a proton gradient across the cellular membrane (modified from Saidijam et al. 2006)

(P16B-1, K7A-1, and P10A-3) isolated from the same mining site location, an unknown, high turnover transporter system was assumed to be responsible for the high resistance to more than100 mM $NiSO_4$ and 100 mM $ZnSO_4$ (Schmidt et al. 2009a).

Essential metals necessitate homeostasis in the cell. Fast and unspecific uptake systems for major biometal cations (Mg^{2+}, Ca^{2+}, Zn^{2+}, Ni^{2+}, Mn^{2+}) or inducible, high specific, but slow and energy-expensive uptake ATPases are antagonized by inducible efflux pumps. For *E. coli,* four zinc uptake systems (P-type ATPase ZntA, CDF proteins ZntB, ZitB and also the iron-dependent FieF) are known. The actinobacterium *M. tuberculosis* codes for P-type export ATPases with Zn^{2+} as possible substrate and a ZitB/ZitF ortholog, ZitA, related to FieF. Overexpression of ZitA in *M. smegmatis* mc^2155 revealed slightly, but not significant lower accumulation of $^{65}Zn^{2+}$ compared to an empty vector control, which can be seen as a hint for the importance of P-type ATPase-dependent efflux (Riccardi et al. 2008). In *Bacillus subtilis*, *M. tuberculosis* and *S. coelicolor*, Zur represses the expression of genes encoding zinc uptake systems. Like with *Xanthomonas campestris,* genes of the operon *zur* balance import and efflux in *M. tuberculosis* (Huang et al. 2008; Riccardi et al. 2008). Expression of efflux systems to decrease intracellular concentrations of nonessential metals for homeostasis maintenance of essential metals are important strategies of resistance, and the presence of these proteins, as well as their regulation need to be investigated in more detail, especially in heavy metal-resistant streptomycetes.

9.4.5 Intracellular Sequestration

Intracellular sequestration or bioaccumulation of metals intracellularly (Acheampong et al. 2009) necessitates metal-binding proteins within the cytosol of resistant strains. For plants and yeasts, e.g., *Schizosaccharomyces pombe* and *Saccharomyces cerevisiae*, two types of cystein-rich, metal-binding peptides, the phytochelatins ((γ-GluCys)$_n$Gly where $n = 2$–11) and the metallothioneins (small peptides of up to 100 amino acid residues of high cystein content) store certain essential trace metals and detoxify excessive concentrations of metals (Van de Weghe and Ow 2001; Hall 2002; Shanker et al. 2005; Yang et al. 2005; Pal and Rai 2010; Xu et al. 2010). In contrast, bacterial cells have only recently been reported to express metallothioneins, like the cyanobacterial SmtA and BmtA (Blindauer et al. 2002). The chromium-reducing *Streptomyces* sp. MC1 accumulated Cr(III) after reduction. Analysis of these metal deposits supported the idea of storage by chromium binding proteins (Polti et al. 2010). For the gram-negatives *Pseudomanonas aeruginosa* and *P. putida*, proteins able to bind multiple zinc ions have been shown (Blindauer et al. 2002). To test metal sequestration, overexpression of metallothioneins was performed showing significantly increased intracellular metal concentrations, although no uniform effect on resistance was observed. Only the expression of human MT-II increased resistance of the cells. Expression of a repetitive, cysteine-rich motif (CGCCG)$_3$ as fusion with the maltose-binding protein in *E. coli* caused ten times higher binding of Cd^{2+} and Hg^{2+} without enhancing metal resistance (Shi et al. 1992; Mejáre and Bülow 2001). Since for function, metal chelating cystein and histidine motifs are the relevant portion of peptides, the screening of genomes for predicted short peptides with high cystein and histidine contents is feasible. Schmidt et al. (2010) analyzed 73 actinobacterial genomes for peptides with high contents of cystein and histidine and identified 103 putative metallothioneins and metallohistines. The putative peptides showed different evolutionary origin, including indications of heterologous gene transfer. The analysis also supports the idea of underestimations of the number of metallothioneins occurring in bacteria. However, much more research is needed concerning the identification and application of metallothioneins in bacteria.

In addition to peptides, intracellular phosphates can sequester metals by ionic interactions. The fungus *Cladosporium cladosporioides*, e.g., revealed intracellular crystals of manganese and phosphor (1.6:1) on 15 mM $MnSO_4$of a size ranging from few nm to 200 nm (Shao and Sun 2007). *Pseudomonas aeriginosa* was able to accumulate nickel as phosphide and carbide crystals in the perimplasm (56%), in the membrane (30%) and in cytoplasm (11%) on 0.4 mmol/l Ni (Sar et al. 2001). Within the cytosol of the heavy metal-resistant *S. acidiscabies* E13, the main amount of nickel was eluted with almost no protein but high phosphorous content indicating an intracellular binding of nickel to reduce metal toxicity (Schmidt et al. 2007). In general, intracellular accumulation of heavy metals is widespread in bacteria and could also be shown for streptomycetes, especially by intracellular phosphates, which could have special relevance to remediation directly of contaminated environments.

9.4.6 Biomineralization

Biomineralization is defined as the formation of minerals by complex processes directly or indirectly driven by living organisms, associated with metabolic activity due to the impact of compounds produced. The organisms provide ions for crystal formation and determine size, shape, and structure, as well as physical characteristics of the biominerals (Bäuerlein 2003; Sigel et al. 2008). By X-ray absorption spectroscopy and transmission electron microscopy, *Myxococcus xanthus* was shown to produce biominerals in uranium-supplemented medium, including organic phosphates at the cell surface and meta-autunite-like phases at the cell wall and within the extracellular polysaccharides (Jroundi et al. 2007). Three actinobacterial strains of the species *Microbacterium oxydans* isolated from a groundwater well at a Siberian radioactive waste deposit were able to form similar uranium phosphates and meta-autunite-like phases (Nedelkova et al. 2007). A lack of intracellular uranium contents was seen as indicator for successful application of bacterial resistance mechanisms. The only accumulation of biocrystals by streptomycetes was reported for the multiple metal-resistant *S. acidiscabies* E13, which was able to mineralize nickel struvite on viable colony surfaces. This mechanism is supposed to decrease toxic nickel concentrations in the vicinity of the cells (Haferburg et al. 2008). However, if sufficient magnesium is present in soil microcosms, struvite formation without the incorporation of other metals could be seen, indicating a different role for these biominerals as extracellular storage components. Although there is not much known on mechanisms involved in bacterial crystal formation, the ecological relevance for streptomycetes, as well as relevance for bioremediation approaches seems high enough to warrant future research in this area.

9.4.7 Superoxide Dismutase

The formation of reactive oxygen species (ROS) resulting from Fenton reactions of heavy metals in soil and from metal-loaded humic substances, as well as during chromium reduction (Paciolla et al. 1999; Schützendübel and Polle 2002; Ramírez-Díaz 2008) is one of the toxic effects of heavy metals. Aerobic bacteria express protective systems, including superoxide dismutase (SOD), catalase and peroxidase, which can thus be seen as additional resistance mechanisms. On the one hand, SODs buffer metal concentration by incorporating specific metal ions into their active center and on the other hand, they catalyze the disproportionation of superoxide anions to dioxygen and hydrogen peroxide (Tarhan et al. 2007; Geslin et al. 2001). For *Schizosaccharomyces pombe*, it could be shown that CuZnSOD activity of the wildtype strain increased when cells were treated with elevated concentrations of zinc (1.5 mM $ZnCl_2$) and that CuZnSOD deletion mutants showed slower growth and higher sensitivity toward zinc. This indicated that

SOD is an important factor for zinc detoxification in eukaryotes (Tarhan et al. 2007). The metal chaperone-like protein Pccs, which maintains copper homeostasis and is encoded by *pccs* I to IV was used for expression studies in the yeast *S. cerevisiae*. It could be shown that mutants expressing the entire gene are resistant to copper, whereas mutants lacking domains *pccs* I to III showed increased sensitivity to copper and cadmium. Additionally, overexpression of *pccs* domain IV enhanced copper resistance by one order of magnitude (Laliberté et al. 2004). The authors suggested that Pccs activates apo-CuZnSOD under copper-limiting conditions by its first three domains and protects cells against metal toxicity by its fourth domain. Taken together, CuZnSOD, as well as its activator Pccs, are involved in metal resistance in eukaryotes.

In bacteria, a similar effect could be shown by Geslin et al. (2001) and Eitinger (2004). *E. coli* expressing two highly homologous SODs, a MnSOD and a FeSOD, was used for heavy metal resistance tests. It could be shown that deletion of one of both genes had no significant effect on growth in cadmium, cobalt, nickel, copper and zinc-containing media, whereas FeSOD- and MnSOD-deficient double mutants exhibited hypersensitivity toward these five metals (Geslin et al. 2001). The gene *sodN* and the downstream open reading frame *sodX* of the cyanobacterium *Prochlorococcus marinus*, when expressed in SOD-negative, oxygen-sensitive *E. coli* restored oxygen tolerance in the presence of copper, cobalt, iron, manganese, nickel and zinc (Eitinger 2004). For the lead-tolerant *S. subrutilus* P5, an extracellular protein with high *N*-terminal amino acid sequence homology to bacterial SOD was found to precipitate extracellularly as a result of lead adsorption, which was considered to be a major mechanism of lead tolerance in this streptomycete. After addition of lead ions to the culture filtrate at any growth phase, an immediate protein precipitation was found. This protein showed SOD activity and EPR spectra similar to FeZnSOD from *S. griseus*, but no identical amino acid sequence to any other known *Streptomyces*-SOD, differing especially for its high tryptophan and low tyrosine contents (So et al. 2001). The lead adsorption capacity of this SOD was measured in protein–lead mixtures after protein purification and showed binding of 1,100 lead atoms per subunit at a concentration of 0.075 mM lead ions. Above this concentration, the protein became saturated. For copper, iron and cadmium, a much lower precipitation efficiency was detected which indicates a specific binding of lead. The mechanism of binding cannot be explained yet, but the authors suggest not only chelation to acidic amino acids, but also other mechanisms like, e.g., similarity to multiple copper motifs or iron nucleation of ferritin. This SOD was constitutively expressed and might reduce the intracellular amount of lead by extracellular precipitation and thereby reduce intracellular generation of ROS caused by metal toxification (So et al. 2001).

For the multiple metal-resistant strain *S. acidiscabies* E13, it was shown that nickel, as well as other metals, is able to induce expression of *sodN* and reduce expression of *sodF* indicating that NiSOD is one of the heavy metal resistance mediating proteins (Schmidt et al. 2007). Overexpression of NiSOD from *S. acidiscabies* E13 in *E. coli* in heavy metal-containing medium showed significantly better growth rates for NiSOD-expressing strains compared to an empty vector control (unpublished). *In silico* analysis revealed a high number of proteins with sequence similarity to NiSOD from actinobacteria, proteobacteria, chlamydiae and green algae, with highly conserved nickel ligands and maturation signals for N-terminal proteolysis. Heterologous gene transfer, most likely from actinobacteria to other taxa, has been postulated for these *sodN* genes, which fall into four clusters not consistent with phylogeny of the species (Schmidt et al. 2009b). In general SOD, and especially NiSOD expression in streptomycetes, could be seen as yet another resistance factor for coping with heavy metal stress.

9.5 Conclusions

It has been shown that streptomycetes are a dominant group of bacteria in heavy metal-contaminated soil (Haferburg et al. 2007b; Kothe et al. 2010) and that growth of soil bacteria, especially streptomycetes, had positive effects on bioremediation, as well as on bioavailability of metals in soil (Kuffner et al. 2008; Dimkpa et al. 2009a, b; Rajkumar et al. 2010; Haferburg and Kothe 2010). Additionally, their influence on biogeochemical cycles and hydrogeological parameters warrants more research on this group of soil bacteria. Particularly, isolation of heavy metal-resistant *Streptomyces* strains, a detailed understanding of their resistance mechanisms and detection of relevant processes directly in soil is necessary, and a combination of (micro)biological and (hydro)geological approaches seems warranted for application in bioremediation. Only if a deeper understanding is reached, a transfer of methods and techniques from basic research into applied science and optimized on-site remediation of heavy metal-contaminated areas will be possible.

Acknowledgments We would like to thank the Helmholtz Interdisciplinary Graduate School for Environmental Research (HIGRADE) for scholarship funding, André Schmidt, René Phieler, Michael Klose and Jens Schumacher for their help and support, Sandor Nietzsche for scanning electron microscopy, as well as Dirk Merten for sequential extraction and metal analyses. Petra Mitscherlich is thanked for technical assistance and DFG-GRK1257 and JSMC for support.

References

Acheampong MA, Meulepas RJW, Lens PNL (2009) Removal of heavy metals and cyanide from gold mine wastewater. J Chem Technol Biotechnol 85:590–613

Albarracín VH, Avila AL, Amoroso MJ, Abate CM (2008) Copper removal ability by *Streptomyces* strains with dissimilar growth patterns and endowed with cupric reductase activity. FEMS Microbiol Lett 288:141–148

Amoroso MJ, Schubert D, Mitscherlich P, Schumann P, Kothe E (2000) Evidence for high affinity nickel transporter genes in heavy metal resistant *Streptomyces* sp. J Basic Microbiol 40:295–301

Amoroso MJ, Castro GR, Duran A, Peraud O, Oliver G, Hill RT (2001) Chromium accumulation by two *Streptomyces* spp. isolated from riverine sediments. J Ind Microbiol Biotechnol 26:210–215

Banks D, Younger PL, Arnesen RT, Iversen ER, Banks SB (1997) Mine-water chemistry: the good, the bad and the ugly. Environ Geol 32:157–174

Barkay T, Turner R, Saouter E, Horn J (1992) Mercury biotransformations and their potential for remediation of mercury contamination. Biodegradation 3:147–159

Bäuerlein E (2003) Biomineralization of unicellular organisms: an unusual membrane biochemistry for the production of inorganic nano- and microstructures. Angew Chem Int Ed 42:614–641

Bentley SD, Chater KF, Cerdeno-Tarraga AM, Challis GL, Thomson NR, James KD, Harris DE, Quail MA, Kieser H, Harper D, Bateman A, Brown S, Chandra G, Chen CW, Collins M, Cronin A, Fraser A, Goble A, Hidalgo J, Hornsby T, Howarth S, Huang CH, Kieser T, Larke L, Murphy L, Oliver K, O'Neil S, Rabbinowitsch E, Rajandream MA, Rutherford K, Rutter S, Seeger K, Saunders D, Sharp S, Squares R, Squares S, Taylor K, Warren T, Wietzorrek A, Woodward J, Barrell BG, Parkhill J, Hopwood DA (2002) Complete genome sequence of the model actinomycete *Streptomyces coelicolor* A3(2). Nature 417:141–147

Bibb M, Hesketh A (2009) Analyzing the regulation of antibiotic production in *Streptomycetes* sp. In: David Hopwood (Ed) Complex enzymes in microbial natural product biosynthesis, part A: overview articles and peptides, vol 458, pp 93–116

Blindauer CA, Harrison MD, Robinson AK, Parkinson JA, Bowness PW, Sadler PJ, Robinson NJ (2002) Multiple bacteria encode metallothioneins and SmtA-like zinc fingers. Mol Microbiol 45:1421–1432

Chater KF, Biro S, Lee KJ, Palmer T, Schrempf H (2010) The complex extracellular biology of *Streptomyces*. FEMS Microbiol Rev 34:171–198

de Jonge LW, Moldrup P, Schjonning P (2009) Soil infrastructure, interfaces and translocation processes in inner space ("soil-it-is"): towards a road map for the constraints and crossroads of soil architecture and biophysical processes. Hydrol Earth Syst Sci 13:1485–1502

Dimkpa CO, Svatos A, Dabrowska P, Schmidt A, Boland W, Kothe E (2008) Involvement of siderophores in the reduction of metal-induced inhibition of auxin synthesis in *Streptomyces* spp. Chemosphere 74:19–25

Dimkpa CO, Merten D, Svatos A, Büchel G, Kothe E (2009a) Metal-induced oxidative stress impacting plant growth in contaminated soil is alleviated by microbial siderophores. Soil Biol Biochem 41:154–162

Dimkpa CO, Merten D, Svatos A, Büchel G, Kothe E (2009b) Siderophores mediate reduced and increased uptake of cadmium by *Streptomyces tendae* F4 and sunflower (*Helianthus annuus*), respectively. J Appl Microbiol 107:1687–1696

Eitinger T, Mandrand-Berthelot MA (2000) Nickel transport systems in microorganisms. Arch Microbiol 173:1–9

Eitinger T (2004) *In vivo* production of active nickel superoxide dismutase from *Prochlorococcus marinus* MIT9313 is dependent on its cognate peptidase. J Bacteriol 186:7821–7825

Flärdh K, Buttner MJ (2009) *Streptomyces* morphogenetics: dissecting differentiation in a filamentous bacterium. Nat Rev Microbiol 7:36–49

Geslin C, Llanos J, Prieur D, Jeanthon C (2001) The manganese and iron superoxide dismutases protect *Escherichia coli* from heavy metal toxicity. Res Microbiol 152:901–905

Haferburg G, Kothe E (2007) Microbes and metals: interactions in the environment. J Basic Microbiol 47:453–467

Haferburg G, Merten D, Büchel G, Kothe E (2007a) Biosorption of metal and salt tolerant microbial isolates from a former uranium mining area. Their impact on changes in rare earth element patterns in acid mine drainage. J Basic Microbiol 47:474–484

Haferburg G, Reinicke M, Merten D, Büchel G, Kothe E (2007b) Microbes adapted to acid mine drainage as source for strains active in retention of aluminum or uranium. J Geochem Explor 92:196–204

Haferburg G, Klöss G, Schmitz W, Kothe E (2008) "Ni-struvite" – a new biomineral formed by a nickel resistant *Streptomyces acidiscabies*. Chemosphere 72:517–523

Haferburg G, Groth I, Möllmann U, Kothe E, Sattler I (2009) Arousing sleeping genes: shifts in secondary metabolism of metal tolerant actinobacteria under conditions of heavy metal stress. Biometals 22:225–234

Haferburg G, Kothe E (2010) Metallomics: lessons for metalliferous soil remediation. Appl Microbiol Biotechnol 87:1271–1280
Hall JL (2002) Cellular mechanisms for heavy metal detoxification and tolerance. J Exp Bot 53:1–11
Hider RC, Kong XL (2010) Chemistry and biology of siderophores. Nat Prod Rep 27:637–657
Hopwood DA (2006) Soil to genomics: the *Streptomyces* chromosome. Ann Rev Genet 40:1–23
Huang P-M, Wang M-K, Chiu C-Y (2005) Soil mineral-organic matter-microbe interactions: impacts on biogeochemical processes and biodiversity in soils. Pedobiologia 49:609–635
Huang DL, Tang DJ, Liao Q, Li HC, Chen Q, He YQ, Feng JX, Jiang BL, Lu GT, Chen BS, Tang JL (2008) The Zur of *Xanthomonas campestris* functions as a repressor and an activator of putative zinc homeostasis genes *via* recognizing two distinct sequences within its target promoters. Nucleic Acids Res 36:4295–4309
Jabeen R, Ahmad A, Iqbal M (2009) Phytoremediation of heavy metals: physiological and molecular mechanisms. Bot Rev 75:339–364
Jroundi F, Merroun ML, Arias JM, Rossberg A, Selenska-Pobell S, Gonzalez-Munoz MT (2007) Spectroscopic and microscopic characterization of uranium biomineralization in *Myxococcus xanthus*. Geomicrobiol J 24:441–449
Kamaludeen SPB, Ramasamy K (2008) Rhizoremediation of metals: harnessing microbial communities. Indian J Microbiol 48:80–88
Khamna S, Yokota A, Lumyong S (2009) Actinomycetes isolated from medicinal plant rhizosphere soils: diversity and screening of antifungal compounds, indole-3-acetic acid and siderophore production. World J Microbiol Biotechnol 25:649–655
Kim YJ, Song JY, Hong SK, Smith CP, Chang YK (2008) Effects of pH shock on the secretion system in *Streptomyces coelicolor* A3(2). J Microbiol Biotechnol 18:658–662
Kindler R, Miltner A, RichnowH-H KM (2006) Fate of gram-negative bacterial biomass in soil-mineralization and contribution to SOM. Soil Biol Biochem 38:2860–2870
Kindler R, Miltner A, Thullner M, Richnow H-H, Kästner M (2009) Fate of bacterial biomass derived fatty acids in soil and their contribution to soil organic matter. Org Geochem 40:29–37
Kothe E, Bergmann H, Büchel G (2005) Molecular mechanisms in bio-geo-interactions: from a case study to general mechanisms. Chem Erde-Geochem 65:7–27
Kothe E, Dimkpa C, Haferburg G, Schmidt A, Schmidt A, Schütze E (2010) Streptomycete heavy metal resistance: extracellular and intracellular mechanisms. In: Sherameti I, Varma A (eds) Soil biology. Springer, Heidelberg
Kuffner M, Puschenreiter M, Wieshammer G, Gorfer M, Sessitsch A (2008) Rhizosphere bacteria affect growth and metal uptake of heavy metal accumulating willows. Plant Soil 304:35–44
Laliberté J, Whitson LJ, Beaudoin J, Holloway SP, Hart PJ, Labbe S (2004) The *Schizosaccharomyces pombe* Pcs protein functions in both copper trafficking and metal detoxification pathways. J Biol Chem 279:28744–28755
Lamparter A, Bachmann J, Goebel MO, Woche SK (2009) Carbon mineralization in soil: impact of wetting-drying, aggregation and water repellency. Geoderma 150:324–333
Liang HC, Thomson BM (2009) Minerals and mine drainage. Water Environ Res 81:1615–1663
Mejáre M, Bülow L (2001) Metal-binding proteins and peptides in bioremediation and phytoremediation of heavy metals. Trends Biotechnol 19:67–73
Merroun ML, Selenska-Pobell S (2008) Bacterial interactions with uranium: an environmental perspective. J Contam Hydrol 102:285–295
Miltner A, Kindler R, Knicker H, Richnow H-H, Kästner M (2009) Fate of microbial biomass-derived amino acids in soil and their contribution to soil organic matter. Org Geochem 40:978–985
Morales DK, Ocampo W, Zambiano MM (2007) Efficient removal of hexavalent chromium by a tolerant *Streptomyces* sp affected by the toxic effect of metal exposure. J Appl Microbiol 103:2704–2712
Nedelkova M, Merroun ML, Rossberg A, Hennig C, Selenska-Pobell S (2007) *Microbacterium* isolates from the vicinity of a radioactive waste depository and their interactions with uranium. FEMS Microbiol Ecol 59:694–705

Nogueira MA, Nehls U, Hampp R, Poralla K, Cardoso E (2007) Mycorrhiza and soil bacteria influence extractable iron and manganese in soil and uptake by soybean. Plant Soil 298:273–284

Onaka H, Nakagawa T, Horinouchi S (1998) Involvement of two A-factor receptor homologues in *Streptomyces coelicolor* A3(2) in the regulation of secondary metabolism and morphogenesis. Mol Microbiol 28:743–53

Paciolla MD, Davies G, Jansen SA (1999) Generation of hydroxyl radicals from metal-loaded humic acids. Environ Sci Technol 33:1814–1818

Pal R, Rai JPN (2010) Phytochelatins: peptides involved in heavy metal detoxification. Appl Biochem Biotechnol 160:945–963

Polti MA, Amoroso MJ, Abate CM (2010) Intracellular chromium accumulation by *Streptomyces* sp. MC1. Water Air Soil Pollut 214:49–57

Rajkumar M, Ae N, Prasad MNV, Freitas H (2010) Potential of siderophore-producing bacteria for improving heavy metal phytoextraction. Trends Biotechnol 28:142–149

Ramírez-Díaz MI, Diaz-Perez C, Vargas E, Riveros-Rosas H, Campos-Garcia J, Cervantes C (2008) Mechanisms of bacterial resistance to chromium compounds. Biometals 21:321–332

Rawlings DE (2002) Heavy metal mining using microbes. Ann Rev Microbiol 56:65–91

Riccardi G, Milano A, Pasca MR, Nies DH (2008) Genomic analysis of zinc homeostasis in *Mycobacterium tuberculosis*. FEMS Microbiol Lett 287:1–7

Saidijam M, Benedetti G, Ren QH, Xu ZQ, Hoyle CJ, Palmer SL, Ward A, Bettaney KE, Szakonyi G, Meuller J, Morrison S, Pos MK, Butaye P, Walraven K, Langton K, Herbert RB, Skurray RA, Paulsen IT, O'Reilly J, Rutherford NG, Brown MH, Bill RM, Henderson PJF (2006) Microbial drug efflux proteins of the major facilitator superfamily. Curr Drug Targets 7:793–811

Salomé C, Nunan N, Pouteau V, Lerch TZ, Chenu C (2010) Carbon dynamics in topsoil and in subsoil may be controlled by different regulatory mechanisms. Global Change Biol 16:416–426

Sar P, Kazy SK, Singh SP (2001) Intracellular nickel accumulation by *Pseudomonas aeruginosa* and its chemical nature. Lett Appl Microbiol 32:257–261

Scherr N, Nguyen L (2009) *Mycobacterium* versus *Streptomyces* – we are different, we are the same. Curr Opinion Microbiol 12:699–707

Schmidt A, Haferburg G, Sineriz M, Merten D, Büchel G, Kothe E (2005) Heavy metal resistance mechanisms in actinobacteria for survival in AMD contaminated soils. Chem Erde-Geochem 65:131–144

Schmidt A, Schmidt A, Haferburg G, Kothe E (2007) Superoxide dismutases of heavy metal resistant streptomycetes. J Basic Microbiol 47:56–62

Schmidt A, Gube M, Schmidt A, Kothe E (2009a) *In silico* analysis of nickel containing superoxide dismutase evolution and regulation. J Basic Microbiol 49:109–118

Schmidt A, Haferburg G, Schmidt A, Lischke U, Merten D, Ghergel F, Büchel G, Kothe E (2009b) Heavy metal resistance to the extreme: *Streptomyces* strains from a former uranium mining area. Chem Erde-Geochem 69:35–44

Schmidt A, Hagen M, Schütze E, Schmidt A, Kothe E (2010) *In silico* prediction of potential metallothioneins and metallohistins in actinobacteria. J Basic Microbiol 50:562–569

Schützendübel A, Polle A (2002) Plant responses to abiotic stresses: heavy metal-induced oxidative stress and protection by mycorrhization. J Exp Bot 53:1351–1365

Shanker AK, Cervantes C, Loza-Tavera H, Avudainayagam S (2005) Chromium toxicity in plants. Environ Int 31:739–753

Shao ZZ, Sun FQ (2007) Intracellular sequestration of manganese and phosphorus in a metal-resistant fungus *Cladosporium cladosporioides* from deep-sea sediment. Extremophiles 11:435–443

Shao HB, Chu LY, Ruan CJ, Li H, Guo GD, Li WX (2010) Understanding molecular mechanisms for improving phytoremediation of heavy metal-contaminated soils. Crit Rev Biotechnol 30:23–30

Shi JG, Lindsay WP, Huckle JW, Morby AP, Robinson NJ (1992) Cyanobacterial metallothionein gene expressed in *Escherichia-coli* – metal-binding properties of the expressed protein. FEBS Lett 303:159–163

Sigel A, Sigel H, Sigel RKO (2008) Metal ions in life sciences. In: Biomineralization: from nature to application. Wiley, Chichester

So NW, Rho JY, Lee SY, Hancock IC, Kim JH (2001) A lead-absorbing protein with superoxide dismutase activity from *Streptomyces subrutilus*. FEMS Microbiol Lett 194:93–98

Tarhan C, Pekmez M, Karaer S, Arda N, Sarikaya AT (2007) The effect of superoxide dismutase deficiency on zinc toxicity in *Schizosaccharomyces pombe*. J Basic Microbiol 47:506–512

van de Weghe JG, Ow DW (2001) Accumulation of metal-binding peptides in fission yeast requires *hmt2*$^{(+)}$. Mol Microbiol 42:29–36

Xu J, Tian YS, Peng RH, Xiong AS, Zhu B, Hou XL, Yao QH (2010) Cyanobacteria MT gene SmtA enhance zinc tolerance in *Arabidopsis*. Mol Biol Reports 37:1105–1110

Yang X, Feng Y, He ZL, Stoffella PJ (2005) Molecular mechanisms of heavy metal hyperaccumulation and phytoremediation. J Trace Elements Med Biol 18:339–353

Chapter 10
Role of Mycorrhiza in Re-forestation at Heavy Metal-Contaminated Sites

Felicia Gherghel and Katrin Krause

10.1 Introduction

Re-forestation programs have been shown to substantially profit from the symbiotic association with ectomycorrhizal (ECM) fungi (Nara 2006a, b). ECM associations play an active role, particularly in forests where low litter quality and low decomposition and mineralization rates cause N and P limitation (van Scholl et al. 2008). ECM link trees through common mycorrhizal networks and are involved in biotransformations and biogeochemical cycling (Colpaert 2008; Finlay 2008) by decomposition (Courty et al. 2005), bioweathering (van Scholl et al. 2008), or metal and mineral transformations (Fomina et al. 2008). The formation of an ECM symbiosis depends on host species, age and vigor of the trees, edaphic and environmental conditions, availability of fungal inoculum, competition, microflora and microfauna (Rajala 2008). Mycorrhizal fungi compete for two general classes of resources: host-derived carbon and soil or detritus-derived mineral nutrients. Both types of resources are heterogeneously distributed in space (e.g., soil depth, distance from tree) and time (e.g., season, host successional series; Rajala 2008).

Unreclaimed mine tailing sites are a worldwide problem with thousands of unvegetated, exposed tailing piles presenting a source of contamination for nearby communities (Kothe et al. 2005). In addition, elevated concentrations of Al in soils are a problem arising from acidification. Such soils can also be adversely affected by poor drainage, low organic matter content, and diminished populations of indigenous microbes that cycle nutrients. Revegetation is essential to limit soil erosion by wind and water, including runoff of metallic sediments and initiates

F. Gherghel
Department of Mycology, Philipps-University of Marburg, Karl-von-Frisch Str. 8, 35043 Marburg, Germany

K. Krause (✉)
Institute of Microbiology, Friedrich Schiller University, Neugasse 25, 07743 Jena, Germany
e-mail: katrin.krause@uni-jena.de

E. Kothe and A. Varma (eds.), *Bio-Geo Interactions in Metal-Contaminated Soils*, Soil Biology 31, DOI 10.1007/978-3-642-23327-2_10,

primary stage succession (Kothe et al. 2005; Staudenrausch et al. 2005). On the other site, forest management and harvesting operations often provide disturbances, and those sites show secondary stage succession (Smith and Read 1997; van Scholl et al. 2008).

Although some metal ions (Zn, Cu, Mn, Ni, and Co) are micronutrients essential for plant growth, others are toxic, even at low concentrations without known biological functions (e.g., Pb, Hg, and the semimetal As). The data concerning the effect of heavy metal pollution on ECM fungal communities are controversial: bioavailable toxic metals can lead to a reduction of the abundance and diversity of species, as well as to a selection of resistant, or at least tolerant, population (Colpaert et al. 2004; Gebhardt et al. 2007), but also they may lead to relatively high abundances of these species in metal-polluted soils (Gadd 1993; Meharg 2003; Krpata et al. 2008; Plassard and Fransson 2009).

In the following, we will review the biodiversity of ECM fungi in heavy metal-contaminated ecosystems, the mechanisms of ECM fungi involved in preventing cells from toxicity and their role in phytoremediation. We will include interpretations from studying ECM communities at a former uranium mining area and a control, both 10–15-year-old oak forest sites. While Kanigsberg was remediated after the removal of the uranium mining heap, covered with 30–40 cm top soil from uncontaminated sites and revegetated by different trees like oak, ash, beech and larch resulting in a primary stage forest succession, the control site near Greiz, Germany, is an unpolluted site reforested after a clear-cut of spruce and by planting oak resulting in a secondary stage succession.

10.2 Mycorrhiza

Mycorrhiza is the dominant fungal plant interaction in terrestrial ecosystems which is formed by an estimate of 86% of plant species (Brundrett 2009). Over 400 Ma, fungi started to adapt to plants and it is suggested that different mycorrhizal types, endomycorrhiza, including arbuscular mycorrhiza (AM), orchid and ericoid mycorrhiza and ectomycorrhiza (ECM), both monotropoid and arbutoid types, had arisen independently (Smith and Read 2008; Tedersoo et al. 2010). AM fungi are the most widely spread mycorrhizal fungi, which are, however, of lower importance to forests as there ECM fungi dominate. While AM fungi penetrate the cells of the plant roots forming typical structures, the arbuscules and vesicles, ECM fungi form a hyphal mantle around the plant root, where hyphae and rhizomorphs extend into the soil and transport water and nutrients to the plant interface at the intercellular spaces in the cortex of the plant root, where the Hartig' net allows transfer of nutrients between both partners (Fig. 10.1a).

In temperate and boreal habitats, ECM is the characteristic symbiosis with wooded plants. Approximately 6,000 basidiomycetes, some ascomycetes and few zygomycetes, form this interaction with about 3% of plant species (Smith and Read 2008). This sounds low, but the ecologically and economically most important trees

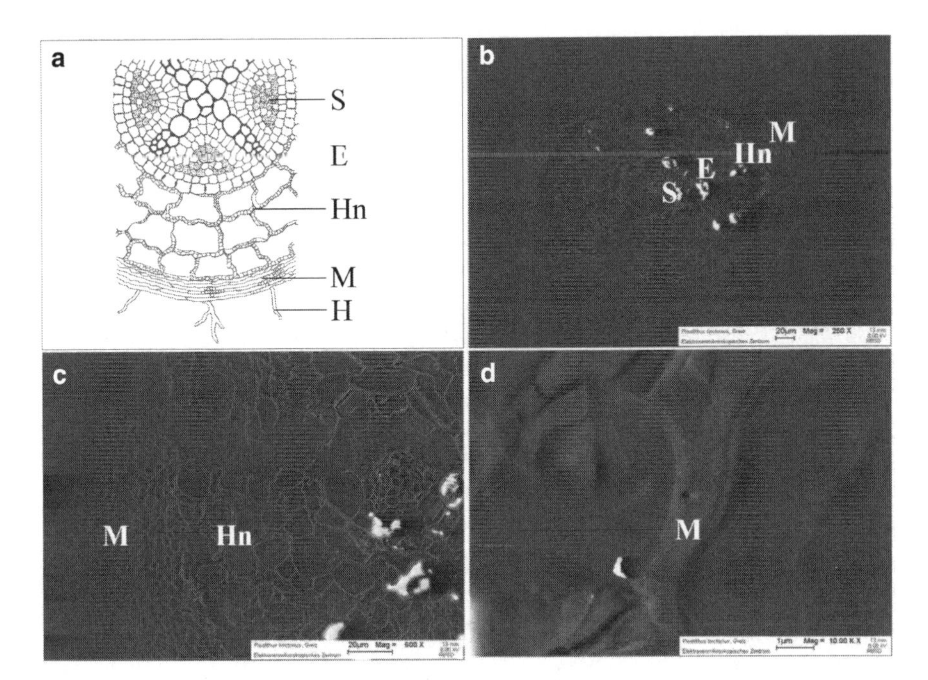

Fig. 10.1 Accumulated heavy metals were detected in the cell walls of the hyphal mantle of *Pisolithus tinctorius*-oak mycorrhiza using EDX scanning electron microscopy. (**a**) Schematic drawing of an ectomycorrhizal root cross section with stele (S), endodermis (E), Hartig' net (Hn), hyphal mantle (M) and extrametrical hyphae (H). (**b**) Cross section with hyphal mantle, Hartig' net, endodermis and stele. (**c**) Hyphal mantle with 8–10 cell layers. (**d**) Hyphal cell walls of the mantle show accumulated heavy metals

of Pinaceae (pines), Fagaceae (beeches and oaks), and Betulaceae (birches and alder), among others, belong to this group, so that ECM fungi play a superior role in seedling establishment and tree growth in forest ecosystems (Tedersoo et al. 2010).

10.3 Biodiversity of ECM Fungi in Heavy Metal-Contaminated Ecosystems

The description of the biodiversity of ECM fungi on metal-polluted soils was assessed by various methods and in studies with diverse host plants and tree ages, but also with different stages of succession. Traditionally, counting fungal fruiting bodies was performed, and there a decrease of fruiting body-producing species was reported with increasing soil metal concentrations. The basidiomycete ECM fungus *Suillus luteus* was identified as a typical mycobiont of young pine in primary succession (Rühling et al. 1984; Rühling and Söderström 1990). It has been shown, however, that there exists a considerable lack of correspondence between above- and below-ground communities of ECM colonizers (Gardes and Bruns 1996). Pritsch et al. (1997) observed fruiting bodies from eight species out of a total of 16 ECM species associated with black alder (*Alnus glutinosa*) collecting fruit bodies monthly over an entire growing season. Similarly, Cripps (2004) found half of the mycorrhizal fungi associated with *Populus tremuloides* when collecting fruiting bodies above-ground. The correspondence between detected fungal fruiting bodies and ECM taxa often depends on the sampling effort and the available taxonomical expertise. Fruiting bodies of some basidiomycete taxa, that have been frequently found on heavily polluted soils, include *Hebeloma* spp., *Pisolithus tinctorius*, *Paxillus involutus*, *Rhizopogon* spp., *Scleroderma* spp. or *Amanita muscaria* (Jones and Hutchinson 1986; Kalač et al. 1991; Turnau et al. 1988). Thus, this traditional identification method can be used for taxonomic identification, but a large proportion of fungi forming ECMs do not fructify, while many others form inconspicuous fruiting bodies.

The ECM root tips can be morphologically characterized (Agerer 2006) and classified into different exploration types based on the amount of emanating hyphae or the presence and differentiation of rhizomorphs of ECM (Agerer 2001). At one extreme, the surface area of the mantle or a few short hyphae provide direct contact to the substrate, while at the other extreme, highly differentiated rhizomorphs, producing vessel-like hyphae with partially or completely dissolved septa, are expected to be relevant for long-distance exploration of nutrient sources (Agerer 2001). Even if the fungal species of ECM root tips cannot be identified directly, this method gives information on functional differences of ectomycorrhizae in the exploration of the substrate by extramatrical mycelium. During investigations in a former uranium mining area at Kanigsberg, Germany below-ground ECM in a primary stage of succession were observed (Iordache et al. 2009). For long-distance exploration types, basidiomycetous fungi like *Tomentella sublinacina*, *Pisolithus tinctorius*, *Paxillus involutus* and *Scleroderma* sp. were found. These fungi have been shown to be predominantly sampled from mining areas in cases of collection of above-ground fruiting bodies. Sequencing of the internal transcribed spacer (ITS) region of rDNA also is commonly used for identification of ECM fungi (Gardes and Bruns 1993; Horton and Bruns 1998). PCR amplification, sequencing, and comparison with the sequences of databases (NCBI: http://www.ncbi.nlm.nih.gov/; UNITE: http://unite.zbi.ee; Kõljalg et al. 2005) allowed the discovery of numerous abundant ECM fungi (Tedersoo et al. 2003, 2006). For metal-rich soils, Staudenrausch et al. (2005) found 23 ECM types associated with *Betula pendula* at an uranium mining heap with the species diversity being lower on polluted land. Mleczko (2004) investigated mycorrhizal and saprobic macrofungi of two zinc waste sites in Southern Poland. In contrast to other data (Krpata et al. 2008; Meharg 2003), he described only 14 ECM types associated with *Pinus sylvestris* and *Betula pendula*. Krpata et al. (2008) reported a high ECM fungal diversity with 54 fungal taxa as mycobionts at a heavily contaminated site associated with *Populus tremula*. The species composition of the contaminated tree stand was comparable to that of a mature forest (Krpata et al. 2008). Many of the taxa detected in this contaminated stand, like *Cenococcum*, *Hebeloma*, *Inocybe*, *Scleroderma*, *Tomentella*, *Tuber*,

Wilcoxina, were also identified in our studies with *Quercus robur* and are typically observed in early successional or disturbed habitats (Muehlmann et al. 2008; Nara et al. 2003; Rudawska et al. 2006). There are indications that in particular ascomycetes are more stress-resistant and thus are more frequent in mycorrhizal communities facing harsh environmental conditions, or after severe disturbances (Baar et al. 1999; Trowbridge and Jumpponen 2004). Colpaert et al. (2004) focused on the occurrence of ECM fungi in pioneer pine forests along a Zn pollution gradient and found only four ECM morphotypes on roots of 25-year-old pine trees on the most polluted area. They also observed the frequent occurrence of morphotypes of the dark septated ascomycete *Hymenoscyphus ericae*. Similar ascomycetes with the same symbiont were reported colonizing a Cu mine spoil in Norway (Vrålstad et al. 2002). It seems ascomycetes are selected for in metal-rich soils. However, it has proven difficult to unequivocally show that metal toxicity exhibited selection pressure on fungal communities and populations.

10.4 Characterization of the Functional Diversity in Contaminated Area

The recent advances in sequencing technologies, the increase of sequence data and the progress in database searches allow assessing ECM diversity in many forest ecosystems. The rapid and relatively inexpensive sequencing technology, like pyrosequencing or 454 technology (Margulies et al. 2005; Schuster 2008) has been used by Jumpponen and Jones (2010), who discuss that the decline of P, Cd, and Pb during the growing season may be correlated with fungal community dynamics, as the fungal communities are sensitive to macronutrients and heavy metal enrichment.

It is well known that, within a fungal species, populations or individuals display a huge genetic diversity (Gryta et al. 2001). Moreover, the life cycle of ECM fungi is driven by adaptation and survival in the soil for the vegetative mycelium and by sexual reproduction continuously producing new genotypes with variable relative impact depending on the fungal species in question (Fiore-Donno and Martin 2001) or on the site conditions (Gryta et al. 1997). Highly variable molecular markers, especially microsatellites, are available and facilitate studying fungal population genetics gene flow (Kretzer et al. 2003; Hogberg et al. 2009).

Taxonomic diversity is relevant to ecosystem function only if it reflects functional diversity. Functional differences between ECM fungi include differences in nutrient cycling, symbiotic abilities, proliferation rates, and to stress tolerance, specifically with respect to drought and heavy metal stress. Different morphologies and physiologies of ECM fungi add up to the total benefit of this specific symbiosis for the host plant, depending on the mycorrhization rates over the entire root system. The interpretation of functional significance of a change in ECM community or diversity of this community to the host plant and to the ecosystem is currently constrained by a lack of data on functional capabilities under field conditions for most ECM taxa. More studies under field conditions thus seem warranted. To identify the function of a species within a community, the fungus has to be isolated and microcosm experiments have to be performed, mostly with a low number of organisms. So far, such experiments are rare in ECM research (Sell et al. 2005).

10.5 Mechanisms to Prevent Cells from Heavy Metal Toxicity

Extracellular and intracellular mechanisms are involved in heavy metal tolerance for fungi. Mostly such mechanisms have not explicitly been studied in mycorrhizal fungi, but it appears likely that data from non-mycorrhizal basidiomycetes (or ascomycetes) can be translated to ECM since they use the same mechanisms (Fig. 10.2). Extracellular chelation by excreted ligands, such as citrate and oxalate, as well as cell-wall binding or an enhanced efflux reduce the amount of heavy metals in the cell. Intracellular chelation by metallothioneins/phytochelatins or

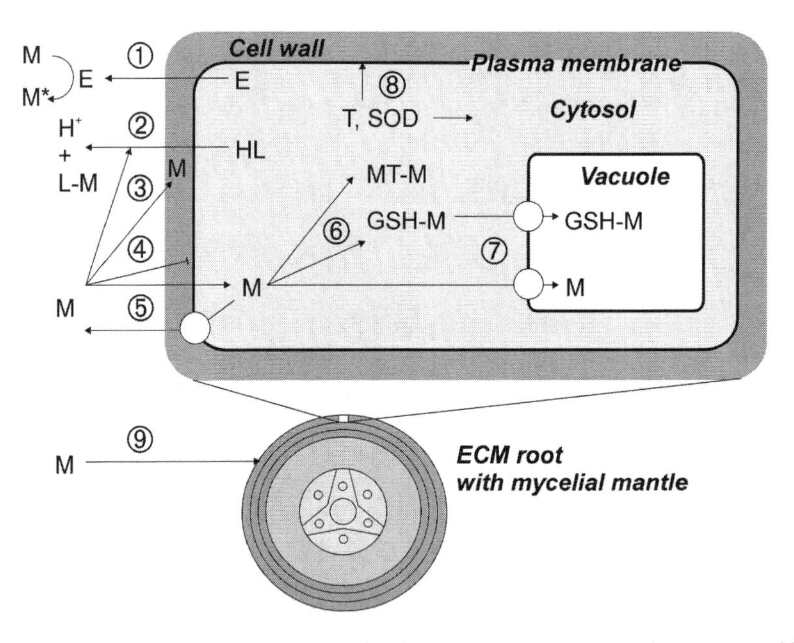

Fig. 10.2 Potentially involved elements of heavy metal stress response in ectomycorrhiza: (1) extracellular modification of metal ions (M) by excreted enzymes (E), (2) extracellular chelation by excreted ligands (L) like low molecular weight organic acids, which may reduce extracellular pH, (3) cell wall binding, (4) reduced influx across plasma membrane, (5) enhanced efflux, (6) intracellular chelation by metallothioneins (MT) or glutathione (GSH), (7) transport into subcellular compartments like the vacuole, (8) protection against toxic metal-induced oxidative stress by thioredoxins (T) and superoxide dismutases (SOD) and (9) filter function of the mycelial mantle (modified after Bellion et al. 2006)

glutatione and transport into subcellular compartments like the vacuole play a role in metal homeostasis of fungi including ECM fungi (see reviews Bellion et al. 2006; Colpaert 2008).

In natural systems, low molecular weight organic acids (LMWOAs) are involved in a wide range of different processes, from nutrient mobilization, mineral weathering and metal detoxification to wood-decay processes and plant pathogenesis because of their dual acidifying and chelating abilities (Landeweert et al. 2001).

However, LMWOAs like malate, citrate and oxalate represent only a small proportion of the exudates released by fungi, amounting to approx. 0.2% of the total carbon fixed by the tree (van Hees et al. 2006). Laboratory experiments have demonstrated that mycorrhizal tree seedlings often exude more LMWOAs than non-mycorrhizal seedlings (Ahonen-Jonnarth et al. 2000), and both plants and ECM fungi have been shown to have significant effects on the concentration of organic acids in soil solution (Rosling et al. 2004). Some ECM fungi form dense mats of hyphae, in which the pH of the bulk soil solution is lower by more than one pH unit (Cromack et al. 1979). Also, a shift in exudates composition is seen, with nonmycorrhizal seedlings exuding mainly malonate, whereas seedlings colonized by *P. involutus* mainly exuded oxalate (van Scholl et al. 2006). It has been shown that excreted natural compounds may influence the mobility of metals, e.g., Cu, from waste-disposal sites, even under the relatively low nutrient fluxes that dominate subsurface systems (Boult et al. 2006). Fomina et al. (2005) could show that in general, metal-tolerant ericoid and ECM fungi grew and solubilized toxic metal minerals better than non-tolerant isolates. Both oxalic acid and chelation are involved in the dissolution of depleted uranium corrosion products and transformation of metallic uranium into meta-autunite minerals, which are capable of long-term uranium retention (Fomina et al. 2008).

Siderophores are the largest class of compounds known to bind, transport or shuttle Fe (but also other metals like actinides), specifically, Fe which is abundant in soils, but highly insoluble (Dancis et al. 1990). The low-molecular-mass coordination molecules are excreted to aid Fe^{3+} assimilation essential for various cellular processes such as respiration, DNA synthesis and metabolism (Neilands 1995). Because of such metal-binding abilities, there are potential applications for siderophores in medicine, reprocessing of nuclear fuel, bioremediation of metal-contaminated sites and treatment of industrial wastes (Gadd 2010). Siderophores secretion is repressed by high ferric iron concentration and induced by iron starvation (Neilands 1995) and can solubilize minerals by protonating or chelating cations (Landeweert et al. 2001). For bacteria, it has been shown that a high concentration of (heavy) metals overrides the repression of siderophore synthesis by iron (Dimkpa et al. 2008). Whether a similar case is found with (mycorrhizal) fungi remains to be investigated.

In addition to mycorrhizal fungi, associated bacteria produce siderophores and together they promote weathering of iron-bearing silicates such as biotite (Frey-Klett et al. 2007). Rineau et al. (2008) demonstrated that *Lactarius subdulcis* was less efficient than *Xerocomus* sp. for accessing free or complexed iron, but instead produced 100 times more oxalate. These preliminary results open the way to study the contribution of ECM communities to nutrient cycling in forest ecosystems.

Toxic metals may cause oxidative stress and several studies on mycorrhizal fungal responses suggested that the fungi may be able to regulate genes providing protection against reactive oxygen species (ROS). In *Paxillus involutus*, thioredoxins (small heat stable oxidoreductases), superoxide dismutases and strongly induced glutathione synthesis are involved in protection against toxic metal-induced oxidative stress (Bellion et al. 2006; Schützendübel and Polle 2002). Such mechanisms are important components of natural biogeochemical cycles for metals as well as associated elements in biomass, soil, rocks and minerals, e.g., sulfur and phosphorus, and metalloids, actinides and metal radionuclides. ECM interact with metals and minerals altering their physical and chemical state, while metals and minerals are also able to affect microbial growth, activity and survival (Gadd 2010). Metal–mineral–microbe interactions are of key importance within the framework of geomicrobiology and also fundamental to microbial biomineralization processes (Kothe et al. 2005).

10.6 Role of Mycelium in Heavy Metal Stress Alleviation

More than 95% of ECM trees are ensheathed by a mycelial mantle around the short root. Nutrients, water and elements released from the soil to the plant are transported by the fungus, including ECM fungi rather than root tips responsible for contact to soil and nutrient equisition (Smith and Read 2008). As shown in Fig. 10.1, the mycelial mantle can play a role as a filter accumulating heavy metals and protecting plants from toxic effects.

Fungal hyphae of extramatrical mycelium replace root hairs in the symbiosis. They are more effective in exploring and penetrating the pores in the soil, because they are, with only approx. 2–6 μm, smaller and much longer than root hairs. Furthermore, hyphae are often ramified exploiting large areas of soil. At the same time, the small diameter allows access to water and nutrients in pores inaccessible for roots and root hairs.

The abundance of the extramatrical mycelium was shown to be important for heavy metal (HM) binding by the fungus. Most of the HMs were demonstrated to be bound to cell wall components like chitin, cellulose, cellulose derivatives, and melanin. The chemical nature of HM binding within the fungal cells is less clear. Polyphosphate granula, which were proposed to have this function, seem to be artifacts of specimen preparation. The high N concentrations associated with the polyphosphate granules rather indicate the occurrence of HM-thiolate binding by metallothionein-like peptides (Galli et al. 1994). Concentrations of heavy metals were usually found to be little altered in roots of mycorrhizal birch, pine and spruce, but were high in extramatrical hyphae of the symbionts *Amanita*, *Paxillus*, *Pisolithus*, *Rhizopogon*, *Scleroderma* and *Suillus* ssp. (Wilkins 1991). The development and differentiation of the extramatrical mycelium may represent important predictive features relevant to the ecological classification of ECM (Agerer 2001).

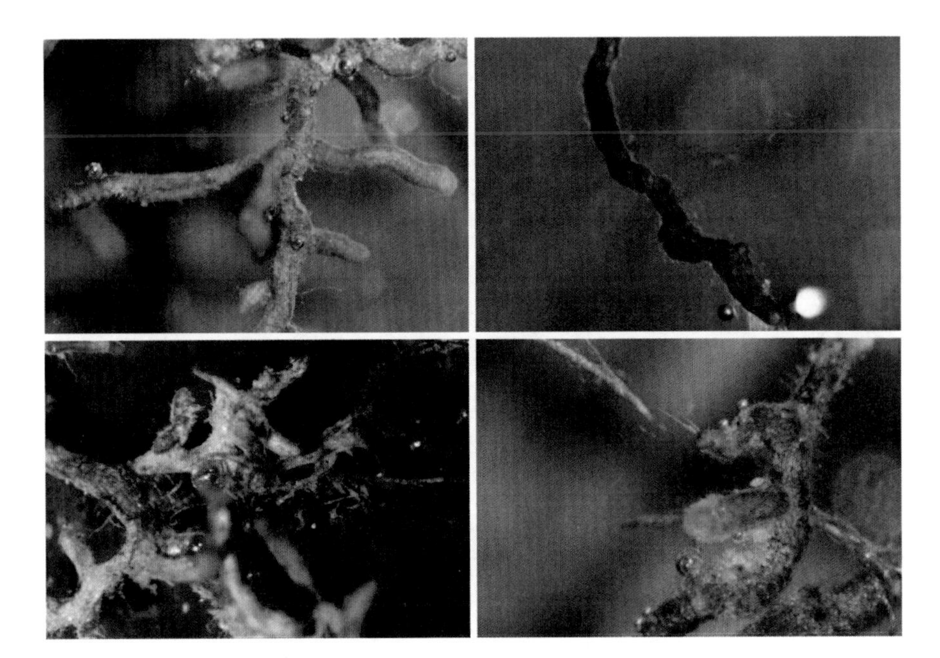

Fig. 10.3 Mycorrhizae on oak roots at a former uranium site showed predominantly long-distance exploration types

This goes along with the finding that in a former uranium mining area most mycorrhizae were long-rooted and showed curved growth and belonged to the long-distance exploration type of mycorrhizae. Rhizomorphs were dominant, and sometimes the hyphal mantle was lacking likely induced by toxic effects of heavy metals (Fig. 10.3; Iordache et al. 2009). Long-distance ECM fungi can alleviate low nutrient stress by scavenging a large area for nutrients while avoiding hot spots of contamination with a well adapted transport system. Mycorrhizal networks (MNs), or the mycorrhizal fungal mycelia that connect two or more plants, are recognized increasingly as mediators of interactions among trees through their effects on tree survival, growth and competitive abilities (Selosse et al. 2006; Simard and Durall 2004). MNs provide a source of mycorrhizal fungal inoculum for establishing seedlings (Nara 2006a), and a potential conduit for interplant transfer of water, carbon and nutrients (Smith and Read 2008).

10.7 Heavy Metal Effects on Transcriptomics of ECM Fungi

Detoxification mechanisms in the mycobiont are probably achieved via several physiological processes rather than through a single mechanism (Hartley et al. 1997). This is seen in studies with EST (expressed sequence tags) data of Cd-treated mycelium of *Paxillus involutus* (Jacob et al. 2004) and microarrays of

Table 10.1 Differential gene expression was observed for gene fragments by Northern blot analyses after growth of the ECM fungus *Tricholoma vaccinum* in 0.5 mM $NiCl_2$-containing medium

Accession no.	Function	Expression (%)
DQ267672	Aldehyde dehydrogenase	87.4
DQ267681	Alcohol dehydrogenase	112.4
DQ267677	APS kinase	100.0
DQ267673	MATE transporter	157.1
DQ267674	MATE transporter	138.4
DQ267676	β-1,4-Glucosidase	115.8
DQ267679	*Gypsy*-like retrotransposon	113.0
AY048578	Hydrophobin	108.7
DQ267682	Lys, Arg and Orn binding protein	96.3
DQ267675	Phosphatase	117.3
DQ267678	Ubiquitin binding enzyme-like MMS2	138.1

Cadophora finlandica (Gorfer et al. 2009) where expression changes are observed for a broad range of genes involved in transport processes, in nutrition, signaling and in stress response. These results were confirmed with a study using fungal probes from *Tricholoma vaccinum–Picea abies* interaction of differential display analyses (Krause and Kothe 2006) and hydrophobin *hyd1* (Mankel et al. 2002). Northern blot analyses were performed with cDNA of *T. vaccinum* grown in MMNb-medium with 0.5 mM $NiCl_2$ as labeled probe (Table 10.1). Among the genes tested, a MATE transporter, a sequence of ubiquitin-conjugating enzyme variant MMS2 and some hypothetical proteins showed the highest expression in $NiCl_2$-containing medium, for others a slight change was observed.

The genomic sequences of the ECM fungi *Laccaria bicolor* and *Tuber melanosporum* have been recently released (Martin et al. 2008, 2010). This allows and will force more detailed expression analyses.

10.8 Effect of Metals on ECM Communities

Major factors responsible for patterns of successional stages and distribution of ECM roots are forest management or metal contamination. This is reflected in discussion in terms of managing systems in order to maximize tree growth and health while effectively restoring soil water balance.

Severe Zn pollution can trigger the evolution of an increased Zn tolerance in ECM fungi as was shown by Colpaert et al. (2004) and Adriaensen et al. (2006) for *Suillus* isolates. However, the influence of abiotic and biotic soil parameters on the spatial structure of ECM communities has only been fragmentarily analyzed to date, and the results are inconclusive (Gebhardt et al. 2007).

The colonization of ECM is completely different in secondary successions where tree seedlings rapidly recruit ECM fungi, more often specialists, from dormant spore banks or other resistant propagules (Izzo et al. 2005; Nara 2006b). Mycorrhizal fungi that colonize podzolic acidic soils can be exposed to high levels of toxic metals such as Al, Fe, and Mn. For a successful symbiosis, both partners must be able to withstand the metal toxicity during all stages of colonization.

We assessed the distribution of ECM in a 15-year-old oak (*Quercus robur*) mixed forest on podzol with acidic pH of 2.85–3.40 near Greiz, Germany. The results show that the observed abundance of ECM types was lower than on alkaline or slightly acidic substrates. The statistical analysis by canonical correspondence analysis indicated that a group of heavy metals, including Al, Cr, Fe, and Pb, strongly contributes to the reduced abundance of ECM fungi *Laccaria amethystina*, *Paxillus involutus*, or brown rhizomorphs. *Russula ochroleuca*, *Lactarius quietus*, and *Pisolithus tinctorius* were in sharp disagreement with these heavy metals but correlated positively with Mn, Cd, Zn, Sr and with the essential element P. *Cenococcum geophilum* was strongly correlated with Ni, Co, and Cd, followed by a cluster of Fe, Pb, and Cu, and also by the essential elements P and S. These findings indicate that the toxicity of heavy metals may be ameliorated by the availability of P. At the same time, not a single element is responsible for the pattern of ECM fungi distribution, but rather multiple stressors determine ECM community structure.

The succession of ectomycorrhizae in contaminated areas was recently analyzed in a more critical manner and an improved theoretical and methodological framework for research on ECM succession was developed taking into account the hierarchical structure of the ecosystem (Iordache et al. 2009, 2011). This led to association of certain fungi with primary succession, while others are dominating secondary succession, as introduced above. It also became obvious that the exploration types should be considered for description of ECM community structure on metal-contaminated soils.

10.9 Application in Phytoremediation

Mycorrhiza in association with fast growing trees is a very useful tool in phytoremediation which has been proposed as an environmentally beneficial and cost-efficient treatment technique for the remediation of heavy metal (HM)-contaminated sites in recent years. Such phytoextraction strategies necessitate tolerance and high HM accumulation of the mycorrhiza, because the aim is uptake and concentration of metals from the contaminated environment into harvestable plant biomass. In contrast, phytostabilization may profit from using metal-tolerant mycorrhiza with low HM accumulation, thus focusing on long-term stabilization and containment of the pollutant, without introduction into harvestable biomass. Such an approach would, e.g., allow for production of lignocelluloses for bioethanol or wood for heat production.

AM (e.g., Joner and Leyval 2001; Bissonnette et al. 2010) and ECM fungi (Marschner et al. 1996; Baum et al. 2006) were investigated to assess the potential of phytoremediation for HM-contaminated soils. An increased uptake of Cd into the leaves of *Populus canadensis* in association with *Paxillus involutus* was reported by Sell et al. (2005), whereas Baum et al. (2006) identified the increased biomass production as the main reason for the increased phytoextraction of Cd, Zn, and Cu in the *Salix* X *dasyclados–P. involutus* interaction. These reports show that despite of the filter function described, mycorrhizal plants can accumulate higher amounts of HM. In few cases fungi may enhance toxicity for plants (Godbold 1994), or no significant effect on HM extraction is seen (Bissonnette et al. 2010). Phytostabilization is often the main benefit of using trees on polluted sites, because they reduce erosion by wind and water erosion and the leaf fall, dead tree roots and root exudates add significantly to increase soil organic matter, promoting nutrient cycling (Pulford and Watson 2003). The positive effect of mycorrhiza to the plant health – especially in degraded and contaminated soils – can be used by companies producing and setting AM and ECM fungi and mycorrhized trees, which can be used in large-scale re-forestation programs on degraded and metal-containing soils.

10.10 Conclusions

ECM fungi are known to affect soil bioavailability and plant uptake of heavy metals (Jentschke and Godbold 2000; Rineau et al. 2008). Therefore, they are predicted to play an important role in phytoremediation involving mycorrhizal hosts (Perotto and Martino 2001). As reviewed in detail by Colpaert (2008), many woody plants may not be considered primary colonizers of metal-polluted soils. However, a number of pioneer tree species, such as willows, poplars, birches, and pines, are able to form pioneer populations. In trees with their long reproductive cycles, the adaptive potential for metal tolerance seems to be low (Meharg and Cairney 2000) and even if there is a selection for individuals with a higher tolerance, it may take many decades before a reasonable tolerant population establishes. Therefore, woody pioneers rely much more on their ECM fungi than herbaceous pioneer plants would rely on their AM mycobionts, irrespective of soil pollution (Ashkannejhad and Horton 2006).

After severe disturbances, when mycorrhizal propagules are scarce, ECM plants are slow colonizers. Therefore, it seems specifically advisable for re-forestation programs to add ECM fungal inoculums to help establish trees that resist extreme metal toxicity through a large phenotypic plasticity as well as by their association with a small guild of well-adapted ECM fungi (Wilkinson and Dickinson 1995; Colpaert 2008). The presence of ECM fungi in (former) mining sites suggests that they play a role for much underestimated heavy metal uptake or avoidance of the plants. In addition to their role as symbionts providing water and nutrients, fungal immobilization of heavy metals could be a potential mechanism by which mycorrhizal fungi alleviate metal toxicity to their hosts (Joner and Leyval 1997).

However, interactions are very complex as they vary in a fungal species and metal-specific manner: Cd, e.g., was almost exclusively localized in fungal cell walls of the Hartig' net, whereas Zn was accumulated in both cell walls and the fungal cytoplasm (Frey et al. 2000); suilloid mycorrhizae significantly enriched Pb and Zn in their mantle, although no such biofiltering effect was observed for mycorrhizae formed by other taxa (Turnau et al. 2002).

In conclusion, fungi have capabilities for the disposal of recalcitrant soil contaminants and their fruitbodies are often found in mining areas indicating potential metal resistance trails. Increased attention to ecological interactions in soil could therefore reduce costs and use of improving the efficacy of restoring a vegetation cover to land impacted by heavy metals or other disturbances. This is not only true for grassland ecosystems relaying on AM fungi, but also for forests with specific, well-adapted ECM communities.

References

Adriaensen K, Vangronsveld J, Colpaert JV (2006) Zinc-tolerant *Suillus bovinus* improves growth of Zn-exposed *Pinus sylvestris* seedlings. Mycorrhiza 16:553–558

Agerer R (2001) Exploration types of ectomycorrhizae – a proposal to classify ectomycorrhizal mycelial systems according to their patterns of differentiation and putative ecological importance. Mycorrhiza 11:107–114

Agerer R (2006) Colour atlas of ectomycorrhizae. Einhorn-Verlag, Schwäbisch Gmünd, Germany

Ahonen-Jonnarth U, van Hees PAW, Lundstrom US, Finlay RD (2000) Organic acids produced by mycorrhizal *Pinus sylvestris* exposed to elevated aluminium and heavy metal concentrations. New Phytol 146:557–567

Ashkannejhad S, Horton TR (2006) Ectomycorrhizal ecology under primary succession on coastal sand dunes: interactions involving *Pinus contorta*, suilloid fungi and deer. New Phytol 169:345–354

Baar J, Horton TR, Kretzer AM, Bruns TD (1999) Mycorrhizal colonization of *Pinus muricata* from resistant propagules after a stand-replacing wildfire. New Phytol 143:409–418

Baum C, Hrynkiewicz K, Leinweber P, Meißner R (2006) Heavy-metal mobilization and uptake by mycorrhizal and nonmycorrhizal willows (*Salix* x *dasyclados*). J Plant Nutr Soil Sci 169:516–522

Bellion M, Courbot M, Jacob C, Blaudez D, Chalot M (2006) Extracellular and cellular mechanisms sustaining metal tolerance in ectomycorrhizal fungi. FEMS Microbiol Lett 254 (2):173–181

Bissonnette L, St-Arnaud M, Labrecque M (2010) Phytoextraction of heavy metals by two *Salicaceae clones* in symbiosis with arbuscular mycorrhizal fungi during the second year of a field trial. Plant Soil 332:55–67

Boult S, Hand VL, Vaughan DJ (2006) Microbial controls on metal mobility under the low nutrient fluxes found throughout the subsurface. Sci Total Environ 372(1):299–305

Brundrett MC (2009) Mycorrhizal associations and other means of nutrition of vascular plants: understnding global diversity of host plants by resolving conflicting information and developing reliable means of diagnosis. Plant Soil 320:37–77

Colpaert JV (2008) Heavy metal pollution and genetic adaptations in ectomycorrhizal fungi. In: Avery S, Stratford M, van West P (eds) Stress in yeasts and filamentous fungi. Elsevier, Amsterdam, pp 157–173

Colpaert JV, Muller LAH, Lambaerts M, Adriaensen K, Vangronsveld J (2004) Evolutionary adaptation to Zn toxicity in populations of *Suilloid* fungi. New Phytol 162:549–559

Courty PE, Pritsch K, Schloter M, Hartmann A, Garbaye J (2005) Activity profiling of ectomycorrhiza communities in two forest soils using multiple enzymatic tests. New Phytol 167:309–319

Cripps CL (2004) Ectomycorrhizal fungi above and below ground in a small, isolated aspen stand: a simple system reveals fungal fruiting strategies and an edge effect. In: Cripps (ed) Fungi in Forest Ecosystems: systematics, diversity and ecology, New York Botanical Garden Press, New York, pp 249–265

Cromack K; Sollins P; Graustein WC; Speidel, K, Todd, AW, Spycher, G, Li CY, Todd RL (1979) Calcium-oxalate accumulation and soil weathering in mats of the hypogeous fungus *Hysterangium-crassum*. Soil Biol Biochem 11(5):463–468

Dancis A, Klausner RD, Hinnesbuch AG, Barriocanal JG (1990) Genetic evidence that ferric reductase is required for iron uptake in *Saccharomyces cerevisiae*. Mol Cell Biol 10: 2294–2301

Dimkpa CO, Svatoš A, Dabrowska P, Schmidt A, Boland W, Kothe E (2008) Involvement of siderophores in the reduction of metal-induced inhibition of auxin synthesis in *Streptomyces* spp. Chemosphere 74:19–25

Finlay RD (2008) Ecological aspects of mycorrhizal symbiosis: with special emphasis on the functional diversity of interactions involving the extraradical mycelium. J Exp Bot 59(5):1115–1126

Fiore-Donno AM, Martin F (2001) Populations of ectomycorrhizal *Laccaria amethystina* and *Xerocomus* spp. show contrasting colonization patterns in a mixed forest. New Phytol 152(3): 533–542

Fomina MA, Alexander IJ, Colpaert JV, Gadd GM (2005) Solubilization of toxic metal minerals and metal tolerance of mycorrhizal fungi. Soil Biol Biochem 37:851–866

Fomina M, Charnock JM, Hillier S, Alvarez R, Livens F, Gadd GM (2008) Role of fungi in the biogeochemical fate of depleted uranium. Curr Biol 18(9):R357–R377

Frey B, Zierold K, Brunner I (2000) Extracellular complexation of Cd in the Hartig net and cytosolic Zn sequestration in the fungal mantle of *Picea abies–Hebeloma crustuliniforme* ectomycorrhizas. Plant Cell Environ 23:1257–1266

Frey-Klett P, Garbaye J, Tarkka M (2007) The mycorrhiza helper bacteria revisited. New Phytol 176(1):22–36

Gadd GM (1993) Interactions of fungi with toxic metals. New Phytol 124:25–60

Gadd GM (2010) Metals, minerals and microbes: geomicrobiology and bioremediation. Microbiology 156:609–643

Galli U, Schüepp H, Brunold C (1994) Heavy metal binding by mycorrhizal fungi. Physiol Plant 92:364–368

Gardes M, Bruns TD (1993) ITS primers with enhanced specificity for basidiomycetes: application to the identificcation of mycorrhizae and rusts. Mol Ecol 2:113–118

Gardes M, Bruns TD (1996) Community structure of ectomycorrhizal fungi in a Pinus muricata forest: Above- and below-ground views. Can J Bot 74:1572–1583

Gebhardt S, Neubert K, Wöllecke J, Münzenberger B, Hüttl RF (2007) Ectomycorrhiza communities of red oak (*Quercus rubra* L.) of different age in the Lusatian lignite mining district, East Germany. Mycorrhiza 17:279–290

Godbold DL (1994) Aluminium and heavy metal stress: from the rhizosphere to the whole plant. In: Gobold DL, Hüttermann A (eds) Effects of acid rain on forest processes. Wiley-Liss, New York, pp 231–264

Gorfer M, Persak H, Berger H, Brynda S, Bandian D, Strauss J (2009) Identification of heavy metal regulated genes from the root associated ascomycete *Cadophora finlandica* using a genomic microarray. Mycol Res 113:1377–1388

Gryta H, Debaud JC, Marmeisse R (2001) Population dynamics of the symbiotic mushroom *Hebeloma cylindrosporum*: mycelial persistence and inbreeding. Heredity 84:294–302

Gryta H, Debaud JC, Effosse A, Gay G, Marmeisse R (1997) Fine-scale structure of populations of the ectomycorrhizal fungus *Hebeloma cylindrosporum* in coastal sand dune forest ecosystems. Mol Ecol 6(4):353–364

Hartley J, Cairney JWG, Meharg AA (1997) Do ectomycorrhizal fungi exhibit adaptive tolerance to potentially toxic metals in the environment? Plant Soil 189:303–319

Hogberg N, Guidot A, Jonsson M, Dahlberg A (2009) Microsatellite markers for the ectomycorrhizal basidiomycete *Lactarius mammosus*. Mol Ecol Resour 9(3):1008–1010

Horton TR, Bruns TD (1998) Multiple host fungi are the most frequent and abundant ectomycorrhizal types in a mixed stand of Douglas fir (*Pseudotsuga menziesii*) and bishop pine (*Pinus muricata*). New Phytol 139:331–339

Iordache V, Gherghel F, Kothe E (2009) Assessing the effect of disturbances on ectomycorrhiza diversity. Int J Environ Res Public Health 6:416–422

Iordache V, Kothe E, Neagoe A, Gherghel F (2011) A conceptual framework for up-scaling ecological processes and application to ectomycorrhizal fungi. In: Rai M, Varma A (eds) Diversity and biotechnology of ectomycorrhizae. Springer, Berlin, pp 255–300

Izzo A, Agbowo J, Bruns TD (2005) Detection of plot-level changes in ectomycorrhizal communities across years in an old-growth mixed-conifer forest. New Phytol 166:619–630

Jacob C, Courbot ML, Martin F, Brun A, Chalot M (2004) Transcriptomic responses to cadmium in the ectomycorrhizal fungus *Paxillus involutus*. FEBS Lett 576(3):423–427

Jentschke G, Godbold DL (2000) Metal toxicity and ectomycorrhizas. Physiol Plant 109:107–116

Joner EJ, Leyval C (1997) Uptake of 109Cd by roots and hyphae of a *Glomus mossae/Trifolium subterraneum* mycorrhiza from soil amended with high and low concentrations of cadmium. New Phytol 135:353–360

Joner EJ, Leyval C (2001) Time-course of heavy metal uptake in maize and clover as affected by root density and different mycorrhizal inoculation regimes. Biol Fertil Soils 33(5):351–357

Jones MD, Hutchinson TC (1986) The effect of mycorrhizal infection on the response of *Betula papyrifera* to nickel and copper. New Phytol 102:429–442

Jumpponen A, Jones KL (2010) Massively parallel 454 sequencing indicates hyperdiverse fungal communities in temperate *Quercus macrocarpa* phyllosphere. New Phytol 184:438–448

Kalač P, Burda J, Staskova I (1991) Concentrations of lead, cadmium, mercury and copper in mushrooms in the vicinity of a lead smelter. Sci Total Environ 105:109–119

Kõljalg U, Larsson KH, Abarenkov K, Nilsson RH, Alexander I, Eberhardt U, Erland S, Høiland K, Kjøller R, Larsson E, Pennanen T, Sen R, Taylor AFS, Tedersoo L, Vrålstad T, Ursing BM (2005) UNITE: a database providing web-based methods for the molecular identification of ectomycorrhizal fungi. New Phytol 166:1063–1068

Kothe E, Bergmann H, Buchel G (2005) Molecular mechanisms in bio-geo-interactions: from a case study to general mechanisms. Chem Erde 65(S1):7–27

Krause K, Kothe E (2006) Use of RNA fingerprinting to identify fungal genes specifically expressed during ectomycorrhizal interaction. J Basic Microbiol 46(5):387–399

Kretzer AM, Dunham S, Molina R, Spatafora JW (2003) Microsatellite markers reveal the below ground distribution of genets in two species of *Rhizopogon* forming tuberculate ectomycorrhizas on Douglas fir. New Phytol 161:313–320

Krpata D, Peintner U, Langer I, Fitz JW, Schweiger P (2008) Ectomycorrhizal communities associated with *Populus tremula* growing on a heavy metal contaminated site. Mycol Res 112: 106–1079

Landeweert R, Hoffland E, Finlay RD, Kuyper TW, van Breemen N (2001) Linking plants to rocks: ectomycorrhizal fungi mobilize nutrients from minerals. Trends Ecol Evol 16:248–254

Mankel A, Krause K, Kothe E (2002) Identification of a hydrophobin gene that is developmentally regulated in the ectomycorrhizal fungus *Tricholoma terreum*. Appl Environ Microbiol 68(3): 1408–1413

Margulies M, Egholm M, Altaian WE, Attiya S, Bader JS, Bemben LA, Berka J, Braverman MS, Chen YJ, Chen Z, Dewell SB, Du L, Fierro JM, Gomes XV, Godwin BC, He W, Helgesen S, Ho CH, Irzyk GP, Jando SC, Alenquer ML, Jarvie TP, Jirage KB, Kim JB, Knight JR, Lanza JR, Leamon JH, Lefkowitz SM, Lei M, Li J, Lohman KL, Lu H, Makhijani VB, McDade KE, McKenna MP, Myers EW, Nickerson E, Nobile JR, Plant R, Puc BP, Ronan MT, Roth GT, Sarkis GJ, Simons JF, Simpson JW, Srinivasan M, Tartaro KR, Tomasz A, Vogt KA, Volkmer GA, Wang SH, Wang Y, Weiner MP, Yu P, Begley RF, Rothberg JM (2005) Genome sequencing in microfabricated high-density picolitre reactors. Nature 437:376–380

Marschner P, Godbold DL, Jentschke G (1996) Dynamics of lead accumulation in mycorrhizal and non-mycorrhizal Norway spruce (*Picea abies* (L) Karst). Plant Soil 178(2):239–245

Martin F, Aerts A, Ahren D, Brun A, Danchin EGJ, Duchaussoy F, Gibon J, Kohler A, Lindquist E, Pereda V, Salamov A, Shapiro HJ, Wuyts J, Blaudez D, Buee M, Brokstein P, Canback B, Cohen D, Courty PE, Coutinho PM, Delaruelle C, Detter JC, Deveau A, DiFazio S, Duplessis S, Fraissinet-Tachet L, Lucic E, Frey-Klett P, Fourrey C, Feussner I, Gay G, Grimwood J, Hoegger PJ, Jain P, Kilaru S, Labbe J, Lin YC, Legue V, Le Tacon F, Marmeisse R, Melayah D, Montanini B, Muratet M, Nehls U, Niculita-Hirzel H, Oudot-Le Secq MP, Peter M, Quesneville H, Rajashekar B, Reich M, Rouhier N, Schmutz J, Yin T, Chalot M, Henrissat B, Kues U, Lucas S, Van de Peer Y, Podila GK, Polle A, Pukkila PJ, Richardson PM, Rouze P, Sanders IR, Stajich JE, Tunlid A, Tuskan G, Grigoriev IV (2008) The genome of *Laccaria bicolor* provides insights into mycorrhizal symbiosis. Nature 452(7183):88–92

Martin F, Kohler A, Murat C, Balestrini R, Coutinho PM, Jaillon O, Montanini B, Morin E, Noel B, Percudani R, Porcel B, Rubini A, Amicucci A, Amselem J, Anthouard V, Arcioni S, Artiguenave F, AuryJ-M BP, Bolchi A, Brenna A, Brun A, Buée M, Cantarel B, Chevalier G, Couloux A, Da Silva C, Denoeud F, Duplessis S, Ghignone S, Hilselberger B, Iotti M, Marçais B, Mello A, Miranda M, Pacioni G, Quesneville H, Riccioni C, Ruotolo R, Splivallo R, Stocchi V, Tisserant E, Viscomi AR, Zambonelli A, Zampieri E, Henrissat B, Lebrun M-H, Paolocci F, Bonfante P, Ottonello S, Wincker P (2010) Périgord black truffle genome uncovers evolutionary origins and mechanisms of symbiosis. Nature 464:1033–1038

Meharg AA (2003) The mechanistic basis of interactions between mycorrhizal associations and toxic metal cations. Mycol Res 107:1253–1265

Meharg AA, Cairney JWG (2000) Co-evolution of mycorrhizal symbionts and their hosts to metal-contaminated environments. Adv Ecol Res 30:69–112

Mleczko P (2004) Mycorrhizal and saprobic macrofungi of two zinc wastes in southern Poland. Acta Biol Cracov Ser Bot 46:25–38

Muehlmann O, Bacher M, Peintner U (2008) *Polygonum viviparum* mycobionts on an alpine primary successional glacier forefront. Mycorrhiza 18:87–95

Nara K (2006a) Ectomycorrhizal networks and seedling establishment during early primary succession. New Phytol 169:169–178

Nara K (2006b) Pioneer dwarf willow may facilitate tree succession by providing late colonizers with compatible ectomycorrhizal fungi in a primary successional volcanic desert. New Phytol 171:187–198

Nara K, Nakaya H, Wu BY, Zhou ZH, Hogetsu T (2003) Underground primary succession of ectomycorrhizal fungi in a volcanic desert on Mount Fuji. New Phytol 159:743–756

Neilands JB (1995) Siderophores – structure and function of microbial iron transport compounds. J Biol Chem 270:26723–26726

Perotto S, Martino E (2001) Molecular and cellular mechanisms of heavy metal tolerance in mycorrhizal fungi: what perspectives for bioremediation? Minerva Biotechnologica 13:55–63

Plassard C, Fransson P (2009) Regulation of low molecular weight organic acid production in fungi. Fungal Biol Rev 23:30–39

Pritsch K, Munch JC, Buscot F (1997) Morphological and anatomical characterisation of black alder *Alnus glutinosa* (L.) Gaertn. Ectomycorrhizas. Mycorrhiza 7(4):201–216

Pulford ID, Watson C (2003) Phytoremediation of heavy metal-contaminated land by trees – a review. Environ Int 29:529–540

Rajala T (2008) Responses of soil microbial communities to clonal variation of Norway spruce. Dissertation, University of Helsinki, Finland

Rineau F, Courty PE, Uroz S, Buée M, Garbaye J (2008) Simple microplate assays to measure iron mobilization and oxalate secretion by ectomycorrhizal tree roots. Soil Biol Biochem 40: 2460–2463

Rosling A, Lindahl BD, Taylor AFS, Finlay RD (2004) Mycelial growth and substrate acidification of ectomycorrhizal fungi in response to different minerals. FEMS Microbiol Ecol 47(1): 31–37

Rudawska M, Leski T, Trocha LK, Gornowicz R (2006) Ectomycorrhizal status of Norway spruce seedlings from bare-root forest nurseries. Ecol Manag 236:375–384
Rühling A, Söderström B (1990) Changes in fruitbody production of mycorrhizal and litter decomposing macromycetes in heavy metal polluted coniferous forests in north Sweden. Water Air Soil Pollut 49:375–387
Rühling A, Bååth E, Nordgren A, Söderström B (1984) Fungi in metal-contaminated soil near the Gusum brass mill, Sweden. Ambio 13:34–36
Schuster SC (2008) Next-generation sequencing transforms today's biology. Nat Methods 5:16–18
Schützendübel A, Polle A (2002) Plant responses to abiotic stresses: heavy metal-induced oxidative stress and protection by mycorrhization. J Exp Bot 53:1351–1365
Sell J, Kayser A, Schulin R, Brunner I (2005) Contribution of ectomacorrhizal fungi to cadmium uptake of poplars and willows from a heavily polluted soil. Plant Soil 277:245–253
Selosse MA, Richard F, He X, Simard SW (2006) Mycorrhizal networks: des liaisons dangereuses? Trends Ecol Evol 21:621–628
Simard SW, Durall DM (2004) Mycorrhizal networks: a review of their extent, function and importance. Can J Bot 82(8):1140–1165
Smith SE, Read DJ (1997) Mycorrhizal symbiosis. Academic, London
Smith SE, Read DJ (2008) Mycorrhizal symbiosis, 3rd edn. Academic, London
Staudenrausch S, Kaldorf M, Renker C, Luis P, Buscot F (2005) Diversity of the ectomycorrhiza community at a uranium mining heap. Biol Fertil Soils 41:439–446
Tedersoo L, Kõljalg U, Hallenberg N, Larsson KH (2003) Fine scale distribution of ectomycorrhizal fungi and roots across substrate layers including coarse woody debris in a mixed forest. New Phytol 159:153–165
Tedersoo L, Suvi T, Larsson E, Koljalg U (2006) Diversity and community structure of ectomycorrhizal fungi in a wooded meadow. Mycol Res 110:734–748
Tedersoo L, May TW, Smith ME (2010) Ectomycorrhizal lifestyle in fungi: global diversity, distribution, and evolution of phylogenetic lineages. Mycorrhiza 20(4):217–263
Trowbridge J, Jumpponen A (2004) Fungal colonization of shrub willow roots at the forefront of a receding glacier. Mycorrhiza 14:283–293
Turnau K, Gucwa E, Mleczko P, Godzik B (1988) Metal content in fruit-bodies and mycorrhizas of *Pisolithus arrhizus* from zinc wastes in Poland. Acta Mycologica 33:59–67
Turnau K, Mleczko P, Blaudez D, Chalot M, Botton B (2002) Heavy metal binding properties of *Pinus sylvestris* mycorrhizas from industrial wastes. Acta Societatis Botanicorum Poloniae 71:253–261
van Hees PAW, Rosling A, Lundstrom US, Finlay RD (2006) The biogeochemical impact of ectomycorrhizal conifers on major soil elements (Al, Fe, K and Si). Geoderma 136(1–2): 364–377
van Scholl L, Hoffland E, van Breemen N (2006) Organic anion exudation by ectomycorrhizal fungi and Pinus sylvestris in response to nutrient deficiencies. New Phytol 170(1):153–163
van Scholl L, Kuyper TW, Smits MM, Landeweert R, Hoffland E, van Breemen N (2008) Rock-eating mycorrhizas: their role in plant nutrition and biogeochemical cycles. Plant Soil 303: 35–47
Vrålstad T, Myhre E, Schumacher T (2002) Molecular diversity and phylogenetic affinities of symbiotic root-associated ascomycetes of the *Helotiales* in burnt and metal polluted habitats. New Phytol 155:131–148
Wilkins DA (1991) The influence of sheathing (ecto-) mycorrhizas of trees on the uptake and toxicity of metals. Agr Ecosyst Environ 35:245–260
Wilkinson DM, Dickinson NM (1995) Metal resistance in trees: the role of mycorrhizae. Oikos 72:298–300

Chapter 11
Historic Copper Spoil Heaps in Salzburg/Austria: Geology, Mining History, Aspects of Soil Chemistry and Vegetation

Wolfram Adlassnig, Stefan Wernitznig, and Irene K. Lichtscheidl

11.1 Introduction

Within the Austrian province of Salzburg, heavy metal (defined according to Appenroth 2010) ores are found in two areas, in the Greywacke zone between the Central and the Northern Alps, and in the Tauernfenster within the Central Alps. In both regions, deposits have been exploited in historical or even prehistorical times. In the Tauernfenster, mining goes back to medieval times, in the Greywacke zone even to the early Bronce Age. All mining districts were closed long ago, partly even in pre-Christian times but spoil heaps and other remnants are still visible and carry characteristic vegetation (for an overview, compare Punz 2008). Therefore, the historical mining districts in Salzburg offer a possibility to study the ecological effects of metal contamination on a different time scale compared to the majority of sites that go back to industrial times only.

This study focuses on three historical with copper contaminated ecosystems. Their geographical location is shown in Fig. 11.1.

a. Schwarzwand: The Grossarl valley cuts through a large ore body and thus hosts a series of Cu deposits (Derkmann 1976). At the Schwarzwand (about 1,550–1,700 m. a.s.l.) next to the village of Hüttschlag, ore veins were exploited on a steep slope until about 1,800. Spoil heap material and Cu-rich mine drainage created an area below the mine, which was gradually covered by blue precipitations rich in Cu (Fig. 11.2). The surrounding vegetation is a subalpine forest consisting of *Picea abies* and *Larix decidua*; on the Schwarzwand itself, only shrubs, herbs, lichens, and mosses are found.

W. Adlassnig (✉) • S. Wernitznig • I.K. Lichtscheidl
Core facility Cell Imaging and Ultrastructure Research, University of Vienna, Althanstrasse 14, Vienna A-1090, Austria
e-mail: wolfram.adlassnig@univie.ac.at

E. Kothe and A. Varma (eds.), *Bio-Geo Interactions in Metal-Contaminated Soils*, Soil Biology 31, DOI 10.1007/978-3-642-23327-2_11,

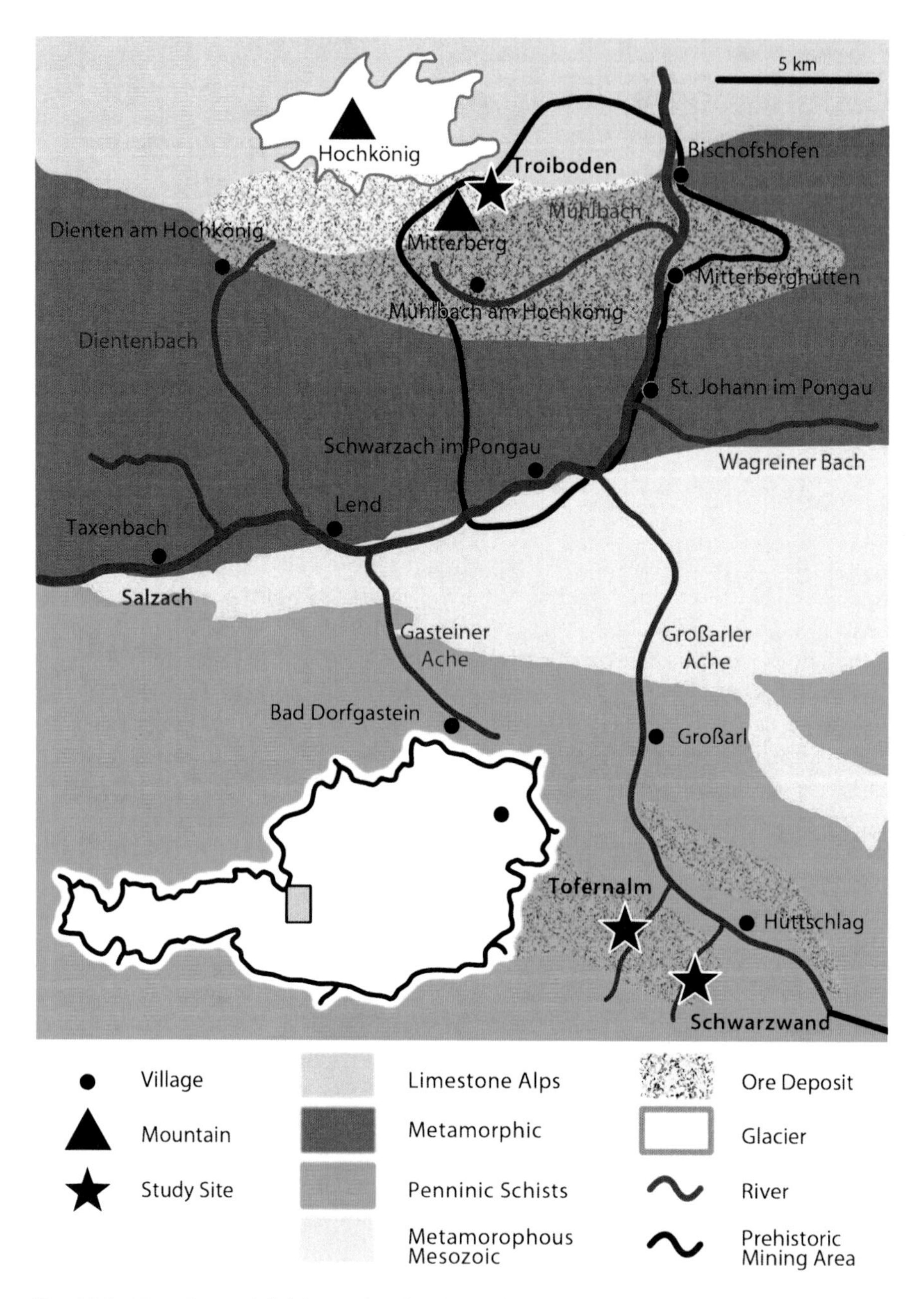

Fig. 11.1 Map of central Salzburg, showing the positions of the Schwarzwand, Tofernalm, and Mitterberg/Hochkönig. The outline of the prehistoric Cu mining district around Mitterberg is shown according to (Stöllner et al. 2006), the outlines of the ore deposits (*red*) are based on (Bernhard 1965; Derkmann 1976), geological background information is from Lechner et al. (1964)

Fig. 11.2 View of the Schwarzwand; the creek precipitating blue copper minerals are clearly visible

b. Tofernalm: At the Tofernalm (1,600 m.a.s.l.), the same ore-bearing layer was exploited as at the Schwarzwand from the sixteenth to the eighteenth century. Today, a spoil heap is the last visible remaining. Dry conditions dominate as mine drainage is lacking. The vegetation differs both from the geological similar habitat at the Schwarzwand and from the surrounding vegetation (Fig. 11.3).
c. Troiboden: The Hochkönig (2,941 m.a.s.l.) is the highest and most prominent peak of the Berchtesgarden Alps. Although the Hochkönig itself consists of limestone (Trias and Dachstein) (Tollmann 1985), the southern foothills around the Mitterberg are ore-bearing greywacke (Bernhard 1965). The greywacke in Salzburg is rich in metal deposits, e.g., Cu, Fe, Ni , U, and Co. Around Mitterberg, Cu was mined in pre-Indoeuropean and Illyrian times (4,000–300 BC), leading to the contamination of several sites in the surroundings. At Troiboden (1,500 m.a.s.l.), after 2,300 years still bare patches without vegetation are found (Fig. 11.4).

In all these mining areas, ore production and the smelting industry have long since come to an end and the local communities have returned to rural farming (Krausmann 2008). As a result of the importance of the area for tourism, environmental conservation has increased. Nevertheless, traces of former mining still can be recognized from their distinct floral diversity: while the original vegetation is characterized by a rich cover of woods and meadows, remnants from heavy metal dumps and smelters are barren. On a close look, plants that have adapted to toxic conditions of such habitats (i.e., heavy metal tolerant plants, "metallophytes") can be found.

Fig. 11.3 View of the Tofernalm; at the left, the spoil heaps and, at their bottom, the swampy area is visible

Fig. 11.4 The Troiboden; the close interlock of bog vegetation and bare smelting places in the background are clearly visible

11.2 Previous Studies

The steep walls of the Grossarl valley host several prominent geological outcrops exhibiting high petrologic diversity. The geology of this area has been studied in detail in spite of the limited economic importance of the ore deposits. Their formation was a matter of long and controversial discussion, a summary was presented by Derkmann (1976).

The moist and cool local climate of the Schwarzwand promotes the development of a rich moss flora; therefore, the first botanists were attracted rather by the bryophytes. In 1815, the Oberbergrath (supreme mining officer) Mathias Mielichhofer described the first copper-tolerant mosses, *Mielichhoferia mielichhoferi* and *M. elongata* from this habitat (Österreichische Akademie der Wissenschaften 1975). After the Second World War, the Institute of Plant Physiology of the University of Vienna made numerous excursions to the Schwarzwand. As a result, the diversity of lichens (Poelt 1955), mosses, and vascular plants (Saukel 1980; Sissolak 1984; Körber-Ulrich 1996) were surveyed. Gams (1966) states that the Schwarzwand is a hotspot of the diversity of copper-resistant lichens and bryophytes. Besides this focus on biodiversity, physiological studies revealed new aspects of Cu tolerance (e.g., Url 1957). For many species of the Schwarzwand, protoplasmatic Cu tolerance and Cu uptake have been quantified (Sissolak 1984; Stummerer 1970; Körber-Ulrich 1996; Url 1956). In some species accumulating significant amounts of Cu, the sites of subcellular deposition have been localized (Hörmann et al. 2001b; Hus 2008). Recent and ongoing studies aim for mapping and modeling Cu distribution within the ecosystem.

The Tofernalm is much smaller than the Schwarzwand and most studies focus on comparing the wet Schwarzwand with the dry Tofernalm, e.g. (Sissolak 1984).

Due to the intensive mining in the nineteenth and twentieth century of the last millennium, the geology and mineralogy of the Mitterberg/Hochkönig area were studied in much detail. Decades of archeological studies on prehistoric mining resulted in a detailed mapping of all preserved spoil heaps (Zschocke and Preuschen 1932) as well as the characterization of slag and spoil heap material (Herdits 1997). Botanical investigations were far less numerous than in the Schwarzwand, and most studies focused on the uptake of Cu into plants (Mutsch 1980; Stummerer 1970).

A classical textbook on the ecology and vegetation of heavy metal contaminated sites (Ernst 1974) presented the Cu tolerant mosses of the Schwarzwand to the global scientific community. The vast majority of studies, however, remain poorly known. Most of these are in German, published in journals of only regional importance or even have been submitted only as unpublished theses. This study aims to summarize the knowledge about these intensively studied, but still little recognized habitats for the international scientific community.

11.3 Geological Background

The Eastern Alps are characterized by a large number of small ore deposits. During the crystallization of granite and other plutonites, metal-rich solutions remained fluid and were pressed into disruptions during several succeeding phases of orogenesis. There, they crystallized in numerous joints. In general, gold is found in the Central Alps, surrounded by a belt of copper and iron deposits. In the periphery, lead and zinc were found (Holzer 1966). General features of the ore formation from magma and hydrothermal solutions are discussed by Petraschek and Pohl (1982).

The three sites discussed in this study belong to two different geological zones: Schwarzwand and Tofernalm are part of a nappe (a continuous sheet of rock) of the Tauernfenster, whereas Mitterberg/Hochkönig is part of the Greywacke zone (Tollmann 1977; Weber et al. 1972) and more in the periphery of the Alps. Thus, the Grossarl valley hosts almost exclusively copper and iron ores, whereas in Mitterberg/Hochkönig a much greater mineral diversity is found (Table 11.1). Concerning the water regime, however, a different classification is evident: The Schwarzwand is wet due to several rivulets, whereas the habitat Tofernalm is only supplied by rain water and rather dry due to the low water holding capacity of the degraded soil. On the Troiboden, parts of the spoil heaps are dry as well, but others are now covered by a peat bog that has developed from a prehistoric pond (Zschocke and Preuschen 1932).

11.3.1 Ore Deposits of the Glockner Facies

The sites of Schwarzwand and Tofernalm are located in the Grossarl valley. The ore-bearing layers were cut open by a glacier. Ore deposits are therefore found at both sides of the valley: in the East at Karteis, Kreemahd, Wasserfallscharte, and Jagerspitzl, in the West, at Aigenbachtal, Tofern, Schattbachalm, and Schwarzwand. In the East, the layers are flat and undisturbed, in the West, S-like folding leads to up to five repetitions of the ore-bearing layers (in both mining sites described here, three repetitions are found (Wiebols 1949). The habitats described in this study belong to two different deposits: (1) the Schwarzwand deposit, which reaches over 8 km from the mine Hubalm (1,311 m) in the Hubtal through the major mining area of the Schwarzwand and further East toward Schappachalm (1,971 m). (2) The Tofern deposit ranges from the Tofernalm to the Tofererscharte or Gamskarscharte, which leads to the neighboring valley of Gastein. The Tofern deposit was exploited by the mines Astentofern and Alttofern North and South of the Tofern creek. All these sites belong to a long series of deposits that continues west into the Gastein valley and further into the Pinzgau. The Glockner Facies is part of the permo-mesozoic layers of the Penninic. It is characterized by layers of Bündner schists rich in carbonates but poor in quartzites, breccia, and a-calcareous phyllites. The dominant rocks are prasinite, black phyllite, and calcareous phyllite. Between these layers, marine effusive rocks are found that are metamorphized to Green Slate and Black Slate. Remains of the former oceanic crust are present as layers of serpentinite. Dolomite and various types of limestone are abundant as well (Peer and Zimmer 1980; Tollmann 1977). The geology and petrology of the region is discussed in detail by Peer and Zimmer (1980).

A description of the ore deposits of the Glockner facies is given by Wiebols (1949) and Derkmann (1976). The facies consists of repeated layers of green slate, black slate, calcareous mica slate, and serpentinite; the most relevant ore bodies are restricted to the black slate and especially to the green slate next to the calcareous

Table 11.1 Heavy metal minerals in the areas of this study.The diversity of minerals and the number of elements is greater in the Mitterberg area. In the Grossarl valley, more secondary minerals, especially sulfates, are found

Grossarl valley	Mitterberg/Hochkönig
Azurite [$Cu_3(CO_3)_2(OH)_2$], Bornite (Cu_5FeS_4], Chalkanthite [$CuSO_4 \cdot 5H_2O$], Chalkopyrite [$CuFeS_2$], Chrysokolla [$(Cu, Al)_2H_2Si_2O_5(OH)_4 \cdot nH_2O$], Goethite [FeO(OH)], Goslarite [$ZnSO_4 \cdot 7H_2O$], Halotrichite [$FeAl_2(SO_4)_4 \cdot 22H_2O$], Hematite [$Fe_2O_3$], Limonite [$FeO(OH) \cdot nH_2O$], Malachite [$Cu_2CO_3(OH)_2$], Melanterite [$FeSO_4 \cdot 7H_2O$], Magnetite [$Fe_3O_4$], Pentlandite [$(Fe,Ni)_9S$], Pyrite [$FeS_2$], Pyrrhotite [$Fe_{0,85-1}S$], Rutile [$TiO_2$], Sampleite [$NaCaCu_5(PO_4)_4Cl \cdot 5H_2O$], Sphalertite [ZnS], Titanite [$CaTiSiO_5$]	Altaite [PbTe], Anatase [TiO_2], Ankerite [$Ca(Fe(II),Mg,Mn)(CO_3)_2$], Annabergite [$Ni_3(II)(AsO_4)_2 \cdot 8H_2O$], Arsenic [As], Arsenopyrite [FeAsS], Azurite [$Cu_3(CO_3)_2(OH)_2$], Becquerelite [$Ca(UO_2)_6O_4(OH)_6 \cdot 8H_2O$], Bournonite [$PbCuSbS_3$], Brannerite [$(U,Ca,Ce)(Ti,Fe)_2O_6$], Bravoite [$(FeNi)S_2$], Brochantite [$CuSO_4 \cdot 3Cu (OH_3)$], Calaverite [$AuTe_2$], Celestine [$SrSO_4$], Chalkanthite [$CuSO_4 \cdot 5H_2O$], Chalkopyrite [$CuFeS_2$], Cinnabar [HgS], Cobaltite [CoAsS], Colloradoite [HgTe], Copiapite [$Fe(II)Fe_4(III) (SO_4)_6(OH_2) \cdot 20H_2O$], Cubanite [$CuFe_2S_3$], Copper [Cu], Covellite [CuS], Erythrite [$Co_3(AsO_4)_2 \cdot 8H_2O$], Fahlore [47% Cu, 14% As, 5% Sb, traces of Hg, V, Zn, Ag, Au], Galenite [PbS], Gersdorffite [NiAsS], Gold [Au], Hematite [Fe_2O_3], Hessite [Ag_2Te], Jamesonite [$Pb_4FeSb_6S_{14}$], Kermesite [Sb_2S_2O], Langite [$Cu_4(SO_4)(OH)_6 \cdot 2H_2O$], Leuchtenbergite [$(Mg, Fe(II))_5Al(Si_3Al)O_{10}OH_8$], Limonite [$FeO(OH) \cdot nH_2O$], Magnetite [$Fe_3O_4$], Malachite [$Cu_2CO_3(OH)_2$], Marcasite [$FeS_2$], Maucherite [$Ni_{11}As_8$], Metacinnabar [HgS], Millertite [NiS], Molybdenite [MoS_2], Morenosite [$NiSO_4 \cdot 7H_2O$], Nickeline [NiAs], Parasymplesite [$Fe_3(II) (AsO_4)_2 \cdot 8H_2O$], Pentlandite [$(Fe,Ni)_9S_8$], Pharmacosiderite [$KFe_4(AsO_4)_3(OH)_4 \cdot 6\text{–}7H_2O$], Pitticite [$Fe_2(III) OHSO_4AsO_4 \cdot nH_2O$], Posnjakite [$Cu_4(SO_4) (OH_6) \cdot H_2O$], Pyrargyrite [$Ag_3SbS_3$], Pyrite [$FeS_2$], Pyrrhotite [$Fe_{0,85-1}S$], Realgar [$As_4S_4$], Retgersite [$NiSO_4 \cdot 6H_2O$], Rutile [$TiO_2$], Skutterudite [$(Co,Ni)As_3$], Sphalertite [ZnS], Skorodite [$FeAsO_4 \cdot 2H_2O$], Strontianite [$SrCO_3$], Studtite [$(UO_2)O_2(H_2O)_2 \cdot 2H_2O$], Titanite [$CaTiSiO_5$], Ullmannite [NiSbS], Uraninite [UO_2], Uranospinite [$Ca(UO_2)_2(AsO_4)_2 \cdot 10H_2O$], Zeunerite [$Cu(UO_2)_2(AsO_4)_2 \cdot 10\text{–}16H_2O$], Zippeite [$K_4(UO_2)_6(SO_4)_3(OH)_{10} \cdot 4H_2O$], Zircon [$ZrSiO_4$]
Derkmann (1976), Friedrich (1936), Günther (2006), Saukel (1980), own observation	Bernhard (1965), Eibner (1993), Nowak (1933), Paar (1993), Siegl (1972), Weber et al. (1972)

mica slate, where a typical pattern can be observed with a clearly distinguished layer of pyrite and magnetite with a thickness of 5–80 cm ("rich ore") followed by a more diffuse layer with a thickness of 50–200 cm consisting of chlorite, quartz, and pyrite ("disseminated ore"). The ores have their origin probably in heavy metal-bearing solutions that remained fluid after the crystallization of granite and were formed simultaneously with the surrounding rocks. Later, both were subjected to metamorphosis and folding during the alpine orogenesis (Tollmann 1977). There is no evidence for unconformities or later translocations of the heavy metals. The primary ores were mainly pyrite and chalcopyrite now showing signs of deformation and recrystallization; furthermore, magnetite and pyrrhotite were formed during metamorphosis (Derkmann 1976). Finally, numerous fissures were formed, which are unrelated to the distribution of ores (Wiebols 1949). Table 11.1 shows heavy metal minerals described so far from the Grossarl valley. The mineral diversity is rather low; only four heavy metals are found, i.e., Cu, Fe, Zn, and Ti. The effects on the vegetation seem to be caused exclusively by Cu (Feitzinger et al. 1998). Besides the sulfidic or oxidic ores, several secondary sulfates are found, probably due to the long-lasting contact of the ore outcrops with atmospheric oxygen. Under aerobic conditions, pyrite is oxidized and dissolved; one product of the oxidation is limonite, which is frequently found (Günther 2006).

11.3.2 Ore Deposits of Mitterberg/Hochkönig

Most of the Hochkönig mountain consists of different limestones; the ore deposits, however, are located in the Greywacke zone, which is closely interlocked with the Northern Limestone Alps at the southern foothills of the Hochkönig (Tollmann 1985). In more detail, the deposits belong to the Wildschönauer Slates of the Mitterberger Layers, the oldest deposits of the Greywacke zone (Ordovician). Similar to the Glockner Facies, the Wildschönauer Slates host a series of deposits 300 km in length with similar geological settings, e.g., in Schwaz, Kitzbühel, Zell am See, Mitterberg/ Hochkönig or Thumersbach (Nowak 1933; Tollmann 1977). Before exploitation, the Mitterberg deposits were the largest of their kind in Austria (Tollmann 1977). Origin and mineralogy of the Mitterberg deposits are described in detail by Bernhard (1965); the stratigraphy of the region is described by Del-Negro (1983). The heavy metal-bearing minerals from Mitterberg are enumerated in Table 11.1.

The ore deposits form a nonconformity, embedded in two petrographic layers, the Gray Series (paleozoic phyllites, serizite quartzite, diabas, green slate, and black slate) and the Violet Series (reddish clay slate, phyllite, quartzite, graphitic slate) (Paar 1993; Weber et al. 1972). Three phases of ore deposition can be distinguished at Mitterberg/Hochkönig: (1) The southern part of the deposit formed simultaneously with the surrounding quartzite layers of Greywacke during the Ordovician. These layers mainly host pyrite and chalcopyrite. (2) During the Permian, possibly due to tectonic processes of the Variscic orogeny, part of the ores were mobilized and translocated to metamorphous clastic rocks of the Fellersbach Layers (Sordian 1983). First, mainly nickel and arsenic ores were deposited at relatively low temperatures, including gersdorffite, ullmannite, pentlandite, or arsenopyrite. Later, chalcopyrite and ankerite were translocated at higher temperatures. (3) During the Alpine orogenesis, heavy metals were mobilized again and deposited in joints as chalcopyrite, fahlore (Cu–As ores) and ankerite. These mobilization processes also enriched uranium next to the joints, which had its origin in the neighboring rocks. The northern deposits (2) and (3) are partly covered by limestones of the Hochkönig. Younger ore joints frequently penetrate ones (Bernhard 1965; Siegl 1972; Tollmann 1977).

The ecosystems described in this chapter resulted from the exploitation of the deposits (2) and (3) at the southern slopes of the Hochkönig. These deposits had a length of 7–10 km and a width of 0.5–12 m. As shown in Fig. 11.1, most of the ore bodies were exploited already in prehistoric times.

11.4 Mining Activities

In all three areas, mining was once the key industry for the region but the mines were abandoned centuries ago. People, techniques and products differed significantly between the Grossarl Valley and Mitterberg/Hochkönig. For an overview of the historical mining in Salzburg compare with Feitzinger et al. (1998).

11.4.1 Historic Mining in the Grossarl Valley

Copper was mined in prehistoric times around Salzburg. However, for the Grossarl valley, this is unknown. Only in 1,010 AD, a small chapel is reported in the village of Hüttschlag, the later mining center (Winkelhofer 1813). Probably, small-scale copper mining started around the same time at the Schwarzwand (Feitzinger et al. 1998). After 1350, mining became an economic factor of regional, and after 1500, of supraregional importance (Drapela et al. 2001; Günther 2006).

At the start, the various small ore deposits were exploited by seven family-owned enterprises (Hübner 1796), virtually each side valley of the Grossarl River hosted at least one mine (for an enumeration of the mines, see Derkmann (1976) and Günther (2006)). Between 1535 and 1594, the dukes of Bavaria made large investments and about 100 people were employed at the mines. Hüttschlag became the center of smelting, which was partly managed by the priest of the village (Günther 2006; Winkelhofer 1813). After privatization (1594), insolvency and formation of a rescue company (1614), all mines and plants were bought by Paris Lodron, the prince-bishop of Salzburg, in 1622. From this time on, the government supported the mining industry in various ways: manpower requirements, for instants, were met by immigration from Carinthia (Hübner 1796). A significant part of the forests of the province was reserved for the needs of the mines (von Hornmayer 1807). Sulfur smelting was restricted to the winter months in order to reduce damage to the forests because of sulfur-oxide emissions. This law, however, was frequently avoided resulting in serious damage to the forest vegetation of the valley. Destructive exploitation led to a fast depletion of the deposits. In the eighteenth century, mining passed its peak, when more than 200 people were employed at the mines (Drapela et al. 2001). All industrial activities were seriously affected by the expulsion of the protestants from Salzburg after 1733 (von Hornmayer 1807). In the following decades, parts of the galleries had to be closed due to the depletion of ores (Hübner 1796). The incorporation of Salzburg into Bavaria in 1810 and into Austria in 1816 led to a cut of investments and as a result prospection for new ore deposits was abandoned. Ore depletion as well as falling copper prices led to the closure of the mines in 1848. Between 1849 and 1947, especially during the World Wars, several enterprises made attempts to resume mining but without any success (Günther 2006). In 1951, the last mining facilities were removed (Pabinger w.y.). At the end of the eighteenth century, when most mines were still working, the Grossarl valley hosted a population of 2,700 people (Hübner 1796), which accounts to roughly 60% of the population of today (Statistik Austria 2010). After 1800, the population decreased quickly (Winkelhofer 1813), and in 1830, only 30% were left and the economic importance of mining was negligible (Veichtlbauer et al. 2006).

During the 300 years of mining activity, a total production of 6,500 tons Cu was achieved. Due to the depletion of copper deposits as well as to changes in ore processing, copper production decreased whereas the amounts of sulfur grew. In 1637, 6.5 tons Cu and 4.5 tons S were produced, whereas in 1816, only 4 tons Cu but 14 tons S were mined. However, due to its higher price, copper always remained the cash cow of the mines (Wiebols 1949). Furthermore, sulfidic copper ores were traded without processing and used as a flux for gold smelting in the gold mines of the Hohe Tauern (Feitzinger et al. 1998). Small amounts of silver, lead, and arsenic were mined as well (Günther 2006).

The Schwarzwand was the oldest, largest, and most important mining district. First evidence for mining goes back to 1521 (Ludwig 1979). At its peak in the seventeenth century, about 100 miners were employed. In the eighteenth century, most existing galleries were abandoned, however. Despite the fact that numerous new galleries were built, only few reached ore deposits. In 1796, only three galleries were active (Günther 2006; Hübner 1796). The search for new deposits continued until 1815 and again between 1849 and 1863. At the end of World War II, mining was resumed by the *Studiengesellschaft Kiesbergbau* (translated: Experimental Company for Sulphidic Ore Mining) but without economic success (Derkmann 1976). In 300 years of mining, a total of 2,700 tons copper had been produced, i.e., more than 40% of the total production of the Grossarl Valley. Today, the ruins of a smelter are still visible, as well as about 15 adits. One gallery ("*Unserer lieben Frau Hauptstollen*", translated: Our Lady's Main Gallery) is still accessible. Mine drainage from the adits feeds the rivulets that precipitate copper (Günther 2006).

At the Tofern Alm, mining activities date back to 1521 or 1534. Large-scale exploitation, however, started in 1627 when a team of up to 25 workers constructed two mines (Alttofern and Astentofern) (Hübner 1796). Depletion of the deposit led to the closure of the mine in 1752. In the following decades, several attempts to find new deposits in 1845 and 1946 were unsuccessful (Feitzinger et al. 1998). A total of about 4,000,000 tons of material was extracted, yielding about 1,000 tons copper, i.e., 15% of the total production of the valley (Günther 2006). Besides the spoil heaps, a cabin built in the sixteenth century and several ruins are present today.

11.4.2 Prehistoric Mining at Mitterberg/Hochkönig

The Eastern Alps were a hotspot of prehistoric Cu mining, with a large number of mining sites excavated in the valleys of the Salzach, the Tiroler Achen, and the Enns. The mining district of Mitterberg/Hochkönig is located in the center of this region and had the greatest number of smelters (Stöllner et al. 2006). In 1828, the first remnants of prehistoric mining were found at Troiboden (Bernhard 1965). Archeological research has been carried out since 1877 (Urban 2002). Summaries of the abundant archeological studies are given by Zschocke and Preuschen (1932), Günther et al. (1993) and Stöllner et al. (2006). The supraregional importance of the Mitterberg mining site was confirmed by the Nebra Sky disk (about 1,600 BC), an artifact, which although found in Sachsen-Anhalt (Germany), has its copper produced in the Eastern Alps, most probably at the Hochkönig (Pernicka et al. 2008).

In the Eastern Alps, techniques for the mining and smelting of copper were imported from the Mediterranean or the Carpathians at about 4,000 BC (Lippert 1993). Ore outcrops were probably identified by the apparent coloration of secondary ore minerals (Eibner 1993). In the first phase of mining (4,000–2,000 BC), only ore veins above the ground water level, i.e., to a depth of about 2 m, were exploited, which means that mainly secondary Cu minerals were mined. After depletion of the surface deposits, mine shafts were dug down to a depth of 70–120 m from 2,000 to 300 BC, and the galleries reached a length of up to 100 m (Eibner 1993). The bearers of this "*Younger East Alpine Technique*" (Eibner 1993) were probably Indoeuropean Illyrians who exported the copper throughout Europe (Eibner 1993; Tollmann 1977; Kirnbauer 1968; Suhling 1983). The much larger, deeper deposits contained both native copper and sulfidic ores, which required elaborate roasting and smelting technologies (Eibner 1993; Lippert 1993). At this time, chalcopyrite was probably the most important ore mineral; especially in the Early Bronze Age. Fahlore (fahlerz) for the production of arsenical bronze was of some importance as well. Tools from the north alpine region made of arsenical bronze show a similar hardness compared to the later tin bronze but were more difficult to produce in a reproducible quality (Kienlin et al. 2006; Stöllner et al. 2006).

After intensive archeological studies, the mining and smelting techniques were reconstructed in detail: ore was loosened by tools made from stone or bronze, but especially by heating the stone with fire, followed by sudden chilling with cold water. Ore was hand-picked from the excavated rocks *in situ*; waste rock was discarded and deposited. Metal extraction was incomplete and as a result, spoil contained high concentrations of heavy metals, and those spoil heaps have remained virtually barren ever since. Ore was processed at Troiboden and in several other places (Henselig 1981; Sperl 1987). Remains of the furnaces and intermediate products enabled the reconstruction of the metallurgical process (Herdits 1997): roasting of the sulfidic ores, predominantly chalcopyrite and fahlore, resulted in the formation of matte (Cu_2S and different Fe oxides). Matte was oxidized by the blast, resulting in the emission of SO_2, the formation of reduced copper and slag containing FeO and silicate flux. Thus, smelting occurred under oxidizing conditions (Moesta 1989), resulting in easier process control than smelting by reduction. The remaining slags are still rich in heavy metals ($3.4 \pm 4.6\%$ Cu, $0.1 \pm 0.2\%$ Ni, $0.1 \pm 0.1\%$ Sn and traces of Zn, Pb, As, Sb, Co and Ag) (Herdits 1997) and contributed to the establishment of contaminated ecosystems. Altogether, a total of 7,000 tons (Eibner 1993) to 17,000 tons (Holzer 1966) of Cu has been mined.

After 1,300 BC, in the late bronze age, mining activities were less extensive and completely abandoned at 300 BC. Remarkably, many intact tools, partly made from valuable bronze, were left behind. Part of the mining equipment was deconstructed. The reasons for the closure of the mines are not clear. In spite of the huge amounts of woods that were needed for smelting, Zschocke and Preuschen (1932) calculated that the wood supply would not have been the limiting factor. Possibly all rich veins in an exploitable depth were depleted, or the miners switched to the more profitable mining of iron (which was discovered around that time) or salt (Eibner 1993). Other theories include chilling of the climate (Tinner et al. 2003) or expulsion of the Illyrians by invading Celts. The latter theory, however, does not explain why the Celts did not resume mining.

In 1829, the modern mining era started with the *Kupfergewerkschaft Mitterberg* when 2 years earlier, the ore deposits had been found again by accident (Günther 1993). After 150 years, the galleries had a total length of more than 17 km (Günther 2006). The modern mine frequently struck prehistoric galleries and revealed the extent and technology of ancient mining (Stöllner et al. 2006). Finally, as a result of declining copper prices, mining was abandoned in 1977 (Günther 1993). Besides copper, uranium was found in 1968. Prospection and experiments concerning mining and smelting were started but came to an end due to the closure of the mine (Günther 1993). Finally, it has to be mentioned that only the construction of modern galleries enabled the mapping and understanding of the prehistoric mines (Stöllner et al. 2006).

11.5 Aspects of Soil Chemistry and Copper-Tolerant Vegetation Today

In all three ecosystems, a range of metals, predominately copper, zinc, and lead are found in metal-rich rocks, soil, and water. Table 11.2 shows metal concentrations observed by different researchers. In all three habitats, heavy metal concentrations

Table 11.2 Heavy metal concentrations [mg kg^{-1}] found in different parts of the three habiats

	Schwarzwand	Tofernalm	Troiboden
Rocks	Green slate: Cu: 162 ± 186; Zn: 261 ± 290; Ni: 54 ± 31; Cr: 314 ± 96; V: 263 ± 39[a]	Green slate: Cu: 66; Zn 243; Ni: 123; Cr: 330[a]	Slag: Cu: 33,580 ± 45,850; Ni: 793 ± 169; As: 151 ± 339; Sn: 536 ± 798[b]
	Chlorite slate: Cu: 621 ± 1,270; Zn: 251 ± 154; Ni: 114 ± 52; Cr: 330 ± 97[a]	Chlorite slate: Cu: 119; Zn 885; Ni: 85; Cr: 225[a]	
	Disseminated ore: Cu: 585 ± 849; Zn: 452 ± 400; Ni: 105 ± 33; Cr: 308 ± 53[a]	Disseminated ore: Cu: 4,372 ± 7,757; Zn: 46 ± 51; Ni: 338 ± 524; Cr: 295 ± 43[a]	
	Rich ore: Cu: 3,410 ± 4,257; Zn: 51 ± 18; Ni: 30 ± 9; Cr: 273 ± 166[a]		
	Quarzite + chalcopyrite: Cu: 32,000 ± 18,248; Zn: 572 ± 229; Ni: 67 ± 25; Cr: 273 ± 166[a]		
Water	Cu: 0.39 ± 0.19; Zn: 0.23 ± 0.03; Mn: 0.04 ± 0.01[c]	Cu: 7–14; Zn: 8–31[d]	Cu: ≤0.6[e]
	Cu: 16.5–24.5; Zn: 15–17; Mn: 10; Fe: 12–13; Cr: 4–6[f]		
	Cu: 0.4[d]		
	Cu: 0.2–0.7; Zn: 0.2–0.3[g]		
	Cu: 0.7–1.5[h]		
Substrate		Soil	
soil	Cu: 3,760–6,910 (550–830)[c]	Cu: 2,200–4,600[d]	
	Cu: 3,200 ± 2,080; Zn: 129 ± 97; Mn: 144 ± 70; Cr: 22 ± 9; Fe: 7,210 ± 3,820[f]		
	Cu: (51 ± 65); Zn: (2 ± 3); Mn: (2 ± 2); Fe: (10 ± 8)[d]		
		Spoil heap material	
	Cu: 290 ± 194; Zn: 267 ± 90; Mn: 172 ± 84; Cr: 39 ± 2; Fe: 8,240 ± 622[f]	Cu: (89 ± 93); Zn: (5 ± 4); Mn: (50 ± 91); Fe: (29 ± 34)[d]	Cu: 1,066–8,100[g]
		Cu: 850–3,200 (62)[g]	
	Creek sediment: Cu: 16,170 ± 5,470; Zn: 701 ± 213; Mn: 310 ± 57; Cr: 45 ± 14; Fe: 38,680 ± 12,060[f]		
	Cu: 0.6–9.5[g]		

Total heavy metal concentrations are shown without brackets, extractable heavy metals are within brackets. Data after (a) Derkmann (1976), (b) Herdits (1997), (c) Hus (2008), (d) Sissolak (1984), (e) Ledl et al. (1980), (f) Saukel (1980), (g) Körber-Ulrich (1996), and (h) Stummerer (1970)

differ extremely between different microhabitats. The bioavailability and the water solubility are always low as far as it was tested yet (e.g., Sissolak 1984). Nothing else could be expected after exposure to intense rain (precipitations > 1,000 mm $year^{-1}$) in the whole region (Zentralanstalt für Meteorologie und Geodynamik 2010) for hundreds or even thousands of years, which had washed out all soluble compounds. Feitzinger et al. (1998) investigated historical mining sites from the viewpoint of protection of people and environment. Although significant contamination of soils was found in the sites described in this study, no remediation measures have been recommended, as all three sites are in great distance to settlements. However, due to the great number of ore deposits, copper concentrations are elevated all over central Salzburg (Zechmeister 1997). Table 11.2 gives an overview on the metal concentrations found in rocks, water, and soil.

Depending on the copper concentration in the soil, the vegetation is clearly different from the surroundings in all three ecosystems. However, in spite of the similar altitude and climate and in spite of the predominance of copper as the main contaminant, floristic similarities between the three sites are limited. Table 11.3 enumerates the plant species described for the three sites.

11.5.1 Schwarzwand

At the Schwarzwand, an area of about 100,000 m^2 is contaminated with metals (Feitzinger et al. 1998). The Schwarzwand is well known as a center of plant and especially bryophyte diversity. Saukel (1980) lists 80 Jungermanniopsida, 177 Bryopsida and 122 species of vascular plants; some more are reported by Punz (1999). Furthermore, 52 species of lichens were described (Poelt 1955). The microclimate of the Schwarzwand shows some specific features supporting the formation of a rich moss flora. The region is covered by snow for 135 ± 21 days $year^{-1}$ (Saukel 1980). Due to the Northern exposition, solar irradiation is extremely low (160–290 W m^{-2} in summer) (Sissolak 1984). Therefore, and because of the high altitude, temperatures are rather low. The mean temperature is 5.5°C (−4.9°C in January; 15.2°C in July). From June to August, Sissolak (1984) measured air temperatures of only 1–25°C, and soil temperatures of 2–20°C. Within cushions of the mosses *Mielichhoferia mielichhoferi*, *Merceya ligulata*, and *Nardia scalaris*, a rather balanced temperature regime was found (daily changes $<$15°C), whereas cushions of *Marsupella emarginata*, *Pleurozium schreberi*, and *Cladonia rangifera* showed changes of $>$20°C (Saukel 1980). Accordingly, several species of the nival zone such as *Gymnomitrium coralloides* and *Anthelia julacea* are found in the Schwarzwand (Poelt 1955). Microbial, fungal, and algal diversity are virtually unknown so far, as well as any faunistic peculiarities.

The Schwarzwand as a whole is covered by a *Larix decidua*-woodland, with large open areas around the creeks and spoil heaps. The surrounding uncontaminated areas at the same altitude and with the same exposition are covered by much denser *Picea abies*-forests. However, it has to be stressed that only a part of the Schwarzwand is exposed to severe concentrations of metals. The following paragraphs focus on these metal-rich sites and their flora, which is rather poor in species. A detailed description and classification of the topography is given by

Table 11.3 Diversity of plants at the three habitats described in this study (Hus 2008; Poelt 1955; Punz and Orasche 1995; Repp 1963; Saukel 1980; Sissolak 1984; Zechmeister and Punz 1990)

Lichens	
Cetraria hepatizon S	*Microglaena gelatinosa* S
Cetraria islandica H	*Polyblastia rivalis* S
Cladonia arbuscula H	*Rhizocarpon furfurosum* S
Cladonia alpestris H	*Rhizocarpon geograficum* S
Cladonia verticillata H	*Rhizocarpon polycarpum* S
Cladonia macroceras H	*Stereocaulon alpinum* SH
Cladonia spp. S	*Stereocaulon botryosum* S
Diploschistes scruposus S	*Stereocaulon tyroliense* S
Haematomma ventosum S	*Stereocaulon vesuvianum* S
Ionaspis suaveolens S	*Stereocaulon* sp. S
Lecanora spp. S	*Thamnolia vermicularis* T
Lecidia spp. S	
Bryophyta	
Gymnomitrium corallioides S	*Mielichhoferia mielichhoferi* S
Cephaloziella massalongoi S	*Narida scalaris* S
Cephaloziella phyllacantha S	*Oligotrichum hercynicum* S
Grimmia atrata S	*Pohlia drummondii* S
Marsupella emarginata S	*Pohlia* sp. H
Merceya ligulata S	*Sacapania obcordata* S
Mielichhoferia elongata S	*Sacapania undulata* S
Vascular plants	
Agrostis alpina T	*Hieracium alpinum* S
Agrostis schraderana ST	*Homogyne alpina* S
Alnus incana T	*Juncus trifidus* SH
Alnus viridis T	*Larix decidua* ST
Arabis alpina T	*Luzula sylvatica* S
Arabis hirsuta T	*Moeringia ciliata* T
Arabis soyeri T	*Moeringia muscosa* T
Atropis tenuis H	*Oxalis acetosella* T
Avenella flexuosa S	*Picea abies* STH
Betula pubescens T	*Poa alpina* T
Calamogrostis villosa S	*Rhododendron ferrugineum* ST
Calluna vulgaris ST	*Rumex acetosella* TH
Campanula scheuchzeri T	*Salix nigricans* T
Cardamine amara T	*Saxifraga aizoides* T
Carex nigra T	*Saxifraga stellaris* ST
Cerastium fontanum T	*Silene inflata* H
Chaerophyllum hirsutum T	*Silene rupestris* T
Cirsium palustre T	*Silene vulgaris* TH
Deschampsia cespitosa HT	*Taraxacum officinale* H
Dryopteris villarii H	*Tussilago farfara* TH
Equisetum palustre T	*Vaccinium myrtillus* STH
Euphrasis minima T	*Vaccinium vitis-idea* S
Galium anisophyllum H	*Veronica alpina* T
Geum montanum T	

Saukel (1980) (supplemented by observations of the authors), who distinguishes seven types of habitats:

1. In the uppermost part of the Schwarzwand above the adits, soil depth and soil conditions vary considerably, but no evidence for extreme metal concentrations is found. The vegetation is closed but the tree layer consists of *Larix decidua* instead of *Picea abies*, the more typical tree at this altitude. Acidophilic mosses such as *Pleurozium schreberi* or *Hylocomium splendens* cover most of the soil.
2. The adits themselves and the surrounding rocks form the second type of habitats. Some of the adits produce mine drainage with secondary copper minerals frequently occurring. The rocks host a vegetation of copper mosses, especially *Mielichhoferia elongata*.
3. Below the adits, a steep (25–45°) slope is found. No fine soil is retained by the smooth rocks, accordingly, no vascular plants occur. Isolated mosses and lichens grow attached to the rocks but it is not clear whether they are affected by metal concentrations, as the petrological situation at this slope has not been studied in detail. Only in some patches, limited amounts of fine soil enable the growth of *Agrostis schraderana*, *Juncus trifidus,* or *Rhododendron ferrugineum.* In Scandinavia, *Juncus trifidus* occurs in heavy metal contaminated meadows (*Lacnis alpina–Arenaria norvegica*) (Ernst 1974), but as a pioneer on highly Cu contaminated mineral soil, it was found at the Schwarzwand for the first time (Saukel 1980).
4. At the base of this slope, a large and steep spoil heap was formed by mine waste produced by the adits above. The spoil heap is watered by two permanent and several temporal creeks and therefore partly saturated. The substrate consists of coarse rock material (i.e., a mixture of quartz, prasinite, and green slate, usually with traces of pyrite weathering; pers. com. M. Bichler). The rocky substrate is partly covered or intermixed with duff and overgrown by small trees of *Larix decidua*, shrubs of *Picea abies,* and various subshrubs such as *Vaccinium myrtillus* and *Rhododendron ferrugineum*, especially at the periphery of the spoil heap. *Silene vulgaris*, a well-known copper-tolerant species, is also found, but only in microhabitats with low contamination.
5. Below this spoil heap, a flat area is situated. Most of the soil consists of leaf mold; little evidence for elevated metal concentrations was found. The vegetation is similar to habitat (1). In some parts, however, highly metal-enriched soils without vegetation occur. This has its origin not in spoil heap material but in precipitations of metals from the creeks (see below).
6. Below area (5), a second spoil heap is found. Compared to (4), it has less organic constituents and more clay minerals. Most of the soil is permanently saturated with water and kept open by constant movements of the soil. The vegetation is dominated by large cushions of mosses such as *Scapania undulata* and other pioneers. In greater distance from the creeks, the spoil heaps host trees, such as *Larix dedicua,* which are up to 300 years old in spite of their small size (Saukel 1980).

7. The lowest part of the Schwarzwand below (6) is covered by a *Larix decidua* wood, which is much denser than in (1) and (5). Little evidence for elevated metal concentrations was found. Due to the moist and shadow conditions, a high diversity of mosses is found. Ferns are virtually restricted to such habitats. Accordingly, Saukel (1980) found no evidence for metal accumulation in *Blechnum spicant* and *Thelypteris limbosperma*.

Recent findings on the chemistry of soils and water by the authors will require a re-evaluation of this classification. In contrast to most other metal-enriched ecosystems, no pH extremes are found in the Schwarzwand. The water of the creeks measures a pH of 5.8–7.4 The only exception is water dripping over ore rocks with an acidic pH of about 3 (Sissolak 1984).

The creeks of the Schwarzwand are a unique feature of this ecosystem. From area (3) to (6), they produce a bluish precipitation that is clearly visible in Fig. 11.2. This substance, probably a calcium carbonate sinter stained by copper minerals (Günther 2006), contains up to 2% Cu (Saukel 1980). Accordingly, the creeks constantly loose copper, in one specific rivulet, the Cu concentrations falls from 1,460 mg l^{-1} Cu at area (2) to 700 mg l^{-1} at area (6) (Stummerer 1970). Only very few plant species are able to cope with the copper-rich water and the precipitations (Fig. 11.5): *Pohlia drummondi* forms extensive cushions; the older parts of the plants are usually embedded into the copper precipitations, whereas the tips emerge above the surface. *Saxifraga stellaris* ssp. *alpigena* grows frequently in the precipitations but strictly avoids the acidic parts of the creek. *Agrostis schraderana* is found at the banks of the creeks (Sissolak 1984). The lichen *Polyblastia rivalis* also settles in the wet precipitations (Poelt 1955).

Fig. 11.5 *Saxifraga stellaris*, the most abundant vascular plant species in the *bluish copper* precipitations, associated with the moss *Pohlia drummondi* growing in and partly covered by the precipitation

The course of the creeks changes frequently, in some parts of the Schwarzwand, sub-fossil layers of the blue precipitations are found in some distance from the creeks, which are now covered by a layer of leaf mold having formed obviously only after the rivulet had changed its course. Consequently, these soils have high copper concentrations (own observation). Table 11.2 gives on overview on the metal concentrations found in rocks, water, and different types of substrate. Besides the metal concentration, lichens from the whole region were reported to exhibit enhanced concentrations of radionuclides such as ^{137}Cs (Eckl et al. 1984).

11.5.2 Tofernalm

As the mine waste of the Tofernalm has its origin in the same ore deposit as on the Schwarzwand, both sites show some similarities: The copper concentrations are more or less the same in rocks, soils, and water. pH values are moderate at the Tofernalm as well, with values of 5.4–5.8 (Sissolak 1984). The bioavailability of metals is therefore low (Plette et al. 1999). On some places, however, seepage water exhibits a pH of 4.0–4.4 leading to an elevated Cu availability (Plette et al. 1999; Marschner 1995).

Similar to the Schwarzwand, several microhabitats can be distinguished (Körber-Ulrich 1996; Sissolak 1984):

1. A spoil heap rich in coarse material lacking fine soil results in low water retention capacity and therefore to high relative abundance of mosses (Punz 1999) and lichens including *Stereocaulon alpinum* and *Cetraria islandica*. In addition, *Thamnolia vermicularis* was found as well (observation of the authors). Among vascular plants, *Silene rupestris* and rarely *S. vulgaris* are found. *Saxifraga stellaris* occurs as well but shows reduced growth due to the xeric conditions (Körber-Ulrich 1996; Sissolak 1984). *Rumex acetosella* is abundant, but remains small (own observations).
2. A spoil heap watered by a creek originating 150 m above, the water is contaminated as well (7–10 mg Cu l^{-1}). In spite of the wet conditions, the vegetation is very similar to (1) and contains *Stereocaulon alpinum*, *Cetraria islandica*, *Silene vulgaris*, and *Silene rupestris*. This region is under constant influence of the creek and often flooded and altered in its appearance.
3. A swampy ore outcrop; the central part is completely devoid of vegetation. In 2010, a little pond developed. In the periphery, the typical copper-tolerant vegetation is supplemented by *Caltha palustris*, *Carex nigra*, and *Equisetum palustre*.
4. Scree formed by rocks originating from the slopes above the Tofernalm. This area is also under constant change. Here, only little copper is found but the material still exceeds the "normal" copper content as expressed by the legal thresholds for soil (500 mg kg^{-1}) (Glaeser 2001). In this habitat, *Silene vulgaris* shows extremely strong growth due to the good water supply.

11.5.3 Troiboden (Mitterberg/Hochkönig)

According to the great number of ancient and modern mining sites, several metal-enriched ecosystems have been described from the Mitterberg area. Recent and very high metal contamination is found at Mitterberghütten (Feitzinger et al. 1998). This study focuses on the Troiboden, a complex of metal-enriched ecosystems at about 1,600–1,800 m close to the timberline of the Hochkönig. Zschocke and Preuschen (1932) have given a detailed map of prehistoric mining residues in this area. Contamination is not restricted exclusively to the prehistoric mining area. In a meadow about 150 m below the Troiboden, a slightly increased copper concentration of 420 ppm is found as well.

Ecologically relevant remains of prehistoric mining form different habitats. Although the copper-tolerant vegetation shows certain differences, it is always dominated by *Silene vulgaris* and is allocated to the alliance Galio anisophylli-Minuartion vernae (Ernst 1974; the validation of this syntaxon was later questioned by Punz and Mucina 1997). *Silene vulgaris* was shown to accumulate up to 235 mg kg^{-1} Cu (compare 0), whereas *Vaccinium myrtillus* or *Picea abies* exclude heavy metals virtually completely (Mutsch 1980). Four types of habitats can be distinguished and were described in detail by Körber-Ullrich 1996:

1. At least three spoil heaps are located in the surroundings of the Troiboden. Findings of stone hammers and other tools give evidence that these heaps were not only the place for the deposition of residues but also the place for the processing of the ores. Furthermore, crater-shaped pits caused by mining activities close to the surface (*Pinge*, glory hole) remained from prehistoric mining. Geoelectric mapping gives evidence that ore minerals have been exploited completely in a depth of 20 m (Stöllner et al. 2006). The open vegetation is dominated by the lichens *Cladonia verticillata*, *Cladonia alpestris*, *Stereocaulon nanodes*, the vascular plants *Silene vulgaris*, *Galium anisophyllum*, *Minuartia verna*, and *Thymus polytrichus*, and especially by the moss *Pohlia* sp. which covers most of the soil (Punz 1999).
2. Two smelting sites are characterized by a red substrate clearly distinguished from the peaty soil in the surroundings. At one smelting site, 1,100 mg kg^{-1} Cu is found in the periphery, compared to 4,730 in the center; in the other, the Cu content is rising from 2,700 mg kg^{-1} at the upper to 8,100 mg kg^{-1} at the lower end. The most contaminated areas carry only lichens such as *Stereocaulon alpinum* or *Cladonia macroceras*. *Silene vulgaris* tolerates up to 5,500 mg kg^{-1}. Concentrations of Ni and Co are increased as well. Next to the smelting places in the peaty soil also more than 3,600 mg Cu kg^{-1} were found below a specimen of *Picea abies*; here, however, the vegetation is not visibly affected, thus confirming the high capability of the peaty soil to immobilize heavy metals (data from Körber-Ulrich 1996).
3. A ditch with a length of about 80 m was probably used for drainage by the prehistoric miners. Today, it fills up with water after heavy rain. The ditch cuts through layers of minerals waste from the mining and contains both charcoal and

slags. The concentrations of copper are much higher than at the spoil heaps (up to 1% Cu), and enhanced concentrations of Ni and Co are measured as well (Mutsch 1980). *Silene vulgaris* grows not only at the walls but also at the bottom of the ditch where very wet conditions and even long-term flooding are tolerated. At the highest copper concentrations, *Rumex acetosella* occurs with reduced growth and reddish shoots and leaves (own observations).

4. Adjacent to the mining site (3) is a peat bog. According to Zschocke and Preuschen (1932), this fen has its origin in an artificial pond that was needed in the smelting process. After closure of the mines, the pond was colonized by *Sphagnum* mosses and the resulting peat bog started to overgrow the former mining area. This colonisation of a site with very high metal concentrations by *Sphagnum* mosses (Zhang and Banks 2005) and *Sphagnum* peat (Kalmykova et al. 2008; Ma and Tobin 2004) may be explained by the inherent capacity of these mosses to immobilize metals, especially Cu.

These habitats are not always clearly distinguished. At the end of ditch (3), e.g., several small spoil heaps are closely interlocked with a quaking bog. Remarkably, in this habitat, *Silene vulgaris* is untypically growing as a part of the fen vegetation (Körber-Ulrich 1996).

11.6 Studies on Copper Tolerance and Uptake

Metal tolerance in plants may be estimated using various approaches. One common stress test involves measuring photosynthetic activity, because especially photosystem II is very sensitive to metals, in particular to copper. Using portable chlorophyll fluorimeters, photosynthetic yield can be measured (e.g., Mallick and Mohn 2003). Another method is the root elongation test (Wilkins 1957), where an index of metal tolerance is derived from plants growing in hydroponics. The rate of root growth is measured in the control medium and again after the addition of different concentrations of the metals.

The tolerance of the cells to certain phytotoxic metals has been frequently tested by tolerance experiments of the plant cytoplasm (Höfler 1932; Biebl 1947, 1949, 1950; Punz and Koerber-Ulrich 1993; Ulrich-Körber 1996; Url 1956). Cells (e.g., algae, moss leaves) or tissues from plant organs containing living, undamaged cells are incubated in graded solutions of metals for 24 or 48 h and then tested for their vitality. Sometimes it is difficult to estimate their vitality from mere observation under a light microscope. In such cases, a valuable indicator for the discrimination between live and dead cells is plasmolysis: only vital cytoplasm, whose plasma membrane has maintained semipermeability, is able to shrink in hypertonic solutions and to detach from the cell wall. Iljin (1935) and later Biebl and Rossi-Pillenhofer (1954) and Url (1956) made a notable observation that in some species, e.g., the copper moss *Nardia scalaris*, so-called "death zones" occurred: The highest mortality was found at intermediate concentrations of metals.

In low and also in high concentrations, most of the cells survived. They postulated the formation of a layer of precipitated metals at the plasma membrane, which is impermeable for metal ions. At high concentrations, the formation of this layer is so fast that no harmful metals can enter the cells (Iljin 1935). Ernst (1982b) compares the rooting technique (Wilkins 1957) with the cytoplasmic toxicity test (Höfler 1932) and suggests with the example of *Silene vulgaris* that the cytoplasmic toxicity test gives a more differentiated picture about the resistance of a population. Furthermore, the cytoplasmic toxicity test provides information about the tolerance in the sense of Levitt (1980).

Various species from the ecosystems described here were tested for their metal tolerance using mainly the cytoplasmic toxicity test. Most of these species accumulate metals in relatively high concentrations (Table 11.5). In such plants, the cells of the shoot are in direct contact with metal ions stored in the living tissue, and therefore the tolerance of the living protoplast against metal stress is of crucial importance. The same is true for mosses, which lack a protective cuticle, so that the cells of the leaflets are in direct contact with metals from the environment.

11.6.1 Acetosella vulgaris

Metal-resistant ecotypes of *Acetosella vulgaris* (syn. *Rumex acetosella*, Polygonaceae) occur in many metal-rich sites all over Europe. In the study area, *A. vulgaris* occurs at the Tofernalm and on Hochkönig, and its response to metal stress is discussed in Chap. 12 of this book.

11.6.2 Saxifraga stellaris

Saxifraga stellaris (*Starry Saxifrage*, *Sternblütiger Steinbrech*, *Saxifrage étoilée*) occurs in central and northern Europe, especially in mountainous habitats such as the Alps. It is characterized by a basal rosette of oval leaves up to 10 cm long (Aeschimann et al. 2004). To our knowledge, *Saxifraga stellaris* has been described only from two metal-enriched sites other than the Schwarzwand/ Tofernalm in Tirol (Punz and Koerber-Ulrich 1993). Here, however, it is the most abundant vascular plant rooting in the precipitations of the creeks of the Schwarzwand. It occurs as *Saxifraga stellaris* ssp. *alpigena*. Three different ecotypes of this species may be distinguished in the vicinity of the Schwarzwand: a medium-sized, metal sensitive variety on soils with low metal concentrations, a small, copper and drought tolerant variety on the nearby Tofernalm and a very large, hydrophilic and copper-tolerant variety from the Schwarzwand. Transplantation experiments gave evidence that all three ecotypes are promoted by wet conditions and do not require a metal-rich substrate (Sissolak 1984). Hydroponic experiments confirmed these differences: plants from the Schwarzwand grew without serious toxicity effects in up to 100 mg kg^{-1} Cu EDTA and survived 1,000 mg kg^{-1}, whereas plants from metal-poor sites showed severe toxicity effects in 100 mg kg^{-1} and died in 1,000 mg kg^{-1} (Sissolak 1984).

In metal tolerance experiments, leaf cells from the Schwarzwand ecotype survived Cu concentrations of up to 0.1 M for 24 h, and even 0.5 M Zn (Sissolak 1984). The metal tolerance for metals such as Cr and V is much lower (Sissolak 1984; Nieboer and Richardson 1980; Bagchi et al. 2001; Basiouny 1984). Saukel (1980) found a similar Zn resistance of 0.5 M, a lower Cu resistance of about 0.005–0.05 M, and resistances against Cr and V in the same order of magnitude. Interestingly, he found no differences between the tolerance of plants from the Schwarzwand and from nonmetal-enriched sites. Besides the metal tolerance and the larger growth form, the Schwarzwand ecotype is distinguished by anatomical features including thicker leaves, increased formation of oxalate crystals, and lacking anthocyanins (Sissolak 1984).

Similar to the Bryophytes, *Saxifraga stellaris* stores metals in above-soil parts (Table 11.5). With the exceptions of Fe and Al, the leaves exhibit much higher concentrations than the stems and the flowers (Saukel 1980).

11.6.3 Silene vulgaris

Silene vulgaris (Syn. *S. cucubalus*, Bladder Campion, Aufgeblasenes Leimkraut, Silène enflé) belongs to the subfamily Silenoideae/Caryophyllaceae. *S. vulgaris* occurs all over Europe and has a large ecological amplitude. It prefers, however, light and well-drained rocky or sandy calcareous substrates. It is also able to grow on serpentine and on silicates up to the alpine zone. Various ecotypes have been described that differ in their morphology. According to Körber-Ullrich (1996), the plants from Salzburg belong to the subspecies *S. vulgaris*. It is a characteristic metal tolerant plant that has been observed in almost every type of heavy metal vegetation. Schubert (1953) and Ernst (1974) describe *S. vulgaris* as a pioneer and terms the initial stage of metal-rich mine site succession the "Silene-*Stadium*" (*Silene* state).

According to this wide distribution throughout the mining areas of Europe, *S. vulgaris* has been analyzed for its metal tolerance numerous times. The plant is especially well built for tests on cytoplasmic toxicity, because it has relatively strong stems and leaves and no hairs in most varieties. A review of the relevant literature as well as an experimental screening of different metal-tolerant *Silene vulgaris* ecotypes from various metal-containing sites is given by Körber-Ulrich (1990, 1996) and by Punz and Koerber-Ulrich (1993). Using the cytoplasmic toxicity test (Höfler 1932), they found that *Silene* from Hochkönig and from Tofernalm was more tolerant toward Cu than *Silene* from control meadows; most tolerant were the plants from Tofernalm; when sections of stems and leaves are incubated in graded $CuSO_4$ solutions, the cells tolerated 50 μM Cu. Although there is almost no

Zn in the natural soil, *Silene* plants from Hochkönig and Tofernalm showed, a low level co-tolerance for Zn (Körber-Ulrich 1996).

S. vulgaris Prat (1934) and Schat and Kalff (1992) proved the inheritance of copper resistance in *Silene dioica* and *Silene vulgaris*. Schat et al. (1993) showed in *Silene* from Germany that copper tolerance is controlled by two specific genes.

Physiologic studies showed that Cu stress results in reduction of biomass production and root elongation (Schiller 1974; Wachsmann 1959), reduction of water content and increase of the osmotic value of the stem (Lolkema and Vooijs 1984; Schiller 1974), degeneration of chlorophyll and resulting chlorosis (Lolkema and Vooijs 1984), and disturbance of nitrogen uptake (Weber et al. 1991). According to de Vos et al. (1991, 1992, 1993), the primary effect of Cu toxicity is damage of the membrane permeability barrier in the root tip resulting in membrane lipids becoming oxidized by free radicals. Moreover, copper oxidizes all cellular thioles.

Tolerance mechanisms were suggested to result from reduced copper uptake, from increased translocation into the stem due to more effective copper transport in the xylem, from external detoxification by excretion of detoxifying molecules, and probably also from detoxification within the cytoplasm (Lolkema et al. 1984, 1986; Lolkema and Vooijs 1984; Schiller 1974). Anatomical studies suggest a strong protective effect of the root endodermis (Punz and Sieghardt 1993). Phytochelatins, complex metals in plants, have been suggested as another mechanism to cope with metal stress independent from tolerance (Schat and Kalff 1992). On the other hand, their production requires sulfate reduction and is therefore energetically expensive for the plant (Steffens 1990). Hence, phytochelatins appear to be not crucial for detoxification of metals but are rather synthesized upon oxidative stress (de Vos et al. 1992).

Metal tolerant ecotypes of *Silene vulgaris* have an increased requirement for Cu and Zn, both essential micronutrients for the cellular metabolism, because Cu and Zn become irreversibly immobilized in the course of detoxification (Ernst 1982a, 1983). Thus, the optimal copper supply is shifted toward higher concentrations in copper-tolerant ecotypes.

11.6.4 Copper Mosses of the Schwarzwand

Since Gams (1966), the Schwarzwand is well known as an area of copper mosses, including *Mielichhoferia nitida*, *Mielichhoferia elongata*, *Merceya ligulata*, and *Grimmia atrata*. *Nardia scalaris* (Jungermanniaceae) is represented with a copper-resistant ecotype on the Schwarzwand; plants transplanted from uncontaminated habitats are not able to establish (Sissolak 1984). Zechmeister and Punz (1990) describe this moss community as a new subassociation Rhacomitro-Andreaeetum rupestris mielichhoferietosum.

Brooks (1971) reported that the appearance of *Mielichhoferia* and *Merceya* is an indication of high copper concentrations of the soil. *Mielichhoferia* is listed in the literature as "copper moss" (Shaw 1989) and well known to appear on soils with high copper concentrations in other parts of the world. Besides *Mielichhoferia*, *Pohlia drummondii* is found at the Schwarzwand. Although *Pohlia* has not been listed as a typical "copper moss" so far, the genus is observed occasionally on metal-rich soils (Folkeson and Anderssonbringmark 1988; Zechmeister and Punz 1990). A few studies (Hörmann 2001; Hörmann et al. 2001a; Hus 2008; Weidinger et al. 2001) took a closer look on the metal tolerance of *Mielichhoferia elongata* and *Pohlia drummondii*. It was shown that both species do not only tolerate high metal concentrations (Table 11.4) but also accumulate metals in significant amounts (Table 11.5). The concentrations within the plants are well above the soil values, especially if compared with the bioavailable fraction of copper in soil. Older parts of the mosses accumulate greater amounts of metals than younger parts. *In vitro* grown mosses clearly take up copper from the growth medium. *M. elongata* tolerates higher copper concentrations than *P. drummondii* but shows reduced growth under strong copper stress (Wernitznig et al. 2009). Microscopic studies indicate the exclusion of metals from the protoplast and retention in the cell walls of *P. drummondii* (Wernitznig et al. 2010).

Table 11.4 Resistance of leaf cells of mosses and *Saxifraga stellaris* from the Schwarzwand against heavy metal stress [g l^{-1}]

Species	Cu^{2+}	Zn^{2+}	Mn^{2+}	Cr^{3+}	VO^{3+}	Fe^{2+}
Saxifraga stellaris	10^{-2*b} $10^{-3\dagger b}$	$10^{-1\dagger *b}$		$10^{-4\dagger *b}$	$10^{-3\dagger *b}$	
Grimmia atrata	$10^{-2\dagger *c}$	$>0.5^{\dagger c}$ 10^{-1*c}		$10^{-3\dagger *c}$	$10^{-2\dagger c}$ 10^{-3*c}	
Lophoza wenzelii	$10^{-1\dagger c}$ 10^{-2*c}	$10^{-1\dagger c}$ 10^{-2*c}		$10^{-2\dagger c}$ 10^{-3*c}	$10^{-2\dagger c}$ 10^{-3*c}	
Merceya ligulata	$>0.5^{\dagger *c}$	$>0.5^{\dagger *c}$		$0.5^{\dagger *c}$	$0.5^{\dagger *c}$	
Mielichhoferia mielichhoferi	$>0.5^{\dagger c}$ $0.5–10^{-1}*c^3$ $10^-1^{\dagger d}$	$>0.5^{\dagger 3c}$ $0.5\text{-}10^{-1*3}$		$>0.5^{\dagger *c,}$ d	$>0.5^{\dagger *c}$ $10\text{-}1^{\dagger d}$	
Mielichhoferia elongata	$10^{-2}–10^{-3*a}$ $>0.5^{\dagger c}$ 10^{-1*c}	10^{-2*a} $>0.5^{\dagger c}$ 0.5^{*c}	$10^{-2}–10^{-3*a}$	$>0.5^{\dagger *c}$	$>0.5^{\dagger *c}$	$10^{-1}–10^{-2*a}$
Nardia scalaris	$10^{-2\dagger *b}$ 10^{-1}– $10^{-2\dagger c}$ 10^{-3*c}	$>0.5^{\dagger *b}$ $10^{-3\dagger *c}$		$10^{-2\dagger *b}$ $10^{-2\dagger *c}$	$10^{-1\dagger b}$ 10^{-2*b} $10^{-2\dagger b}$ 10^{-3*b}	
Pohlia drummondii	$10^{-3}–10^{-5*a}$	$10^{-2}–10^{-4*a}$	$10^{-4}–10^{-6*a}$			$10^{-3}–10^{-4*a}$
Scapania undulata	$10^{-2}–10^{-5*a}$	$10^{-2}–10^{-4*a}$	$10^{-2}–10^{-3*a}$			$10^{0}–10^{-2*a}$

Leaflets were exposed to heavy metals for 24 h (†), 48 h
(*) Data after (a) Hus (2008), (b) Sissolak (1984), (c) Saukel (1980), (d) Url (1956). Only species with a remarkable level of resistance were incorporated into the table

Table 11.5 Heavy Metal content of the leaves of selected plants from the Schwarzwand

Species	Cu	Zn	Mn	Cr	Al	Fe
Merceya ligulata	11–17[a] 7[b]					
Mielichhoferia mielichhoferi	5–11[a] 7[b] 4,310[e]					7,150[e]
Pohlia drummondii	3–103[a] 8,540[e]	2–3[a]	≤1[a]	≤0.4[a]	11–998[a]	2–110[a] 6,230[e]
Oligotrichum hercynicum	1–4[a]	≈1[a]	≈1[a]	≤0.1[a]	98–109[a]	9–69[a]
Saxifraga stellaris	2–87[f]	3–7[f]	1–8[f]	≤2[f]	24–1,042[f]	7–101[f]
Silene vulgaris	235[d]					
Juncus trifidus	130[a]	147[a]	75[a]	0[a]	888[a]	708[a]
Rhododendron ferrugineum	95[a]	2[a]	1[a]		7[a]	1[a]
Vaccinium myrtillus	21[a]	2[a]	1[a]		2[a]	1[a]
Larix decidua	296[a]	33[a]	46[a]	0[a]	669[a]	261[a]

Data in mg kg^{-1}, rounded to whole digits, after (a) Saukel (1980), (b) Ernst (1974), (c) Stummerer (1970), (d) Mutsch (1980), (e) Hus (2008), (f) Sissolak (1984)

11.7 Conclusions and Future Aspects

In spite of the small size and the nonexistent economic importance of the three sites, much knowledge and information has been collected over the years. Tofernalm and Troiboden are very similar to metal-rich ecosystems that have resulted from mining activities in industrial times. It may be concluded that the well-known metal tolerant vegetation, which establishes on modern spoil heaps within a relatively short time is already a climax that may be stable for many centuries.

The Schwarzwand, however, is a unique habitat, with respect both to its diverse moss flora and to the constant precipitation of copper in the creeks. Although the weathering of ores in the galleries and possibly also in the spoil heaps constantly frees metals, the creeks are re-colonized and the habitat does not export significant amounts of metals. Obviously, imitation of the processes of the Schwarzwand offers a possibility to remediate other metal-rich or contaminated sites. As a prerequisite for utilization however, a more complete understanding of the metal cycling of the Schwarzwand is needed. This includes additional data on the water and soil chemistry, mathematical modeling of weathering and precipitation, and probably studies on the microflora of different parts of the sites.

Although more similar to more typical copper enriched sites, the Illyrian spoil heaps on Hochkönig exemplify two important features of copper contamination: (1) Copper contamination may last forever. Even after 2,500–4,000 years, the spoil heaps lack dense vegetation. (2) On the other hand, copper contamination can be constantly restricted to even the smallest areas. Even after several thousands of years, only very limited metal concentrations are found in surrounding habitats, which are below the legal limits (Glaeser 2001) and do not exert a visible influence on the vegetation. The importance of these findings for the long-term storage of contaminated material is obvious.

For several reasons, the ecosystems described in this review are a highly attractive model system:

1. Climate, altitude, and use of the landscape are virtually the same in all three areas. Furthermore, they belong to the same floristic realms. Comparison between the habitats is therefore facilitated.
2. Concerning other parameters, a pair-wise comparison is possible, e.g., ore with a high mineralogical diversity at the Troiboden vs. low diversity at the Grossarl Valley, wet and dry spoil heaps, effects of running and still water, effects of heaps containing overburden or slags, etc.
3. In habitats exposed to recent contamination, only hypotheses can be made on long-term effects, and the need for remediation anticipates any test of these hypotheses. On the Troiboden, long-term effects of more than 2,500 years of contamination can be studied.
4. None of the three sites poses a threat to human health (Feitzinger et al. 1998); thus, the natural development and succession of the habitats can be studied without any need to interfere to achieve accelerated remediation.

The present literature, however, deals mainly with the properties of isolated habitats and less with such comparisons. Furthermore, little of this literature is available to international researchers, and many studies are old, when modern techniques were not available. Especially mathematical modeling of habitats such as the Schwarzwand can be expected to yield important results.

References

Aeschimann D, Lauber K, Moser DM and Theurillat J-P (2004) Flora Alpina. Ein Atlas sämtlicher 4500 Gefäßpflanzen der Alpen. Haupt-Verlag, Bern.

Appenroth K-J (2010) Definition of "Heavy Metals" and their role in biological systems. In: Varma A, Sherameti I (eds) Soil heavy metals. Springer, Heidlberg, pp 19–30

Bagchi D, Bagchi M, Stohs SJ (2001) Chromium (VI)-induced oxidative stress, apoptotic cell death and modulation of p53 tumor suppressor gene. Mol Cell Biochem 222:149–158

Basiouny FM (1984) Distribution of Vanadium and its influence on chlorophyll formation and iron-metabolism in tomato plants. J Plant Nutr 7:1059–1073

Bernhard J (1965) Die Mitterberger Kupferkieslagerstätte. Erzführung und Tektonik. Jahrbücher der Geologischen Bundesanstalt 109:3–90

Biebl R (1947) Über die gegensätzliche Wirkung der Spurenelemente Zink und Bor auf die Blattzellen von *Mnium rostratum*. Oesterreichische Botanische Zeitschrift 94:61

Biebl R (1949) Vergleichende chemische Resistenzstudien an pflanzlichen Plasmen. Protoplasma 39:1–13

Biebl R (1950) Über die Resistenz pflanzlicher Plasmen gegen Vanadium. Protoplasma 39:251–259

Biebl R, Rossi-Pillenhofer W (1954) Die Änderung der chemischen Resistenz pflanzlicher Plasmen mit dem Entwicklungszustand. Protoplasma 44:113–135

Brooks RR (1971) Bryophytes as a guide to mineralisation. NZ J Bot 9:674–677

de Vos CHR, Schat H, de Waal MAM, Vooijs R, Ernst WHO (1991) Increased resistance to copper-induced damage of the root cytoplasma in copper tolerant *Silene cucubalus*. Physiol Plant 82:523–528
de Vos CHR, Vonk MJ, Vooijs R, Schat H (1992) Glutathione depletion due to copper-induced phytochelatine synthesis causes oxidative stress in *Silene cucubalus*. Plant Physiol 98:853–858
de Vos CHR, Ten Bookum WM, Vooijs R, Schat H, de Kok LJ (1993) Effect of copper on fatty acid composition and peroxidation of lipids in the roots of copper tolerant and sensitive *Silene cucubalus*. Plant Physiol Biochem 31:151–158
Del-Negro W (1983) Geologie des Landes Salzburg. Amt der Salzburger Landesregierung. Landespressebüro, Salzburg
Derkmann KJ (1976) Geochemisch-lagerstättenkundliche Untersuchungen an Kiesvorkommen in den Metabasiten der oberen Tauern-Schieferhüllte, Ph.D. thesis, Ludwig-Maximilian-Universität, Munich
Drapela J, Jungmeier M, Kircmeir H, Kohlmayr B, Dullnig G, Zollner D (2001) Almwirtschaftliche Nutzungserhebung – Nationalpark Hohe Tauern Salzburg. E.C.O. Institut für Ökologie, Klagenfurt
Eckl P, Türk R, Hofmann W (1984) Natural and man-made radionuclide concentrations in lichens at several locations in Austria. Nord J Bot 4:521–524
Eibner C (1993) Die Pongauer Siedlungskammer und der Kupferbergbau in der Urzeit. In: Günther W, Eibner C, Lippert A, Paar W (eds) 5000 Jahre Kupferbergbau Mühlbach am Hochkönig-Bischofshofen. Gemeinde Mühlbach am Hochkönig, Mühlbach am Hochkönig, pp 11–26
Ernst W (1974) Schwermetallvegetation der Erde. Gustav Fischer Verlag, Stuttgart
Ernst WHO (1982a) Fluor-und Selenpflanzen. In: Kinzel H (ed) Pflanzenökologie und Mineralstoffwechsel. Eugen Ulmer Verlag, Stuttgart, pp 507–519
Ernst WHO (1982b) Schwermetallpflanzen. In: Kinzel H (ed) Pflanzenökologie und Mineralstoffwechsel. Eugen Ulmer Verlag, Stuttgart, pp 472–605
Ernst WHO (1983) Ökologische Anpassungen an Bodenfaktoren. Berichte der Deutschen Botanischen Gesellschaft 96:49–71
Feitzinger G, Günther W, Brunner A (1998) Bergbau- und Hüttenaltstandorte im Bundesland Salzburg. Land Salzburg - Abteilung 16 Umweltschutz, Salzburg
Folkeson L, Anderssonbringmark E (1988) Impoverishment of vegetation in a coniferous forest polluted by copper and zinc. Canadian Journal of Botany-Revue Canadienne de Botanique 66:417–428
Friedrich OM (1936) Zur Geologie der Kieslager des Großarltals. Hölder-Pichler-Tempsky, Wien
Gams H (1966) Erzpflanzen der Alpen. Jahrbuch des Vereins zum Schutze der Alpenpflanzen und -Tiere 31:65–72
Glaeser O (2001) Materialienband zum Salzburger Klärschlammkonzept 2001. Stand Juni 2001. Land Salzburg, Abteilung 16 der Salzburger Landesregierung, Salzburg
Günther W (1993) Von der Mitterberger Kupfergewerkschaft zur Kupferbergbau Mitterberg Ges. m.b.H. in Mühlbach am Hochkönig. Zur neuzeitlichen Entwicklung des Kupferbergbaues in Mühlbach am Hochkönig, St. Johann im Pongau und Bischofshofen 1829–1977. In: Günther W, Eibner C, Lippert A, Paar W (eds) 5000 Jahre Kupferberbau Mühlbach am Hochkönig-Bischofshofen. Gemeinde Mühlbach am Hochkönig, Mühlbach am Hochkönig, pp 57–367
Günther W, Eibner C, Lippert A, Paar W (eds) (1993) 5000 Jahre Kupferbergbau Mühlbach am Hochkönig-Bischofshofen. Gemeinde Mühlbach am Hochkönig, Mühlbach am Hochkönig
Günther W (ed) (2006) Salzburgs Bergbau und Hüttenwesen im Wandel der Zeit. Buntmetalle und stahlveredelnde Metalle. Leoganger Bergbaumuseumsverein, Leogang
Henselig KO (1981) Bronze, Eisen, Stahl. Bedeutung der Metalle in der Geschichte. Rowohlt Taschenbuch Verlag, Reinbek bei Hamburg
Herdits H (1997) Ein bronzezeitlicher Kupferverhüttungsplatz in Mühlbach/Hochkönig (Salzburg). Diploma thesis, Universität Wien, Vienna
Höfler K (1932) Vergleichende Protoplasmatik. Berichte der Deutschen Botanischen Gesellschaft 50:53–67

Holzer H (1966) Erläuterungen zur Karte der Lagerstätten mineralischer Rohstoffe der Republik Österreich. In: Beck-Mannagetta P, Grill R, Holzer H, Prey S (eds) Erläuterungen zur Geologischen und zur Lagerstatten-Karte 1:1,000.000 von Österreich. Geologische Bundesanstalt, Wien, pp 30–57

Hörmann D (2001) Schwermetallanalyse bei Kupfermoosen. Über die Elementverteilung in Moospflanzen. Universität Wien, Vienna

Hörmann D, Lichtscheidl I, Weidinger M, Gruber D, Klepal W (2001a) The localisation of heavy metals in mosses from copper-rich soil by scanning electron microscopy. In: "Dreiländertagung für Elektronenmikroskopie", a Conference for Modern Microscopical Methods, Innsbruck, 20

Hörmann D, Lichtscheidl IK, Weidinger M, Gruber D, Klepal W, Horak O (2001b) Über die Lokalisation von Schwermetallen in Mooszellen. In: 14. Tagung des Arbeitskreises für Pflanzenphysiologie, Neuberg an der Mürz, Österreich

Hübner L (1796) Beschreibung des Erzstiftes und Reichsfürstenthums Salzburg in Hinsicht auf Topographie und Statistik. Zweyter Band. Das Salzburgische Gebirgsland. Pongau, Lungau und Pinzgau. L. Hübner, Salzburg

Hus K (2008) Die Aufnahme und Verteilung von Schwermetallen in Kupfermoosen der Schwarzwand/Salzburg. Diploma thesis, Universität Wien, Vienna

Iljin WS (1935) Das Absterben der Planzenzellen in reinen und balancierten Salzlösungen. Protoplasma 24:409–430

Kalmykova Y, Strömvall A-M, Steenari B-M (2008) Adsorption of Cd, Cu, Ni, Pb and Zn on *Sphagnum* peat from solutions with low metal concetrations. J Hazard Mater 152:885–891

Kienlin TL, Bischoff E, Opielka H (2006) Copper and bronce during the eneolithic and early bronze age: a metallographic examination of axes from the northalpine region. Archaeometry 48:453–468

Kirnbauer F (1968) Historischer Bergbau I und II. Bergbaue, Schmeltzhütten, Hammerwerke und Salinen, Münz- und Prägestätten in der Zeit von 1500 bis 1600. In: Wolfram R (ed) Österreichischer Volkskundeatlas, Kommentar. Verlag Hermann Böhlaus Nachf. Ges.m.b.H. Wien-Köln-Graz, Wien, pp 1–70

Körber-Ulrich SM (1990) Comparative studies on metal tolerance and metal uptake in *Silene vulgaris*. Phyton 30:323–324

Körber-Ulrich SM (1996) Physiologische Untersuchungen zur Schwermetallresistenz von Normal- und Kupferpopulationen des Gemeinen Leimkrautes *Silene vulgaris* (Moench) Garcke im österreichischen und norditalienischen Ostalpenraum. Ph.D. thesis, Universität Wien, Vienna

Krausmann F (2008) Land use and socioeconomic metabolism in pre-industrial agricultural systems: four nineteenth-century Austrian villages in comparison. Soc Ecol Work Paper 72:1–45

Lechner K, Holzer H, Ruttner A, Grill R (1964): Karte der Lagerstätten mineralischer Rohstoffe der Republik Österreich 1:1,000.000. - Geol. B.A., Wien

Ledl G, Horak O, Janauer GA (1980) Mangan, Eisen, Kupfer und andere Schwermetalle in einigen österreichischen Fließgewässern. Österreichische Abwasser-Rundschau 1:28–32

Levitt J (1980) Responses of plants to environmental stresses: water, radiation, salt and other stresses. Academic, New York

Lippert A (1993) Frühe Zeugnisse von Kupfermetallurgie im Raum Mühlbach am Hochkönig-Bischofshofen. In: Günther W, Eibner C, Lippert A, Paar W (eds) 5000 Jahre Kupferbergbau Mühlbach am Hochkönig-Bischofshofen. Gemeinde Mühlbach am Hochkönig, Mühlbach am Hochkönig, pp 27–40

Lolkema PC, Donker MH, Schouten AL, Ernst WHO (1984) The possible role of metallothioneins in copper tolerance of *Silene cucubalus*. Planta 162:174–179

Lolkema PC, Vooijs R (1984) Copper tolerance in *Silene cucubalus*. Subcellular distribution of copper and its effects on chloroplasts and plastocyanin synthesis. Planta 167:30–36

Lolkema PC, Doornhof M, Ernst WHO (1986) Interaction between a copper-tolerant and a copper-sensitive population of *Silene cucubalus*. Physiol Plant 67:654–658

Ludwig K-H (1979) Die Agricola-Zeit im Montangemälde. Frühmoderne Technik in der Malerei des 18. Jahrhunderts. Verlag des Vereins Deutscher Ingenieure, Düsseldorf

Ma W, Tobin JM (2004) Determination and modelling of effects of pH on peat biosorption of chromium, copper and cadmium. Biochem Eng J 18:33–40

Mallick N, Mohn FH (2003) Use of chlorophyll fluorescence in metal-stress research: a case study with the green microalga *Scenedesmus*. Ecotoxicol Environ Saf 55:64–69

Marschner H (1995) Mineral nutrition of higher plants. Academic Press, London

Moesta H (1989) The furnace of Mitterberg. An oxidizing bronze age copper process. Bullet Metals Museum 14:5–16

Mutsch F (1980) Schwermetallanalysen an Freilandpflanzen im Hinblick auf die natürliche Schwermetallversorgung und die Schwermetalltoxikation. Ph.D. thesis, Universität Wien, Vienna

Nieboer E, Richardson DHS (1980) The replacement of the nondescript term "heavy metals" by a biologically and chemically significant classification of metal ions. Environ Pollut (Ser B) 1:3–26

Nowak O (1933) Die Kupferkieslagerstätte von Mitterberg bei Bischofshofen in Salzburg und ihre Beziehung zu den übrigen ostalpinen Erzlagerstätten der nördlichen Grauwackenzone. Ph.D. thesis, Universität Wien, Vienna

Österreichische Akademie der Wissenschaften (ed) (1975) Österreichisches Biographisches Lexikon. vol 6 (Mai – Mus). ÖAW, Wien

Paar WH (1993) Geologische Einführung und Mineralogie der Kupfererzlagestätten im Raum Mühlbach am Hochkönig - Bischofshofen. In: Günther W, Eibner C, Lippert A, Paar W (eds) 5000 Jahre Kupferbergbau Mühlbach am Hochkönig-Bischofshofen. Gemeinde Mühlbach am Hochkönig, Mühlbach am Hochkönig, pp 41–55

Pabinger P (without year) Nationalpark Hohe Tauern. Hüttschlager Anteil. Nationalparkverein Hohe Tauern, Hüttschlag

Peer H, Zimmer W (1980) Geologie der Nordrahmenzone der Hohen Tauern (Gasteiner Ache bis Saukarkopf-Großarltal). Jahrbücher der Geologischen Bundesanstalt 123:411–466

Pernicka E, Wunderlich C-H, Reichenberger A, Meller H, Borg G (2008) Zur Echtheit der Himmelscheibe von Nebra - eine kurze Zusammenfassung der durchgeführten Untersuchungen. Archäologisches Korrespondenzblatt 38:331–352

Petraschek WE, Pohl W (1982) Lagerstättenlehre. Eine Einführung in die Wissenschaft von den mineralischen Bodenschätzen. E. Schweizerbart'sche Verlagsbuchhandlung, Stuttgart

Plette ACC, Nederlof MM, Temminghoff EJM, van Riemsdijk WH (1999) Bioavailability of heavy metals in terrestrial and aquatic systems: a quantitative approach. Environ Toxicol Chem 18:1882–1890

Poelt J (1955) Flechten der Schwarzen Wand in der Großarl. Verhandlungen der Zoologisch-Botanischen Gesellschaft in Österreich 95:107–113

Prat S (1934) Die Erblichkeit der Resistenz gegen Kupfer. Berichte der Deutschen Botanischen Gesellschaft 52:65–67

Punz W (1999) Zur Moosflora auf Bergbauhalden und anderen Schwermetallstandorten im Ostalpenraum – ein Überblick. Abhandlungen der Zoologisch-Botanischen Gesellschaft in Österreich 30:131–140

Punz W (2008) Schwermetallstandorte und deren Vegetation im Land Salzburg. Sauteria 16:375–378

Punz W, Koerber-Ulrich S (1993) Resistenzökologische Befunde zu schwermetallbewohnenden Pflanzen im Ostalpenraum. Verhandlungen der Zoologisch-Botanischen Gesellschaft in Österreich 130:201–224

Punz W, Mucina L (1997) Vegetation on anthropogenic metalliferrous soils in the Eastern Alps. Folia Gebotanica et Phytotaxonomica 32:283–295

Punz W, Sieghardt H (1993) The response of roots of herbaceous plant species to heavy metals. Environ Exp Bot 33:85–98

Punz W, Orasche IC (1995) Pflanzen auf Schwermetallstandorten im Ostalpenraum und deren Häufigkeitsverteilung. Verhandlungen der Zoologisch-Botanischen Gesellschaft in Österreich 132:61–80

Repp G (1963) Die Kuperresistenz des Protoplasmas höherer Pflanzen auf Kupfererzböden. Protoplasma 57:643–659

Saukel J (1980) Ökologisch-soziologische, systematische und physiologische Untersuchungen an Pflanzen der Grube "Schwarzwand" im Großarltal (Salzburg). Ph.D. thesis, Universität Wien, Vienna

Schat H, Kalff MMA (1992) Are phytochelatins involved in differential metal tolerance or do they merely reflect metal-imposed strains? Plant Physiol 99:1475–1480

Schat H, Kuiper E, Ten Bookum WM, Vooijs R (1993) A general model for the genetic control of copper tolerance in *Silene vulgaris*: evidence from crosses between plants from different tolerant populations. Heredity 70:142–147

Schiller W (1974) Versuche zur Kupferresistenz bei Schwermetallökotypen von *Silene cucuballus* Wib. Flora 163:327–341

Schubert R (1953) Die Schwermetallpflanzengesellschaften des östlichen Harzvorlandes. Wiss. Z. Martin-Luther-Univ. Halle-Wittenberg, Math. Nat. Kl. III 1953/54, 51 70

Shaw AJ (1989) Metal tolerance in bryophytes. In: Shaw AJ (ed) Heavy metal tolerance in plants: evolutionary aspects. CRC, Boca Raton, FL, pp 133–152

Siegl W (1972) Die Uranparagenese von Mitterberg (Salzburg, Österreich). Tschmeraks Mineralogisch Petrologische Mitteilungen 17:263–275

Sissolak M (1984) Ökophysiologische Untersuchung an Pflanzen an kupferbelasteten und unbelasteten Standorten im Gebiet von Hüttschlag (Salzburg). Ph.D. thesis, Universität Wien, Vienna

Sordian H (1983) Zur Erfassung der Lockergesteine und zur baugeologischen Detailkartierung des Gebietes Dachegg - Dientner Sattel - Elmau (Dienten - Mühlbach am Hochkönig, Salzburg, Österreich). Archiv für Lagerstättenforschung der geologischen Bundesanstalt 5:157–165

Sperl G (1987) Frühes Hüttenwesen in Niederösterreich. In: Kustering A (ed) Bergbau in Niederösterreich, Vorträge und Diskussionen des sechtsen Symposions des Niederösterreichischen Instituts für Landeskunde, Pitten, 1–3. Juli 1985. Selbstverlag des NÖ Instituts für Landeskunde, Vienna, pp 411–428

Statistik Austria (2010) Endgültige Bevölkerungszahl 31.10.2008 für die Finanzjahre 2009 und 2010 gemäß § 9 Abs. 9 FAG 2008. Vienna

Steffens JC (1990) The heavy metal-binding peptids in plants. Ann Rev Plant Physiol Mol Biol 41:553–575

Stöllner T, Cierny J, Eibner C, Boenke N, Herd R, Mass A, Röttger K, Sormaz T, Steffens G, Thomas P (2006) Der bronzezeitliche Bergbau im Südrevier des Mitterberggebietes. Bericht zu den Forschungen der Jahre 2002 bis 2006. Archaeologia Austriaca 90:87–137

Stummerer H (1970) Kupfer-Analysen an Pflanzen Cu-reicher Standorte. Österreichische Botanische Zeitschrift 118:189–193

Suhling L (1983) Aufschließen, Gewinnen und Fördern. Geschichte des Bergbaus. Rowohlt Taschenbuch Verlag, Reinbek bei Hamburg

Tinner W, Lotter AF, Ammann B, Conedra M, Hubschmid P, van Leeuwen JFN, Wehrli M (2003) Climatic change and contemporaneous land-use phases north and south of the Alps 2300 BC to 800AD. Quat Sci Rev 22:1447–1460

Tollmann A (1977) Geologie von Österreich. Band I. Die Zentralalpen. Franz Deuticke, Vienna

Tollmann A (1985) Geologie von Österreich. Band II. Außerzentralalpiner Anteil. Franz Deuticke, Vienna

Urban OH (2002) ". . . und der deutschnationale Antisemit Dr. Matthäus Much" - der Nestor der Urgeschichte Österreichs? Archaeologia Austriaca 86:7–43

Url W (1956) Über Schwermetall-, zumal Kupferresistenz einiger Moose. Protoplasma 46:768–793

Url W (1957) Zur Kenntnis der Todeszonen im konzentrationsgestuften Resistenzversuch. Physiol Plant 10:318–327

Veichtlbauer O, Zeileis A, Leisch F (2006) The impact of interventions on a pre-industrial Austrian alpine population. Coll Antropol 30:1–11

von Hornmayer J (1807) Historisch-statistisches Archiv für Süddeutschland. Doll, Frankfurt

Wachsmann C (1959) Wasserkulturversuche zur Wirkung von Blei, Kupfer und Zink auf die Gartenform und Schwermetallbiotypen von *Silene inflata* SM. Inaugural-Dissertation, Westfälische Wilhelms-Universität, Münster

Weber L, Pausweg F, Medwenitsch W (1972) Zur Mitterberger Kupfervererzung (Mühlbach/Hochkönig, Salzburg). Mitteilungen der Geologischen Gesellschaft in Wien 65:137–158

Weber MB, Schat H, Ten Bookum WM (1991) The effect of copper toxicity on the contents of nitrogen compounds in *Silene vulgaris* (Moench) Garcke. Plant Soil 133:101–109

Weidinger M., Hörmann D., Gruber D., Lichtscheidl I. and Klepal W. (2001) Differential uptake of heavy metals by bryophytes detected by x-ray microanalysis. In: "Dreiländertagung für Elektronenmikroskopie", a Conference for Modern Microscopical Methods, Innsbruck.

Wernitznig S, Lang I, Weidinger M, Sassmann S, Lichtscheidl I (2009) The heavy metal distribution in two copper tolerant bryophytes *Pohlia drummondii* and *Mielichhoferia elongata*. In: EMC 2009, Graz

Wernitznig S, Sassmann S, Lichtscheidl I, Lang I (2010) Uptake and Cellular Visualization of Zinc in the Moss Pohlia drummondii. In: ATSPB2010. Illmitz

Wiebols J (1949) Zur Tektonik des hinteren Groß-Arl-Tales. Jahrbuch der Geologischen Bundesanstalt 93:37–55

Wilkins DA (1957) A technique for the measurment of lead tolerance in plants. Nature 180:37–38

Winkelhofer A (1813) Der Salzach-Kreis. Geographisch, historisch und statistisch beschrieben. Mayr'sche Buchhandlung, Salzburg

Zechmeister H (1997) Schwermetalldeposition in Österreich, erfaßt durch Biomonitoring mit Moosen (Aufsammlung 1995). Umweltbundesamt, Wien

Zechmeister H, Punz W (1990) Zum Vorkommen von Moosen auf schwermetallreichen Substraten, insbesondere Bergwerkshalden, im Ostalpenraum. Verhandlungen der Zoologisch-Botanischen Gesellschaft in Österreich 127:95–105

Zentralanstalt für Meteorologie und Geodynamik (2010) Klimadaten von Österreich 1971–2000. Online under http://www.zmag.ac.at/fix/klima/oe71-00, Accessed Oct 2010

Zhang Y, Banks C (2005) The interaction between Cu, Pb, Zn and Ni in their biosorption onto polyurethane-immobilised *Sphagnum* moss. J Chem Technol Biotechnol 80:1297–1305

Zschocke K, Preuschen E (1932) Das urzeitliche Bergbaugebiet von Mühlbach-Bischofshofen. Anthropologische Gesellschaft, Wien

Chapter 12
Natural Vegetation, Metal Accumulation and Tolerance in Plants Growing on Heavy Metal Rich Soils

Viera Banásová, Eva Ďurišová, Miriam Nadubinská, Erika Gurinová, and Milada Čiamporová

12.1 Introduction

Heavy metal rich habitats characteristic by extremely high concentrations of elements such as Zn, Pb, Cd, Cu, Ni, Co, Cr, etc. occur worldwide. Except natural ore outcrops there are numerous man-made metalliferous habitats located in the surroundings of mines. The spoil heaps occurring close to mines composed of parent stones and leftovers from rocky ore are the main source of the metals. By weathering of such materials the soils with elevated metal contents develop. They are a distinctive phenomenon in the nature as they are unfavorable for most plant species. Plants that can cope with extraordinary high metal concentrations occur there, developing unique adaptation mechanisms for surviving in such habitats. Plant species, comprising heavy metal plant communities, are genetically altered ecotypes with specific tolerances to, e.g., cadmium, copper, lead, nickel, zinc, and arsenic, adapted through microevolutionary processes (Baker et al. 2010; Ernst 1974). Some of them, called metallophytes are confined only to metalliferous substrates (local endemics), e.g., *Viola lutea* subsp. *calaminaria* (Baker et al. 2010) while the plant species occurring in both metalliferous and non-metalliferous habitats were termed "pseudometallophytes" (Lambinon and Auquier 1963). Metallophytes and pseudometallophytes achieve their resistance by a true tolerance strategy resulting from population differentiation (Baker 1987). The evolution of plant species tolerance occurs within one generation in plant populations as a result of the powerful selective forces of metal toxicity provided that appropriate genetic variability is available on which selection can act (Wu et al. 1975).

V. Banásová • E. Ďurišová • M. Nadubinská • E. Gurinová
Institute of Botany, Slovak Academy of Sciences, Dúbravská cesta 9, 845 23 Bratislava, Slovakia

M. Čiamporová (✉)
Institute of Botany, Slovak Academy of Sciences, Dúbravská cesta 9, 845 23 Bratislava, Slovakia
e-mail: milada.ciamporova@savba.sk

E. Kothe and A. Varma (eds.), *Bio-Geo Interactions in Metal-Contaminated Soils*, Soil Biology 31, DOI 10.1007/978-3-642-23327-2_12,

Based on metal uptake, translocation and accumulation, metal tolerant plants are termed either as excluders, or accumulators. Excluders either block the uptake of metals by roots, or they restrict transport of accepted metal ions to the above-ground tissues, so that majority of the metal remains in the roots. Accumulators, on the contrary, take up readily metals and sequester them. Nevertheless, great accumulation does not necessarily equal great tolerance, since tolerance and accumulation are genetically independent traits (Bert et al. 2003; Macnair et al. 1999). Within the group of accumulating plants, the term "hyperaccumulator" has been proposed by Brooks (1998) for plants which (1) have great metal uptake by roots, (2) translocate the metal ions to above-ground parts, and (3) possess high tolerance to these metals at cellular level. Later, the use of this term was shifted and the main criterion became the metal concentration in the above-ground part; threshold values for different metal ions were set conventionally to earn the assignation "hyperaccumulator". Nevertheless, Baker and Whiting (2002) recommend to take into consideration also plants which accumulate extremely high metal amounts in their shoot/leaves, even if they did not reach the thresholds.

Successful colonization of metal-enriched soils requires adaptation of plants, i.e., genetic modifications leading to physiological and structural/morphological traits enabling the plants to tolerate the specific stress conditions. Basal tolerance to heavy metal excess, present in all plant species uses two main mechanisms: sequestration and efflux. Hypertolerance of metal-tolerant plants may use also other mechanisms, but generally they are connected with either metal inactivation or metal transport, which are interconnected (Clemens 2006). Inactivation of metal ions may occur even before a metal enters the root in the rhizosphere via exudation of protons, organic acids, phytosiderophores or by mycorrhizas. In general, the effort of metal-tolerant plants is to regulate metal uptake by roots, metal transport to the above-ground tissues, to allocate metal surplus to metabolically less active tissues (Chardonnens et al. 1999; Verkleij et al. 2009), or to deciduous organs at the time of senescence, and limitation of metal amount allocated into the seeds (Ernst 1974).

Within the cells, heavy metals can be chelated by diverse ligands such as organic acids (malate, citrate), amino acids (histidine, nicotianamine), and two classes of peptides, the metallothioneins and phytochelatins (reviewed by Hall 2002). Zinc was shown to be chelated also by phytic acid ligands (Van Steveninck et al. 1994; Vollenweider et al. 2011). Calcium oxalate crystals present in the cells of various plants and tissues represent structures that may incorporate toxic metal ions (reviewed by Franceschi and Nakata 2005; Nakata 2003). Important mechanisms for heavy metal tolerance and accumulation of these elements in hyperaccumulating plants are the compartmentation of metal surplus in the vacuole (Frey et al. 2000; Küpper et al. 1999; Marques et al. 2004) and binding of toxic ions to cell walls (Bringezu et al. 1999; Frey et al. 2000; Memon and Schröder 2009). These mechanisms provide removal of surplus metals from the cytosol to avoid interference with metabolic processes.

The metal-tolerant plants have evolved adaptations which employ modulation of expression of metal transporter genes, regulation of synthesis of metal-binding substances, or substances related to antioxidative defense mechanisms (Luque-

Garcia et al. 2011; Yadav 2010). Today it is not possible to declare particular genes as responsible for hyperaccumulation or tolerance, but several genes are known to contribute. Generally, these are the genes of metal homeostasis regulation and stress responses. Up to now, investigations have not detected specific metabolites in metal-tolerant ecotypes (Clemens 2006; Weber et al. 2004).

There are a number of metal transporter proteins and presumably only part of them has already been identified. In eukaryotic cells, transporters from three families have been proved. ZIP transporters bring metal ion into the cytoplasm, either from extracellular space or from different cell compartments; CDF transporters work in the opposite direction, i.e., from cytosol into the lumen of organelles or out from the cell; P-type ATPases have been localized on plasma membrane and are active in Zn and Cd efflux and xylem loading (Eren and Arguello 2004; Clemens 2006).

The phenomenon of heavy metal tolerance in plants has been attracting interest of plant ecologists, plant physiologists, and evolutionary biologists for more than 50 years (for review see Baker 1987; Ernst 1974). In the last decades heavy metal-tolerant plants have been increasingly studied, especially the mechanism of metal uptake, transport and accumulation at both physiological and molecular levels (Baker et al. 2010), mainly because of their potential use in phytoremediation (for review see Masarovičová et al. 2010).

Great amount of literature data are available relating to metal uptake and tolerance in the species like *Arabidopsis thaliana*, *A. halleri*, or *Rumex acetosa*. Poorly known are the strategies to cope with heavy metals of their relatives *Arabidopsis arenosa* and *Rumex acetosella*. In this chapter we extend the knowledge of *A. arenosa* and *R. acetosella* relation to surplus of Zn, Pb, Cd, Cu, and we characterize the interaction of soil properties and natural vegetation of mine habitats rich in Zn–Pb–Cd (region Banská Štiavnica 18°˜51′45″E, 48°˜27′55″N), the site A, and rich in Cu (region Staré Hory, 19°˜07′50″E, 48°˜50′23″N), the site B in Central Slovakia. Archeological findings of hammers and small graphite cans testify that primitive mining of ores had been developing since the Bronze period (third to second century BC). From thirteenth to eighteenth century the ore-fields in Banská Štiavnica and Staré Hory regions belonged to the richest in Europe. In medieval times only metal-rich ore had been mined and the poorer ones were removed into the heaps. Due to this we can find several ore minerals in these habitats and the substrates have extraordinary high contents of metals.

In either of the studied sites A and B (with diverse soil pollution) both plant species *A. arenosa* and *R. acetosella* can grow together under natural field condition forming populations A and B respectively. In these populations, we compared heavy metal (Zn, Pb, Cd, Cu) accumulation in plant roots, their translocation to leaves and the degree of metal (Zn, Cu) tolerance at both cellular level and primary root growth, expressed as tolerance index. Root systems developed in *A. arenosa* and *R. acetosella* seedlings growing in the soils of their natural habitats were also compared.

12.2 Soil Characteristics

As both study sites are situated far from potential sources of air pollution, the heavy metal contamination of their soils originates only from the weathered mine wastes. The mine heaps are composed of the waste rocks excavated by deep underground mining of ores. A primitive soil with coarse texture has slowly developed due to incompletely weathered stones. Particle size analysis confirmed that the fraction of over 2 mm stones exceeded 50% in both study sites. Imperfect, poorly developed upper layer (A horizon) reached 0–5 cm. Parent material including ore minerals were the main component of the C horizon. Such stratification of the soil is typical for the wastes originating from the underground mining. The weathering of minerals such as sulfides induced soil acidification, resulting in low pH values especially in the site A (Table 12.1). The soils are highly permeable and dry. The plants on mine heaps are not only confronted with a surplus of one or more metals, but also with a shortage of macronutrients (Ernst 1974). Our analyses documented the low nutrient status of the soils in mine heaps in both investigated sites (Table 12.1).

The metal contents in soils and in plants were determined using atomic absorption spectrometry (Žemberyová et al. 2007). The soil metal concentrations found in the investigated metalliferous habitats (Table 12.2) exceeded the values of natural soils. Extraordinary high contents of Zn, Pb, and Cd were found in the soil of the site A. By contrast very high Cu content was found in the mine heaps in the site B, whereas the concentrations of Zn and Cd were close to those in the natural soils. Only part of total metal amount in the soil is available for uptake by plants. The assessment of heavy metal phytoavailability is usually based on extraction methods using the EDTA or CH_3COOH (exchangeable fraction). However, Zn availability at the soil–root interface can still be difficult to determine satisfactorily (Broadley et al. 2007).

Table 12.1 Selected soil properties of the upper horizon (0–10 cm) taken from the root space of *A. arenosa* and *R. acetosella* in mine heaps in the mine sites A (Zn–Pb–Cd) and B (Cu)

Site	pH	Proportion of stones > 2 mm (%)	*C* (%)	Total *N* (%)	$CaCO_3$ (%)
A	4.5 ± 0.2	63.8 ± 23	1.2 ± 0.1	0.14 ± 0.03	0.8 ± 0.1
B	5.4 ± 0.2	53 ± 19	2.2 ± 1.4	0.16 ± 0.1	0.11 ± 0.01

Table 12.2 Total metal concentration in soil (mg kg^{-1}), and the exchangeable fraction of metals (in %), extracted in 0.05M EDTA or 0.013M CH_3COOH in the samples taken from the rooting zone of *A. arenosa* and *R. acetosella* (depth 0–10 cm) in the mine sites A and B (four soil samples per site; each sample was assayed in three replicates)

Site	Total metal contents (mg kg^{-1})				Exchangeable fraction in EDTA/ CH_3COOH (%)			
	Zn	Pb	Cd	Cu	Zn	Pb	Cd	Cu
A	2,368 ± 615	3,925 ± 1,495	22 ± 7.2	269 ± 55	30/32	42/10	48/47	38/10
B	124 ± 41	107 ± 68	1 ± 1.2	3,558 ± 2,484	22/34	19/72	25/12	19/13

The amounts of the total soil Zn and Cd as well as those extracted by EDTA or CH_3COOH were higher in the site A than in the site B. High percentage of Pb extracted by CH_3COOH was found in the site B. In the case of Cu, low exchangeable amounts were determined using both extractants in site B (Table 12.2).

The difference in the levels of elements extracted by CH_3COOH and EDTA could be connected with the differences in soil pH between the sites. In the more acidic soil in the site A the exchangeable fraction of metals was higher with the exception of Cu. The retention of metals to soil organic matter is also weaker at low pH, resulting in more available metal in the soil solution for root absorption (Prasad and Freitas 2003).

12.3 Characteristics of Vegetation

Toxic concentrations of heavy metals in the soil, properties of a poorly developed soil as well as low nutrient status, and water-deficiency conditions maintain physiognomy, structure, and floristic composition of vegetation on mine spoil heaps (Banásová et al. 2006; Ernst 1974). It is the characteristic that annual plants show neither vigor nor persistence on metalliferous soils (Baker et al. 2010). The specific feature of plant communities on mine heaps in both sites is the prevalence of perennial vascular plants and non-vascular lichens and mosses. High heavy metal concentrations are too toxic for expanding of trees. Extreme ecological conditions retard natural succession. Sporadically dwarf trees of *Betula pendula*, *Salix caprea* or *Picea excelsa* could be found.

Although the surrounding meadows are species-rich, the number of inhabited species on the heaps has been restricted because of the low potential of most plants to evolve metal resistance and to survive there (Banásová et al. 2006; Ernst 2006). The non-metalliferous meadows in the surroundings of mine heaps in both sites serve as propagule sources for the flora on contaminated habitats. However, many propagules extinct under high metal toxicity, nutrient and water deficiency. Extreme ecological conditions are known to determine viability, density of plants, and vegetation composition (Banásová et al. 2006; Ernst 1974). Such relationships have resulted in the occurrence of heavy metal-tolerant plants, e.g., *Agrostis stolonifera*, *Arabidopsis arenosa* subsp. *borbasii*, *Dianthus carthusianorum*, *Rumex acetosella* agg., *Silene dioica*, *S. vulgaris*, *Thymus pulegioides*, *Thlaspi caerulescens*, and *Viola tricolor* at both study sites while the meadow species such as *Dactylis glomerata*, *Briza media*, *Jacea pseudophrygia*, *Knautia arvensis*, *Leucanthemum vulgare*, and *Salvia pratensis* were absent from the heaps although growing commonly in their vicinity.

The typical feature of the vegetation in both study sites is the formation of open mosaic stands with patchy cover, poor in species number and plants growing in the fissures between stones (e.g., fern *Asplenium septentrionale*). For the site A the occurrence of xerothermophilous grassland species, e.g., *Scleranthus perennis* and *Sedum acre*, and of the species with clonal growth like *Pilosella officinarum*,

Fig. 12.1 Typical features of vegetation containing *A. arenosa* and *R. acetosella* in mine heaps (**a**) open mosaic patches in the Zn–Pb–Cd-rich site A (Banská Štiavnica) (**b**) distinctive plant community where vascular plants are less extensive in the Cu-rich site B (Staré Hory)

Teucrium chamaedrys, *Lotus corniculatus*, is characteristic. Also the hyperaccumulator of Zn and Cd *Thlaspi caerulescens* is abundant there (Banásová et al. 2008). Markedly lower number of vascular plants and a dense cover of lichens and mosses is the typical feature of the site B, where the species capable of evolving metal-tolerant ecotypes, e.g., *Agrostis stolonifera*, *Silene dioica*, and *S. vulgaris* grow.

Some differences between both sites with diverse heavy metal composition were found also in the vegetation structure. The ground layer is sparse in Zn–Pb–Cd heaps (site A, Fig. 12.1a). A few terrestrial lichens *Cetraria islandica*, *Cladonia fimbriata*, *C. pyxidata* and mosses *Ceratodon purpureus*, *Hedwigia albicans* occur there. The mosses can occasionally create a dense cover while the lichens appear in low frequency and coverage. On the contrary, the Cu heaps (site B, Fig. 12.1b) are characteristic by the high diversity and abundance of lichens. Terrestrial lichens and mosses occupy ecological niches between stones and developed a dense carpet. High abundance and cover is reached by the species *Ceratodon purpureus*, *Cetraria islandica*, *Cladonia arbuscula* subsp. *mitis*, *C. fimbriata*, *C. furcata*, *C. cariosa*, *C. coniocraea*, *C. pyxidata*, *Parmelia saxatilis* and *Stereocaulon incrustatum*. Metalliferous sites may harbor important populations of rare lichens (Baker et al. 2010) as has been documented also for our localities (Bačkor and Bodnárová 2002; Pišút 1980).

12.4 Characteristics of *Arabidopsis arenosa*, *Rumex acetosella* and Their Relations to Zn, Pb, Cd, and Cu

Plants growing in metal-enriched substrates take up metals to varying degrees in response to external and internal factors (Baker and Walker 1989). They are mostly characterized by enhanced metal levels in their tissues (Ernst 1974, 2006). Metal uptake can be restricted in plant species living in symbiosis with arbuscular mycorrhizal fungi. The investigated plants *A. arenosa* and *R. acetosella* are non-mycorrhizal (Pawlowska et al. 1996; Sandberg et al. 2009). This might have contributed to their high metal uptake. Both investigated species are members of the orders with a considerable incidence of metal accumulation in plant shoots (Broadley et al. 2007). *A. arenosa* and *R. acetosella* belong to orders Brassicales, and Caryophyllales respectively. It is known that both species often occur in metalliferous sites in Central Europe (e.g., Banásová et al. 2008; Ernst 1974; Przedpełska and Wierzbicka 2007; Reeves et al. 2001) and they both are present also in the investigated metalliferous sites A and B without exhibiting symptoms of toxicity (Banásová et al. 2006).

Plants of metalliferous habitats are typical by their tolerance and sequestration of exceptional quantities of metals in their leaves at concentrations that would be toxic to "normal" plants (Baker and Whiting 2002; Ernst 1974). There are two important approaches in assessing accumulating properties: BCF – the ratio of metal concentration in plant tissues to metal concentration in their rooted soils, and TF – the ratio of metal concentration in shoots (leaves) to metal concentration in roots.

Tolerance of metal toxicity can be demonstrated at different levels: in cells, tissues, and organs of a plant (Ernst et al. 1992). Plant roots are directly exposed to negative effects of toxic heavy metal ions. Changes in their growth is one of the fastest and most obvious responses of plants to toxic concentrations of heavy metals, and they can be measured easily (Appenroth 2010; Macnair 1993). Determining root growth exposed to increasing heavy metal concentrations is an effective method for assessing the degree of tolerance of this kind of stress. However, as the root length is not a direct measure of tolerance and, sensitivity to metal ions can change with plant age, it is reasonable to consider more parameters to characterize plant tolerance.

We studied the tolerance to two biogenic elements, zinc and copper of two pseudometallophytes *A. arenosa* and *R. acetosella,* that can withstand excess of both of these elements. Our experimental design was based on the condition that both species have created populations occurring in natural biotopes contaminated with either of these elements. We estimated cellular tolerance, tolerance index, and root system morphology of such *A. arenosa* and *R. acetosella* populations.

The cellular tolerance was based on searching for the threshold metal concentrations, in which leaf cells were able to survive (Marquès et al. 2004). Briefly, sections were excised from mature leaves of three plants, immersed into graded concentrations of $CuSO_4.5H_2O$ or $ZnSO_4.7H_2O$ and after 24 (Cu) or 48 (Zn) h, cell viability was investigated under microscope. In root growth test, the seedlings of the populations were grown on MS agar medium with graded concentrations of Zn (10 as control, 100, 1,000 $\mu M\ l^{-1}$) and Cu (0.1 as control, 1, 10, 100, 1,000 $\mu M\ l^{-1}$). Length of primary seminal roots was measured on the fourth day after germination and the tolerance index (TI) expressed as ratio of the length of roots exposed to metal to the length of control roots, in % (Macnair 1993; Przedpełska and Wierzbicka 2007). The early stages of root system development were compared in the seedlings of populations A and B of both species growing on their original, well watered soils for 14 days. Lengths of the washed primary seminal and lateral roots (intact, with all tips preserved) were measured to characterize root system architecture from each population. Sum of the lengths of all roots for each root system was calculated and average of these data was calculated as the total root length.

12.4.1 Arabidopsis arenosa (L.) Lawalré

12.4.1.1 Biology and Ecology

A. arenosa (syn. *Cardaminopsis arenosa*) commonly occurs on nutrient-poor, non-metalliferous skeletous soils, e.g., dry grasslands, roadsides, sandy areas, and in heavy metal polluted sites (Banásová et al. 2006). Przedpełska and Wierzbicka (2007) tested *A. arenosa* plants originating from a Zn–Pb–Cd metalliferous site, in laboratory conditions and they found significant heritable morphological differences between the metallicolous and non-metallicolous populations.

12.4.1.2 Metal Accumulation

It seems that the only data on heavy metal contents in the species *A. arenosa* are those published by Reeves et al. (2001). They detected high plant concentrations of Zn (5,180 mg kg^{-1}), Pb (849 mg kg^{-1}), and Cd (25 mg kg^{-1}). Our results showed lower values of such elements in this species. In the leaves we found maximum concentrations 1,861 mg kg^{-1} Zn, 105 mg kg^{-1} Pb, and 24.3 mg kg^{-1} Cd. No relevant information about the relation of *A. arenosa* to Cu could be found. In our samples, maximum concentration of Cu was 66.7 mg kg^{-1} in the site B.

According to our analyses from both sites, *A. arenosa* could take up high quantities of Zn, Pb, Cd, and Cu (Fig. 12.2). The concentrations of metals in the plants reflected those in the soil. In the site A with high soil Zn content the accumulation of Zn in plants was higher than in the site B but the difference between roots and leaves was not significant. In the case of low soil Zn concentration (site B) there was a tendency of *A. arenosa* to accumulate more Zn in the leaves than in the roots. The relative species *A. halleri* is known to hyperaccumulate Zn and Cd in the leaves to concentrate over 10,000 and 100 mg kg^{-1} respectively (e.g., Broadley et al. 2007). The capability of *A. arenosa* population from the site A to accumulate Zn and Cd in the leaves was noteworthy reminding of plants with intermediate but very high Zn concentrations frequently over 1,000 mg kg^{-1} (Baker and Whiting 2002). It is open to discussion if this *A. arenosa* population can be considered as hyperaccumulator.

Differences between populations A and B occurred in BCF and TF (Table 12.3). Although Zn leaf concentration in population A exceeded 1,000 mg kg^{-1}, average concentration was not higher than in the soils and roots (BCF < 1; TF < 1). In the

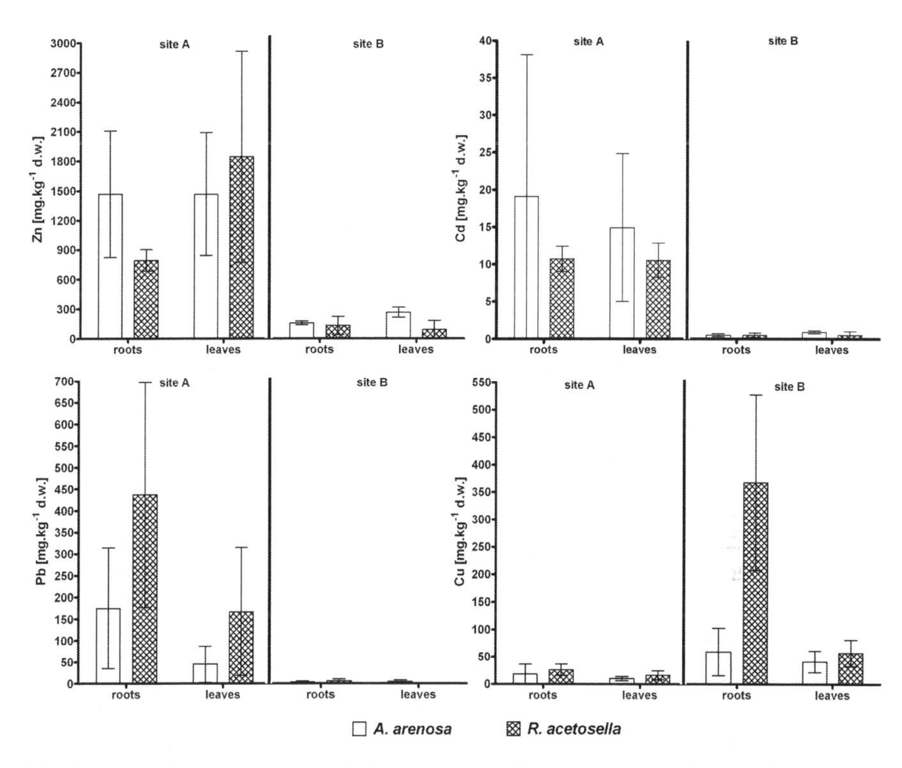

Fig. 12.2 Average metal contents (in logarithmic scale) in the roots and leaves of *Arabidopsis arenosa and Rumex acetosella* from metalliferous sites A (Zn–Pb–Cd-rich soil) and B (Cu-rich soil). Number of samples: four (*A. arenosa*) and five (*R. acetosella*), each analyzed in three replicates; error bars = s.d.

Table 12.3 Bioconcentration factor (BCF) calculated as ratio of average metal concentration in *A. arenosa* and *R. acetosella* leaves to average metal concentration in the soil; translocation factor (TF) calculated as ratio of average metal concentrations in the leaves to average metal concentrations in the roots of *A. arenosa* and *R. acetosella* originating from natural metalliferous habitats

	Site		Zn	Pb	Cd	Cu
A. arenosa	A	BCF	0.6	0.01	0.7	0.04
		TF	0.9	0.2	0.7	0.6
	B	BCF	2.1	0.05	0.9	0.01
		TF	1.7	1.3	1.0	0.7
R. acetosella	A	BCF	0.8	1.6	0.5	0.06
		TF	2.3	0.4	1.7	0.6
	B	BCF	0.7	0	0.5	0.02
		TF	0.7	0	1.0	0.2

population B the leaf Zn concentration was almost ten times lower than in population A. However, the concentration of Zn accumulated in the leaves of *A. arenosa* in the site B exceeded that in the soils and roots as documented by BCF $>$ 1; TF $>$ 1. In the population B the calculated TF $>$ 1 also for Pb and Cd.

12.4.1.3 Metal Tolerance

The cell tolerance test showed that the leaf cells of *A. arenosa* plants from the population A survived in tenfold higher $ZnSO_4$ concentration in comparison with population B, suggesting higher degree of Zn tolerance (Table 12.4). Similarly, using the TI test, a higher level of Zn tolerance (up to 1,000 μM l^{-1}) and even stimulating effect of the concentration 100 μM l^{-1} Zn were found in the population A from the Zn–Pb–Cd site (Table 12.5). In comparison with our results, Przedpełska and Wierzbicka (2007) recorded lower values of TI for Zn in the metallicolous population of *A. arenosa* from the site with elevated Zn and Pb soil concentrations. The difference suggests high intraspecific variability and heterogeneity of this taxon and that the degree of tolerance of a plant population may depend on the characteristics of the habitat.

The Cu tolerance at cellular level was relatively high in both populations but not increased in the population B. According to the tolerance index, the values of *A. arenosa* population A decreased from the concentration 1 μmol l^{-1} onwards, down to extremely strong decrease at 1,000 μmol l^{-1} Cu. On the contrary, in plants from population B the TI values indicated that root growth in concentrations 1 and 10 μmol l^{-1} was even stimulated and the decrease occurred only in the higher Cu concentrations (Table 12.5). These results might be the manifestation of different tolerance in different plant organs and their developmental stage.

The root systems of metallophytes have been little studied outside hydroponic cultures although the root system morphology, particularly its length is important in solute uptake (Fitter 2002) and thus in phytoextraction efficiency. On agar media with increased concentration of Cr(VI) (Castro et al. 2007) or Cu(II) (Lequeux et al. 2010; Pasternak et al. 2005) reorganization of root system architecture resulted

Table 12.4 Threshold metal concentrations, in which mature leaf cells of *A. arenosa* and *R. acetosella* originating from the sites A (Zn–Pb–Cd-rich) and B (Cu-rich) survived for 24 (Cu) or 48 (Zn) h

Population	*A. arenosa*		*R. acetosella*	
	Zn (μM)	Cu (μM)	Zn (μM)	Cu (μM)
A	100	10	10,000	10
B	10	10	100	100

Table 12.5 Tolerance index (in %) of *A. arenosa* and *R. acetosella* seedlings ($n = 25$) from the population A (Zn–Pb–Cd-rich soil) and population B (Cu-rich soil) cultivated under increasing concentrations of $ZnSO_4$ or $CuSO_4$

	Population	Concentration (μM l^{-1})					
		Zn		Cu			
		100	1,000	1	10	100	1,000
A. arenosa	A	115	91	95	80	75	5
	B	91	68	120	128	89	26
R. acetosella	A	197	132	92	98	97	44
	B	132	117	121	108	115	66

from the reduction of primary root growth and stimulation of lateral root formation in *Arabidopsis thaliana*. Various plant species develop different root systems to adapt to their environment and be efficient in element uptake from the soil. High concentration of metal ions in the soil inhibited root growth in non-metalliferous plants like *Cannabis sativa* (Bona et al. 2007), *Vicia sativa* or *Zea mays* (Pečiulytė et al. 2006) but it could also stimulate the development of laterals often leading to a more densely branched root system (Hagemeyer and Breckle 2002).The quantity of available Zn and also the distribution of the metal in the soil influenced the development of root system in metallicolous populations of *Thlaspi caerulescens* (Keller et al. 2003; Schwartz et al. 1999).

In the two differently contaminated soils the seedlings of the respective metallicolous *A. arenosa* populations developed different types of root systems. In the population from the site A (Zn–Pb–Cd) the number of laterals was low. Only 68.6% of primary roots developed laterals. Thus the contribution of the main root to total length of the roots within the root system was dominant (Fig. 12.3a). On the contrary, in the population from the site B (Cu) all seedlings developed lateral roots, and in 28.5% of the seedlings also second order laterals were present (Fig. 12.3b). Their contribution to the total root length within the root system exceeded that of the main root due to both higher number and sum of their lengths. However, the total root length within the root system in population B was lower than that in population A (Fig. 12.3c). Although these results compare only natural populations without experimentally induced non-metal control, they indicate that in *A. arenosa* the enhanced Cu concentration affected particularly root elongation and development of laterals was restricted in the soil enriched in Zn, Pb, and Cd. The diversity of root systems developed by the seedlings of both *A. arenosa* populations confirms that all the important components of root system architecture are susceptible to environmental modifications (Fitter 2002).

12.4.2 *Rumex acetosella* L.

12.4.2.1 Biology and Ecology

R. acetosella occurs in open, non-metalliferous, dry, slightly acid soils, e.g., ruderal sites, but it is also commonly found on the metalliferous wastes in Europe (Babalonas et al. 1997; Banásová 1983; Banásová et al. 2006; Bagatto and Shorthause 1999; Reeves et al. 2001; Rodwell 2000; see also Chap. 11), and can tolerate high metal concentrations also on natural ultramafic soils (Wenzel et al. 2003). According to Baker et al. (2010) and Ernst (1974) this species is tolerant to soil heavy metals and it can be considered as indicator species of the outcrop of ore veins (Banásová et al. 1998; Kelepertsis et al. 1985).

R. acetosella is very good competitor especially on acid substrates which often originate by weathering of sulfides in waste material. It can rapidly occupy the bare space on the contaminated soils where less tolerant plants are not able to grow. Due to its clonal growth it creates extensive and abundant patches. High rate of

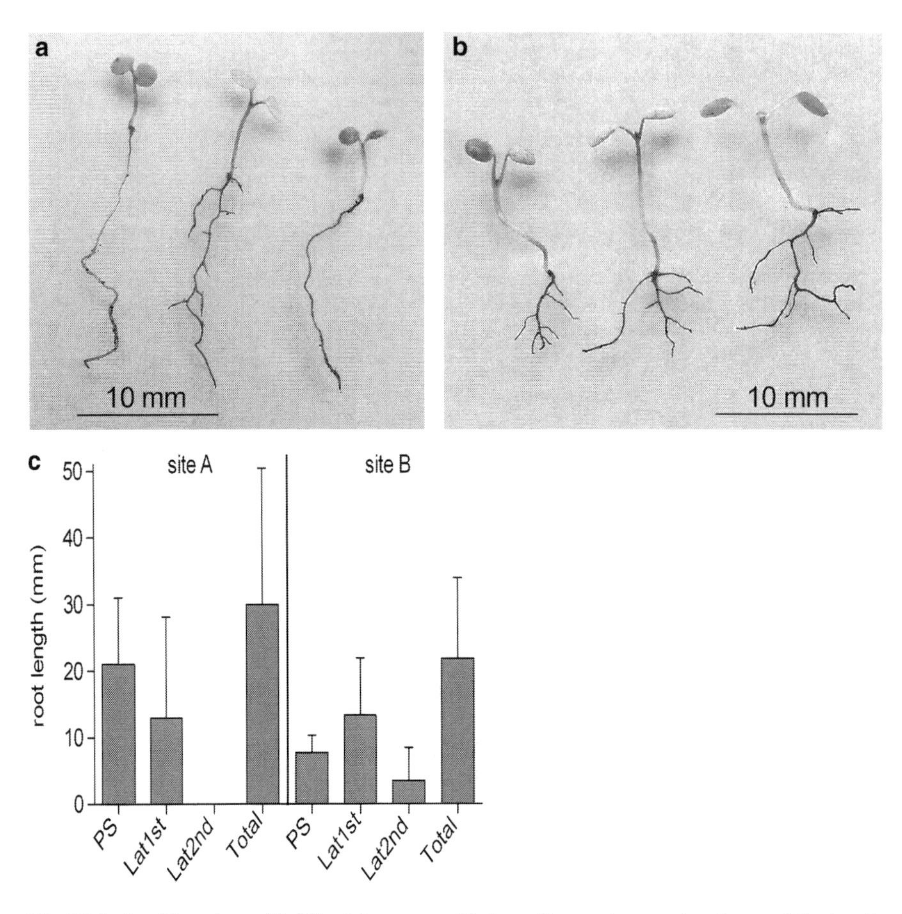

Fig. 12.3 Seedlings (n = 40–50) of *A. arenosa* 14 days after sowing: root system architecture in population A (**a**) and in population B (**b**) and lengths of individual root types (**c**). PS = main (primary seminal) root, (Lat1st) and (Lat2nd) = laterals of first and second order, total = average of the sums of lengths of all root types within each root system (The zero values in case of lateral root absence were omitted, thus the total lengths are not a mere sum of average values of PS + laterals)

R. acetosella root growth, development of numerous new ramets, and formation of extensive and abundant patches are the natural traits of this species on non-metalliferous soils (Lebedev 1993; Röttgerman et al. 2000), and were observed also in the studied metalliferous sites (Banásová 1976).

12.4.2.2 Metal Accumulation

In the shoots of *R. acetosella* from metalliferous sites, high concentrations of Zn 3,100 and 1,390 mg kg^{-1} were reported by Reeves et al. (2001) and Kelepertsis et al. (1985) respectively. We found maximum concentration 3,505 mg kg^{-1} Zn in the leaves of the population A from the soil with high Zn content. Comparing two populations, Zn accumulation was remarkably higher in the site A than in population from the site B. In spite of the high Zn leaf concentration, the values of BCF < 1 were recorded in both sites. Important was, that TF for Zn > 1 was obtained in plants from the habitat A. Remarkable sequestration of Zn into the leaves of this population seems to be its characteristic property. Also our EDX analysis suggested an uneven distribution of Zn in leaf tissues: the highest amount of Zn was detected in subsidiary cells of the stomatal complex, and in the trichomes (unpublished results).

The concentration of Cd and Cu accumulated in *R. acetosella* leaves never exceeded that in the soils of both sites as documented by BCF < 1 (Table 12.3). Our results confirmed also the capability of *R. acetosella* to take up Cd in concentrations higher than those in the majority of plant species (12.9 mg kg^{-1} in population A). The TF > 1 for Cd was recorded in both sites (Table 12.3). *R. acetosella* accumulated higher amount of Cu in the roots than in the leaves (Fig. 12.2) in both localities (maximum 88 mg kg^{-1}), similarly as *R. acetosella* plants grown in hydroponic culture (Masarovičová and Holubová 1999).

The test of cellular tolerance as well as the TI test indicate that *R. acetosella* populations from our study sites are adapted to elevate soil concentrations of respective metal (Table 12.4). The mature leaf cells from the population A were able to survive in extraordinary high Zn concentration (10^{-2} M $ZnSO_4$). The tolerance index also confirmed high Zn tolerance of both *R. acetosella* populations (Table 12.5), with the population A from Zn–Pb–Cd rich habitat being more tolerant than the population B.

At the cellular level the population B tolerated higher Cu concentration in comparison with the population A. Similarly on the basis of TI the higher degree of Cu tolerance and even stimulation of root growth in concentrations 1, 10 and 100 $\mu M\, l^{-1}$ were found in the population B growing on Cu-contaminated soil. The development of tolerant ecotypes in *R. acetosella* from copper mine heaps in the site B was reported by Masarovičová and Holubová (1999).

Similarly as with *A. arenosa*, we found both types of the root system morphology in *R. acetosella* populations, depending on the type of soil contamination (Fig. 12.4a, b). In the population A, the root system of the seedlings consisted of main root and only 32.5% of seedlings possessed also lateral roots of first order but no second order laterals were present. The lengths of the laterals contributed only with 9.6% to the total root lengths within the system. On the contrary, all plants from the population B possessed first order lateral roots and a 12.5% of seedlings have formed also laterals of the second order; all laterals together contributing with 39.6% to the total root length within the root system. Also, the sum of lengths of the main and lateral roots within the root system was almost three times higher in the population B growing on Cu-contaminated soil compared with the population A growing on Zn–Pb–Cd enriched soil (Fig. 12.4c), indicating considerable tolerance of soil Cu toxicity in this *R. acetosella* population. To our best knowledge, there is lack of data on differences in root system architecture, between metallophyte populations growing on soils with different kinds of metal contamination. Although our data were recorded with 14-day-old seedlings, they are important as the early seedling root growth and development can determine the optimum root system throughout the entire life of a plant (Nicola 1998).

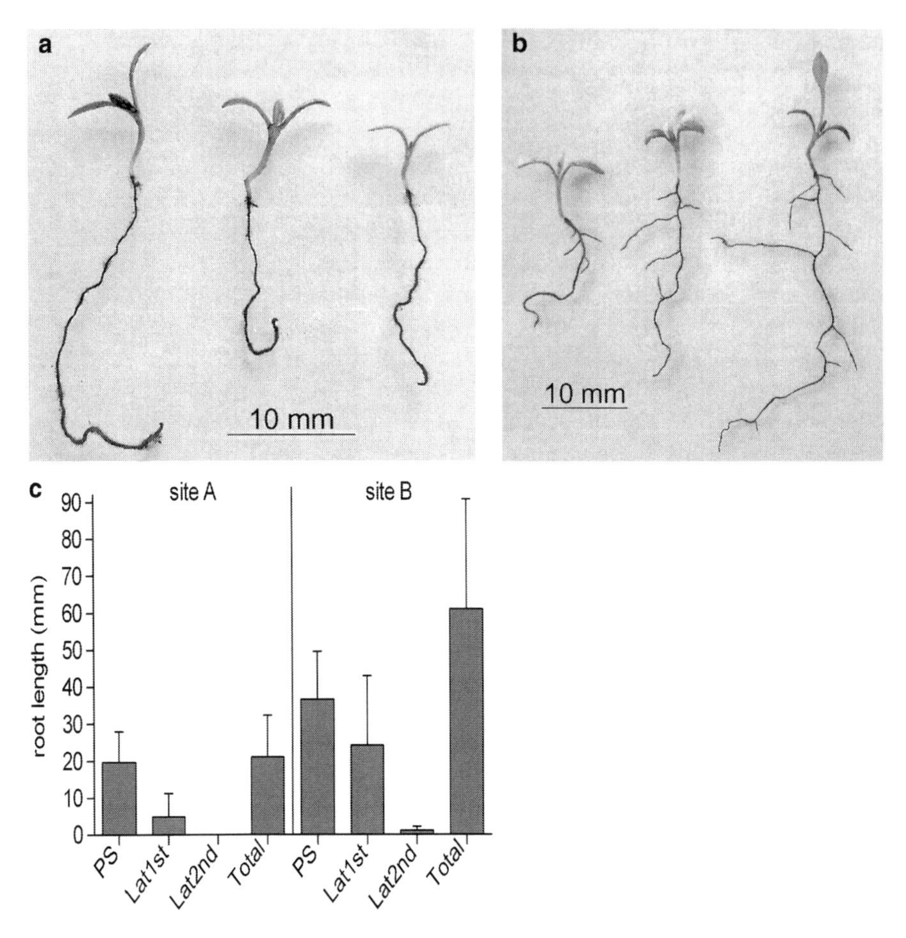

Fig. 12.4 Seedlings ($n = 40$) of *R. acetosella* 14 days after sowing: root system architecture in population A (**a**) and in population B (**b**) and lengths of individual root types (**c**). PS = main (primary seminal) root, (Lat1st) and (Lat2nd) = laterals of first and second order, total = average of the sums of lengths of all root types within each root system (The zero values in case of lateral root absence were omitted, thus the total lengths are not a mere sum of average values of PS + laterals)

12.5 Conclusions

The soil heavy metal content is one of the most important edaphic factors for plant life affecting the development of plant cover with specific species composition. An important aspect of plant strategies under conditions of heavy metal pollution is the development of metal-tolerant ecotypes capable of surviving and reproducing in the contaminated habitats. The two selected former mine habitats in Central Slovakia with soil structure typical for heaps differed in heavy metal contamination: the site A with high soil concentrations of Zn–Pb–Cd, and the site B with soil rich in Cu. In both sites, species poor vegetation with specific structure and with dominating metal-tolerant plants has developed. Especially in the site B the phytotoxicity of high doses of Cu resulted in a low number of vascular plants.

Arabidopsis arenosa and *Rumex acetosella* are abundant in these two mine sites showing no symptoms of injury. Both species take up Zn, Pb, Cd, and Cu into their roots and accumulate Zn in their leaves. Unlike the constitutive trait of Zn and Cd hyperaccumulation known in the species *Arabidopsis halleri*, the metal accumulation was not a constitutive trait in the field populations of *A. arenosa* and *R. acetosella*. The metal accumulation in these populations was connected with the pattern of metal contamination in the soil. Noticeable accumulation of Zn preferably in the leaves was found in *R. acetosella* population growing on the substrate rich in Zn. Similar tendency was found also in *A. arenosa*. Concerning Cu, the accumulation by plants of all populations exceeded that in "normal" plants.

The ecotypes more tolerant to Zn (site A) accumulated more Zn in their tissues when grown in the soil rich in Zn. Ecotypes tolerant to Cu (site B) accumulated more Cu in their tissues when grown in the soil with high Cu concentrations. There were differences in the translocation of Zn into the leaves expressed using TF. The more tolerant *A. arenosa* (population A) showed lower Zn concentration in the leaves than in the roots. The more tolerant *R. acetosella* (population A) translocated Zn into the leaves in higher concentrations than in the roots.

Comparing the two species, data on cellular tolerance and TI provide evidence for a higher Zn and Cu tolerance in *R. acetosella* than in *A. arenosa*. Within each of the species, the populations differ in tolerance: populations originating from Zn-rich habitats are more tolerant to Zn and populations originating from Cu-rich habitats are more tolerant to Cu. The differences in most data on metal uptake and translocation, metal tolerance, and root system morphology recorded in the plant populations from two different habitats suggest that distinct ecotypes have evolved in either of the species *A. arenosa* and *R. acetosella*, as a result of interactions of metal concentrations in the soil and the plants.

Acknowledgments We are especially grateful to Irene Lichtscheidl and Othmar Horak (University of Vienna) for fruitful discussions and co-operation, and to Ivan Pišút (Institute of Botany Bratislava) for lichen determination. The work was supported by the Grants VEGA No. 2/0149/08, 2/0200/10, and APVV-0432-06.

References

Appenroth K-J (2010) Definition of "heavy metals" and their role in biological systems. In: Sherameti I, Varma A (eds) Soil heavy metals. Soil biology, vol 19. Springer, Berlin, pp 19–29

Babalonas D, Mamolos AP, Konstantinou M (1997) Spatial variation in a grassland on soil rich in heavy metals. J Veg Sci 8:601–604

Bačkor M, Bodnárová M (2002) Additions to lichen flora of Slovak Republic I. Thaiszia – J Bot 12:173–178

Bagatto G, Shorthause D (1999) Biotic and abiotic characteristics of ecosystems on acid metalliferous mine tailings near Sudbury, Ontario. Can J Bot 77:410–425

Baker AJM (1987) Metal tolerance. New Phytol 106(Suppl):93–111
Baker AJM, Walker PL (1989) Ecophysiology of metal uptake by tolerant plants. In: Shaw AJ (ed) Heavy metal tolerance in plants: evolutionary aspects. CRC, Boca Baton, FL, pp 156–193
Baker AJM, Whiting SN (2002) In search of the Holy Grail – a further step in understanding metal hyperaccumulation? New Phytol 155:1–7
Baker AJM, Baker M, Ernst WHO, Ent A, Malaisse F, Ginocchio R (2010) Metallophytes: the unique biological resource, its ecology and conservational status in Europe, central Africa and Latin America. In: Batty LC, Hallberg KB (eds) Ecology of industrial pollution. Cambridge University Press, Cambridge, pp 7–40
Banásová V (1976) Vegetation of copper and antimony mine heaps. Biol Práce 22:1–109
Banásová V (1983) Die Vegetation auf Pyrithalden und der Gehalt an Cu, Pb, Zn, As, Ag, Fe und S in den Pflanzen und im Boden. Biológia, Bratislava 38:469–480
Banásová V, Danáková A, Križáni I (1998) Specifity of the vegetation on the ore body Terézia in the Banská Štiavnica region. Bull Slov Botan Spoločn 20:166–172
Banásová V, Horak O, Čiamporová M, Nadubinská M, Lichtscheidl I (2006) The vegetation of metalliferous and non-metalliferous grasslands in two former mine regions in Central Slovakia. Biologia 61:433–439
Banásová V, Horak O, Nadubinská M, Čiamporová M, Lichtscheidl I (2008) Heavy metal content in *Thlaspi caerulescens* J. et C. Presl growing on metalliferous and non-metalliferous soils in Central Slovakia. Int J Environ Pollut 33:133–145
Bert V, Meerts P, Saumitou-Laprade P, Salis P, Gruber W, Verbruggen N (2003) Genetic basis of Cd tolerance and hyperaccumulation in *Arabidopsis halleri*. Plant Soil 249:9–18
Bona E, Marsano F, Cavaletto M, Berta G (2007) Copper stress in *Cannabis sativa* roots: morphological and proteomic analysis. Caryologia 60:96–101
Bringezu K, Lichtenberger O, Leopold I, Neumann D (1999) Heavy metal tolerance of *Silene vulgaris*. J Plant Physiol 154:536–546
Broadley MR, White PJ, Hammond JP, Zelko I, Lux A (2007) Zinc in plants. New Phytol 173:677–702
Brooks RR (ed) (1998) Plant that hyperaccumulate heavy metals. CAB International, Wallingford, CT
Castro RO, Trujillo MM, Bucio JL, Cervantes C, Dubrovsky J (2007) Effects of dichromate on growth and root system architecture of *Arabidopsis thaliana* seedlings. Plant Sci 172:684–691
Chardonnens AN, Ten Bookum WM, Vellinga S, Schat H, Verkleij JAC, Ernst WHO (1999) Allocation patterns of zinc and cadmium in heavy metal tolerant and sensitive *Silene vulgaris*. J Plant Physiol 155:778–787
Clemens S (2006) Toxic metal accumulation, responses to exposure and mechanisms of tolerance in plants. Biochimie 88:1707–1719
Eren E, Arguello JM (2004) Arabidopsis HMA2, a divalent heavy metal-transporting PIB-type ATPase, is involved in cytoplasmic Zn^{2+} homeostasis. Plant Physiol 136:3712–3723
Ernst WHO (1974) Schwermetallvegetation der Erde. Gustav Fischer Verlag, Stuttgart
Ernst WHO (2006) Evolution of metal tolerance in higher plants. For Snow Landsc Res 80:251–274
Ernst W, Verkleij JAC, Schat H (1992) Metal tolerance in plants. Acta Bot Neerl 41:229–248
Fitter A (2002) Characteristics and functions of root systems. In: Waisel Y, Eshel A, Kafkafi L (eds) Plant roots – the hidden half. Dekker, New York, pp 15–32
Franceschi VR, Nakata PA (2005) Calcium oxalate in plants: formation and function. Annu Rev Plant Biol 56:41–71
Frey B, Keller C, Zierhold K, Schulin R (2000) Distribution of Zn in functionally different leaf epidermal cells of the hyperaccumulator *Thlaspi caerulescens*. Plant Cell Environ 23:675–687
Hagemeyer J, Breckle SW (2002) Trace elements stress in roots. In: Waisel Y, Eshel A, Kafkafi L (eds) Plant roots – the hidden half. Dekker, New York, pp 763–785
Hall JL (2002) Cellular mechanisms for heavy metal detoxification and tolerance. J Exp Bot 53:1–11

Kelepertsis AE, Andrulakis I, Reeves RD (1985) *Rumex acetosella* L. and *Minuartia verna* (L.) Hiern. as geobotanical and biochemical indicators for ore deposits in northern Greece. J Geochem Explor 23:203–212
Keller C, Hammer D, Kayser A, Richner W, Brodbeck M, Sennhauser M (2003) Root development and heavy metal phytoextraction efficiency: comparison of different plant species in the field. Plant Soil 249:67–81
Küpper H, Zhao FJ, McGrath SP (1999) Cellular compartmentation of zinc in leaves of the hyperaccumulator *Thlaspi caerulescens*. Plant Physiol 119:305–311
Lambinon J, Auquier P (1963) La flore et la végétation des terraine calaminaires de la Wallonie septentrionale et de la Rhénnanio arxoist. Nat Mosana 16:113–130
Lebedev VP (1993) Population structure of some vegetative mobile weeds in ecotype-based vegetation groupings. Bot Zh 78:29–34
Lequeux H, Hermans C, Lutts S, Verbruggen N (2010) Response to copper excess in *Arabidopsis thaliana*: impact on the root system architecture, hormone distribution, lignin accumulation and mineral profile. Plant Physiol 48:673–682
Luque-Garcia JL, Cabezas-Sanchez P, Camara C (2011) Proteomics as a tool for examining the toxicity of heavy metals. Trends Anal Chem 30:703–716
Macnair MR (1993) The genetics of metal tolerance in vascular plants. New Phytol 124:541–559
Macnair MR, Bert V, Huitson SB, Saumitou-Laprade P, Petit D (1999) Zinc tolerance and hyperaccumulation are genetically independent characters. Proc R Soc Lond B 266:2175–2179
Marques L, Cossegal M, Bodin S, Czernic P, Lebrun M (2004) Heavy metal specificity of cellular tolerance in two hyperaccumulating plants, *Arabidopsis halleri* and *Thlaspi caerulescens*. New Phytol 164:289–295
Masarovičová E, Holubová M (1999) Intraspecific tolerance of some herbs to the copper effect. Rostl Výroba 45:473–476
Masarovičová E, Kráľová K, Kummerová M (2010) Principles of classification of medicinal plants as hyperaccumulators or excluders. Acta Physiol Plant 32:823–829
Memon AR, Schröder P (2009) Implications of metal accumulation mechanisms to phytoremediation. Environ Sci Pollut Res 16:162–175
Nakata PA (2003) Advances in our understanding of calcium oxalate crystal formation and function in plants. Plant Sci 164:901–909
Nicola S (1998) Transplant production and performance: understanding root systems to improve seedling quality. Hort Technol 8:544–549
Pasternak T, Rudas V, Potters G, Jansen MAK (2005) Morphogenetic effects of abiotic stress: reorientation of growth in *Arabidopsis thaliana* seedlings. Environ Exp Bot 53:299–314
Pawlowska TE, Baszkowski J, Rühling Å (1996) The mycorrhizal status of plants colonizing a calamine spoil mound in southern Poland. Mycorrhiza 6:499–505
Pečiulytė D, Repečkienė J, Levinskaitė L, Lugauskas A, Motuzas A, Prosučevas I (2006) Growth and metal accumulation ability of plants in soil polluted with Cu, Zn and Pb. Ekologija 1:48–52
Pišút I (1980) Lichenes Slovakiae exsiccati editi a Museo Nationali Slovaco. Bratislava Fasc XII (276–300):1–7
Prasad MNV, Freitas HMO (2003) Metal hyperaccumulation in plants – biodiversity prospecting for phytoremediation technology. Electronic J Biotechnol 6(3):276–321
Przedpełska E, Wierzbicka M (2007) *Arabidopsis arenosa* (Brassicaceae) from a lead–zinc waste heap in southern Poland – a plant with high tolerance to heavy metals. Plant Soil 299:43–53
Reeves RD, Schwartz Ch, Morel JL, Edmondson J (2001) Distribution of metal-accumulating behaviour of *Thlaspi caerulescens* and associated metallophytes in France. Intern J Phytoremed 3:145–172
Rodwell JS (2000) British plant communities, vol 5, Maritime communities and vegetation of open habitats. Cambridge University Press, Cambridge
Röttgerman M, Steinlein T, Beyschlag W, Dietz H (2000) Linear relationships between aboveground biomass and plant cover in low open herbaceous vegetation. J Veg Sci 11:145–148

Sandberg J, Jones DL, Fransson AM (2009) Despite high uptake efficiency, non-mycorrhizal *Rumex acetosella* increases available phosphorus in the rhizosphere soil, whereas *Viscaria vulgaris*, *Plantago lanceolata* and *Achillea millefolium* does not. Nordic J Bot 27:444–448

Schwartz C, Morel JL, Saumier S, Whiting SN, Baker AJM (1999) Root development of the zinc-hyperaccumulator plant *Thlaspi caerulescens* as affected by metal origin, content and localization in soil. Plant Soil 208:103–115

Van Steveninck RFM, Babare A, Fernando DR, Van Steveninck ME (1994) The binding of zinc but not cadmium, by phytic acid in roots of crop plants. Plant Soil 167:157–164

Verkleij JAC, Golan-Goldhirshb A, Antosiewiszc DA, Schwitzguebel J-P, Schroder P (2009) Dualities in plant tolerance to pollutants and their uptake and translocation to the upper plant parts. Environ Exp Bot 67:10–22

Vollenweider P, Bernasconi P, Gautschi H-P, Menard T, Frey B, Günthardt-Goerg MS (2011) Compartmentation of metals in foliage of *Populus tremula* grown on soils with mixed contamination. II. Zinc binding inside leaf cell organelles. Environ Pollut 159:337–347

Weber M, Harada E, Vess C, von Roepenack-Lahaye E, Clemens S (2004) Comparative microarray analysis of *Arabidopsis thaliana* and *Arabidopsis halleri* roots identifies nicotianamine synthase, a ZIP transporter and other genes as potential metal hyperaccumulation factors. Plant J 37:269–281

Wenzel WW, Bunkowski M, Puschenreiter M, Horak O (2003) Rhizosphere characteristics of indigenously growing nickel hyperaccumulator and excluder plants on serpentine soil. Environ Pollut 123:131–138

Wu L, Bradshaw AD, Thurman DA (1975) The potential for the evolution of heavy metal tolerance in plants. III. The rapid evolution of copper tolerance in *Agrostis stolonifera*. Heredity 34:165–187

Yadav SK (2010) Heavy metals toxicity in plants: an overview on the role of glutathione and phytochelatins in heavy metal stress tolerance of plants. South Afr J Bot 76:167–179

Žemberyová M, Barteková J, Závadská M, Šisoláková M (2007) Determination of bioavailable fractions of Zn, Cu, Ni, Pb and Cd in soils and sludges by atomic absorption spectrometry. Talanta 71:1661–1668

Chapter 13
Hyperaccummulation: A Key to Heavy Metal Bioremediation

Ileana Cornelia Farcasanu, Mihaela Matache, Virgil Iordache, and Aurora Neagoe

13.1 Introduction

Industrialization, along with numerous benefits, brought up important issues such as environment awareness and environment protection. Strict regulations compel industries to find ways to limit the discharge of pollutants into the environment or to use eco-friendly approaches to clean up contaminated sites. Heavy metals are challenging pollutants, as they are natural components of the earth's crust, they are persistent in the environment and are nondegradable. Regarding the interaction with the living organisms, they have a dualistic behavior. On the one hand, many of the heavy metals are essential in minute amounts for the normal metabolism, binding to and stabilizing biomolecules, or acting as cofactors for various enzymatic processes. On the other hand, heavy metals can be toxic in high concentrations, mainly by nonspecific binding to biomolecules or by intereference to other metals' metabolism. The sources of heavy metal pollution can be industrial effluents, automobile exhaustion, mining industry, leaching of metal ions from the soil into lakes, rivers and aquifers by acid rain, etc. Heavy metal contamination of soils and waters is alarming due to potential penetration through the food chain leading to serious health problems at all levels (Duruibe et al. 2007; Fraústo da Silva and Williams 2001; Sánchez 2008). The term "heavy metal" is ill defined and has raised controversy (Duffus 2002), but it is still commonly used when referring to any metallic chemical element that has a relatively high density and is toxic or poisonous at low concentrations. The term is usually applied not only to common transition metals, such as Cd, Co, Cu, Hg, Mn, Ni, Pb, Zn, but also to metalloids such as As or Se. Among these, Co, Cu, Mn, Ni, Zn, and Se are elements that are essential for life in

I.C. Farcasanu (✉) • M. Matache
University of Bucharest, Sos Panduri 90-92, Bucharest 050665, Romania
e-mail: ileana.farcasanu@g.unibuc.ro

V. Iordache • A. Neagoe
University of Bucharest, Spl Independentei 91-95, Bucharest 050089, Romania

E. Kothe and A. Varma (eds.), *Bio-Geo Interactions in Metal-Contaminated Soils*, Soil Biology 31, DOI 10.1007/978-3-642-23327-2_13,

very low amounts (essential trace elements). These are taken up by cells via intricate active transport systems and are maintained at relatively constant levels by means of strictly regulated homeostasis mechanisms. Other heavy elements such as Cd, Hg, Pb, As are not essential for life, but they can compete with the essential trace elements either for the transport systems or in binding to various biomolecules. Essential or not, when present in high concentrations, the heavy metals become toxic, causing serious damages to organisms.

Although avoidance of pollution should be a pre-requisite for any anthropomorphic activity, this is not always achieved. Cleaning polluted or contaminated sites is never an easy task, especially when classical physicochemical approaches are expensive or inefficient, produce secondary pollution, or simply fail. As pollution-related issues become more stringent, cost-effective, sustainable, and environmentally friendly methods for removal of hazardous substances are more and more needed. Bioremediation by microorganisms, algae or plants is often considered as inexpensive, safe and efficient way of cleaning up wastes, sediments, and soils. Heavy metal bioremediation may be regarded as the removal of the excess hazardous substance following the interaction between the pollutant and (1) one or more living species; (2) dead or inactivated biomass. Bioremediation can be achieved through a plethora of mechanisms, but the most actions against heavy metals involve biosorption and bioaccumulation, actions that can take place in situ or in bioreactors designed both for the hazardous substance and for the bioremediator. Biosorption implies binding of the heavy metal ions to the cell surface, usually by noncovalent, electrostatic forces and is considered a cost-effective biotechnology for the treatment of high-volume-and-low-concentration complex wastewater. Most times, biosorption refers to a property of certain types of inactive, dead, microbial biomass to bind and concentrate heavy metals from dilute aqueous solutions. The biomass exhibits this property, acting as an ion exchanger of biological origin, and it is particularly the cell wall structure of certain algae, fungi and bacteria, which was found responsible for this phenomenon. Bioaccumulation is another attractive possibility to remove heavy metals from the environment. In contrast to biosorption, bioaccumulation can be achieved only by living cells, requiring metabolically driven continuous uptake that needs to occur with a rate greater than the excretion rate. The retention of a chemical within the living cells is a key step for bioremediation and obtaining resistant strains is vital for such a process.

13.2 Hyperaccumulation as a Primary Tool for Heavy Metal Bioremediation

Bioaccumulation has bioremediation significance only when correlated with increased (gained) tolerance to the pollutant and easy separation of the bioremediator from the site to be decontaminated. Thus, an ideal heavy metal bioremediator would have the following characteristics: tolerance to nonphysiological metal concentrations, abundant growth on/in the contaminated site, hyperaccumulating capacity, and facile separation from the bioremediated site. Nevertheless, heavy metals inhibit very often the biological remediation processes due to metal sensitivity of most organisms. Under such circumstances, strategies for efficient operation have to be considered and heavy metal hyperaccummulating plants seem to be the best models to follow when designing or developing a suitable heavy metal bioremediator. Hyperaccumulation was a term first used by Brooks et al. (1977) for plants that are endemic to metalliferous soils and are able to tolerate and accumulate metals in their above-ground tissues. Metal hyperaccumulator plants are naturally capable of accumulating trace elements, in their above-ground tissues, without developing any toxicity symptoms (Baker 2002; Baker and Brooks 1989). The concentrations of these elements in dry leaf biomass are usually up to 100-fold higher than the concentrations in the soil (McGrath and Zhao 2003). Natural hyperaccumulator species are often an indication of elevated soil heavy metal concentrations, and hence they can function as bioindicators of contamination; their potential role in phytoremediation, phytoextraction, and phytomining has been extensively studied (for review see Boyd 2010; Chaney et al. 2007; Cheng 2003; Krämer 2005; Lone et al. 2008). Although metal hyperaccumulator plants seem very promising, most produce little biomass and are therefore used mainly as model organisms for research purposes (Krämer 2005). In this paper, studies concerning the use of hyperaccumulating plants for understanding the hyperaccumulation and tolerance mechanisms, the use of transgenic approaches to obtain new hyperaccumulators, as well as the possibility to extent the plant hyperaccumulation concept to other organisms are reviewed.

13.3 Metals Commonly Hyperaccumulated by Plants

Most plants, when exposed to potentially toxic metals in their growing medium, take up the metal into the root, but restrict its further translocation to the shoot. A very rare class of plants, named hyperaccumulators, translocate substantial amounts to their shoots, so that shoot:root ratios exceed unity (Baker 1981; Macnair 2003). Most hyperaccumulators that have been identified so far hyperaccumulate nickel, but hyperaccumulators of cadmium, arsenic and zinc are also well characterized (Macnair 2003). To be considered a genuine hyperaccumulator, plants need to accumulate high concentrations of metals in any part that grows above the ground (Baker and Brooks 1989; Baker et al. 2000) in concentrations 10–500 times higher than the same plant species from nonpolluted environments (Yanqun et al. 2005). In addition, the shoot-to-root concentrations ratios must be higher than one (McGrath and Zhao 2003; Yanqun et al. 2005), meaning higher concentrations in the plant than in the soil. Most metal hyperaccummulators are endemic to soils with high concentrations of metal (Baker and Brooks 1989; Pollard et al. 2002). Table 13.1 summarizes the most studied heavy metal plant hyperaccumulators. In particular,

Table 13.1 Some heavy metal plant accumulators

Plant name	Metal accumulated	Possible tolerance mechanism	Comments	References (selected)
Alyssum sp.	Ni	Chelated by free L-histidine Ni^{2+}/H^{+} antiport V-ATPase at the tonoplast that can drive vacuolar accumulation of Ni through a secondary active transport mechanism	*Alyssum* accounts for 48 of the 318 known Ni hyperaccumulator species known	Baker et al. (1994, 2000), Brooks et al. (1977), Corem et al. (2009), Ingle et al. (2005b, 2008), Krämer et al. (1996), Reeves and Baker (2000)
		Slow vacuolar (SV) channel activity	Proteomic analysis in *A. lesbiacum*	
Arabidopsis halleri	Cd, Ni, Zn	Metal chelation Metal sequestration *AhHMA4* increased activity	One of the closest wild relatives of *A. thaliana* Transcriptomic studies Proteomic analysis of shoots in response to Cd, Zn and rhizosphere microorganisms Field and soil studies	Becher et al. (2004), Farinati et al. (2009), Hanikenne et al. (2008), Küpper et al. (2000), Marquès et al. (2004), McGrath et al. (2006), Talke et al. (2006), Vera-Estrella et al. (2009), Weber et al. (2004), Zhao et al. (2006)
Athyrium yokoscense	As, Cd		Fern common in metal-contaminated areas in Asia	Morishita and Boratynski (1992), Van et al. (2006)
Avena strigosa	Cd	High activities of antioxidative enzymes High amounts of total soluble phenolics	Crop plant	Uraguchi et al. (2006)
Crotalaria juncea	Cd	High activities of antioxidative enzymes High amounts of total soluble phenolics	Crop plant	Uraguchi et al. (2006)
Fagopyrum esculentum	Pb	Organic acid chelation		Tamura et al. (2005)
Gossia sp.	Mn/foliar	–	One of the species co-accumulates Ni	Fernando et al. (2007, 2009)

Maytenus founieri ssp.	Mn	Sequestration primarily in dermal tissues	Tree species, difficult to follow on short periods	Fernando et al. (2007)
Phytolacca acinosa	Mn	Sequestration of Mn in leaf epidermis	Herbaceous Mn hyperaccumulator, with high biomass and fast growth	Xue et al. (2004, 2005), Xu et al. (2006)
Pteris vitata	As	Suppression of endogenous arsenate reduction in roots may serve to enhance root-to-shoot translocation of As Chelation by phytochelatins	The plant transports 95% of the absorbed arsenic from roots to shoots Applicability in the field	Bondada et al. (2004), Caille et al. (2004), Dhankher et al. (2002), Ma et al. (2001), Tu and Ma (2002), Wei et al. (2006, 2007), Zhao et al. (2003)
Sedum alfredii	Cd, Pb, Zn	Enhanced root-to-shoot translocation Glutathione, rather than phytochelatins involved in Zn and Pb transport	Transcriptomic analysis under Zn induction	Chao et al. (2010), Long et al. (2009), Lu et al. (2008), Ni and Wei (2003), Sun et al. (2005, 2007), Yang et al. (2004)
Thlaspi caerulescens	Cd, Ni, Zn	Metal chelation with organic ligands Increased expression of *ZNT1*, *MTP1*, and *HMA4* genes	One of the most studied model organisms for hyperaccumulation mechanisms Variations between populations and ecotypes Stimulated shoot metal accumulation Transcriptome studies Field and soil studies	Baker et al. (2000), Basic et al. (2006), Hammond et al. (2006), Küpper et al. (2004), McGrath et al. (2006), Milner and Kochian (2008), van de Mortel et al. (2006), Papoyan et al. (2007), Rigola et al. (2006), Ueno et al. (2005), Yanai et al. (2006)
Thlaspi goesingense	Ni	Enhanced ability to compartmentalize Ni in shoot vacuoles	Vacuolar metal ion transport proteins, termed *m*etal *t*olerance *p*roteins (TgMTPs) described	Krämer et al. (2000), Persans et al. (2001)

Thlaspi caerulescens and *Arabidopsis halleri*, have been studied extensively for their ability to hyperaccumulate several heavy metals, mainly Zn, Cd, and Ni, as well as the species from *Alissum* genus, known for their ability to hyperaccumulate Ni. To be considered a hyperaccumulator, the plants need to store more than 1,000 mg kg^{-1} (dry weight biomass) of metal for Ni, Cu, Co, Cr, or Pb, or 10,000 mg kg^{-1} for Zn or Mn (Baker and Brooks 1989). Some of the heavy metals more commonly stored by hyperaccumulator plants are presented below.

13.3.1 Cadmium

Cd is nonessential but a major pollutant that is extremely toxic to organisms. Soils can be contaminated with Cd as a result of fertilization with phosphates as well as from mining and smelting industries (McGrath et al. 2001; Sanità di Toppi and Gabbrielli 1999). Daily consumption of Cd-contaminated foods poses a risk to human health (Watanabe et al. 2000). The principal plant used for Cd phytoextraction is *Thlaspi caerulescens*, a Zn/Cd hyperaccumulator (Baker et al. 2000) owing its potential to the exceeding bioconcentration factor of Cd and Zn in the shoots that enables a remarkable yield of both metals from contaminated soil (McGrath and Zhao 2003; McGrath et al. 2001). Usually, Cd overlaps the Zn tolerance and hyperaccumulation (Uraguchi et al. 2006; Yang et al. 2004). Apart from *T. caerulescens*, *Arabidopsis halleri* (Küpper et al. 2000), *Sedum alfredii* (Yang et al. 2004; Zhao et al. 2006), and *Athyrium yokoscense*, (Morishita and Boratynski 1992) have been reported as evident Cd-hyperaccumulator plants. These are all wild metal-accumulators, growing slowly in the field. Uraguchi et al. (2006) studied the behavior of crop species and found two Cd hyperaccumulators, *Avena strigosa* and *Crotalaria juncea*. In this case, the Cd tolerance seems to be the result of enhanced activities of antioxidative enzymes and of augmentation in total soluble phenolics, both well-known antioxidant defense strategies. In plant cells, Cd tends to be stored in the apoplast and in vacuoles. In addition to the metal compartmentation mechanism, antioxidative ability might play an important role in the tolerance. Cd is a redox-inactive metal that is incapable of producing reactive oxygen species (ROS) directly, but can indirectly promote oxidative stress by disrupting physiological processes. Also, phytochelatins (PCs) and other thiol (SH)-containing compounds have been proposed to play an important role in the detoxification and tolerance of some heavy metals. For instance, glutathion may be responsible for Cd and tolerance in mine population of *Sedum alfredii* (Sun et al. 2007).

13.3.2 Copper

Cu is an essential element, and Cu poisoning is rarely an issue of environmental risk. Nevertheless, sources of Cu contamination are the electroplating industry, smelting and refining, mining, biosolids (Liu et al. 2005). There are not many Cu hyperaccumulators described, and most of them are not Cu-specific, like some varieties of *Thlaspi caerulescens*, or *Athyrum yokoscense* (Cheng 2003; Honjo et al. 1984).

13.3.3 Manganese

Manganese, an essential trace element that is found in varying amounts in all tissues, is one of the most widely used metals in industry. Exposure to excess manganese results in manganese toxicity, including Parkinson-like symptoms (Chan et al. 2000; Erikson and Aschner 2003; Gerber et al. 2002), and abnormalities of the immune system (Vartanian et al. 1999). Manganese hyperaccumulation has been arbitrarily defined by a threshold foliar concentration of 10,000 mg kg^{-1} dry weight (Baker and Brooks 1989). Plants that hyperccumulate Mn are predominantly woody and hence unsuited to short-term controlled study (Fernando et al. 2008). The number of species has varied with taxonomic changes, and currently nine are recognized worldwide (Bidwell et al. 2002; Reeves and Baker 2000; Xue et al. 2004). The heterogeneity of the Mn-hyperaccumulative trait was studied in *Phytolacca acinosa* a herbaceous species, under controlled conditions (Xue et al. 2005) and in natural populations of the tree *Gossia bidwillii* (Fernando et al. 2007). Recently, Fernando et al. (2009) demonstrated up to seven new Mn hyperaccumulators, mostly tropical rainforest species from the *Gossia* genus, one of them exhibiting also elevated foliar Ni concentrations. Among the Mn hyperaccumulating plants, *P. acinosa* is a good candidate for phytoremediation of Mn polluted soil for its high biomass and fast growth. Xu et al. (2006) determined Mn distribution within the hyperaccummulating plant and found that the highest Mn content was in the vascular tissues of root, stem, petiole and midrib and that, Mn content in leaf epidermis was higher than that in mesophyll, which suggested that the sequestration of Mn in leaf epidermis might be one of the detoxification mechanisms of *P. acinosa*.

13.3.4 Nickel

Approximately 450 species of metal-hyperaccumulating plants are currently known, of which more than 330 are Ni-hyperaccumulators (Reeves 2003; Reeves and Adigüzel 2004; Reeves and Baker 2000). Ni is ubiquitously distributed in nature and constitutes a trace element in most living cells. In high concentrations, it is toxic to most cells and is also listed as a possible human carcinogen (group 2B) and associated with reproductive problems and birth defects. Ni-hyperaccumulators are able to take up more than 1,000 mg kg^{-1} dry weight (Baker et al. 1994) without significant detrimental effect on plant survival or health. In terms of remediation using classical chemical approaches, Ni is one of the most recalcitrant pollutants; therefore, Ni hyperaccumulating plants have received special attention. Ni-hyperaccumulators belong to a group of taxonomically diverse plants that can accumulate Ni to concentrations in excess of 0.1% shoot dry biomass (Baker and Brooks 1989; Reeves and Baker 2000). The genus *Alyssum* accounts for 48 of the known Ni hyperaccumulator species, and includes *Alyssum lesbiacum* which is capable of accumulating Ni to over 3% shoot dry biomass (Baker et al. 2000; Reeves and Baker 2000). Ni accumulates predominantly in the aerial tissues of hyperaccumulators, and the shoot epidermal cells are a major site of Ni deposition in hyperaccumulator plants, including several *Alyssum* species, *Cleome heratensis*, *Hybanthus floribundus* and *Senecio coronatus* (Asemaneh et al. 2006; Bidwell et al. 2004; de la Fuente et al. 2007; Küpper et al. 2001; Marquès et al. 2004; Mesjasz-Przybyłowicz et al. 1994).

13.3.5 Lead

Pb is a nonessential heavy metal discharged in the environment from mining and smelting of metalliferous ores, burning of leaded gasoline, municipal sewage, industrial wastes enriched in Pb, paints (Gispert et al. 2003; Seward and Richardson 1990). There are certain cultivars of Indian mustard (*Brassica juncea*) (Kumar et al. 1995), the fern *Athyrium yokoscense* (Honjo et al. 1984), or the leguminous shrub *Sesbania drummondii* (Sahi et al. 2002) that have been reported as Pb hyperaccumulators. Tamura et al. (2005) found that common buckwheat (*Fagopyrum esculentum* Moench), known as an aluminum (Al) accumulator (Ma and Hiradate 2000; Ma et al. 1997; Shen et al. 2002), can accumulate a high concentration of lead (Pb) in the shoot and especially in the leaf. Since buckwheat can grow with relatively high biomass productivity this plant may prove to be a useful phytoremediator of Pb-contaminated soils around the world.

13.3.6 Zinc

Zn is another essential trace element that has deleterious effect on organsism when present in excess. Zn pollution sources can be electroplating industry, smelting and refining, mining, biosolids (Liu et al. 2005). Around 14 species have been described as Zn hyperaccumulators defined as containing more than 10 g kg^{-1} shoot dry weight (Baker et al. 2000). *Thlaspi caerulescens* has been widely studied for its remarkable properties to tolerate toxic levels of Zn as well as Cd, and sometimes Ni (reviewed by Milner and Kochian 2008). Along with *Arabidopsis halleri*, *T. caerulescens* has been in the primelight for studies concerning the hyperaccumulation process. Because *T. caerulescens* is a slow-growing plant species that does not generate significant shoot biomass, it has been used primarily as a model system for the investigation and identification of the molecular and physiological mechanisms of hyperaccumulation, with the ultimate goal of transferring these mechanisms to higher biomass plant species. *A. halleri* has been described as constitutively zinc Zn-tolerant as well as Zn-hyperaccumulating

(Bert et al. 2000, 2002, 2003; Macnair 2002; Pauwels et al. 2006). The Zn-accumulating capacity of this plant is quite remarkable (Kashem et al. 2010). Another plant reported to accumulate Zn is *Sedum alfredii* Hance. When grown on an old lead/zinc mining site, Long et al. (2009) found this plant to significantly hyperaccumulate zinc and cadmium under field conditions.

13.3.7 Arsenic

Arsenic (As) has been categorized as a toxic and carcinogenic element and contribute to environmental and human health problems worldwide; the highest number of cases has been reported in South-East Asian countries (Mandal and Suzuki 2002). Drinking water from wells located in areas with As-rich underground sediments has been shown to be the source of arsenocosis (Patel et al. 2005; Yang et al. 2002). Ma et al. (2001) discovered the first known vascular plant, *Pteris vittata* L. commonly known as Chinese brake fern to hyperaccumulate arsenic. The brake fern takes up high concentrations of arsenic (as high as 2.3%) from soil and allocated most of it to the above-ground pars for final storage (Tu and Ma 2002). Moreover, the hyperaccumulation of arsenic is accompanied by an increased biomass of the above-ground plant parts, an important characteristic, which is indispensable for phytoremediation (Ma et al. 2001; Tu and Ma 2002). Other desirable characters permitting brake fern as an ideal plant for phytoremediation include its perennial growth habit, disease and pest resistance, fast vigorous growth, and its preference for soils with high pH where arsenic exists in high abundance (Bondada and Ma 2003; Bondada et al. 2004). Studies on arsenic hyperaccumulation by *P. vittata* were diverse (Caille et al. 2004; Wei et al. 2006, 2007) and other species of ferns, *including Pteris cretica* and *Pityrogramma calomelanos*, have also been determined to be As-hyperaccumulators and show great potential in phytoremediation (Francesconi et al. 2002; Visoottiviseth et al. 2002; Wei et al. 2006). The root systems of the As hyperaccumulating fern *P. vittata* possess a higher affinity for arsenate uptake than those of a related nonaccumulator fern species, and a suppression of endogenous arsenate reduction in roots may serve to enhance root-to-shoot translocation of As (Dhankher et al. 2002). Phytochelatins, metal-chelating molecules synthesized by the ubiquitous plant enzyme phytochelatin synthase are known to contribute to As detoxification in As hyperaccumulator plants (Zhao et al. 2003).

13.4 What is Responsible for the Heavy Metal Tolerance of the Hyperaccumulating Plants?

At least three processes appear to make a major contribution to the ability of certain species to hyperaccumulate metals: enhanced uptake, root-to-shoot translocation, and detoxification via chelation and sequestration (Clemens et al. 2002; Pollard et al. 2002). In addition, hyperaccumulator plants appear to have highly effective antioxidant systems to protect against the potentially damaging effects of metal-induced ROS (Freeman et al. 2004).

13.4.1 Metal Chelation

Regarding the deleterious effects of heavy metals on the living organisms, it is considered that mainly the osmotically free forms of the metal ions are genuinely toxic, as they are prone to bind nonspecifically to biomolecules or to interfere with the essential metals' metabolism. To avoid the growth impairments caused by the potentially toxic heavy metals, hyperaccummulating plants must possess biochemical defense mechanisms. Plants developed a number of strategies to resist this toxicity, including active efflux, sequestration, and binding of heavy metals inside the cells by strong ligands. The primary antidote against the osmotically free ions may be the presence of chelating agents that form easily nontoxic complexes. Thus, an essential component of tolerance is the buffering of free metal ions in the cytoplasm via chelation with high-affinity ligands. The response of plants is complex with considerable variation between species. Several detoxification strategies are known to occur and different effects are observed with different metals and metal concentrations. Still and Williams (1980) first proposed that Ni hyperaccumulation might involve a ligand containing two nitrogen donor centers and one oxygen donor center because this would exhibit a sufficiently high affinity for Ni and could account for the observed preference for Ni over Co in these plants. This was demonstrated by Krämer et al. (1996), who observed a linear relationship between the concentrations of Ni and free histidine appearing in the xylem of hyperaccumulating species of *Alyssum* exposed to a range of Ni concentrations. Thus, it was found that the concentration of free histidine in the roots of the hyperaccumulator species *A. lesbiacum* could be several fold higher than in the nonaccumulator *Brassica juncea*, even in the absence of Ni (Kerkeb and Krämer 2003). Hisidine forms soluble complex compounds with Ni, which are completely nontoxic. To determine the molecular basis of the histidine response and its contribution to Ni tolerance, Ingle et al. (2005a, b) analyzed the transcripts of the enzymes involved in histidine biosynthesis and found that the transcript levels were constitutively higher in the hyperaccumulating *A. lesbiacum* compared to the congeneric nonaccumulator *A. montanum*. This was noted especially for the first enzyme in the biosynthetic pathway, ATP-phosphoribosyltransferase (ATP-PRT); comparison with the weak hyperaccumulator *A. serpyllifolium* revealed a close correlation between Ni tolerance, root histidine concentration, and ATP-PRT transcript abundance. Thus, it seems that *ATP-PRT* expression plays a major role in regulating the pool of free histidine and contributes to the exceptional Ni tolerance of hyperaccumulator *Alyssum* species. Free histidine seems to be involved in Ni tolerance in other hyperaccumulators as well. Thus, histidine concentration in the roots is 17-fold higher in the Ni hyperaccumulator *Thlaspi goesingense* than in the nonaccumulator *T. arvense* (Persans et al. 1999). Nevertheless, it seems that the Ni hyperaccumulating phenotype in *T. goesingense* is not determined by the overproduction of His in response to Ni (Persans et al. 1999). Coordination with histidine is not an universal determinant for heavy metal hyperaccumulation, not even for Ni. Phytochelatins are other plant chelators that might be worth considering when metal ion buffering is desired. Complexation with phytochelatin peptides synthesized from glutathione has been identified as an important mechanism for detoxifying metals such as Cd, Pb, and Zn in nonhyperaccumulator plants and plant cell cultures (Cobbett 2000; Zenk 1996), but phytochelatins do not appear to be responsible for hyperaccumulation or metal tolerance in hyperaccumulator species (Ebbs et al. 2002; Krämer et al. 1996, 2000; Shen et al. 2000). Sun et al. (2005) investigated whether phytochelatins were differentially produced in mine populations of *Sedum alfredii* compared with a nonmine control of the same species and found that phytochelatins and cysteine were not responsible for Zn and Pb tolerance in the mine population; instead, Zn and Pb treatments resulted in the increase of glutathione, suggesting that glutathione, rather than phytochelatins, may be involved in Zn and Pb transport, hyperaccumulation/accumulation and tolerance in mine population of *S. alfredii*. There are many indications that organic acids are involved in heavy metal tolerance, transport and storage in plants, including for Cd, Ni, and Zn (Godbold et al. 1984; Krotz et al. 1989; Nigam et al. 2001; Yang et al. 1997). In hyperaccumulators, the levels of citric, malic, malonic, and oxalic acids have been correlated with elevated concentrations of Ni or Zn in the biomass (Li et al. 2010; Tolrá et al. 1996). Determining the distribution of Cd and Ni in hairy roots of the Cd hyperaccumulator, *Thlaspi caerulescens* and the Ni hyperaccumulator, *Alyssum bertolonii* that contained high constitutive levels of citric, malic and malonic acids, Boominathan and Doran (2003) found that about 13% of the total Cd in *T. caerulescens* roots and 28% of the total Ni in *A. bertolonii* were associated with organic acids, while the hairy roots remained healthy and grew well. Nicotianamine (NA), a nonproteinogenous amino acid synthesized in all plants through the activity of the enzyme nicotianamine synthase (NAS) is another chelator that can be involved in heavy metal tolerance (Higuchi et al. 1994). Based on yeast screens, Mari et al. (2006) demonstrated root-to-shoot long-distance circulation of nicotianamine and nicotianamine–nickel chelates in the metal hyperaccumulator *Thlaspi caerulescens*, while NA–Ni chelates are absent in the nontolerant nonhyperaccumulator-related species *T. arvense*. Furthermore, *Arabidopsis thaliana* lines overexpressing Tc*NAS* cDNA produce a large amount of NA, correlated with a better resistance to the toxicity of this metal (Pianelli et al. 2005). An intriguing fact is that the Cd/Zn hyperaccumulator *Thlaspi caerulescens* is sensitive toward Cu, which may be a problem for phytoremediation of soils with mixed contamination, demonstrating that hyperaccumulation and metal resistance are highly metal specific. A few individuals of *T. caerulescens* that were more Cu resistant revealed that a large proportion of Cu in *T. caerulescens* leaves is bound by sulfur ligands, in contrast to the known binding environment of cadmium and zinc in the same species, which is dominated by oxygen ligands (Mijovilovich et al. 2009). This observation suggests that hyperaccumulators detoxify hyperaccumulated metals differently compared with nonaccumulated metals. Thus, it seems that hyperaccumulators have different strategies of detoxification for metals that are hyperaccumulated compared with nonhyperaccumulated metals. For the hyperaccumulated metals, detoxification is mainly based on active sequestration into the vacuoles of the epidermis, where they are stored only loosely associated with organic acids that are abundant in this organelle (Küpper et al. 2004). Strong ligands such as the phytochelatins and metallothioneins that detoxify heavy metals in nonaccumulator plants do not play a major role in the detoxification of hyperaccumulated metals (Küpper et al. 2009; Mijovilovich et al. 2009). At the same time, Ni and Zn are possibly transported by similar ligands and thus competition for binding sites is likely to be a factor in the relationship between uptake of Zn and Ni when levels of both are high in the soil (Assunção et al. 2003).

13.4.2 Metal Transport, Distribution, Compartmentalization

Plants developed different strategies to grow on soils rich in heavy metals. Many of them, the excluders, are able to restrict root uptake, and in particular, root-to-shoot translocation of heavy metals (Baker 1981; Küpper et al. 2000, 2001). On the opposite side, hyperaccumulators store high amounts of heavy metals in the above-ground parts. Understanding how metal transport and accumulation processes differ between normal and hyperaccumulator plants is important. "Normal" nonhyperaccumulator plants tend to store the absorbed heavy metals in the roots, whereas hyperaccumulator plants are capable of transporting most of the accumulated heavy metals to the shoots (Lasat et al. 1998, 2000). Hyperaccumulator plants exhibit stronger influx of heavy metals into the roots than the nonaccumulator species do (Lasat et al. 1996). Studies examining the competitive effects between Zn and Cd transport in *Thlaspi caerulescens* suggested that, at least in the leaf, the heavy metal Cd is transported via cellular Zn transporters (Cosio et al. 2004). While Zn and Cd are transported via the same transporter in the case of the Prayon ecotype, Cd transport in the Ganges ecotype (which hyperaccumulates Cd to a higher degree) takes place through a separate transporter (Lombi et al. 2001). In contrast, in nonaccumulator plants, Zn and Cd uptake and accumulation are negatively correlated (Hart et al. 2002; Wu et al. 2003). Papoyan et al. (2007) suggested that xylem loading may be one of the key sites responsible for the hyperaccumulation of Zn and Cd accumulation in *Thlaspi caerulescens*. Interestingly, the expression of one of the metal transporter involved, (HMA4) in *Arabidopsis thaliana* is downregulated upon exposure to heavy metals (Mills et al. 2003), while its expression in *T. caerulescens* is upregulated upon exposure to high concentrations of Cd and Zn (Papoyan and Kochian 2004). In a quantitative study of cell compartmentation, Küpper et al. (2000) showed that the major storage site for Zn and Cd hyperaccumulation in *Arabidopsis halleri* are the leaf mesophyll cells. In *Sedium alfredii* Hance, altered Zn transport across the tonoplast in the root stimulate Zn uptake in the leaf cells, being the major mechanisms involved in the strong Zn hyperaccumulation observed (Yang et al. 2006). At the cellular level, a major role has been ascribed to vacuolar compartmentalization of excess cytosolic metals. Likewise, hyperaccumulation is connected to the ability to transport large amounts of metals into leaf vacuoles (Krämer et al. 2000; Persans et al. 2001). Shira et al. (2009) investigated the transport properties of the Slow Vacuolar (SV) channel identified in leaf vacuoles of *Alyssum bertolonii* Desv. Accumulation of Ni in epidermal cells seemed to be a common feature in the leaves of Ni-hyperaccumulators, such as *Thlaspi goesingense*, *Alyssum* species and *Berkheya coddii* (Broadhurst et al. 2004; Kerkeb and Krämer 2003; Krämer et al. 1996; Küpper et al. 2000, 2001; McNear et al. 2005). The sequestration of toxic metals in leaf epidermis is not a universal detoxification mechanism in all hyperaccumulators. Cu in *Elsholtzia splendens* (Shi et al. 2004) and Mn in *Gossia bidwillii* (Fernando et al. 2006) seem to be more abundant in the mesophyll than in the epidermis of leaves. The cellular distribution of Pb in leaves is less studied. The Pb-As co-hyperaccumulator *Viola principis* H. de Boiss has similar Pb and As compartmentalization patterns in the leaves, as both elements accumulate in the bundle sheath and the palisade mesophyll, while the vascular bundle and the epidermis contained lower levels of Pb and As. The palisade enrichment of Pb and As indicates that *V. principis* H. de Boiss. may have a special mechanism on detoxification of toxic metals within the mesophyll cells (Lei et al. 2008). Apart from complexation with organic ligands, cellular compartmentation is likely to be involved in heavy metal tolerance of the hyperaccumulators. Krämer et al. (2000) estimated that approximately 70% of Ni in *Thlaspi goesingense* was associated with the apoplast, while Küpper et al. (2001) reported that the majority of Ni was localized in the vacuoles in leaf tissue of *Alissum lesbiacum* and *Thlaspi goesingense*. This apparent discrepancy may be due in part to the different methods used to study Ni localization (Smart et al. 2007). The vacuole is the largest subcellular compartment, occupying 70–80% of the total volume of mature parenchymatous plant cells, and plays an important role in the storage of inorganic ions (Martinoia et al. 2007). It is rich in carboxylic acids, such as citric and malic acid, which may serve to chelate Ni with moderately high affinity (Saito et al. 2005; Smart et al. 2007). If the vacuole is a site of Ni sequestration in hyperaccumulators, some form of active transport will be required at the tonoplast membrane to move Ni into the vacuole against its electrochemical gradient. Ingle et al. (2008) highlighted a Ni^{2+}/H^{+} antiport system at the tonoplast of *Alissum lesbiacum* that can drive vacuolar accumulation of Ni via a secondary active transport mechanism.

13.5 Molecular Studies

Most hyperaccumulators are slow-growing plant species that do not generate significant shoot biomass. This is why they are used mainly as model systems for the investigation and identification of the underlying molecular and physiological mechanisms of hyperaccumulation, with the ultimate goal of transferring these mechanisms to higher biomass plant species. A number of genes involved in the hyperaccumulating phenotype were identified. Among them, a major role for *AhHMA4* in naturally selected Zn hyperaccumulation and associated Cd and Zn hypertolerance in *Arabidopsis halleri* was demonstrated (Hanikenne et al. 2008). The function of identified genes was also checked by complementation analysis in model organisms such as *Arabidopsis thaliana* or *Saccharomyces cervisiae* (Table 13.2). The molecular basis of the widely used hyperaccumulator *Thlaspi caerulescens* was thoroughly reviewed by Milner and Kochian (2008). More widely, systemic studies such as transcriptomic and proteomic studies are expected to shed light on the intricate mechanisms that lead to hyperaccumulation.

13.5.1 Transcriptomic Studies

A comparative transcriptome analysis between the Zn/Cd hyperaccumulator *Arabidopsis halleri*, and the nonaccumulator *Arabidopsis thaliana* using the gene chip arrays revealed that hyperexpression may be a general property of metal hyperaccumulators, as a large number of genes were found to be more highly expressed in *A. halleri* compared with *A. thaliana* (Becher et al. 2004; Weber et al. 2004). In a separate study, a set of candidate genes for Zn hyperaccumulation, Zn and Cd hypertolerance, and the adjustment of micronutrient homeostasis in *A. halleri* was identified using a combination of genome-wide cross species microarray analysis and real-time PCR (Talke et al. 2006). In this study, 18 putative metal homeostasis genes were newly identified to be more highly expressed in *A. halleri* than in *A. thaliana*, and 11 previously identified candidate genes were confirmed (Talke et al. 2006). Interestingly, the study showed that in the steady state, *A. halleri* roots, but not the shoots, act as physiologically Zn-deficient under conditions of moderate Zn supply. An analysis of the first EST collection from *Thlaspi caerulescens* obtained from Zn-exposed roots and shoots revealed that *T. caerulescens* expresses a relatively large number of genes, which are expressed at a very low level in *A. thaliana*, and 8% of the total set of expressed uni-genes did not have an *Arabidopsis* ortholog (Rigola et al. 2006). Interestingly, some of the genes shown to be more highly expressed in *A. halleri* also exhibited elevated expression in *T. caerulescens*, suggesting this set of genes might be important for hyperaccumulation in both plant species. Also, cDNA-AFLP analysis of inducible gene expression in the Zn hyperaccumulator *Sedum alfredii* Hance under zinc induction was reported (Chao et al. 2010).

Table 13.2 Transgenic organisms expressing genes isolated from hyperaccumulator plants

Target organism	Gene/source	Gained phenotype	Function	Comments	References
Arabidopsis thaliana	*TcNAS1*/*Thlaspi caerulescens*	Ni tolerance	Nicotianamine synthase	Ni tolerance accumulation in the shoot	Pianelli et al. (2005)
	ATP-PRT cDNA/ *Alissum lesbiacum*	Ni tolerance, but no Ni hyperaccumulation	Histidine synthesis pathway	Overexpression increased the pool of free His up to 15-fold	Ingle et al. (2005a)
Saccharomyces cerevisiae	*AhHMA4*/ *Arabidopsis halleri*	Zn and Cd tolerance	CPx-ATPase	Metal substrates: Zn and Cd	Talke et al. (2006)
	PvPCS1/*Pteris vittata*	Cd tolerance	Phytochelatin synthase		Dong et al. (2005)
	TcHMA4/*Thlaspi caerulescens*	Cd tolerance	CPx-ATPase	Expression in *Tc* is induced by both Zn-deficiency and high-Zn, as well as by high Cd	Bernard et al. (2004), Papoyan and Kochian (2004)
	TcMT3 (metallothionein)/ *Thlaspi cerulescens*	Cd and Cu tolerance	Cysteine-rich, low-molecularweight, metal-binding protein	Maintains normal Cu homeostasis under high cytoplasmic Cd and Zn	Roosens et al. (2004)
	TcNAS1 cDNA/ *Thlaspi caerulescens*	Ni tolerance	Synthesis of nicotianamine, a nonproteinaceous amino acid, capable of chelating metal ions	Nicotianamine synthase	Vacchina et al. (2003)
	TcZNT1/*Thlaspi caerulescens*	Mediate Zn and Cd uptake	Plasma membrane-localized transporter	In yeast acts as a high affinity Zn and low-affinity Cd uptake transporter	Pence et al. (2000)

(continued)

Table 13.2 (continued)

Target organism	Gene/source	Gained phenotype	Function	Comments	References
	TgMTP1/Thlaspi goesingense	Tolerance to Cd, Co, and Zn	Vacuolar metal ion transporter Cation-efflux family member	*TgMTP1* transcripts are highly expressed in *T. goesingense* compared with orthologs in the nonaccumulators *Arabidopsis thaliana*, *Thlaspi arvense*, and *Brassica juncea*	Persans et al. (2001)
	TgMTP2/Thlaspi goesingense	Tolerance to Ni	Vacuolar metal ion transporter Cation-efflux family member		Persans et al. (2001)

13.5.2 Proteomic Studies

Broader molecular characterization of hyperaccumulators has been performed mainly by transcriptomics. In proteomics, even though large and hydrophobic transporters and low abundant or small polypeptides may remain undetected in 2-DE-based proteomics, many other proteins, such as regulatory proteins and those contributing to stress protection that appear to have importance in the hyperaccumulation phenotype might be detected (Farinati et al. 2009; van de Mortel et al. 2006; Verbruggen et al. 2009). In addition, posttranscriptional regulation could be very important, and there is often no proportionality between the transcript and protein abundance. Different profiling techniques are thus clearly complementary, and the proteomics approach is of increasing interest in exploring the hyperaccumulation phenomenon. A proteomic analysis of the Ni hyperaccumulator plant *Alyssum lesbiacum* was carried out to identify proteins that may play a role in the Ni tolerance and accumulation characteristic. As very few polypeptides were found to change in abundance in root or shoot tissue after plants were exposed to conditions representing the optimum for growth and hyperaccumulation of Ni in the shoot, it was concluded that constitutively expressed genes may be sufficient to allow for effective chelation and sequestration of Ni without the need for additional protein synthesis (Ingle et al. 2005b). Tuomainen et al. (2006) identified differences in protein intensities among three *Thlaspi caerulescens* accessions with pronounced differences in tolerance, uptake and root-to-shoot translocation of Zn and Cd, noticing clearest differences mainly among the *Thlaspi* accessions, while the effects of metal exposures were less pronounced. When looking at protein profiles of *Thlaspi caerulescens*, two accessions, and lines derived from the two accession proteins that showed co-segregation with high or low Zn accumulation were manganese superoxide dismutase, glutathione S-transferase, S-formyl glutathione hydrolase (Tuomainen et al. 2010).

13.6 The Hyperaccumulation Concept Applied to Organisms Other than Plants

Searching through literature uncovers the fact that the heavy metal hyperaccumulation seems to be restricted to plants, as seen from the number of studies or from the unequivocal definition of a hyperaccumulator. Hyperaccumulation is a phenomenon that generated the idea of phytoremediation and phytoextraction, immediately related to eco-friendly bioremediation processes. Even if the large-scale application of hyperaccumulating plants in bioremediation is in its infancy, it is undeniable that this is somehow restricted to soils. Phycoremediation, or the use of algae in bioremediation is another increasing possibility that would enlarge the applicability to contaminated waters (Olguin 2003). Marine organisms are of increasing interest and exceptionally high levels of trace metals have been reported in specific tissues of certain *Polychaetes* (Gibbs et al. 1981; Ishii et al. 1994; Fattorini et al. 2005, 2010; Sandrini et al. 2006). In a remarkable review, Gifford et al. 2007 introduced the concept of zooremediation and defined an animal heavy metal hyperaccumulator by similitude with plants as animal species known to accumulate >100 mg kg^{-1} Cd, Cr, Co or Pb; or >1,000 mg kg^{-1} Ni, Cu, Se, As or Al; or >10,000 mg kg^{-1} Zn or Mn. Even though this field would probably be limited to invertebrates for ethical reasons, emerging data are tempting and self-financing zoo remediation models such as pearl oysters, sponges are proposed as models (Gifford et al. 2007). Work concerning the use of zooremediation is less intense than in plants, however one system is already in use for the recovery of Cd in waste scallop tissue (Seki and Suzuki 1997; Shiraishi et al. 2003, http://www.unirex-jp.com/engcadmium/engcadmium.htm). Although microorganisms are the most common group of organisms used for bioremediation, biosorption is the main governing process used to remove heavy metals from contaminated sites. Manipulating heavy metal resistance of bacteria by overexpressing genes from hyperacummulating plants has been reported (Freeman et al. 2005), but the term hyperaccumulation may be awkward to apply in this case. Molecular mechanisms concerning heavy metal metabolism are widely studied in bacteria, and expressing various bacterial genes into plants to obtain hyperaccumulating transgenics is a promising approach (Dhankher et al. 2002, 2003; Rugh et al. 1996). The genus *Saccharomyces* looks like the group to which the term is worth extrapolating, mainly because of the general tendency to extend plant molecular processes to model organisms such as *Saccharomyces cerevisiae* (Table 13.2). The ease of growth and the exceptional elegance of genetic manipulation make the yeast cells a versatile tool for biotechnology. *S. cerevisiae* is not a heavy metal accumulator, and attempts to obtain tolerant hyperaccumulating in *S. cerevisiae* yeast strains failed, noticing a biunivocal relationship tolerance-exclusion, hyperaccumulation-sensitivity. Nevertheless, there are studies indicating that heavy metal sensitive but accumulating *kamikaze* strains could be used for heavy metal bioremediation through bioaccumulation (Ruta et al. 2010). Moreover, the cell surface of yeast cells can be engineered for heavy metal increases absorption capacity using the molecular display (arming) technology in which heterologous proteins that are expected to offer novel functionality to cell wall can be expressed on the surface of the cell (Georgiou et al. 1997; Murai et al. 1997; Shibasaki et al. 2009). Using such technology, cells with improved heavy metal biosorption abilities were obtained (Kambe-Honjoh et al. 2000; Nakajima et al. 2001; Kotrba and Rumi 2010). Combining improved biosorbents with bioaccumulative induction may be the basis of obtaining hyperaccumulating yeast strains.

13.7 Conclusions

As heavy metal pollution poses serious problems and needs to be overcome in a friendly, noninvasive way, the extraordinary phenotype of hyperaccumulating plants seems to be the appropriate tool to design bioremediation systems capable of removing the contaminating heavy metals from various sites. In recent years, major scientific progress has been made in understanding the physiological and molecular mechanisms of metal uptake and transport in these plants. General metal hyperaccumulators, however, may be inadequate for bioremediation technology because of their small size and slow growth rates. At present, therefore, the focus is on searching for new metal hyperaccumulators or on trying to improve metal uptake in plants and other organisms using metal chelators and on producing transgenic organisms that show metal hyperaccumulation.

Acknowledgments This review was done in the Romanian Consortium for the Biogeochemistry of Trace Elements with financing from National University Research Council (CNCSIS) by projects 176 and 291/2007 (codes ID 965 and 1006) and within the frame of European project UMBRELLA (FP7_ENV-2008-1 no. 226870).

References

Asemaneh T, Ghaderian SM, Crawford SA, Marshall AT, Baker AJM (2006) Cellular and subcellular compartmentation of Ni in the Eurasian serpentine plants *Alyssum bracteatum*, *Alyssum murale* (*Brassicaceae*) and *Cleome heratensis* (*Capparaceae*). Planta 225:193–202

Assunção AGL, Schat H, Aarts MGM (2003) *Thlaspi caerulescens*, an attractive model species to study heavy metal hyperaccumulation in plants. New Phytol 159:351–360

Baker AJM (1981) Accumulation and excluders – strategies in the response of plants to heavy metals. J Plant Nutr 3:643–654

Baker AJM (2002) The use of tolerant plants and hyperaccumulators. In: Wong MH, Bradshaw AD (eds) Restoration and management of derelict land: modern approaches, [derived from an Advanced Study Institute], Kowloon, China, Nov 2000, pp 138–148

Baker AJM, Brooks RR (1989) Terrestrial higher plants which hyperaccumulate metallic elements – a review of their distribution, ecology and phytochemistry. Biorecovery 1:81–126

Baker AJM, McGrath SP, Sidoli CMD, Reeves RD (1994) The possibility of in situ heavy metal decontamination of polluted soils using crops of metal-accumulating plants. Resour Conservat Recycl 11:41–49

Baker AJM, McGrath SP, Reeves RD, Smith JAC (2000) Metal hyperaccummulator plants: a review of the ecology and physiology of a biological resource for phytoremediation of metal-polluted soils. In: Terry N, Baelos G (eds) Phytoremediation of contaminated soil and water. Lewis, Boca Raton, FL, pp 85–107

Basic N, Salamin N, Keller C, Galland N, Besnard G (2006) Cadmium hyperaccumulation and genetic differentiation of *Thlaspi caerulescens* populations. Biochem Syst Ecol 34:667–677

Becher M, Talke IN, Krall L, Krämer U (2004) Cross-species microarray transcript profiling reveals high constitutive expression of metalhomeostasis genes in shoots of the zinc hyperaccumulator *Arabidopsis halleri*. Plant J 37:251–268

Bernard C, Roosens N, Czernic P, Lebrun M, Verbruggen N (2004) A novel CPx-ATPase from the cadmium hyperaccumulator *Thlaspi caerulescens*. FEBS Lett 569:140–148

Bert V, Macnair MR, De Lague rie P, Saumitou-Laprade P, Petit D (2000) Zinc tolerance and accumulation in metallicolous and nonmetallicolous populations of *Arabidopsis halleri* (*Brassicaceae*). New Phytol 146:225–233

Bert V, Bonnin I, Saumitou-Laprade P, de Laguerie P, Petit D (2002) Do *Arabidopsis halleri* from nonmetallicolous populations accumulate zinc and cadmium more effectively than those from metallicolous populations? New Phytol 155:47–57

Bert V, Meerts P, Saumitou-Laprade P, Salis P, Gruber W, Verbruggen N (2003) Genetic basis of Cd tolerance and hyperaccumulation in *Arabidopsis halleri*. Plant Soil 249:9–18
Bidwell SD, Woodrow IE, Batianoff GN, Sommer-Knusden J (2002) Hyperaccumulation of manganese in the rainforest tree Austromyrtus bidwillii (Myrtaceae) from Queensland, Australia. Funct Plant Biol 29:899–905
Bidwell SD, Crawford SA, Woodrow IE, Sommer-Knudsen J, Marshall AT (2004) Sub-cellular localization of Ni in the hyperaccumulator, *Hybanthus floribundus* (Lindley) F. Muell. Plant Cell Environ 27:705–716
Bondada BR, Ma LQ (2003) Tolerance of heavy metals in vascular plants: arsenic hyperaccumulation by Chinese brake fern (*Pteris vittata* L.). In: Chandra S, Srivastava M (eds) Pteridology in new millenium. Kluwer, The Netherlands, pp 397–420
Bondada BR, Tu S, Ma LQ (2004) Absorption of foliar-applied arsenic by the arsenic hyperaccumulating fern (*Pteris vittata* L.). Sci Total Environ 332:61–70
Boominathan R, Doran PM (2003) Organic acid complexation, heavy metal distribution and the effect of ATPase inhibition in hairy roots of hyperaccumulator plant species. J Biotechnol 101:131–146
Boyd RS (2010) Heavy metal pollutants and chemical ecology: exploring new frontiers. J Chem Ecol 36:46–58
Broadhurst CL, Chaney RL, Angle JS, Maugel TK, Erbe EF, Murphy CA (2004) Simultaneous hyperaccumulation of nickel, manganese, and calcium in *Alyssum Leaf* Trichomes. Environ Sci Technol 38:5797–5802
Brooks RR, Lee J, Reeves R, Jaffre T (1977) Detection of nickeliferous rocks by analysis of herbarium specimens of indicator plants. J Geochem Explor 7:49–58
Caille N, Swanwick S, Zhao FJ, McGrath SP (2004) Arsenic hyperaccumulation by *Pteris vittata* arsenic contaminated soils and the effect of liming and phosphate fertilization. Environ Pollut 132:113–120
Chan DW, Son SC, Block W, Ye R, Douglas P, Pelley J, Goodarzi AA, Khanna KK, Wold MS, Taya Y, Lavin MF, Lees-Miller SP (2000) Purification and characterization of ATM from human placenta, a manganese-dependent, wortmanninsensitive serine/threonine protein kinase. J Biol Chem 275:7803–7810
Chaney RL, Angle JS, Broadhurst CL, Peters CA, Tappero RV, Sparks DL (2007) Improved understanding of hyperaccumulation yields commercial phytoextraction and phytomining technologies. J Environ Qual 36:1429–1443
Chao Y, Zhang M, Feng Y, Yang X, Islam E (2010) cDNA-AFLP analysis of inducible gene expression in zinc hyperaccumulator *Sedum alfredii Hance* under zinc induction. Environ Exp Bot 68:107–112
Cheng S (2003) Heavy metals in plants and phytoremediation: a state-of-the-art report with special reference to literature published in Chinese journals. Environ Sci Pollut Res Int 10:335–340
Clemens S, Palmgren M, Krämer U (2002) A long way ahead: understanding and engineering plant metal accumulation. Trends Plant Sci 7:309–315
Cobbett CS (2000) Phytochelatin biosynthesis and function in heavy-metal detoxification. Curr Opin Plant Biol 3:211–216
Corem S, Carpaneto A, Soliani P, Cornara L, Gambale F, Scholz-Starke J (2009) Response to cytosolic nickel of Slow Vacuolar channels in the hyperaccumulator plant *Alyssum bertolonii*. Eur Biophys J 38:495–501
Cosio C, Martinoia E, Keller C (2004) Hyperaccumulation of cadmium and zinc in *Thlaspi caerulescens* and *Arabidopsis halleri* at the leaf cellular level. Plant Physiol 134:716–725
de la Fuente V, Rodriguez N, Diez-Garretas B, Rufo L, Asensi A, Amils R (2007) Nickel distribution in the hyperaccumulator *Alyssum serpyllifolium* Desf. spp. from the Iberian Peninsula. Plant Biosyst 141:170–180
Dhankher OP, Li Y, Rosen BP, Shi J, Salt D, Senecoff JF, Sashti NA, Meagher RB (2002) Engineering tolerance and hyperaccumulation of arsenic in plants by combining arsenate reductase and g-glutamylcysteine synthetase expression. Nat Biotechnol 20:1140–1145

Dhankher OP, Shasti NA, Rosen BP, Fuhrmann M, Meagher RB (2003) Increased cadmium tolerance and accumulation by plants expressing bacterial arsenate reductase. New Phytol 159:431–441

Dong R, Formentin E, Losseso C, Carimi F, Benedetti P, Terzi M, Lo Schiavo F (2005) Molecular cloning and characterization of a phytochelatin synthase gene, *PvPCS1*, from *Pteris vittata* L. J Ind Microbiol Biotechnol 32:527–533

Duffus JH (2002) "Heavy metals" a meaningless term? (IUPAC Technical Report). Pure Appl Chem 74:793–807

Duruibe JO, Ogwoegbu MOC, Egwurugwu JN (2007) Heavy metal pollution and human biotoxic effects. Int J Phys Sci 2:112–118

Ebbs S, Lau I, Ahner B, Kochian L (2002) Phytochelatin synthesis is not responsible for Cd tolerance in the Zn/Cd hyperaccumulator *Thlaspi caerulescens* (J. & C. Presl). Planta 214:635–640

Erikson KM, Aschner M (2003) Manganese neurotoxicity and glutamate-GABA interaction. Neurochem Int 43:475–480

Farinati S, DalCorso G, Bona E, Corbella M, Lampis S, Cecconi D, Polati R, Berta G, Vallini G, Furini A (2009) Proteomic analysis of *Arabidopsis halleri* shoots in response to the heavy metals cadmium and zinc and rhizosphere microorganisms. Proteomics 9:4837–4850

Fattorini D, Notti A, Halt MN, Gambi MC, Regoli F (2005) Levels and chemical speciation of arsenic in polychaetes: a review. Mar Ecol 26:255–264

Fattorini D, Notti A, Nigro M, Regoli F (2010) Hyperaccumulation of vanadium in the Antarctic polychaete *Perkinsiana littoralis* as a natural chemical defense against predation. Environ Sci Pollut Res 17:220–228

Fernando DR, Batianoff GN, Baker AJM, Woodrow IE (2006) In vivo localization of manganese in the hyperaccumulator *Gossia bidwillii* (Benth.) N. Snow & Guymer (Myrtaceae) by cryo-SEM/EDAX. Plant Cell Environ 29:1012–1020

Fernando DR, Baker AJM, Woodrow IE, Batianoff GN, Bakkaus EJ, Collins RN (2007) Variability of Mn hyperaccumulation in the Australian rainforest tree *Gossia bidwillii* (Myrtaceae). Plant Soil 293:145–152

Fernando DR, Woodrow IE, Jaffré T, Dumontet V, Marshall AT, Baker AJM (2008) Foliar manganese accumulation byn*Maytenus founieri* (Celastraceae) in its native New Caledonian habitats: populational variation and localization by X-ray microanalysis. New Phytol 177:178–185

Fernando DR, Guymer G, Reeves RD, Woodrow IE, Baker AJ, Batianoff GN (2009) Foliar Mn accumulation in eastern Australian herbarium specimens: prospecting for 'new' Mn hyperaccumulators and potential applications in taxonomy. Ann Bot 103:93–939

Francesconi K, Visoottiviseth P, Sridokchan W, Goessler W (2002) Arsenic species in an As hyperaccumulating fern, *Pityrogramma calomelanos*: a potential phytoremediator of As-contaminated soils. Sci Total Environ 284:27–35

Fraústo da Silva JJR, Williams RJP (2001) The biological chemistry of the elements: the inorganic chemistry of life, 2nd edn. Oxford University Press, Oxford

Freeman JL, Persans MW, Nieman K, Albrecht C, Peer W, Pickering IJ, Salt DE (2004) Increased glutathione biosynthesis plays a role in nickel tolerance in *Thlaspi* nickel hyperaccumulators. Plant Cell 16:2176–2191

Freeman JL, Persans MW, Nieman K, Salt DE (2005) Nickel and cobalt resistance engineered in *Escherichia coli* by overexpression of serine acetyltransferase from the nickel hyperaccumulator plant *Thlaspi goesingense*. Appl Environ Microb 71:8627–8633

Georgiou G, Stathopoulos C, Daugherty PS, Nayak AR, Iverson BL, Curtiss R 3rd (1997) Display of heterologous proteins on the surface of microorganisms: from the screening of combinatorial libraries to live recombinant vaccines. Nat Biotechnol 15:29–34

Gerber GB, Leonard A, Hantson P (2002) Carcinogenicity, mutagenicity and teratogenicity of manganese compounds. Crit Rev Oncol Hematol 42:25–34

Gibbs PE, Bryan GW, Ryan KP (1981) Copper accumulation by the polychaete *Melinna palmata*: an antipredation mechanism? J Mar Biol Ass UK 61:707–722

Gifford S, Dunstan RH, O'Connor W, Koller CE, MacFarlane GR (2007) Aquatic zooremediation: deploying animals to remediate contaminated aquatic environments. Trends Biotechnol 25:60–65

Gispert C, Ros R, de Haro A, Walker DJ, Pilar Bernal M, Serrano R, Avino JN (2003) A plant genetically modified that accumulates Pb is especially promising for phytoremediation. Biochem Biophys Res Commun 303:440–445

Godbold DL, Horst WJ, Collins JC, Thurman DA, Marschner H (1984) Accumulation of zinc and organic acids in roots of zinc tolerant and non-tolerant ecotypes of *Deschampsia caespitosa*. J Plant Physiol 116:59–69

Hammond JP, Bowen HC, White PJ, Mills V, Pyke KA, Baker AJ, Whiting SN, May ST, Broadley MR (2006) A comparison of the *Thlaspi caerulescens* and *Thlaspi arvense* shoot transcriptomes. New Phytol 170:239–260

Hanikenne M, Talke IN, Haydon MJ, Lanz C, Nolte A, Motte P, Kroymann J, Weigel D, Krämer U (2008) Evolution of metal hyperaccumulation required cis-regulatory changes and triplication of HMA4. Nature 453:391–395

Hart JJ, Welch RM, Norvell WA, Kochian LV (2002) Transport interactions between cadmium and zinc in roots of bread and durum wheat seedlings. Physiol Plant 116:73–78

Higuchi K, Kanazawa K, Nishizawa NK, Chino M, Mori S (1994) Purification and characterization of nicotianamine synthase from Fe-deficient barley roots. Plant Soil 165:173–179

Honjo T, Hatta A, Taniguchi K (1984) Characterization of heavy metals in indicator plants – studies on the accumulation of lead and tolerance of gregarious fern, *Athyrium yokoscense*, in the polluted areas from the lead tile of the ruins of Kanazawa Castle, now the campus of Kanazawa University. J Phytogeogr Taxon 32:68–80

Ingle RA, Mugford ST, Rees JD, Campbell MM, Smith JAC (2005a) Constitutively high expression of histidine biosynthetic pathway contributes to nickel tolerance in hyperaccumulator plants. Plant Cell 17:2089–2106

Ingle RA, Smith JAC, Sweetlove LJ (2005b) Responses to nickel in the proteome of the hyperaccumulator *Alyssum lesbiacum*. Biometals 18:627–641

Ingle RA, Fricker MD, Smith JAC (2008) Evidence for nickel/proton antiport activity at the tonoplast of the hyperaccumulator plant *Alyssum lesbiacum*. Plant Biol 10:746–753

Ishii T, Otake T, Okoshi K, Nakahara M, Nakamura R (1994) Intracellular localization of vanadium in the fan worm *Pseudopotamilla occelata*. Mar Biol 121:143–151

Kambe-Honjoh H, Ohsumi K, Shimoi H, Nakajima H, Kitamoto K (2000) Molecular breeding of yeast with higher metal-adsorption capacity by expression of histidine-repeat insertion in the protein anchored to the cell wall. J Gen Appl Microbiol 46:113–117

Kashem MA, Singh BR, Kubota H, Sugawara R, Kitajima N, Kondo T, Kawai S (2010) Zinc tolerance and uptake by *Arabidopsis halleri ssp. gemmifera* grown in nutrient solution. Environ Sci Pollut Res 17:1174–1176

Kerkeb L, Krämer U (2003) The role of free histidine in xylem loading of nickel in *Alyssum lesbiacum* and *Brassica juncea*. Plant Physiol 131:716–724

Kotrba P, Rumi T (2010) Surface display of metal fixation motifs of bacterial P1-type ATPase specifically promotes biosorption of Pb(2+) by *Saccharomyces cerevisiae*. Appl Environ Microbiol 76:2615–2622

Krämer U (2005) Phytoremediation: novel approaches to cleaning up polluted soils. Curr Opin Biotechnol 16:133–141

Krämer U, Cotter-Howells JD, Charnock JM, Baker AJM, Smith JAC (1996) Free histidine as a metal chelator in plants that accumulate nickel. Nature 379:635–638

Krämer U, Pickering IJ, Prince RC, Raskin I, Salt DE (2000) Subcellular localization and speciation of nickel in hyperaccumulator and non-accumulator *Thlaspi* species. Plant Physiol 122:1343–1353

Krotz RM, Evangelou BP, Wagner GJ (1989) Relationships between cadmium, zinc, Cd-peptide, and organic acid in tobacco suspension cells. Plant Physiol 91:780–787

Kumar NPBA, Dushenkov V, Motto H, Raskin I (1995) Phytoextraction: the use of plants to remove heavy metals from soils. Environ Sci Technol 29:1232–1238

Küpper H, Lombi E, Zhao FJ, McGrath SP (2000) Cellular compartmentation of cadmium and zinc in relation to other elements in the hyperaccumulator *Arabidopsis halleri*. Planta 212:75–84

Küpper H, Lombi E, Zhao FJ, Wieshammer G, McGrath SP (2001) Cellular compartmentation of nickel in the hyperaccumulators *Alyssum lesbiacum*, *Alyssum bertolonii* and *Thlaspi goesingense*. J Exp Bot 52:2291–2300

Küpper H, Mijovilovich A, Meyer-Klaucke W, Kroneck PMH (2004) Tissue- and age-dependent differences in the complexation of cadmium and zinc in the Cd/Zn hyperaccumulator *Thlaspi caerulescens* (Ganges ecotype) revealed by x-ray absorption spectroscopy. Plant Physiol 134:748–757

Küpper H, Götz B, Mijovilovich A, Küpper FC, Meyer-Klaucke W (2009) Complexation and toxicity of copper in higher plants. I. Characterization of copper accumulation, speciation, and toxicity in *Crassula helmsii* as a new copper accumulator. Plant Physiol 151:702–714

Lasat MM, Baker A, Kochian L (1996) Physiological characterization of root Zn^{2+} absorption and translocation to shoots in Zn hyperaccumulator and nonaccumulator species of *Thlaspi*. Plant Physiol 112:1715–1722

Lasat MM, Baker A, Kochian L (1998) Altered Zn compartmentation in the root symplasm and stimulated Zn absorption into the leaf as mechanisms involved in Zn hyperaccumulation in *Thlaspi caerulescens*. Plant Physiol 118:875–883

Lasat MM, Pence NS, Garvin DF, Ebbs SD, Kochian LV (2000) Molecular physiology of zinc transport in the Zn hyperaccumulator *Thlaspi caerulescens*. J Exp Bot 51:71–79

Lei M, Chen TB, Huang ZC, Wang YD, Huang YY (2008) Simultaneous compartmentalization of lead and arsenic in co-hyperaccumulator *Viola principis* H. de Boiss.: an application of SRXRF microprobe. Chemosphere 72:1491–1496

Li WC, Ye ZH, Wong MH (2010) Metal mobilization and production of short-chain organic acids by rhizosphere bacteria associated with a Cd/Zn hyperaccumulating plant, *Sedum alfredii*. Plant Soil 326:453–467

Liu XM, Wu QT, Banks MK (2005) Effect of simultaneous establishment of *Sedum alfridii* and *Zea mays* on heavy metal accumulation in plants. Int J Phytoremediation 7:43–53

Lombi E, Zhao F, McGrath S, Young S, Sacchi G (2001) Physiological evidence for a high-affinity cadmium transporter highly expressed in a *Thlaspi caerulescens* ecotype. New Phytol 149:53–60

Lone MI, He Z, Stoffella PJ, Yang X (2008) Phytoremediation of heavy metal polluted soils and water: progress and perspectives. J Zhejiang Univ Sci B 9:210–220

Long XX, Zhang YG, Dai J, Zhou Q (2009) Zinc, cadmium and lead accumulation and characteristics of rhizosphere microbial population associated with hyperaccumulator *Sedum Alfredii Hance* under natural conditions. Bull Environ Contam Toxicol 82:460–467

Lu L, Tian S, Yang X, Wang X, Brown P, Li T, He Z (2008) Enhanced root-to-shoot translocation of cadmium in the hyperaccumulating ecotype of *Sedum alfredii*. J Exp Bot 59:3203–3213

Ma JF, Hiradate S (2000) Form of aluminium for uptake and translocation in buckwheat (*Fagopyrum esculentum* Moench). Planta 211:355–360

Ma JF, Zheng SJ, Matsumoto H, Hiradate S (1997) Detoxifying aluminium with buckwheat. Nature 390:569–570

Ma LQ, Komar KM, Tu C, Zhang W, Cai Y, Kennelley ED (2001) A fern that hyperaccumulates arsenic. Nature 409:579

Macnair MR (2002) Within and between population genetic variation for zinc accumulation in *Arabidopsis halleri*. New Phytol 155:59–66

Macnair MR (2003) The hyperaccumulation of metals by plants. Adv Bot Res 40:63–106

Mandal BK, Suzuki KT (2002) Arsenic round the world: a review. Talanta 58:201–235

Mari S, Gendre D, Pianelli K, Ouerdane L, Lobinski R, Briat JF, Lebrun M, Czernic P (2006) Root-to-shoot long-distance circulation of nicotianamine and nicotianamine-nickel chelates in the metal hyperaccumulator *Thlaspi caerulescens*. J Exp Bot 57:4111–4122

Marquès L, Cossegal M, Bodin S, Czernic P, Lebrun M (2004) Heavy metal specificity of cellular tolerance in two hyperaccumulating plants, *Arabidopsis halleri* and *Thlaspi caerulescens*. New Phytol 164:289–295

Martinoia E, Maeshima M, Neuhaus HE (2007) Vacuolar transporters and their essential role in plant metabolism. J Experim Bot 58:83–102

McGrath SP, Zhao FJ (2003) Phytoextraction of metals and metalloids from contaminated soils. Curr Opin Biotechnol 14:277–282

McGrath SP, Zhao FJ, Lombi E (2001) Plant and rhizosphereprocesses involved in phytoremediation of metal-contaminated soils. Plant Soil 232:207–214

McGrath SP, Lombi E, Gray CW, Caille N, Dunham SJ, Zhao FJ (2006) Field evaluation of Cd and Zn phytoextraction potential by the hyperaccumulators *Thlaspi caerulescens* and *Arabidopsis halleri*. Environ Pollut 141:115–125

McNear DH, Peltier E, Everhart J, Chaney RL, Sutton S, Newville M, Rivers M, Sparks DL (2005) Application of quantitative fluorescence and absorption-edge computed microtomography to image metal compartmentalization in *Alyssum murale*. Environ Sci Technol 39:2210–2218

Mesjasz-Przybyłowicz J, Balkwill K, Przybyłowicz WJ, Annegarn HJ (1994) Proton microprobe and X-ray fluorescence investigations of nickel distribution in serpentine flora fromSouth Africa. Nucl Instrum Meth Phys Res B 89:208–212

Mijovilovich A, Leitenmaier B, Meyer-Klaucke W, Kroneck PMH, Goötz B, Küpper H (2009) Complexation and toxicity of copper in higher plants. II. Different mechanisms for copper versus cadmium detoxification in the copper-sensitive cadmium/zinc hyperaccumulator *Thlaspi caerulescens* (Ganges ecotype). Plant Physiol 151:715–731

Mills R, Krijger G, Baccarini P, Hall J, Williams L (2003) Functional expression of AtHMA4, a P1B-type ATPase of the Zn/Co/Cd/Pb subclass. Plant J 35:164–176

Milner MJ, Kochian LV (2008) Investigating heavy-metal hyperaccumulation using *Thlaspi caerulescens* as a model system. Ann Bot Lond 102:3–13

Morishita T, Boratynski K (1992) Accumulation of Cd and other metals in organs of plants growing around metal smeltersin Japan. Soil Sci Pl Nutr 38:781–785

Murai T, Ueda M, Yamamura M, Atomi H, Shibasaki Y, Kamasawa N, Osumi M, Amachi T, Tanaka A (1997) Construction of a starch-utilizing yeast by cell surface engineering. Appl Environ Microbiol 63:1362–1366

Nakajima H, Iwasaki T, Kitamoto K (2001) Metalloadsorption by *Saccharomyces cerevisiae* cells expressing invertase-metallothionein (Suc2-Cup1) fusion protein localized to the cell surface. J Gen Appl Microbiol 47:47–51

Ni TH, Wei YZ (2003) Subcellular distribution of cadmium in mining ecotype *Sedum alfredii*. Acta Bot Sin 45:925–928

Nigam R, Srivastava S, Prakash S, Srivastava MM (2001) Cadmium mobilisation and plant availability: the impact of organic acids commonly exuded from roots. Plant Soil 230:107–113

Olguin EJ (2003) Phycoremediation: key issues for cost-effective nutrient removal processes. Biotechnol Adv 22:81–91

Papoyan A, Kochian LV (2004) Identification of *Thlaspi caerulescens* genes that may be involved in heavy metal hyperaccumulation and tolerance. Characterization of a novel heavy metal transporting ATPase. Plant Physiol 136:3814–3823

Papoyan A, Pineros M, Kochian LV (2007) Plant Cd^{2+} and Zn^{2+} status effects on root and shoot heavy metal accumulation in *Thlaspi caerulescens*. New Phytol 175:51–58

Patel KS, Shrivas K, Brandt RN, Jakubowski WC, Hoffmann P (2005) Arsenic contamination in water, soil, sediment and rice of central India. Environ Geochem Health 27:131–145

Pauwels M, Frérot H, Bonnin I, Saumitou-Laprade P (2006) A broadscale study of population differentiation for Zn-tolerance in an emerging model species for tolerance study: *Arabidopsis halleri* (*Brassicaceae*). J Evol Biol 19:1838–1850

Pence NS, Larsen PB, Ebbs SD, Letham DL, Lasat MM, Garvin DF, Eide D, Kochian LV (2000) The molecular basis for heavy metal hyperaccumulation in *Thlaspi caerulescens*. Proc Natl Acad Sci USA 97:4956–4960

Persans MW, Yan X, Patnoe J-MML, Krämer U, Salt DE (1999) Molecular dissection of the role of histidine in nickel hyperaccumulation in *Thlaspi goesingense* (Hálácsy). Plant Physiol 121:1117–1126

Persans MW, Nieman K, Salt DE (2001) Functional activity and role of cation-efflux family members in Ni hyperaccumulation in *Thlaspi goesingense*. Proc Natl Acad Sci USA 98:9995–10000

Pianelli K, Mari S, Marques L, Lebrun M, Czernic P (2005) Nicotianamine over-accumulation confers resistance to nickel in *Arabidopsis thaliana*. Transgenic Res 14:739–748

Pollard AJ, Powell KD, Harper FA, Smith JAC (2002) The genetic basis of metal hyperaccumulation in plants. Crit Rev Plant Sci 21:539–566

Reeves RD (2003) Tropical hyperaccumulators of metals and their potential for phytoextraction. Plant Soil 249:57–65

Reeves RD, Adigüzel NN (2004) Rare plants and nickel accumulators from Turkish serpentine soils, with special reference to *Centaurea* species. Turk J Bot 28:147–153

Reeves RD, Baker AJM (2000) Metal-accumulating plants. In: Raskin I, Ensley BD (eds) Phytoremediation of toxic metals: using plants to clean up the environment. Wiley, New York, NY, pp 193–221

Rigola D, Fiers M, Vurro E, Aarts MGM (2006) The heavy metal hyperaccumulator *Thlaspi caerulescens* expresses many species-specific genes, as identified by comparative expressed sequence tag analysis. New Phytol 170:753–766

Roosens N, Bernard C, Leplae R, Verbruggen N (2004) Adaptive evolution of metallothionein 3 in the Cd/Zn hyperaccumulator *Thlaspi caerulescens*. Z Naturforsch 60:224–228

Rugh CL, Wilde HD, Stack NM, Thompson DM, Summers AO, Meagher RB (1996) Mercuric ion reduction and resistance in transgenic *Arabidopsis thaliana* plants expressing a modified bacterial *merA* gene. Proc Natl Acad Sci USA 93:3182–3187

Ruta LL, Paraschivescu CC, Matache M, Avramescu S, Farcasanu IC (2010) Removing heavy metals from synthetic effluents using "kamikaze" *Saccharomyces cerevisiae* cells. Appl Microbiol Biotechnol 85:763–771

Sahi SV, Bryant NL, Sharma NC, Singh SR (2002) Characterization of a lead hyperaccumulator shrub, *Sesbania drummondii*. Environ Sci Technol 36:4676–4680

Saito A, Higuchi K, Hirai M, Nakane R, Yoshiba M, Tadano T (2005) Selection and characterization of a nickel-tolerant cell line from tobacco (*Nicotiana tabacum* cv. bright yellow-2) suspension culture. Physiol Plant 125:441–453

Sánchez ML (ed) (2008) Causes and effects of heavy metal pollution. Nova Science, Hauppauge

Sandrini JZ, Regoli F, Fattorini D, Notti A, Inacio AF, Linde-Arias AR, Laurino J, Bainy AC, Marins LF, Monserrat JM (2006) Short-term responses to cadmium exposure in the estuarine polychaete *Laeonereis acuta* (polychaeta, Nereididae): subcellular distribution and oxidative stress generation. Environ Toxicol Chem 25:1337–1344

Sanità di Toppi LS, Gabbrielli R (1999) Response to cadmium in higher plants. Environ Exp Bot 41:105–130

Seki H, Suzuki A (1997) A new method for the removal of toxic metal ions from acid-sensitive biomaterial. J Colloid Interface Sci 190:206–211

Seward MRD, Richardson DHS (1990) Atmospheric sources of metal pollution and effects on vegetation. In: Shaw AJ (ed) Heavy metal tolerance in plants: evolutionary aspects. CRC, Florida, pp 75–92

Shen ZG, Li XD, Chen HM (2000) Comparison of elemental composition and solubility in the zinc hyperaccumulator *Thlaspi caerulescens* with the non-hyperaccumulator *Thlaspi ochroleucum*. Bull Environ Contam Toxicol 65:343–350

Shen R, Ma JF, Kyo M, Iwashita T (2002) Compartmentation of aluminium in leaves of an Al-accumulator, *Fagopyrum esculentum* Moench. Planta 215:394–398

Shi JY, Chen YX, Huang YY, He W (2004) SRXRF microprobe as a technique for studying elements distribution in *Elsholtzia splendens*. Micron 35:557–564
Shibasaki S, Maeda H, Ueda M (2009) Molecular display technology using yeast–arming technology. Anal Sci 25:41–49
Shira C, Carpaneto A, Soliani P, Cornara L, Gambale F, Scholz-Starke J (2009) Response to cytosolic nickel of Slow Vacuolar channels in the hyperaccumulator plant *Alyssum bertolonii*. Eur Biophys J 38:495–501
Shiraishi T, Tamada M, Saito K, Sugo T (2003) Recovery of cadmium from waste of scallop processing with amidoxime adsorbent synthesized by graftpolymerization. Radiat Phys Chem 66:43–47
Smart KE, Kilburn MR, Salter CJ, Smith JAC, Grovenor CRM (2007) NanoSIMS and EPMA analysis of nickel localisation in leaves of the hyperaccumulator plant *Alyssum lesbiacum*. Int J Mass Spectrom 260:107–114
Still ER, Williams RJP (1980) Potential methods for selective accumulation of nickel(II) ions by plants. J Inorg Biochem 13:35–40
Sun Q, Ye ZH, Wang XR, Wong MH (2005) Increase of glutathione in mine population of *Sedum alfredii*: a Zn hyperaccumulator and pb accumulator. Phytochem 66:2549–2556
Sun Q, Ye ZH, Wang XR, Wong MH (2007) Cadmium hyperaccumulation leads to an increase of glutathione rather than phytochelatins in the cadmium hyperaccumulator *Sedum alfredii*. J Plant Physiol 164:1489–1498
Talke I, Hanikenne M, Krämer U (2006) Zinc-dependent global transcriptional control, transcriptional deregulation, and higher gene copy number for genes in metal homeostasis of the hyperaccumulator Arabidopsis halleri. Plant Physiol 142:148–167
Tamura H, Honda M, Sato T, Kamachi H (2005) Pb hyperaccumulation and tolerance in common buckwheat (*Fagopyrum esculentum* Moench). J Plant Res 118:355–359
Tolrá RP, Poschenrieder C, Barceló J (1996) Zinc hyperaccumulation in *Thlaspi caerulescens*. II. Influence on organic acids. J Plant Nutr 19:1541–1550
Tu C, Ma LQ (2002) Effects of arsenic concentrations and forms on arsenic uptake by the hyperaccumulator Ladder Brake. J Environ Qual 31:641–647
Tuomainen MH, Nunan N, Lehesranta SJ, Tervahauta AI, Hassinen VH, Schat H, Koistinen KM, Auriola S, McNicol J, Kärenlampi SO (2006) Multivariate analysis of protein profiles of metal hyperaccumulator *Thlaspi caerulescens* accessions. Proteomics 6:3696–3706
Tuomainen M, Tervahauta A, Hassinen V, Schat H, Koistinen KM, Lehesranta S, Rantalainen K, Häyrinen J, Auriola S, Anttonen M, Kärenlampi S (2010) Proteomics of *Thlaspi caerulescens* accessions and an interaccession cross segregating for zinc accumulation. J Exper Bot 61:1075–1087
Ueno D, Ma JF, Iwashita T, Zhao FJ, McGrath SP (2005) Identification of the form of Cd in the leaves of a superior Cd-accumulating ecotype of *Thlaspi caerulescens* using ^{113}Cd-NMR. Planta 221:928–936
Uraguchi S, Watanabe I, Yoshitomi A, Kiyono M, Kuno K (2006) Characteristics of cadmium accumulation and tolerance in novel Cd-accumulating crops, *Avena strigosa* and *Crotalaria juncea*. J Exp Bot 57:2955–2965
Vacchina V, Mari S, Czernic P, Marques L, Pianelli K, Schaumloeffel D, Lebrun M, Lobinski R (2003) Speciation of nickel in a hyperaccumulating plant by high-performance liquid chromatography-inductively coupled plasma mass spectrometry and electrospray MS/MS assisted by cloning using yeast complementation. Anal Chem 75:2740–2745
Van de Mortel JE, Almar Villanueva L, Schat H, Kwekkeboom J, Coughlan S, Moerland PD, Loren V, van Themaat E, Koornneef M, Aarts MGM (2006) Large expression differences in genes for iron and zinc homeostasis, stress response, and lignin biosynthesis distinguish roots of *Arabidopsis thaliana* and the related metal hyperaccumulator *Thlaspi caerulescens*. Plant Physiol 142:1127–1147

Van TK, Kang Y, Fukui T, Sakurai K, Iwasaki K, Aikawa Y, Phuong NM (2006) Arsenic and heavy metal accumulation by *Athyrium yokoscense* from contaminated soils. Soil Sci Plant Nutr 52:701–710

Vartanian JP, Sala M, Henry M, Hobson SW, Meyerhans A (1999) Manganese cations increase the mutation rate of human immune deficiency virus type 1 ex vivo. J Gen Virol 80: 1983–1986

Vera-Estrella R, Miranda-Vergara MC, Barkla BJ (2009) Zinc tolerance and accumulation in stable cell suspension cultures and in vitro regenerated plants of the emerging model plant *Arabidopsis halleri* (*Brassicaceae*). Planta 229:977–986

Verbruggen N, Hermans C, Schat H (2009) Molecular mechanisms of metal hyperaccumulation in plants. New Phytol 181:759–776

Visoottiviseth P, Francesconi K, Sridokchan W (2002) The potential of Thai indigenous plant species for the phytormediation of As contaminated land. Environ Pollut 118:453–461

Watanabe T, Moon CS, Zhang ZW, Shimbo S, Nakatsuka H, Matsuda-Inoguchi N, Higashikawa K, Ikeda M (2000) Cadmium exposure of women in general populations in Japan during 1991–1997 compared with 1977–1991. Int Arch Occup Environ Health 73:26–34

Weber M, Harada E, Vess C, von Roepenack-Lahaye E, Clemens S (2004) Comparative microarray analysis of *Arabidopsis thaliana* and *Arabidopsis halleri* roots identifies nicotianamine synthase, a ZIP transporter and other genes as potential metal hyperaccumulation factors. Plant J 37:269–281

Wei CY, Wang C, Sun X, Wang WY (2006) Factors influencing arsenic accumulation by *Pteris vittata*: a comparative field study at two sites. Environ Pollut 141:488–493

Wei CY, Wang C, Sun X, Wang WY (2007) Arsenic accumulation by ferns: a field survey in southern China. Environ Geochem Health 29:169–177

Wu F, Zhang G, Yu J (2003) Interaction of cadmium and four microelements for uptake and translocation in different barley genotypes. Commun Soil Sci Plant Anal 34:2003–2020

Xu XH, Shi JY, Chen YX, Xue SG, Wu B, Huang YY (2006) An investigation of cellular distribution of manganese in hyperaccumulator plant *Phytolacca acinosa* Roxb. Using SRXRF analysis. J Environ Sci (China) 18:746–751

Xue SG, Chen YX, Reeves RD, Baker AJM, Lin Q, Fernando DR (2004) Manganese uptake and accumulation by the hyperaccumulator plant *Phytolacca acinosa* Roxb. (Phytolaccaceae). Environ Pollut 131:393–399

Xue SG, Chen YX, Baker AJM (2005) Manganese uptake and accumulation by two populations of *Phytolacca acinosa* Roxb. (*Phytolaccaceae*). Water Air Soil Pollut 160:3–14

Yanai J, Zhao FJ, McGrath SP, Kosaki T (2006) Effect of soil characteristicson Cd uptake by the hyperaccumulator *Thlaspi caerulescens*. Environ Pollut 139:167–175

Yang XE, Baligar VC, Foster JC, Martens DC (1997) Accumulation and transport of nickel in relation to organic acids in ryegrass and maize grown with different nickel levels. Plant Soil 196:271–276

Yang LS, Peterson PJ, Williams WP, Wang WY, Hou SF, Tan JA (2002) The relationship between exposure to arsenic concentrations in drinking water and the development of skin lesions in farmers from Inner Mongolia, China. Environ Geochem Health 24:293–303

Yang XE, Long XX, Ye HB, He ZL, Calvert DV, Stoffella PJ (2004) Cadmium tolerance and hyperaccumulation in a new Zn-hyperaccumulating plant species (*Sedum alfredii Hance*). Plant Soil 259:181–189

Yang X, Li T, Yang J, He Z, Lu L, Meng F (2006) Zinc compartmentation in root, transport into xylem, and absorption into leaf cells in the hyperaccumulating species of *Sedum alfredii* Hance. Planta 224:185–195

Yanqun Z, Yuan L, Jianjun C, Haiyan C, Li Q, Schvartz C (2005) Hyperaccumulation of Pb, Zn and Cd in herbaceous grown on lead-zinc mining area in Yunnan, China. Environ Int 31:755–762

Zenk MH (1996) Heavy metal detoxification in higher plants: a review. Gene 179:21–30

Zhao FJ, Wang JR, Barker JHA, Schat H, Bleeker PM, McGrath SP (2003) The role of phytochelatins in arsenic tolerance in the hyperaccumulator *Pteris vittata*. New Phytol 159: 403–410

Zhao FJ, Jiang RF, Dunham SJ, McGrath SP (2006) Cadmium uptake, translocation and tolerance in the hyperaccumulator *Arabidopsis halleri*. New Phytol 172:646–654

Chapter 14
Nickel Hyperaccumulating Plants and *Alyssum bertolonii*: Model Systems for Studying Biogeochemical Interactions in Serpentine Soils

Alessio Mengoni, Lorenzo Cecchi, and Cristina Gonnelli

14.1 The Serpentine Factor as a Tool for Studying Biogeochemical Interactions

Serpentine rocks (or ophiolites) derive their name from the olive greenish-gray color, striped in different shades, that looks like the skin of a snake (*serpens* in latin, *οφίς – ophis* in Greek). They originate from metamorphic alterations of peridotites with water and may form near the Earth's surface or in the upper part of the Earth's mantle during subduction events. In a wider concept, the same term is extended to all substrates which are derived from the weathering of ultramafic (igneous or metamorphic) rocks that contain at least 70% hydrous magnesium – iron phyllosilicates such as antigorite and chrysotile, minerals with the general formula $(Mg, Fe^{II})_3Si_2O_5(OH)_4$ (Brooks 1987; Kruckeberg 2002). Serpentine outcrops are spread worldwide within 22 of the 35 floristic regions (as defined by Takhtajan 1986), from sea level up to 2,000–3,000 m, ranging from 0 to 70 latitude degrees but cover no more than 1% of total Earth's surface (Fig. 14.1).

Serpentine soils have such extreme chemical and physical properties to render them potentially toxic and unsuitable for most plant species (Brooks 1987; Brady et al. 2005; Chiarucci et al. 1998) and for many microorganisms (Mengoni et al. 2010) (Table 14.1).

As a general rule, in comparison with other rock types, ultramafites are strongly enriched in elements such as iron, nickel, cobalt, and chromium, whereas they present much lower abundance of plant nutrients, such as calcium, nitrogen, phosphorus, and potassium. The relatively high concentrations of nickel and cobalt in serpentines largely depends on the fact that the ionic radii of their divalent states are very close to that of Mg^{2+} so that ionic substitution readily takes place into magnesium-rich minerals, which are dominant in ophiolitic rocks. Chromium is

A. Mengoni (✉) • L. Cecchi • C. Gonnelli
Department of Evolutionary Biology, University of Firenze, via Romana 17, I-50125, Firenze, Italy
e-mail: alessio.mengoni@unifi.it

E. Kothe and A. Varma (eds.), *Bio-Geo Interactions in Metal-Contaminated Soils*, Soil Biology 31, DOI 10.1007/978-3-642-23327-2_14,

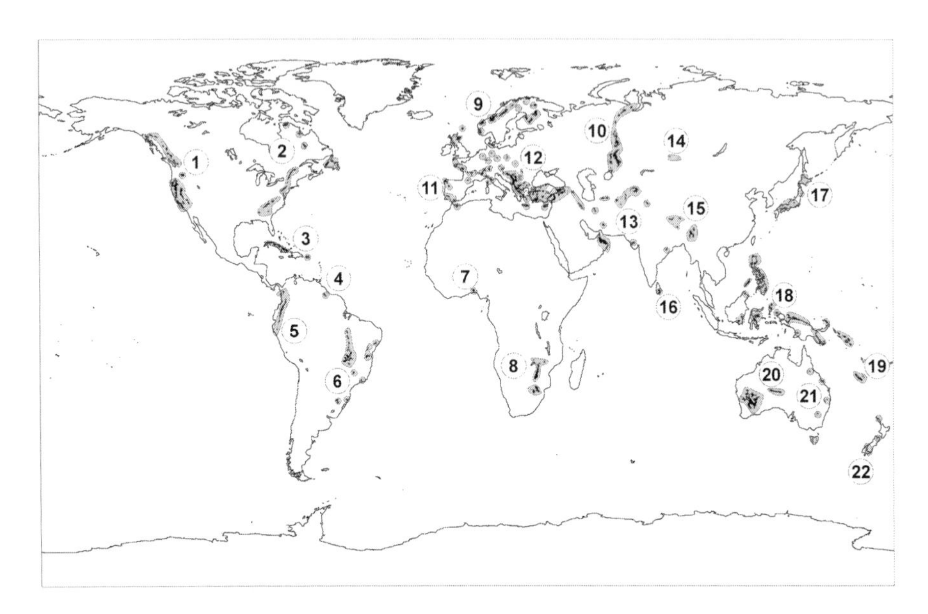

Fig. 14.1 Worldwide distribution of ultramafic outcrops. 1, Western North America (from Alaska to northern California); 2, eastern North America (from Labrador and Newfoundland to South Carolina); 3, Caribbean (Cuba and Puerto Rico); 4, Guyana; 5, Andes; 6, Brazil; 7, Ghana; 8, southern Africa (from Zaire to South Africa); 9, northern Europe (Scotland and Fennoscandia); 10, western Europe and northern Morocco; 11, eastern Europe, Anatolia, and Cyprus; 13, Oman and south-western Asia; 14. central Asia; 15, north-western India and Bangladesh; 16, Ceylon; 17, Japan; 18, Indopacific Islands (from Malay archipelago to Solomon Islands); 19, New Caledonia; 20, south-western Australia; 21, eastern Australia; 22, New Zealand. *Black* dots indicate localization of the presence of ultramafic outcrop within wider areas defined by *gray* color

Table 14.1 Ranges of chemical characteristics of serpentine soils in Tuscany (Italy)

Chemical parameters	Range of variation
pH (1:2.5, soil:H_2O)	7.15–7.52
CEC (meq g^{-1} DW)	25.2–32.4
Organic matter (%)	1.7–8.3
Ni	979–2,724
Cr	897–3,342
Co	63–229
Ca	2,015–4,881
Mg	38,785–13,3010
Ca/Mg	0.019–0.239

CEC, cation exchange capacity. Total metal concentrations are expressed in μg g^{-1} DW. Concentrations in surface soils exceeding 100 μg g^{-1} DW for Ni and Cr and 50 μg g^{-1} DW for Co are considered phytotoxic (Kabata-Pendias and Pendias 1991). Data from Galardi et al. (2007a) and Mengoni et al. (2004)

enriched because Cr^{3+} readily substitutes in Fe^{3+} minerals (Brooks 1987). In any event, the cation concentration of these soils is known to vary markedly as they are derived from world-spread rocks that occur under a wide range of climates (Kruckeberg 2002).

Nickel is often believed to play a major role in determining the flora and vegetation in many serpentine areas (Brooks 1987; Vergnano Gambi 1992; Robinson et al. 1997). Nickel has a relatively high availability in the range of pH values of serpentine soils and values of bioavailable nickel in serpentine soils are often significantly higher than the toxicity threshold (as defined for common plants; see Kabata-Pendias and Pendias 1991). However, not all serpentine soils are ever nickel toxic as shown by early experiments (Slingsby and Brown 1977), suggesting that serpentine adaptation is not always linked to the presence of heavy metals such as Ni. The discovery of a large number of taxa that accumulate Ni in their tissues (Brooks 1987; Bani et al. 2007) may be a further evidence of the high selective pressure that this element exerts on serpentine plants. However, clear evidence for nickel toxicity in any serpentine is sadly lacking in the literature. Although relatively high concentrations of Co are available in plants in ultramafic soils, its accumulation in plant tissues is rare (Robinson et al. 1997). On the other hand, Cr has very low exchangeable concentrations in the soil and few species are known that truly hyperaccumulate this element (Brooks 1987; Robinson et al. 1997; Chiarucci 2003; Zhang et al. 2007). Another possible selective factor is the high concentration of Mg and/or the deficiency of Ca, i.e., the unfavorable ratio of Mg to Ca in serpentine soils. Strong effects of the Mg/Ca quotient (Brooks 1987; Proctor and Woodell 1975; Kruckeberg 2002; Roberts and Proctor 1992) and the toxic influence of Mg (Proctor 1971; Brooks and Yang 1984; Bani et al. 2007) have been found in several studies, and the addition of Ca to serpentine soils may reverse the unfavorable conditions of these soils, at least to some extent (Proctor 1971; Brooks 1987; Brady et al. 2005). Moreover, nonserpentine soils with high Mg concentrations share several floristic elements with ultramafic environments (Mota et al. 2008), suggesting that serpentine adaptation may often be explained as a mere tolerance to the "magnesium factor."

Another problem can be represented by the low nutrient content of ophiolites (Brooks 1987; Proctor and Nagy 1992). Fertilization with P, K, or N enhanced cover and productivity and resulted in a change in the floristic composition of serpentine plant communities (Huenneke et al. 1990; Proctor and Nagy 1992; Chiarucci et al. 1999; Chiarucci and Maccherini 2007; Bani et al. 2007).

The relative available concentrations of all the above-mentioned elements in water matrices are mutually influenced, as a consequence of their direct chemical interactions and their indirect contribution to the soil organic matter content and pH values. As a consequence, the effects of fertilization with a certain nutrient elements may lead to the misinterpretation of its actual role in producing the "serpentine factor" (Brooks 1987).

The physical conditions of serpentine are also hostile to many plants. Serpentine outcrops are often steep and relatively rocky, making them particularly vulnerable to erosion, which results in shallow freely draining soils. In addition, they generally have negligible contents of silt and clay. Combined, these factors yield an environment with little moisture and depressed nutrient levels (Kruckeberg 2002; Proctor and Woodell 1975; Walker 1954). Furthermore, the scarce plant cover also promotes erosion and elevated soil thermal excursions (Kruckeberg 2002). Each of these factors poses an additional stress to plant life. As a collective result, three traits can be identified as strictly characteristic of serpentine environments: poor plant productivity, high rates of endemism, and vegetation types distinct from those of neighboring areas (Whittaker 1954).

Jointly in the still elusive "serpentine factor," the chemical, physical, and biotic components of such soils produce what Jenny (1980) defined as the "serpentine syndrome," i.e., the cumulative effect of these components on plant form, development, and distribution. Such a "syndrome" is the key for the evolution of endemic taxa (Pichi Sermolli 1948; Kruckeberg 1954; Kruckeberg and Kruckeberg 1990), and the reason is that serpentine outcrops have to be considered as "ecological islands" (Lefèbvre and Vernet 1990), taking also into account that they are ubiquitous but patchily distributed. Because of all the above-mentioned reasons, the linkage between the ophiolites and their flora generates an extremely valuable and irreplaceable tool for studying bio–geo interactions.

14.2 Plants that "Like" Metals

The notion that species are indicators of particular environments is a time-honored one in plant science (Kruckeberg 2002). This concept was widely exploited even in mineral exploration, so that the first practical geobotanists can really be identified in the mediaeval miners and metallurgists. In fact, the biological method for prospecting (geobotany), depending only on visual observation of vegetation cover, has a very long history dating back at least to Roman times, whereas biogeochemical procedures, depending on advances in analytical chemistry, date back only to the last century (Brooks 1998). As a consequence, the connection of a specific flora to a specific environment seemed to be so strong to have allowed a whole profit-based discipline, such as the mineral exploration itself, to develop; this depends on the amazing ability of some plants to evolve tolerance to unfavorable substrates like the metal-enriched ones. Metal-adapted genotypes are the result of the Darwinian natural selection of metal-tolerant individuals selected from surrounding nonmetallicolous populations (Antonovics et al. 1971; Baker 1987; Ernst 2006). Once tolerance evolved, a tight link between plants and the metal-rich environment is established, depending on the fact that these plant populations are competitive only in such environments, where the fitness takes advantage from the acquired tolerance mechanisms. Such selection can lead ultimately to speciation and the evolution of endemic taxa.

These unique plants with an ability to tolerate metal toxicities and survive and reproduce on metalliferous soils are called metallophytes. The majority of them are able to tolerate specific metals in the substrate by physiologically restricting the entry of metals into the root and/or their transport to the shoot (termed "excluders" by Baker 1981). A few species, however, have extremely specialized biological mechanisms in that they are able to accumulate, or even "hyperaccumulate," metals in their shoots at concentrations that can exceed 2% of their dry weight (DW) (Baker et al. 2000). These latter plants are the so-called metal "hyperaccumulators," a term first coined by Brooks et al. (1977) to define particular plants living on nickel-rich serpentine substrates with nickel concentrations >1,000 $\mu g\ g^{-1}$ DW in their above-ground parts. Hyperaccumulators may be at a disadvantage when resources are abundant, but thrive in disturbed habitats because, for example, the high concentration of metals in their organs deters some animals from grazing upon them (Pollard and Baker 1997; Boyd 2004; Jiang et al. 2005). Among the metallophytes, metal hyperaccumulators can represent the most indicative case of the linkage between a certain plant and a certain soil, considering also that for some hyperaccumulators the presence of the metal at high concentration in the soil is essential for a normal growth (Küpper et al. 2001). Hyperaccumulators have therefore been studied as peculiar interesting examples of evolution and adaptation, and as useful indicators in prospecting for metals (Brooks 1983). Recently, the development of new plant-based technologies for the remediation of polluted sites (Vassilev et al. 2004) has stimulated new research interest on metallophytes (Whiting et al. 2004), and on the underlying physiological mechanisms that enable some of these plants to take up such extraordinary amounts of metals.

14.3 Evolution of Serpentine Plants

Serpentine habitats are geologic islands in a "sea" of other soil types. When these rocks were exposed, new species spread on to them from the surrounding substrates. In due course, those that could colonize and survive on serpentines evolved on a different route from their nonserpentine relatives. In several cases, the new species survived on a patch of serpentine because they were poor competitors on other substrates. As a result of this "island" effect, serpentine soils show a large number of species that are found only on such substrates and have highly restricted geographical ranges.

The subject of adaptation of plant species to the total environment of serpentine soils has occupied scientist for many years. Plant biologists have studied in depth the ecology, physiology, phylogeny, and taxonomy of plants occurring on serpentine soils, the so-called serpentinophytes (for a review see Brady et al. 2005). The ecological island model has boosted much research on evolution and adaptation and provoked discussion on the microevolutionary dynamics of metal tolerance in plants, from the population to the single-gene level (for examples see Nyberg Berglund et al. 2004; Kazakou et al. 2010; Quintela-Sabarís et al. 2010; Mengoni et al. 2000, 2001, 2003a, b, c; Rajakaruna et al. 2003; Vekemans and Lefèbvre 1997).

The tight bond between these plants and their environment has been proved from a long time ago by Kruckeberg (1950, 1954), through experiments showing that serpentine-tolerant species and races are limited to serpentine soils because of their inability to compete in nonserpentine environments. This suggests that along the evolutionary trajectory toward serpentine tolerance, genetic trade-offs occur, rendering the serpentine-adapted plant species or ecotypes unable to re-colonize their parental habitat. Moreover, the self-fertility of metal-tolerant populations has proved to be usually much greater than that of nontolerant taxa, presumably as a strategy to reduce reduction of tolerance by flow of nontolerant genes from the surrounding populations (Brady et al. 2005; Brooks 1987) or could be a side effect of previous history of higher self-pollination rates because of the expected low number of first colonizers. Thus, serpentine-tolerant taxa are often endemic to serpentine regions (Brady et al. 2005) and, indeed, the occurrence of plant species restricted to serpentine substrates was documented as long ago as the sixteenth century (Vergnano Gambi 1992).

Serpentinophytes comprise facultative taxa, plants that will grow quite well on serpentine soils without having a specific requirement for any of the edaphic or physical properties of the substrate and obligate taxa that are presumed to grow on serpentine because of a specific nutritional or other requirement which only such soils can provide, mainly the protection from biotic factors present in nonserpentine substrates (Brooks 1987; Boyd 2004). Widespread serpentine endemics can act as flag species, because they are loyal to the substrate; they are thus good indicator plants for serpentine. Furthermore, these species often display unusual and characteristic features in their habitus. In fact, studies of serpentine floras have noted the so-called serpentinomorphoses (Novák 1928), morphological differences between populations or taxa growing on serpentine and nonserpentine soils (Kruckeberg 2002) that concur to plant adaptation to the serpentine factor. The most frequent serpentinomorphoses are xeromorphic foliage, including reduced leaf size and sclerophylls, development of a large root system, dwarfism, plagiotropism, glabrescence or pubescence, glaucescence, and erythrism (Vergnano Gambi 1993; Brady et al. 2005).

Serpentine plants also show a wide range of physiological strategies to adapt to the particular substrate they colonize. Indeed, the most intriguing ones are those related to overcoming the often high heavy metal concentrations present. In relation to such strategies, the two main categories, excluders and accumulators *sensu* Baker (1981), can be found. Tolerant plants are often excluders, limiting the entry and root-to-shoot translocation of trace metals. Differential uptake and transport between root and shoot in excluders leads to more-or-less constant low shoot levels over a wide range of external concentrations. On the other hand, accumulators concentrate metals in plant parts from low or high background levels. Among the latter, a class of rare plants shows extreme behavior in metal uptake and translocation to the shoots are the so-called hyperaccumulators (Brooks et al. 1977) as mentioned above. Inevitably, metal hyperaccumulation is associated with a strongly enhanced ability to detoxify the metal accumulated in above-ground tissues, and thus with metal hypertolerance (Krämer 2010; Rascio and Navari-

Izzo 2011). Metal hyperaccumulation requires complex alterations in the plant metal homeostasis network. Briefly, the main processes supposed to be involved are increased root metal uptake rates, enhanced rates of metal loading from the root symplasm into the apoplastic xylem, highly effective metal detoxification, and sequestration in the leaves (Krämer 2010).

Hyperaccumulators occur in over 54 different families of angiosperms, and very few species among conifers and pteridophytes (see Krämer 2010 for a comprehensive list). Because Ni hyperaccumulation occurs in a broad range of unrelated families, it is certainly of polyphyletic origin (Macnair 2003). The Brassicaceae family is relatively rich in Ni hyperaccumulators, in particular the genera *Alyssum* and *Noccaea* (*Thlaspi* s.l.). In Sect. 14.5 a phylogenetic discussion about the evolution of hyperaccumulation in tribe Alysseae is presented.

The selective factors causing the evolution of hyperaccumulation are unknown and difficult to identify retrospectively. Increased metal tolerance, protection against herbivores or pathogens, inadvertent uptake, drought tolerance, and allelopathy are the different nonmutually exclusive hypotheses formulated so far (Boyd and Martens 1992). Anyway, the supposition of defense against natural enemies is certainly the most accepted one (Boyd 2004, 2007). Whatever the reason of the evolution of this particular phenomenon was, metal hyperaccumulators can surely be the most representative emblem of the link between geology and plant life, thus representing a valuable model system for studying biogeochemical interactions.

14.4 Hyperaccumulation as a Variable Trait

Whatever the physiological strategies for nickel hyperaccumulation are, it is of fundamental importance to investigate bio–geo interactions, i.e., if possible variations in the soil, in terms of chemical characteristics of the substrates and spatial distribution, can affect plant variability both in terms of phenotype and of selective pressure on target genes for tolerance and hyperaccumulation. In the field, individual plants of a metal-hyperaccumulating species exhibit wide phenotypic variation, even within a single population (Boyd et al. 1999; Macnair 2002). Obviously, the two most important determinants are bioavailable soil metal concentration and individual genotype. Plant metal concentrations may be expected to be related to soil metal levels. However, it is also possible that they could be relatively insensitive to those of the soil, especially when the curve relating plant metal uptake to the soil concentrations suggests saturation at quite a low external metal concentration (Baker 1981).

Molecular variability in genetic and biochemical pathways (Krämer 2010; Verbruggen et al. 2009) involved in metal accumulation and metal tolerance can also lead to variation in plant metal concentrations. In *Noccaea* (*Thlaspi*) *caerulescens* and *Cardaminopsis* (*Arabidopsis*) *halleri*, the variability of Zn and Cd accumulation has been widely investigated (Assunçao et al. 2003, 2008; Bert et al. 2002; Macnair, 2002; Taylor and Macnair 2006), showing that there is heritable variation in degree of metal accumulation between local populations and that microevolutionary adaptation plays important role on the onset of the enhanced tolerance in metallicolous populations. In particular, it has been reported that metallicolous populations are of polyphyletic origin (Verbruggen et al. 2009). In other hyperaccumulators such as *Sedum alfredii* (Crassulaceae), both Zn and Cd hyperaccumulations are not constitutive at the species level but confined to metallicolous populations (Deng et al. 2007; Yang et al. 2006). The distinct intraspecific variations in *S. alfredii* provide very useful potential material for genetic and physiological dissection of the hyperaccumulation trait in a species not belonging to the Brassicaceae family. In *N. caerulescens* the variation in Cd accumulation among populations is correlated with the variation in Zn accumulation suggesting the hypothesis of common determinants for Cd and Zn hyperaccumulation (Verbruggen et al. 2009). However, *N. caerulescens* populations from southern France do not show such a correlation (Escarré et al. 2000), indicating that molecular mechanisms correlating Zn and Cd accumulation are variable among populations and no simple conclusions can be drawn about the hyperaccumulating phenotype even in a constitutive hyperaccumulator.

Ni hyperaccumulation in *N. caerulescens* also shows considerable variation and seems to be confined to serpentine populations. Moreover, Ni tolerance and Ni accumulation are not correlated (Richau and Schat 2009). Concerning the preference for Zn and Ni, the Turkish serpentine endemics *Masmenia rosularis*, *Noccaea violascens*, and *Thlaspiceras oxyceras* (all species formerly included in *Thlaspi* s.l.), contrarily to *N. caerulescens*, do not seem to take up Zn over Ni, suggesting that different strategies for Ni hyperaccumulation may have been evolved within the tribe Noccaeeae (Peer et al. 2003). Recently (Kazakou et al. 2010), in an effort to characterize the Ni hyperaccumulation capacity of the serpentine endemic *Alyssum lesbiacum* over all its distribution area, large inter-population differences were recorded and related to soil Ni availability. Extreme intra-specific variation for Ni has also been found in the South African hyperaccumulator *Senecio coronatus* (Boyd et al. 2008) for which differences in elemental content of, e.g., Ca, Fe, Mn, and Zn have recorded. Intra-specific variability in metal uptake has also been shown for other metals such as Mn in *Gossia bidwillii* (Myrtaceae) (Fernando et al. 2007).

14.5 Phylogenetic Pattern of Ni Hyperaccumulation in *Alyssum* and Its Relatives

Hyperaccumulation of nickel is a rare physiological adaptation shared worldwide by a small number of serpentine endemic or subendemic plants (ca. 360 species), especially at tropical and subtropical latitudes. Despite the fact that a large amount of tropical flora is still waiting to be studied, the relatively few Ni hyperaccumulator species we know, even in the richest and best known serpentine floras of Northern hemisphere, suggest this ability is unlikely shared by more than a small percentage of the metallophytes all over the world.

The Brassicaceae is undoubtedly the widest and most diversified group of Ni hyperaccumulators, with up to 83 species distributed in 8–12 genera (see Checchi et al. 2010 and references therin). The traditional morphological classification of this family, mainly based on homoplasic characters such as the fruit shape or dehiscence, was recently shown as widely artificial by a molecular phylogenetic approach (reviewed in Koch and Al-Shehbaz 2009). Following the deep ongoing rearrangement of intrafamilial taxonomy, inspired by a monophyly criterion, hyperaccumulators of Ni in the Brassicaceae can be now referred to only 5 out of the 35 natural tribes (Koch and Al-Shehbaz 2009; Cecchi et al. 2010): Aethionemaeae (1 species), Alysseae (56), Cardamineae (1), Noccaeeae (24), and Schizopetaleae (1). Within the Alysseae they are circumscribed to the genera *Alyssum* (50), *Bornmuellera* (5), and *Leptoplax* (1), and their main specific diversity occurs in Anatolia and the Balkans, which include some of the largest serpentine outcrops in Europe and one of the richest serpentine floras in the world (Brooks 1987; Stefanović et al. 2003).

As already noted above, the unusual behavior of such plants with respect to the presence of Ni does not bear necessarily to a true "dependence" on that metal or tolerance of it, but rather to a facultative advantage in synecological dynamics. Nevertheless, because Ni hyperaccumulators are almost absent on ultramafics which were involved by glacial phenomena during the Quaternary (Proctor and Nagy 1992), it could be suggested that they need a long time to develop either such physiological adaptations or a preadaptive genetic pattern (from which they can easily differentiate when metalliferous soils outcrop). Indeed, one of the most intriguing topics concerning metallophytes is their evolution from nonmetallophyte ancestors, which is also a good starting point to approach the genetic bases of such specialization. The distribution of hyperaccumulators (not only Ni hyperaccumulators) through the angiosperms is highly uneven, with a few groups covering the main percentage of the total alone, but it is unclear how many times hyperaccumulation of a given metal evolved in a given group, and whether this specialization represents a widespread a homoplasic character or is mainly a synapomorphic trait restricted to given lineages. Today, a very few researches are dedicated to the origin of serpentine and heavy metal adaptation at the superspecific level (Broadley et al. 2001; Jansen et al. 2002, 2004; Patterson and Givnish 2004; Cecchi and Selvi 2009), and the only ones regarding Ni hyperaccumulation just deal with *Alyssum* species and their relatives (Mengoni et al. 2003a; Cecchi et al. 2010).

In order to assess the actual relationships among Ni-hyperaccumulating Alysseae and the significance of physiological adaptation from an evolutionary point of view, nuclear ribosomal internal transcribed spacer (ITS nDNA) sequences have been recently obtained for comparison from a wide sampling of species and populations, and their phylogenetic pattern was reconstructed at the tribal, generic, and specific levels (Cecchi et al. 2010). This also allows the development of a clearer and more practical taxonomy of European hyperaccumulators in this group, and the identification of suitable model systems consisting of phylogenetically related taxa for comparative studies of the molecular mechanisms of metal hyperaccumulation, and for practical applications.

Both morphological, caryological, and molecular data agree with a double origin of this specialization within the tribe, namely the clades of *Bornmuellera/Leptoplax* and that of *Alyssum*. Despite the fact that several metal-tolerant species in the latter are able to grow on serpentine soils, hyperaccumulation is restricted to the monophyletic sect. *Odontarrhena*, a widely polymorphic group which accounts some 50 species in the Mediterranean and Irano-Turanian regions. It has been suggested that this should deserve the position of an independent genus because of the paraphyletic structure of *Alyssum* s.l. (Warwick et al. 2008).

In the clade including the Greek endemic *Leptoplax emarginata* and the very closely related west Mediterranean and Irano-Turanian species in the genus *Bornmuellera*, Ni hyperaccumulation must be probably considered as one of the traits they inherited from a common ancestor, thus reducing the total number of natural groups where this physiological character has occurred. By contrast, in Odontarrhena it must have evolved multiple times as a consequence of a complex of preadaptive, genetic traits shared by all the taxa of this group.

Such a different frequency of evolution of metal tolerance, depending on the phylogenetic depth, has been observed for obligate serpentinophytism even in tribe Lithospermeae of Boraginaceae (Cecchi and Selvi 2009), and is in line with the results of similar phylogenetic inferences for serpentine adaptation in *Calochortus* (Patterson and Givnish 2004) or Al accumulation in the Ericales (Jansen et al. 2002, 2004). Obligate serpentine taxa in Lithospermeae and Ni hyperaccumulators in Alysseae also share a similar evolutionary pattern regarding their ancestry among nonserpentine taxa; in both the cases, these sister groups of serpentine endemics are strictly basophilous and xerophytic plants growing on limestone, or even dolomite rocks, with a high magnesium content. Thus, there is evidence that the combined tolerance to an ultrabasic pH value, dry environments, and, especially, high levels of magnesium in the soil may be key factors for the evolution of serpentinophile (then hyperaccumulator) plants.

14.6 *Alyssum bertolonii*

One of the earliest reports in the scientific literature about the strict connection between plants and geology dates back to over four centuries (Cesalpino 1583), when the Italian botanist Andrea Cesalpino observed the crucifer currently known as *A. bertolonii* (Fig. 14.2) growing on black stony soils of the Upper Tiber Valley in Tuscany (Vergnano Gambi 1992). Its discovery as a curious and bizarre case of evolution was in the late 1940s, when Minguzzi and Vergnano (1948) discovered its uncommonly high concentration of nickel in its leaves. Since then, some studies have attempted to investigate the physiological mechanisms of its Ni tolerance and hyperaccumulation. For example, Gabbrielli et al. (1991) found that *A. bertolonii* is characterized by a higher Ni tolerance as compared to other serpentine

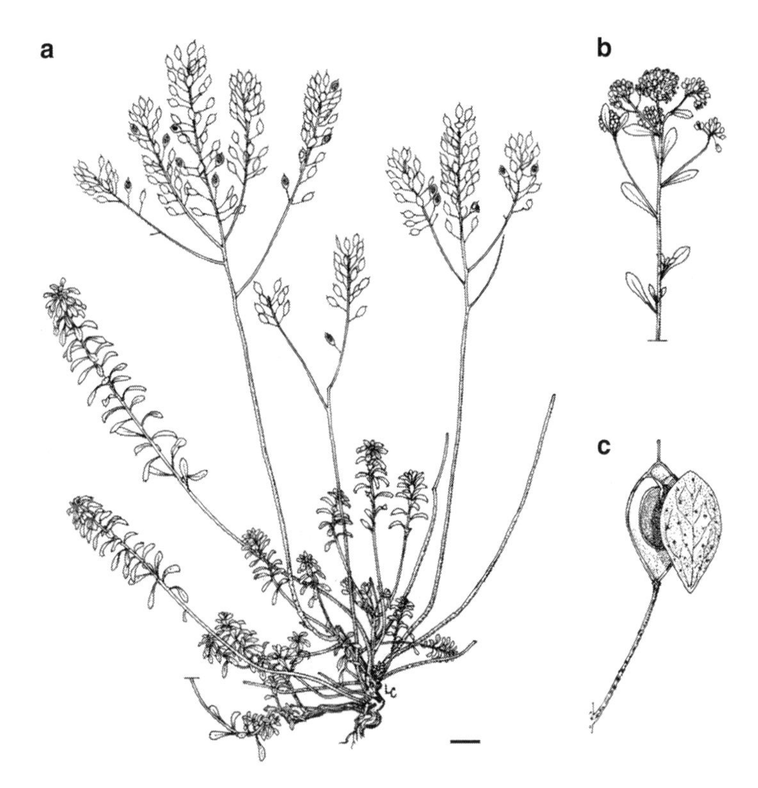

Fig. 14.2 *Alyssum bertolonii*. (**a**) Habit (fruiting plant); (**b**) early inflorescence; (**c**) ripe fruit (silicula) with seed. Scale bar is 1 cm in (**a**) and (**b**) and 1 mm in (**c**). Drawing by L. Cecchi

nonaccumulator species. The hypertolerance strategy confers high costs, but is important for surviving in unfavorable serpentine conditions, thus determining the inseparable connection between hyperaccumulators and serpentine soils. Gabbrielli and Pandolfini (1984) showed, instead, that in *A. bertolonii* the internal Ca and Mg concentrations possibly counteract Ni toxicity or in any case enhance Ni tolerance, through physiological mechanisms still unknown. However, Marmiroli et al. (2004) investigated the Ni distribution in its tissues and found a specific pattern of nickel distribution, with the highest concentrations present in parenchyma and sclerenchyma cells for the roots, while in the shoots, the highest amounts of nickel were found in the stem epidermis, the leaf epidermal surface, and the leaf trichome base.

In terms of biogeochemical interactions, serpentine soils are well known to markedly differ in their cation concentrations as they are derived from rocks occurring under a wide range of climates, from temperate to tropical regions (Kruckeberg 2002). Galardi et al. (2007a) showed that even at the local scale of the distribution of *A. bertolonii*, mainly central Italy, there could be statistically significant heterogeneity in the levels of cations of these soils, probably due both to microclimatic factors and to differences in the composition of the original rocks

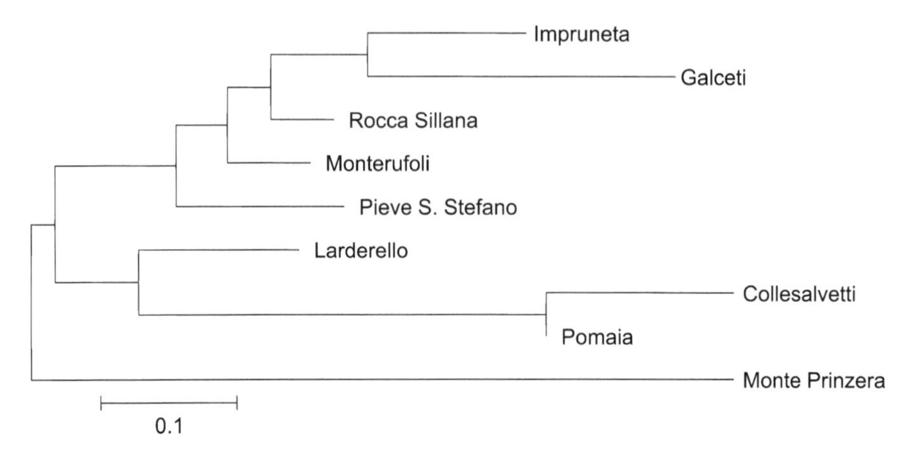

Fig. 14.3 Pattern of genetic relationships among Tuscan *A. bertolonii* populations. The neighbor joining method was applied to an average squared distance matrix among populations. *Scale bar* indicates average squared distance (*Microsat* 1.5). Original data from Mengoni et al. (2003b)

between different outcrops. In that study, for example, Ni concentration spanned a wide range of values, from around 1,000 μg g^{-1} to more than threefold higher, while an Ni mean value of 2,000 μg g^{-1} DW was reported by Brooks (1987) for serpentine outcrops. That heterogeneity in soil Ni concentrations was shown to generate substantial differences in mineral element concentration between *A. bertolonii* populations, as Ni shoot concentrations showed a fivefold range, from 4,000 to 21,000 μg g^{-1} DW. The scale of concentration variation of the other elements was similar to that of nickel, irrespective of their absolute values. Moreover, in the study of Galardi et al. (2007a) it was also demonstrated that *A. bertolonii* was not only a well-known faithful indicator of serpentine soils for geobotanical prospecting but also a useful tool for biogeochemistry as, in the case of nickel and cobalt, it is representative of the degree of mineralization of the soil. A previous population genetic study (Mengoni et al. 2003b) showed that *A. bertolonii* populations are strongly genetically distinct from each other and that a relatively high genetic heterogeneity does exist within the same population (Fig. 14.3). Furthermore, in the same study a clear relationship between geographical isolation and genetic differentiation of populations has been found. Evaluating the relationship between soil and plant metal concentration differences among outcrops and population genetic diversity, at the intra-population level, a hypothetical edaphic effect on the genetic diversity of populations was suggested, i.e., the more variable the soil Ni concentrations were, the more genetically variable were the plant populations. Thus, Ni concentration variability of soil seems to be an important factor shaping *A. bertolonii* genetic diversity. Considering the geographical distribution of the outcrops (Fig. 14.4), Galardi et al. (2007a) suggested also that the center of diversity, then possibly the center of origin, of *A. bertolonii* was in the outcrop with the

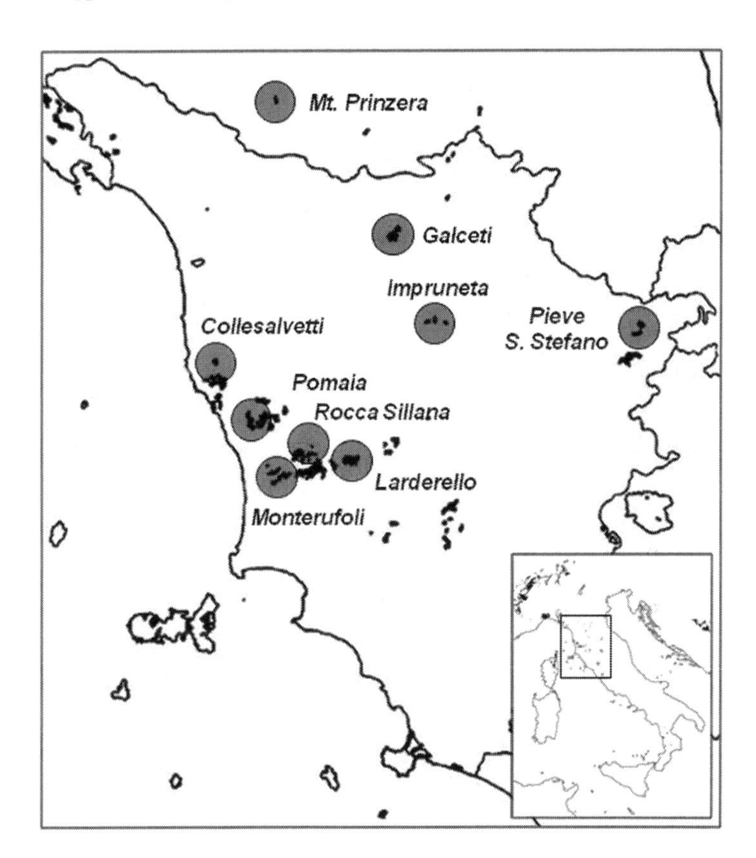

Fig. 14.4 The patchy distribution of serpentine outcrops in Tuscany (Central Italy). *Highlighted areas* and names in *bold* indicate localities where *A. bertolonii* populations were sampled in the study by Mengoni et al. (2003b)

lowest Ni concentration and that from there, plants might have diffused into the other outcrops with higher Ni concentrations.

After having demonstrated for the first time that there can be significant variation in Ni tolerance and hyperaccumulation in populations of a species endemic to metalliferous soils, such as *A. bertolonii*, and that their relationship was positive, Galardi et al. (2007b) compared data obtained in hydroponic tests with data on metal concentration collected in the field, in order to assess the effects of local soil and plant metal concentration on Ni tolerance and accumulation. In the field, a positive correlation has been found between soil Ni concentration and shoot Ni concentration (Galardi et al. 2007a), but neither of these measures seemed to be related to the considerable differences in Ni tolerance and accumulation levels measured under controlled conditions. Hence, in contrast to the general notion that the least tolerant populations are found on the least metalliferous soils, with tolerance being a result of adaptive evolution in response to soil toxicity (Pollard et al. 2002), *A. bertolonii* populations do not show this particular feature. Variation in Ni tolerance and accumulation also shows no relationship to the variation in genetic diversity that was found by Mengoni et al. (2003b) for *A. bertolonii* populations. So genotypic differences in Ni tolerance and accumulation do not seem to be the main cause for generating the differences in shoot Ni concentrations shown by *A. bertolonii* populations in the field (Galardi et al. 2007a), whereas the nickel "serpentine factor" has been demonstrated to play a significant role (Galardi et al. 2007a).

14.7 Conclusions

Recent studies on the special features of interactions played between serpentine hyperaccumulators and soil substrate are highlighting more and more the role of genetic background, physiological constraints, and facilitated variation in the evolution of metal hyperaccumulation from nonaccumulating relatives. Very recently, evidence has shown that Cd tolerance and accumulation are not independent in *Cardaminopsis halleri* (Willems et al. 2010), as well as the important role played by both genes (sets of genes) and environmental interactions in the evolution of Zn tolerance and hyperaccumulation (Frérot et al. 2010). Similar evidence has also been suggested by field and population studies on *A. bertolonii* which have shown a high degree of heterogeneity of population metal concentrations, Ni tolerance, and hyperaccumulation capacities as well as a strong positive linear relationship between Ni tolerance and hyperaccumulation and Ni in plants and soils. These features render *A. bertolonii* an attractive model for studying evolution, both physiological and molecular, of the most striking feature produced by the interactions of biological systems with the geological substrate–metal hyperaccumulation.

References

Antonovics J, Bradshaw AD, Turner RG (1971) Heavy metal tolerance in plants. Adv Ecol Res 7:1–85

Assunçao AGL, ten Bookum WM, Nelissen HJM, Vooijs R, Schat H, Ernst WHO (2003) Differential metal-specific tolerance and accumulation patterns among *Thlaspi caerulescens* populations originating from different soil types. New Phytol 159:411–419

Assunçao AGL, Bleeker P, Ten Bookum WM, Vooijs R, Schat H (2008) Intraspecific variation of metal preference patterns for hyperaccumulation in *Thlaspi caerulescens*: evidence from binary metal exposures. Plant Soil 303:289–299

Baker AJM (1981) Accumulators and excluders – strategies in the response of plants to heavy metals. J Plant Nutr 3:643–654

Baker AJM (1987) Metal tolerance. New Phytol 106:93–111

Baker AJM, McGrath SP, Reeves RD, Smith JAC (2000) Metal hyperaccumulator plants: a review of the ecology and physiology of a biological resource for phytoremediation of metal-polluted

soils. In: Terry N, Bañuelos G (eds) Phytoremediation of contaminated soil and water. Lewis, Boca Raton, FL, pp 85–107
Bani A, Echevarria G, Sulçe S, Morel JL, Mullai A (2007) In-situ phytoextraction of Ni by a native population of *Alyssum murale* on an ultramafic site (Albania). Plant Soil 293:79–89
Bert V, Bonnin I, Saumitou-Laprade P, De Laguérie P, Petit D (2002) Do *Arabidopsis halleri* from nonmetallicolous populations accumulate zinc and cadmium more effectively than those from metallicolous populations? New Phytol 155:47–57
Boyd RS (2004) Ecology of metal hyperaccumulation. New Phytol 162:563–567
Boyd RS (2007) The defense hypothesis of elemental hyperaccumulation: status, challenges and new directions. Plant Soil 293:153–176
Boyd RS, Davis MA, Balkwill K (2008) Elemental patterns in Ni hyperaccumulating and non-hyperaccumulating ultramafic soil populations of *Senecio coronatus*. S Afr J Bot 74:158–162
Boyd RS, Jaffré T, Odom JW (1999) Variation in nickel content in the nickel-hyperaccumulating shrub *Psychotria douarrei* (Rubiaceae) from New Caledonia. Biotropica 31:403–410
Boyd RS, Martens SN (1992) The raison d'être for metal hyperaccumulation by plants. In: Baker AJM, Proctor J, Reeves RD (eds) The vegetation of ultramafic (serpentine) soils. Intercept, Andover, pp 279–289
Brady KU, Kruckeberg AR, Bradshaw HD (2005) Evolutionary ecology of plant adaptation to serpentine soils. Ann Rev Ecol Evol Syst 36:243–266
Broadley MR, Willey NJ, Wilkins JC, Baker AJM, Mead A, White PJ (2001) Phylogenetic variation in heavy metal accumulation in angiosperms. New Phytol 152:9–27
Brooks RR (1983) Biological methods of prospecting for minerals. Wiley, New York
Brooks RR (1998) Plants that hyperaccumulate heavy metals. CAB International, Wallingford
Brooks RR (1987) Serpentine and its vegetation. A multidisciplinary approach. Dioscorides, Portland
Brooks RR, Lee J, Reeves RD, Jaffré T (1977) Detection of nickeliferous rocks by analysis of herbarium specimens of indicator plants. J Geochem Explor 7:49–57
Brooks RR, Yang XH (1984) Elemental levels and relationships in the endemic serpentine flora of the Great Dyke, Zimbabwe and their significance as controlling factors for this flora. Taxon 33:392–399
Cecchi L, Gabbrielli R, Arnetoli M, Gonnelli C, Hasko A, Selvi F (2010) Evolutionary lineages of nickel hyperaccumulation and systematics in European Alysseae (Brassicaceae): evidence from nrDNA sequence data. Ann Bot 106(5):751–767
Cecchi L, Selvi F (2009) Phylogenetic relationships of the monotypic genera *Halacsya* and *Paramoltkia* and the origins of serpentine adaptation in circummediterranean Lithospermeae (Boraginaceae): insights from ITS and matK DNA sequences. Taxon 58:700–714
Cesalpino A (1583) De plantis libri XVI, Florentiae
Chiarucci A (2003) Vegetation ecology and conservation on Tuscan ultramafic soils. Bot Rev 69(3):252–268
Chiarucci A, Maccherini S (2007) Long-term effects of climate and phosphorus fertilisation on serpentine vegetation. Plant Soil 293:133–144
Chiarucci A, Maccherini S, Bonini I, De Dominicis V (1999) Effects of nutrient addition on community productivity and structure of serpentine vegetation. Plant Biol 1:121–126
Chiarucci A, Robinson BH, Bonini I, Petit D, Brooks RR, De Dominicis V (1998) Vegetation of tuscan ultramafic soils in relation to edaphic and physical factors. Folia Geobot 33:113–131
Deng DM, Shu WS, Zhang J, Zou HL, Lin Z, Ye ZH, Wong MH (2007) Zinc and cadmium accumulation and tolerance in populations of *Sedum alfredii*. Environ Pollut 147:381–386
Ernst WHO (2006) Evolution of metal tolerance in higher plants. For Snow Landsc Res 80:251–274
Escarré J, Lefèbvre C, Gruber W, Leblanc M, Lepart J, Rivière Y, Delay B (2000) Zinc and cadmium hyperaccumulation by *Thlaspi caerulescens* from metalliferous and nonmetalliferous sites in the Mediterranean area: implications for phytoremediation. New Phytol 145:429–437

Fernando DR, Woodrow IE, Bakkaus EJ, Collins RN, Baker AJM, Batianoff GN (2007) Variability of Mn hyperaccumulation in the Australian rainforest tree *Gossia bidwillii* (Myrtaceae). Plant Soil 293:145–152

Frérot H, Faucon M-P, Willems G, Godé C, Courseaux A, Darracq A, Verbruggen N, Saumitou-Laprade P (2010) Genetic architecture of zinc hyperaccumulation in *Arabidopsis halleri*: the essential role of QTL x environment interactions. New Phytol 187:355–367

Gabbrielli R, Mattioni C, Vergnano O (1991) Accumulation mechanisms and heavy metal tolerance of a nickel hyperaccumulator. J Plant Nutr 14:1067–1080

Gabbrielli R, Pandolfini T (1984) Effect of Mg_2 and Ca_2 on the response to nickel toxicity in a serpentine and nickel accumulating species. Physiol Plant 62:540–544

Galardi F, Corrales I, Mengoni A, Pucci S, Barletti L, Arnetoli M, Gabbrielli R, Gonnelli C (2007a) Intra-specific differences in nickel tolerance and accumulation in the Ni-hyperaccumulator *Alyssum bertolonii*. Environ Exp Bot 60:377–384

Galardi F, Mengoni A, Pucci S, Barletti L, Massi L, Barzanti R, Gabbrielli R, Gonnelli C (2007b) Intra-specific differences in mineral element composition in the Ni-hyperaccumulator *Alyssum bertolonii*: a survey of populations in nature. Environ Exp Bot 60:50–56

Huenneke LF, Hamburg SP, Koide R, Mooney HA, Vitousek PM (1990) Effects of soil resources on plant invasion and community structure in Californian serpentine grassland. Ecology 71:478–491

Jansen S, Broadley M, Robbrecht E, Smets E (2002) Aluminium hyperaccumulation in angiosperms: a review of its phylogenetic significance. Bot Rev 68:235–269

Jansen S, Watanabe T, Caris P, Geuten K, Lens F, Pyck N, Smets E (2004) The distribution and phylogeny of aluminium accumulating plants in the Ericales. Plant Biol 6:498–505

Jenny H (1980) The soil resource: origin and behavior. Ecol Stud 37:256–259

Jiang RF, Ma DY, Zhao FJ, McGrath SP (2005) Cadmium hyperaccumulation protects *Thlaspi caerulescens* from leaf feeding damage by thrips (*Frankliniella occidentalis*). New Phytol 167:805–814

Kabata-Pendias A, Pendias H (1991) Trace elements in soils and plants, 2nd edn. CRC, Boca Raton, FL

Kazakou E, Adamidis GC, Baker AJM, Reeves RD, Godino M, Dimitrakopoulos PG (2010) Species adaptation in serpentine soils in Lesbos Island (Greece): metal hyperaccumulation and tolerance. Plant Soil 332:369–385

Koch M, Al-Shehbaz IA (2009) Phylogeny of *Brassica* and wild relatives. In: Gupta SK (ed) Biology and breeding of crucifers. Taylor & Francis, Boca Raton, FL, pp 1–19

Krämer U (2010) Metal hyperaccumulation in plants. Ann Rev Plant Biol 61:517–534

Kruckeberg AR (1950) An experimental inquiry into the nature of endemism on serpentine soils. Ph.D. thesis. University of California, Berkeley, p 154

Kruckeberg AR (1954) The ecology of serpentine soils: a symposium III. Plant species in relation to serpentine soils. Ecology 35:267–274

Kruckeberg AR (2002) Geology and plant life. University Press, Washington

Kruckeberg AR, Kruckeberg AL (1990) Endemic metallophytes: their taxonomic, genetic and evolutionary attributes. In: Shaw AJ (ed) Heavy metal tolerance in plants: evolutionary aspects. CRC, Boca Raton, FL, pp 301–312

Küpper H, Lombi E, Zhao F-J, Wieshammer G, McGrath SP (2001) Cellular compartmentation of nickel in the hyperaccumulators *Alyssum lesbiacum*, *Alyssum bertolonii* and *Thlaspi goesingense*. J Exp Bot 52:2291–2300

Lefèbvre C, Vernet P (1990) Microevolutionary processes on contaminated deposits. In: Shaw AJ (ed) Heavy metal tolerance in plants: evolutionary aspects. CRC, Boca Raton, FL, pp 286–297

Macnair MR (2002) Within and between population genetic variation for zinc accumulation in *Arabidopsis halleri*. New Phytol 155:59–66

Macnair MR (2003) The hyperaccumulation of metals by plants. Adv Bot Res 40:63–105

Marmiroli M, Gonnelli C, Maestri E, Gabbrielli R, Marmiroli N (2004) Localisation of nickel and mineral nutrients Ca, K, Fe, Mg with scanning electron microscopy microanalysis in tissues of the nickel-hyperaccumulator *Alyssum bertolonii* Desv. and the non-accumulator *Alyssum montanum* L. Plant Biosyst 138:231–243

Mengoni A, Baker AJM, Bazzicalupo M, Reeves RD, Adigüzel N, Chianni E, Galardi F, Gabbrielli R, Gonnelli C (2003a) Evolutionary dynamics of nickel hyperaccumulation in Alyssum revealed by ITS nrDNA analysis. New Phytol 159:691–699

Mengoni A, Barabesi C, Gonnelli C, Galardi F, Gabbrielli R, Bazzicalupo M (2001) Genetic diversity of heavy metal tolerant populations of *Silene paradoxa* L.: a chloroplast microsatellite analysis. Mol Ecol 10:1909–1916

Mengoni A, Gonnelli C, Brocchini E, Galardi F, Pucci S, Gabbrielli R, Bazzicalupo M (2003b) Chloroplast genetic diversity and biogeography in the serpentine endemic Ni-hyperaccumulator *Alyssum bertolonii*. New Phytol 157:349–356

Mengoni A, Gonnelli C, Galardi F, Gabbrielli R, Bazzicalupo M (2000) Genetic diversity and heavy metal tolerance in populations of *Silene paradoxa* L. (Caryophyllaceae): a RAPD analysis. Mol Ecol 9:1319–1324

Mengoni A, Gonnelli C, Hakvoort HW, Galardi F, Bazzicalupo M, Gabbrielli R, Schat H (2003c) Evolution of copper-tolerance and increased expression of a 2b-type metallothionein gene in *Silene paradoxa* L. populations. Plant Soil 257:451–457

Mengoni A, Grassi E, Barzanti R, Biondi EG, Gonnelli C, Kim CK, Bazzicalupo M (2004) Genetic diversity of bacterial communities of serpentine soil and of rhizosphere of the nickel-hyperaccumulator plant *Alyssum bertolonii*. Microb Ecol 48:209–217

Mengoni A, Schat H, Vangronsveld J (2010) Plants as extreme environments? Ni-resistant bacteria and Ni-hyperaccumulators of serpentine flora. Plant Soil 331:5–16

Minguzzi C, Vergnano O (1948) Il contenuto di nichel nelle ceneri di *Alyssum bertolonii*. Atti Soc Tosc Sci Nat 55:49–74

Mota JF, Medina-Cazorla JM, Navarro FB, Pérez-García FJ, Pérez-Latorre A, Sánchez-Gómez P, Torres JA, Benavente A, Blanca G, Gil C, Lorite J, Merlo ME (2008) Dolomite flora of the Baetic Ranges glades (South Spain). Flora 203:359–375

Novák FA (1928) Quelques remarques relative au problème de la végétation sur les terrains serpentiniques. Preslia 6:42–71

Nyberg Berglund AB, Dahlgren S, Westerbergh A (2004) Evidence for parallel evolution and site-specific selection of serpentine tolerance in *Cerastium alpinum* during the colonization of Scandinavia. New Phytol 161:199–209

Patterson TB, Givnish TJ (2004) Geographic cohesion, chromosomal evolution, parallel adaptive radiations, and consequent floral adaptations in *Calochortus* (Calochortaceae): evidence from a cpDNA phylogeny. New Phytol 161:253–264

Peer WA, Mamoudian M, Lahner B, Reeves RD, Murphy AS, Salt DE (2003) Identifying model metal hyperaccumulating plants: germplasm analysis of 20 Brassicaceae accessions from a wide geographical area. New Phytol 159:421–430

Pichi Sermolli R (1948) Flora e vegetazione delle serpentine e delle altre ofioliti dell'alta valle del Tevere (Toscana). Webbia 17:1–380

Pollard AJ, Baker AJM (1997) Deterrence of herbivory by zinc hyperaccumulation in *Thlaspi caerulescens* (Brassicaceae). New Phytol 135:655–658

Pollard AJ, Dandridge Powell K, Harper FA, Smith JAC (2002) The genetic basis of metal hyperaccumulation in plants. Crit Rev Plant Sci 21:539–566

Proctor J (1971) The plant ecology of serpentine II. Plant responses to serpentine soils. J Ecol 59:397–410

Proctor J, Nagy L (1992) Ultramafic rocks and their vegetation: an overview. In: Baker AJM, Proctor J, Reeves RD (eds) The vegetation of ultramafic (serpentine) soils. Intercept, Andover, pp 469–494

Proctor J, Woodell SRJ (1975) The ecology of serpentine soils. Adv Ecol Res 9:255–365

Quintela-Sabarís C, Vendramin G, Castro-Fernández D, Fraga M (2010) Chloroplast microsatellites reveal that metallicolous populations of the Mediterranean shrub *Cistus ladanifer* L. have multiple origins. Plant Soil 334:161–174

Rajakaruna N, Baldwin BG, Chan R, Desrochers AM, Bohm BA, Whitton J (2003) Edaphic races and phylogenetic taxa in the *Lasthenia californica* complex (Asteraceae: eliantheae): an hypothesis of parallel evolution. Mol Ecol 12:1675–1679

Rascio N, Navari-Izzo F (2011) Heavy metal hyperaccumulating plants: How and why do they do it? And what makes them so interesting? Plant Sci 180:169–181

Richau KH, Schat H (2009) Intraspecific variation of nickel and zinc accumulation and tolerance in the hyperaccumulator *Thlaspi caerulescens*. Plant Soil 314:253–262

Roberts BA, Proctor J (1992) The ecology of areas with serpentinized rocks: a world view. Kluwer, Dordrecht

Robinson BH, Brooks RR, Kirkman JH, Gregg PEH, Alvarez HV (1997) Edaphic influences on a New Zealand ultramafic ("serpentine") flora: a statistical approach. Plant Soil 188:11–20

Slingsby DR, Brown BH (1977) Nickel in British serpentine soils. J Ecol 65:597–618

Stefanović V, Tan K, Iatrou G (2003) Distribution of the endemic Balkan flora on serpentine I. – obligate serpentine endemics. Plant Syst Evol 242:149–170

Takhtajan AL (1986) The floristic regions of the World. University of California Press, Berkeley

Taylor SI, Macnair MR (2006) Within and between population variation for zinc and nickel accumulation in two species of *Thlaspi* (Brassicaceae). New Phytol 169:505–514

Vassilev A, Schwitzguébel JP, Thewys T, Van der Lelie D, Vangronsveld J (2004) The use of plants for remediation of metal-contaminated soils. Sci World J 4:9–34

Vekemans X, Lefèbvre C (1997) On the evolution of heavy metal tolerant populations in *Armeria maritima*: evidence from allozyme variation and reproductive barriers. J Evol Biol 10:175–191

Verbruggen N, Hermans C, Schat H (2009) Molecular mechanisms of metal hyperaccumulation in plants. New Phytol 181:759–776

Vergnano Gambi O (1992) The distribution and ecology of the vegetation of ultramafic soils in Italy. In: Roberts BA, Proctor J (eds) The ecology of areas with serpentinized rocks – a world view. Kluwer, Dordrecht, The Netherlands, pp 217–247

Vergnano Gambi O (1993) Gli adattamenti delle piante. In: Le ofioliti dell'Appennino Emiliano (ed) Regione Emilia-Romagna, pp 103–128

Walker RB (1954) The ecology of serpentine soils: a symposium II. Factors affecting plant growth on serpentine soils. Ecology 35:259–266

Warwick SI, Sauder CA, Al-Shehbaz IA (2008) Phylogenetic relationships in the tribe Alysseae (Brassicaceae) based on nuclear ribosomal ITS DNA sequences. Botany 86:315–336

Whiting SN, Reeves RD, Richards D, Johnson MS, Cooke JA, Malaisse F, Paton A, Smith JAC, Angle JS, Chaney RL, Ginocchio R, Jaffré T, Johns R, McIntyre T, Purvis OW, Salt DE, Schat H, Baker AJM (2004) Research priorities for conservation of metallophyte biodiversity and their potential for restoration and site remediation. Rest Ecol 12:106–116

Whittaker RH (1954) The ecology of serpentine soils: a symposium. I. Introduction. Ecology 35:258–259

Willems G, Frérot H, Gennen J, Salis P, Saumitou-Laprade P, Verbruggen N (2010) Quantitative trait loci analysis of mineral element concentrations in an *Arabidopsis halleri* x *Arabidopsis lyrata petraea* F2 progeny grown on cadmium-contaminated soil. New Phytol 187:368–379

Yang X, Li T, Yang J, He Z, Lu L, Meng F (2006) Zinc compartmentation in root, transport into xylem, and absorption into leaf cells in the hyperaccumulating species of Sedum alfredii Hance. Planta 224:185–195

Zhang X-H, Liu J, Huang H-T, Chen J, Zhu Y-N, Wang D-Q (2007) Chromium accumulation by the hyperaccumulator plant *Leersia hexandra* Swartz. Chemosphere 67:1138–1143

Chapter 15
The Role of Organic Matter in the Mobility of Metals in Contaminated Catchments

Aurora Neagoe, Virgil Iordache, and Ileana Cornelia Fărcăşanu

15.1 Introduction

The concept of "role" is dependent on a theory of systems (the role is what a structural component of a system does in order to support the functioning of the integrated system). Although assumed by default in most of the current scientific literature, the theory of nested hierarchies of systems is unrealistic for the description of environmental systems (Iordache et al. 2011), because there are no true emergent properties associated with each hierarchical level (emergent properties are a conceptual must for the identification of hierarchical levels). A realistic approach is by a pseudo-hierarchy of systems, which claims that simple biological or abiotic environmental objects (and their associated new processes) are perceivable and measurable only at certain scales. There is a continuum of relevant scales of analyses in a complex environmental object (Iordache et al. 2011, 2012). This continuum of relevant scales is discretized for methodological reasons in hierarchical levels, i.e., it is modeled by a limited number of hierarchically organized systems, which has not reality, but has epistemic value.

In Chap. 19 of this book, it will be shown that the overall pattern of metal mobility at larger scale (e.g., catchments) apparently resulting from smaller scale processes (e.g., local sites) cannot be reduced to the small-scale mechanisms because of the intervention of larger scale processes, that are different from the small-scale

The contribution of the first two authors to this chapter is equal.

A. Neagoe (✉) • V. Iordache
Research Centre for Ecological Services (CESEC), Faculty of Biology, University of Bucharest, Spl Independentei 91-95, Bucharest 050089, Romania
e-mail: aurora.neagoe@unibuc.ro

I.C. Fărcăşanu
Research Centre for Applied Organic Chemistry, University of Bucharest, Bucharest 050665, Romania

E. Kothe and A. Varma (eds.), *Bio-Geo Interactions in Metal-Contaminated Soils*, Soil Biology 31, DOI 10.1007/978-3-642-23327-2_15,

processes, a difference that is observed for both abiotic processes and biological processes. Thus, the patterns at large scale, while dependent on smaller scale mechanisms, are *decoupled* in explanatory terms. For example, the quantity of metals leached from a soil column depends not only on small-scale processes occurring at soil aggregate level, but also on the hydrological conductivity partially controlled by the preferential flow, which can be perceived and estimated only at the scale of the whole soil column. As a result, for the particular and simpler case of abiotic processes, hierarchically structured models should be not a way of reducing the large level processes to smaller level processes, but only a convenient representation of the relationships between the objects involved in processes overlapped in space and *observable* by investigation at different space–time scales.

In this context, in this review article we aim to emphasize the role of a small-scale simple environmental object, the presence of organic matter (OM) in the processes supporting the mobility of metals in a large-scale complex environmental object, i.e., a catchment. The studies and works that are reviewed in this paper demonstrate that this topic has both fundamental and applied importance.

From a fundamental perspective, any local environmental entity delineated in space and time in a catchment can be a source of metals. Whether is a primary or a secondary source, or a system of both types of sources depends on the scale of the analyses (scope and resolution). Metals mobilized from environmental entities at a certain scale are buffered (retained temporarily) in parts of a system identifiable at a larger scale. Metals retained can be then remobilized by changing of environmental conditions. A key biogeochemical parameter is the space–time scale of retention. The organic matter is involved both in the retention and in the remobilization of metals at scales from soil aggregate to river system. Nevertheless, the molecular identity of the organic carbon as mobility facilitator very seldom remains the same over long distances, because of its degradability and its interactions with biotic and abiotic compartments. Another point is that the role of OM on metals mobility is controlled by feedbacks resulted from the ecotoxicological effects of metals on OM production. We will explore in this chapter the existing data concerning the role of OM on the mobility of metals, as well as the effects propagated in larger scale systems. Also, this chapter focuses on how many hierarchical levels and at which scales a catchment can be discretized in this kind of research and what kind of proofs (phenomenological correlations, manipulative experiments, models) can be considered relevant at each level/scale?

The applied aspect of this chapter is associated with the management of contaminated sites, considering that managers do not work with hierarchical theories, but with operational concepts. The contamination is managed at "site" scale, and a "contaminated site" is an area perceived as contaminated and in need for management. "Perceiving" contamination at societal level is associated with the management of private and public natural resources and services disturbed by the contamination. Nevertheless, the production of these resources and services occurs at different scales. While for soil management a "site" is only the local area of primary contamination, for water management, a "site" may extend to the surface of land supporting the development of an aquifer. It may even extend to a whole catchment, spotted by contaminated hot spots, whose water quality has to be preserved. Such a site is considered rather a "region." In order to explain and predict the effect of myco-phyto-remediation of upland contaminated sites on downstream water quality (i.e., to integrate soil management with water management), one needs information about the effects of the interaction between microorganisms and plants on the mobility of metals from contaminated sites. What is the effect of organic matter on the mobility of metals in all kinds of contaminated sites, whatever the scale? Answering this question is a matter of biogeochemistry across scales. We will explore to what extent the existing literature supports such an approach.

The red line of the chapter is the following: first, we present general aspects about OM and details about its structure, needed as a background for the other chapters, second the role of OM at specific scale (in mobilizing, immobilizing, or influencing other processes involved in the mobility of metals) is reviewed, and third we screen the literature for the effects of these roles on the fluxes of metals occurring in larger scale entities (soil column, site, catchments of different complexities).

15.2 The Structure of the Organic Matter

Organic matter is a structural component of the ecological systems derived from living biomass, representing a control variable both in models of the effect of metals on biological systems (nutritional, toxicological, and ecotoxicological models) and in models of metal transport by abiotic fluxes. The simplest system structure of interest for understanding the effect of OM on the mobility of metals should include at least three components: mineral particles, organic matter, and microorganisms. The causal effects of OM at contaminated site level is inseparable from understanding the role of minerals (Jianu et al. 2011) and of microorganisms (review in Tabak et al. 2005). The scale in space of such a system is of the order of 10^{-6} m^2. A system with increased complexity would also include plants and its scale in space starts from 10^{-1} m^2.

Natural OM refers to detrital OM, excluding living organisms and manmade compounds (Warren and Haack 2001). The literature concerning the interaction of organic pollutants with metals is so scarce, that practically the natural organic matter will be discussed primarily in this chapter. In some cases, like landfills, OM from anthropogenic sources will also be referred.

OM is present in all phases in soil: solid, liquid, and gaseous. In this chapter, two aspects are of interest: the types of OM in each physical phase classified as a function of the interactions with metals, and the life time of each type of OM. Knowledge about the turnover is crucial because, for instance, a strong chemical immobilization of metals may not be relevant at larger scale if the life time of the organic complex is short because of other processes (like degradation by microorganisms). On the contrary, weak interaction may become relevant if the OM involved is refractory to physical or microbiological interactions.

OM includes many other elements apart from carbon, but in this chapter we will review strictly the role of OM based only on measurements of carbon content. There are two terminological situations in the literature: either the effect of OM is reported as such (the effect is explicitly mentioned in the cited text as due to OM), or it is reported under the term "organic carbon" or OC (in the cited text). Although there is a difference in absolute concentrations between OM and the OC corresponding to it, in this chapter the usual terminological difference between OM (real entity) and organic carbon (analytical result of the measurement of carbon content in OM) will not be used. We keep the original terminology of the citing authors, but refer by both terms, OM and OC, to the same real entity. We do this conventionally because no publication discriminating between the effects of the carbon, nitrogen, or phosphorus included in OM on the mobility of metals, or proving that there are such separate influences of the constituent elements of OM was found. Consequently, the reader should take into consideration that all effects of "organic carbon" reviewed below may be due in principle, at least to some extent, also to other not-measured elements co-occurring with carbon in OM.

OM molecules found in the environment sites are polyelectrolytes, heterogeneous and with no unique structure and mass (Warren and Haack 2001). They can be characterized in the first instance by fractionation methods. The methodology and significance of dissolved OM is a research subfield in itself. For instance, it is documented that if it can pass through a filter with specific pore size (usually around 0.45 μm), it is referred to as dissolved organic matter (DOM) – a generic term which needs to be used with caution, especially when one is concerned with its ecological function (Zsolnay 2003). On the other hand, the particulate organic matter (POM) is comprised of large particles of organic matter (250–2,000 μm, Bronick and Lal 2005). Using this definition, much of the colloidal material from the soil that ranges from 1 nm up to 1 μm (Pédrot et al. 2008) would be considered to be DOM. However, colloids have different attributes compared to the truly dissolved material. Taking into account the dimension the soil particles, the OC can be fractionated by filtration and ultra-filtration methods in particulate organic carbon (POC), colloidal organic carbon (COC), fine colloidal organic carbon (FCOC), and dissolved organic carbon (DOC) (Chow et al. 2005). Organic fragments larger than 2 mm, although not categorized above, are also important in the immobilization of metals directly (especially in the case of litter and atmospheric deposition around smelters) or by their decomposition products.

DOC can be further fractionated in function of the hydrophobic and hydrophilic character (Quails and Haines 1991), with direct relevance for the solubility of respective fractions. Depending on the extraction procedures, other terms for DOC, such as water extractable organic carbon (WEOC), or water soluble organic matter (WSOM) can be used.

Another possibility to group organic compounds in soil is in nonhumic substances (low percent and higher turnover rate) and humic substances (large percent, up to 90%, and lower turnover rate). Aquatic humic and fulvic acids originate in the soil, are operationally defined as refractory fractions, and occur both in dissolved and particulate size classes (Warren and Haack 2001). Humic acids account for about 5–10% of the natural organic matter in freshwaters (Warren and Haack 2001). Fulvic acids are derived from humic acids, are of smaller size, with less aromatic functional groups and account for about 40–80% of the natural OM in freshwaters (Warren and Haack 2001). We will not provide here information about the structure of the rest of OM because it has much higher turnover rate and its role in the transport or immobilization of metals is consequently lower.

The dynamic of OM in soil was reviewed by Kalbitz et al. (2000) and Bolan et al. (2004) and their general appreciation underlined the fragmentary and often inconsistent existing knowledge. The major sources of OM in soil are plants and microorganisms. Plant litter and its particulate form decomposition products (Berg and McCluagherty 2008) play a major role in metal immobilization in top soil. On the other hand, DOM turnover is dominated by microorganisms, directly and by exuded compounds, although in the rhyzosphere the plants seem to play an equally important role by the root exudates. Of the total DOM in a lake, 78% originated from bacteria, compared to 50% in a forest soil (Schulze et al. 2005). Heavy metals in aqueous solution have an influence on the biodegradability of OM, either by toxicity (reduction of degradation rate), or by flocculation (increase of degradation rate by facilitating the attachment of microbial colonies on larger organic structures) (Marschner and Kalbitz 2003).

The result of the interactions between mineral particles, organic matter, and microorganisms is the formation of soil aggregates (Velde and Barré 2010), which play a key role in soil hydrology and the immobilization of metals. Bronick and Lal (2005) review in detail the role in soil structure formation of each molecular type of organic carbon (carbohydrates, polysaccharides, phenols, lignin, lipids, humic substances), and the factors influencing the OM in soil (climate, erosion, texture, porosity). POM exists as free POM light fraction (LF) or embedded with soil particles and having lower turnover rates as a result of this (Bronick and Lal 2005). Soil aggregates are grouped by size in macro-aggregates (>250 μm) and micro-aggregates (<250 μm), which differ in properties such as binding agents and carbon and nitrogen distribution. While micro-aggregates are formed from organic molecules attached to clay and polyvalent cations to form compound particles, macro-aggregates can form around POM, which can be decomposed, and microbial exudates are released. Therefore, macro-aggregates becomes more stable while C:N ratio decreases, and micro-aggregates form inside. The utilization of carbon contaminants by microorganisms leads to a demand for N and P for building cell constituents (the C:N:P ratio for microbial activity is about 100:10:1). Increases in microbial biomass are associated with increases in aggregate stability (Haynes and Beare 1997). The internally formed micro-aggregates contain a larger recalcitrant organic carbon pool. At low OC concentration, macro-aggregate stability is enhanced by carbonates, which is highly relevant for designing the amendment of contaminated and acidified soil, or of tailing substrate. Warmer climate and increased erosion result in lower OM content, increasing clay content and less pore space. This leads to more stable OM by the larger reactive surface and lower spaces for gas diffusion and water transport (Bronick and Lal 2005).

Turnover time of upper (down to 1 m) OC is in the range of years to decades (Jardine et al. 2006), but highly variable by each category of OM. The most labile OC is represented by the small molecules directly usable by organotrophic microorganisms. Lower turnover rates can be associated either with free refractory OM (humic and fulvic type) or with OC incorporated into soil aggregates.

The content of OM is different at the surface soil, the unsaturated (vadose) zone and in the saturated zone; surface soils have more organic matter and, as a result, contribute to a larger microbial biomass compared to the vadose or saturated zones (Leug et al. 2007). The stability of OM usually increases with increasing soil depth, and the stability of OM with depth is decreased by a fresh supply of OC (energy source for soil heterotrophic microorganisms) from upper soil layers (Fontaine et al. 2007).

15.3 Roles of Organic Matter

The role of OM in metal mobility in soils was recently reviewed (Carrillo-González et al. 2006), but the multiscale characteristics and consequences of this role in contaminated sites are not systematically approached in the existing literature. An excellent work investigating the scale-specific mechanisms of DOC mobility was done by Jardine et al. (2006). In the current literature, there is no similar information concerning the role of organic carbon (DOC included) in the mobility of metals.

While the roles of microorganisms, organic carbon or minerals in the mobility of metals in contaminated sites are well documented, the propagation of their effects on metal mobility (in particular the consequences on the fluxes occurring at larger scales) is much less clear. As already explained, understanding these up-scaled effects is crucial because the scale for the management of contaminated areas is not the very small scales specific to microorganisms.

From methodological point of view, the roles of the microorganisms and OC are seldom estimated exactly at their specific scale in environmental studies. This is not only because of instrumental restrictions, but also because what is often at stake is not to understand organism level processes, but population and community ones in heterogeneous media, more appropriate to field extrapolation. This is the reason why the literature concerning the effects of OC and microorganisms in metal mobility relevant for contaminated sites management results from studies at several increasing scales (corresponding to pseudo-hierarchical levels usable for the description of the real environmental objects), from soil aggregate up to an experimental field plot. Nevertheless, the scales of managerial interest are still larger: from site and catchment.

15.3.1 Role in the Immobilization of Metals

The overall effect of OM attachment on mineral–metal interactions depends on the nature of mineral surfaces (Kahle et al. 2004) of OM and of their environmental dependent interaction (Warren and Haack 2001). Reciprocal stabilization of Fe oxides and OM occurs in soil, eventually in ternary association with minerals (Wagai and Mayer 2007). The tertiary structure of humic and fulvic acids depends on pH and ionic strength and implicitly so does their ability to scavenge metals in free form or in interaction with mineral surfaces. There is competition for binding sites, that is dominant elements such as Al and Fe blocking the binding of other metals. Although metals sorbed by OM are more strongly bound than those to oxyhydroxides of Fe, the decomposition of labile OM leads to release of metals (Warren and Haack 2001). This may be one reason for the fact that metal transport from certain types of sites (e.g., landfills) is rather in colloidal form than in organic ligands form (details in part 15.4.6).

Although for analytical reasons the role in immobilization is separately assessed experimentally for minerals or OC, or microorganisms, in nature it occurs on complex surfaces generated by minerals, OM and microorganisms. For micro-organisms, the surface reactivity to metals is much more complex than that of OM or mineral surfaces because the complexity of the functional groups differ between major taxons, species, and individuals compared to structured consortia and biofilms. As a general rule, at low pH the adsorption of metal ions to mineral surfaces is increased by the OM, and at high pH is decreased (Warren and Haack 2001). Details by mineral type and metals can be found in this source. Mineral surface hydroxyl functional groups are effective in the sorption of metals between pHs 6 and 8, while organic surfaces can be effective sorbents at much lower pH because of the pKs of some functional groups, especially carboxylic groups (Warren and Haack 2001).

The reactivity of OC controls the mobility of metals either directly or indirectly, by interaction with mineral surfaces or bacteria. In the pH ranges of natural systems, the OM and bacterial surfaces are negatively charged (Warren and Haack 2001). Due to the polyelectrolyte and complex chemical character, the OM attaches to mineral surfaces, changing their affinity for metals (by ligand exchange, cation bridging, proton exchange, water bridging, hydrogen bonding, and van deer Wails interactions). In particular, net hydrophobic expulsion is involved in the accumulation of humic and fulvic acids at mineral surfaces (Warren and Haack 2001) leading to a strong retention of OC in soil (Jardine et al. 2006), with consequences on the immobilization of metals. Controls over sorption and desorption of DOC in soils, with indirect relevance for the mobility of metals, are summarized by Neff and Asner (2001).

Another mechanism of immobilization is by diffusion to small pores. DOC, and presumably associated metals being solubilized in the near-surface soil during storm events, is transported to deeper profile depths, where it may diffuse into smaller pores where microbial degradation processes are limited (Jardine et al. 2006).

This is dependent on the pore water velocity (it takes place mostly when the residence time of the solution is long), and on the absence of strong preferential flow. Colloidal transport of metals could be facilitated not only because of the larger life time of this form, but also because colloids cannot enter the smallest pores and thus keep the pathways with larger water velocity. Smucker et al. (2007) discuss in detail the micro-pores development in soil macro-aggregates with direct relevance for the transport of soluble carbon.

Relevant information can be extracted also from the experimental literature about phytoremediation. For instance, after a study at pot level, Banks et al. (1994) conclude that the revegetation of mining wastes may increase the leaching of Zn, and this effect is attenuated by the presence of (inoculation of) microbes. The mechanism supporting this phenomenon (in terms of OC speciation in the substrate) was not investigated. The situation seems to be highly metal and plant specific, with interplay between the effects of soil OM and the new DOC exudates by plants. In another study, the organic acids exudated by *Lupinus albus* chelated and solubilized Al, Ca, Mn, and Zn, while Cu and Pb remained bound to soil OM (Dessurealut-Rompre et al. 2008).

OM is an important (even dominant) sorbent for metals in the surface horizons of the soils, but this depends on the type of OM (reactive surface and functional groups) and on the competition with Al and Fe compounds (Gustafsson et al. 2003). Deep soil OM has a very long residence time, and implicitly a large potential in metal retention (Rumpel and Kögel-Knabner 2010). The stabilization of OM can involve interaction with polymeric metal species (Mikutta et al. 2006). For a recent review of stabilizing mechanisms and properties of carbonic, *see* Rumpel and Kögel-Knabner (2010).

15.3.2 Role in the Mobilization of Metals

Molecular account of the interactions between organic matter and metals in the context of aqueous chemistry was given by Dudal and Gérard (2004). Trace element transport and transport pathways in soil, factors influencing mobility and transport models have been reviewed by Carrillo-González et al. (2006). Zhou and Wong (2003) reviewed the effects of DOM on the behavior of heavy metals in soil.

More recently, Degryse et al. (2009) reviewed in detail the problem of metal mobility estimation in soils by the distribution coefficient between solid and liquid phases (K_d). Mechanistic and regression models for predicting K_d are discussed and the influence of OM is considered in both types of models. For instance, the mechanistic model WHAM assumes an affinity of metals to DOM in the following order: Co $<$ Ni $<$ Cd $\approx$ Zn $< <$ Pb $\approx$ Cu. All reported regression models have a linear relationship between K_d and the logarithm of OC (or OM) percent in soil. It is interesting to note that the K_d of elements with large affinity for OM, like Cu, can be reduced to the ratio between the concentrations of TOC and DOC in soil. K_d is of direct use for estimating the metal leaching and the bioavailability of metals to plants. These applications are discussed in detail by the mentioned authors. From a methodological perspective, water or weak salt extracts are not recommended to estimate total dissolved metal concentrations of Cu and Pb in soil because of dilution of DOC and colloids (Degryse et al. 2009).

Zhao Lu et al. (2007) found that the decomposition of OM was the main driver of coupled DOC and metal mobility in the studied soil columns. Cu mobilization depended on DOC, whereas Zn mobilization depended mainly on Ca and to a smaller extent on DOC. The quality of DOC is also an important factor. For instance, the DOC resulted from drying and rewetting cycles has a low affinity for metals (Amery et al. 2007). The decomposition of DOM after drying and rewetting in soil may be related to the toxicity of metal (Merckx et al. 2001), probably as a result of its use by microorganisms leading to a reduction in soil redox potential and farther changes in redox sensitive metal concentrations (Cornu et al. 2007).

Pédrot et al. (2008) investigated the relative importance of OM and Fe–Mn oxyhydroxides as elemental carriers along soil columns using synthetic percolating solutions. Truly dissolved species with fastest leaching properties were the alkaline, alkaline-earth metals, Mn, Si, B, and V. Cu, Zn, Co, and Ni were partly mobilized by colloids (humic substances with weight lower than 5 kDa and fulvic acids). Fully depending on colloids in transport were Al, Cr, U, Mo, Pb, rare earth elements, Fe, Ti and Th. Moreover, it was shown that the fraction ranging between 5 and 30 kDa was the major carrier by an interaction between Fe-oxyhydroxides and complex stabilizing humic acids. The authors also cited literature demonstrating that the export of Co, Ni, Cu, Zn, and V was associated with organic carrier phases. Several other sources are cited to demonstrate the export of rare earth elements, Pb, U, Ti, and Th by iron-rich organic phases. OM was also found to increase the mobility of ^{137}Cs by binding as DOM complexes (Nakamaru et al. 2007).

Uroz et al. (2009) presented microbial molecules acting as mineral-weathering agents (oxalate, lactate, gluconate, citrate, cathecol, and pseudobactin), along with the chemical mechanisms of this process (acidolyses, chelation, and oxido-reduction). According to this source, the organic acids liberated by microorganisms have three actions on mineral weathering: (1) they adhere to mineral surfaces and extract nutrients from mineral particles by electron transfer; (2) they break the oxygen links; and (3) they chelate ions present in solution through their carboxyl and hydroxyl groups, indirectly accelerating the dissolution rate of the mineral by creating an imbalance between cation and anion concentrations in the solution (Uroz et al. 2009).

Chelating agents can be deliberately used for mobilizing metals in soils (Nowack 2005). Some chelating agents used in phytoremediation studies are: ethylenediaminetetraacetic acid (EDTA), *N*-hydroxyethylethylenediaminetriacetic acid (HEDTA), diethylenetriaminepentaacetic acid (DTPA), *trans*-1,2-diaminocyclohexane-*N*,*N*,*N* ¢,*N*¢-tetraacetic acid (CDTA), ethylenebis(oxyethylenetrinitrilo) (EGTA)-*N*,*N*,*N*¢, *N*¢-tetraacetic acid, ethylenediamine-di (*o*-hydroxyphenylacetic acid) (EDDHA), *N*-(2-hydroxyethyl) iminodiacetic acid (HEIDA), ethylenediaminesuccinate

(EDDS), nitrilotriacetic acid (NTA), *N,N*-di(2-hydroxybenzyl)ethyleneamide *N,N* ¢-diacetic acid (HBED), citric acid, malic acid (Alkorta et al. 2004).

Besides the direct effects observed in lab-scale experiments, more subtle effects can be observed in field or lysimeter studies. The heterogeneity of organo-mineral particles in saturated soils and in sediments leads to isolated super-saturation of metals and sulfide (Stockdale et al. 2009). In a phytoremediation lysimeter experiment, Nowack et al. (2006) found that Cu in soil solution was significantly correlated with DOC, but Zn only to phosphate and pH (regression equations are provided, but the statistical relation between DOC and pH is not mentioned). Another interesting point is that the DOC concentration in the contaminated soil was smaller than in the reference soil (probably for ecotoxicological reasons at root and microbial levels), with consequences on metal mobility (Nowack et al. 2006).

At low metal–carbon ratios, the dominant form of Fe and Al in water extracts from a sandy soil was metal–DOC complexes (Jansen et al. 2003). At high metal–carbon ratios, the pH had a strong influence on Al, with insoluble (formerly dissolved) OM–Al complexes at pH 4.5, but soluble Al dominating at pH < 3.5, and insoluble (formerly dissolved) OM–Fe complexes at all pHs (Jansen et al. 2003). In another study, Jansen et al. (2005) showed that the OC in batch experiments was retained mostly in the B horizon as controlled mainly by precipitation in organic–metal complexes and adsorption to Al and Fe phases. Al was mobilized within soil, mainly by acid dissolution, and transported in particulate and organic form, only a small part of it in dissolved (more toxic) form (Driscoll et al. 1994). High total organic matter content in peaty soils led to larger concentration in pore water (Van Gestel 2008), but total metal availability for soil organisms could not be predicted well based only on the metal concentration in pore water.

The role of colloids in the mobility of metals is underlined in recent reviews (Kretzschman and Schäfer 2005). Working with undisturbed soil monolith, Karathanasis and Johnson (2006) showed that eluted metal loads varied with metal (Mo $>$ Cd $>$ Cr), colloid state (increase with OM content), and soil type (increase with soil porosity). In batch experiments, Klitzke and Lang 2007 demonstrated that drying of soil increased the hydrophobicity of the solid phase in soil, but not that of the dispersed colloids. The drying-induced colloid mobilization of metals was not dependent on the properties of the colloids, but on the disintegration of aggregates or increase of share forces (Klitzke and Lang 2007). In another study, Klitzke et al. (2008) showed that liming of contaminated soil, by the increase in soil pH, resulted in an increase in the mobilization of Pb in colloidal form (soil mineral stabilized by organic coatings, and Fe–organic complexes). This was not the case, in similar experiments, for As and Sb (Klitzke and Lang 2009). In the case of As and Sb, the Ca existing in the system strongly diminished the solubility of their complexes with organic carbon.

In a lysimeter experiment with translocated soil, Luster et al. (2008) found a decrease in the DOC that occurred in the second year of the experiment compared with the first year (probably as a result of the depletion of the pools resulted from initial aggregates destruction), but this was not associated with an important change in the concentrations of Zn and Cu in the percolating water. A correlation between

Cu and DOC concentrations in water was found only below calcareous soils, and was missing below acidic soils (Luster et al. 2008).

15.3.3 Influence on Other Processes

The influence on the activity of microorganisms will be detailed in Jianu et al. (2011). As an example, the weathering by certain bacteria was stronger when xylose (wood) or glucose is available, whereas the weathering activity of other bacteria is increased in the presence of lactose or mannitol (herbaceous vegetation, algae and fungi) (Uroz et al. 2009). The use of OC as energy source in acid mine drainage is discussed in part 15.4.3.

A different type of example is provided by Strobel et al. (2005). DOC increased the extractability of Cu by direct interaction, but increased the release of Cd by buffering the solution at more acidic pH.

15.4 Effects of Organic Matter on Processes Occurring at Larger Scales

15.4.1 Effects on Fluxes of Metals Transferred to Plants

The picture of metal mobilization by plants under the influence of organic carbon is complicated by the species specific preferences and mechanisms characterizing the uptake of each metal. It is beyond the purpose of this chapter to go into the physiological and molecular biology details of these processes.

One simple direction of research in this area was to look for the effect of added organic ligands on metal uptake by plants. Seuntjens et al. (2004) propose a model for metal uptake and leaching in the root zone in the presence of organic ligands. The simulations suggested that the stimulation or inhibition of uptake by the addition or exudation of ligands depended on other parameters (especially on pH).

The effect of OC on the transfer of metals to plants was studied also in the context of organic material (e.g., sewage sludge) application in agrosystems, mining dumps, or tailing dams. Iordache et al. (2004) provided a review of the agricultural type of studies performed in Romania. Neagoe et al. (2005, 2009) assess the influence of various organic amendments on the phytoremediation success of a mining dump. In this case, the main effect of organic matter is on plant development by changing the humidity status and the microbial communities, the effect on metal mobility being masked by such important ecological and physiological effects.

15.4.2 *Effects on Fluxes of Metals Transferred to Lower Soil Layers*

This research area developed intensively, either by experiments with soil columns and lysimeters, or by field observations. In a column study, Forsberg et al. (2008) investigate the effect of sewage sludge amendment on tailing material. They found an initial large leaching of Al, Cu, Zn, Ni, and Pb correlated with large DOC concentrations, but following this the amounts of Al, Cu, Mn, and Ni were related to sulfide oxidation in each column as reflected by substrate's pH. The total amounts of metals leached were not statistically different between the variants with and without sewage sludge.

According to Citeau et al. (2003), 75% of the Pb collected by zero-tension lysimeters was in colloidal form, and Cd and Zn were mainly present as free ion or labile complexes. Yin et al. (2010) also found that Pb vertical transport was colloid facilitated. In a sandy soil with low OC, 87% of the leached Pb was in the coarse colloidal fraction (0.45–8 μm), but in sandy soil with high organic carbon 66% of the Pb was in the nano-sized colloid fraction (<0.1 μm) (Yin et al. 2010). Zn in lysimeter water from contaminated soils was not correlated with TOC; biocolloids contained Zn and Pb and mineral colloids (smectites) contained only Zn (Denaix et al. 2001).

Zhang and Zhang (2010) suggest that because of the competition with DOC at adsorption sites, excessive phosphorous application on soil leads to leaching of Cu, Cd, and Zn in organic forms. Interestingly, they found a decrease of the Pb leaching by the added phosphate in the experimental conditions.

After studying a 2 m deep vadose zone, Dudal et al. (2005) found a fivefold decrease of the soluble OM concentration, but the profile of the complexation constant for Cu, Fe, and Mn was homogenous. They interpreted this result as indicating that only a small part of the organic carbon was responsible for metal binding. Strong rainfalls were able to transport only a small amount of metal-binding organic matter, but freeze-thaw cycles release more such OM (Dudal et al. 2005). The concentrations of Cu, Cr, Ni and Cd in the water transferred from the organic rich forest ecosystem soil layer to lower layers were significantly correlated with DOC (that of Zn was not correlated) at 20 cm depth the correlation for Cd disappeared, and at 90 cm depth there was no significant correlation (Huang et al. 2010).

In a soil amended with sewage sludge, the OC had an indirect effect on metal mobility by determining a disaggregation of soil macro-aggregates responsible for the accumulation of heavy metals in the coarsest fraction (Parat et al. 2007). Another interesting finding in field experiments with sewage sludge was the importance of space–time variations of OM distribution on the mobility of Cu, Ni, and Pb, when increase of mobility of these elements especially at high pHs was noticed (Ashworth and Alloway 2008).

The leaching of metals in bioremediation experiments was found to correlate with soil contamination, pH, redox conditions, particulate and DOM, previous washing from the soil, timing of rainfalls with respect to the plants development, and ecological characteristics of the plants and microorganisms (e.g., potential of mycorhization), in a complex manner (Iordache et al. 2006). Careful microcosm experiments involving the full plant cycle development and hydrological conditions reflecting the natural ones were recommended in order to find indications about the effects of a certain bioremediation solution on the export of metals from the targeted contaminated site (Iordache et al. 2006).

In an interesting study, Besser and Rabeni (2009) evaluate the toxicity of leachates resulting from phytoremediation experiments with tailings material on aquatic invertebrates. Bioaccumulation and toxicity were greatest in treatments leading to organic/metal complexes in the leachate. However, the leachate was sampled just below the test plot and it is not clear what its content of organic/metal complexes would have been after traversing the full length of tailing material.

15.4.3 Effects at Site Scale – Site Scale Immobilization of Metals

In metal mining areas, the OM is usually not taken into consideration as a parameter controlling the metal mobility because of the strictly mineral nature of the substrate. However, in restoration situations, the OM may become relevant, either directly by organic amendments or indirectly by the result of plant life cycle in phytoremediation projects (Mendez and Maier 2008). The use of organic covers for acid mine drainage control from reactive sulfide wastes is reviewed by Peppas et al. (2000). More recently, Vangronsveld et al. (2009) also discussed the application of biosolids for site remediation.

An experiment for the stabilization tailing dam with different amendments showed that when the material was covered with sewage sludge, the leachate had the lowest metals concentration, redox potential and highest pH compared to variant without organic matter (Alakangas and Öhlander 2006).

In an attempt to simulate the fate of tailings placed above alluvial soils, Flores and Sola (2010) performed mixed column experiments. Al, Ba, Cd, Cu, Mn, Pb, and Zn were mobilized when the tailing was leached, but they were retained in the soil. Fe and Sr remained mobile through the soil after leaching from the tailing material. The mechanisms involved in the immobilization were primarily related to the pH increase, but sorption (including on organic matter) played also a role for Cu, Pb, and Zn (Flores and Sola 2010). An alternative with strong retardation effect of heavy metals is the use of compacted sewage sludge as bottom barrier to acid mine drainage from tailing dams (Wang et al. 2010).

Another relevant situation is that of acid mine drainage treatment. Reactive layers with organic carbon may be useful for such purposes. For instance, Hulshof et al. (2009) compared the efficiency of different organic carbon materials. Various types of organic carbon have different efficiency as substrate for the bacteria involved in the treatment (Zagury et al. 2006). A review of remediation options for acid mine drainage, including the use of organic matter is provided by Johnson and Hallberg (2005). Recent work in acid mine drainage treatment using OM is, for instance, that of Wu et al. (2010).

Up-scaling the role of OM at site scale based on knowledge at smaller scale is not trivial because of its heterogeneity in space. The variability of OM in space is moderate to high compared to other soil variables, having a coefficient of variation of 21–41% (Keur and Iversen 2006). The autocorrelation length of soil organic matter is reported to be in the range 34–250 m (Keur and Iversen 2006).

15.4.4 Site Scale Mobilization of Metals

Various amendments rich in organic carbon may be added on soils or on contaminated sites like tailing dams (Antoniadis and Alloway 2002), either directly generating pollution with metals or changing the mobility of existing metals. Example of such amendments are: lime stabilized biosolid, anaerobic digested biosolid, aerobically digested biosolid, sewage sludge, cattle manure, poultry manure, pig manure, paper mill sludge, and secondary digested sewage sludge.

At field scale, Burckhard et al. (1994) found that the mobility of Pb and Cd was not affected by organic acids, being probably limited along the vertical profile by heterotrophic bacteria using the chelating organic carbon. Li and Zhou (2010) showed in a field lysimeter study that colloids decreased the leaching depth of Cd by blocking the soil matrix capillary and, in the case of preferential flow, by deposition on the wall of the macropores. But the addition of OC increased the penetration of Cd by 10 cm (Li and Zhou 2010).

Alakangas and Ohlander (2006) found that the amendment with sewage sludge generated a larger export of As from a tailing material than in the variants lacking OM.

15.4.5 Site Scale Transfer to Plants

Although OM is not a factor directly correlated to the distribution of vegetation in mining areas (Guo et al. 2010; statistical correlations being observed mainly for macro-nutrients such as nitrogen), it plays an important role in the primary and secondary succession processes occurring in these areas (Neagoe et al. 2009). A general discussion of the management of metal fluxes in contaminated zones is provided by Robinson et al. (2006). The spatial relationship between metals and OC heterogeneities and the distribution of plants is important for understanding and predicting the phytoextraction remediation solutions (Robinson et al. 2006).

For the reasons mentioned in part 15.4.1 and because certain aspects concerning the plant population and community ecology would have to be presented in detail, this role is not discussed in detail in this chapter.

15.4.6 Site Scale Transfer to Groundwater and Surface Water

The fate of the carbon associated with the erosion of soil OM, and implicitly of the associated metals, is not clarified, and it is an important topic of research (recent reviews, Lal 2009). The proofs for this effect are many: occasionally from experiments with amendments, and dominantly from correlations between metals and organic carbon in water (in some studies manipulated by injection experiments), and from catchment scale studies.

Proofs from experiments with amendments: In an experiment for leaching from tailings covered with different amendments, the lowest concentrations of metals, of redox potential, and highest pH and As concentrations were found in the variant amended with sewage sludge (organic rich material) (Alakangas and Öhlander 2006). Most of the literature in this area is not directly at field scale, but rather at lysimeter or plot scale. The situation is due to the difficulty to organize large-scale experimental settings linking surface soil processes with groundwater quality.

Proofs from correlations between metals and organic carbon in water: A review of colloid associated contaminants transport (Sen and Khilar 2006) underlines the importance of organic coatings and of bacteria (beside mineral and oxyhydroxide colloids) on mineral phases in the transport of metals in saturated environments. An excellent review about the occurrence of inorganic pollutants in ground waters, with source-sink aspects in function of environmental conditions (including the role of DOC for Cu and Pb at neutral conditions) was done by Helvoort et al. (2009).

In seepage water from a landfill, 77% of heavy metals were bound to particles, most of them to humic colloids (other particle types included silicates, insoluble salts, iron (hydro) oxides and carbonates – Klein and Niessner 1998). Baun and Christensen (2004) review the speciation of heavy metals in landfill leachates. They conclude that free ions are less than 30% and, typically less than 10% of the total metal concentration, a more important part being that of colloids and organic complexes. The general topics of landfill leachate plumes biogeochemistry and their long-term composition are reviewed by Christensen et al. (2001) and Kjeldsen et al. (2002). DOC and heavy metals are important categories of landfill leachate pollutants but the export of metals is low and its association with organic carbon is highly variable from site to site, not showing according to models an increasing trend within thousands of years (Kjeldsen et al. 2002). DOC in the leachate from landfills acts as a substrate for microbial redox processes, and heavy metals have low concentrations and a rapid attenuation by sorption and precipitation (Christensen et al. 2001).

Although the role of DOC is important in a landfill context, high concentrations of Pb, Cd, Zn, Cu, and Fe were carried by groundwater at contaminated sites on long distances only when groundwater was highly acidic (pH 3–3.5) (Lee and Saunders 2003). Kjoller et al. (2004) also point out the role of acidification in the mobilization of Ni, Be, Cd, and Co in a sandy aquifer. The acidification front functioned as a geochemical trap for the accumulation of these trace elements (Kjoller et al. 2004). In such cases, DOC has no significant role.

An ash pound had a 600–900 m zone of attenuation for Ba, Fe, Cu, Mn, S, and Zn in groundwater, but Pb did not show any significant attenuation within 1,200 m (Praharaj et al. 2002). It is not clear whether OC played a role in the generation of this pattern. Goldberg et al. (2007) review the adsorption–desorption processes in subsurface reactive transport modeling, with issues directly relevant for the role of DOC in the transport of elements. An experimental proof on the role of OC in the mobility of metals is provided by Muller (2000). He injected metals with and without EDTA and DOC from a lignite pit in a gravel aquifer and studied the mobility at a 10 m scale. The mobility of elements deceased in the following order: As $>$ Se $>$ Ni $>$ Hg $>>$ Zn $>$ Co $>$ Cr $>>$ Sb $>$ Cu $\approx$ Cd $\approx$ Pb. EDTA decreased the mobility of As, Sb, Cr, Hg, and Se, and increased the mobility of Pb, Cd, Cu, Zn, Co, and Ni. Lignite pit DOC decreased the mobility of As, Sb, Cd, Co, Ni, and Cr, and increased the mobility of Pb, Hg, Cu, Se, and Zn. One can conclude that the effect of DOC on the mobility of metals in groundwater is element specific and for some elements is DOC type specific. Tang and Weisbord (2009) also showed experimentally that organic–mineral colloidal Pb was mobile in fractured rocks, while lead carbonate particles were not mobile. By spiking a landfill-leachate polluted groundwater with Cd, Ni, Zn, Cu, and Pb, Jensen et al. (1999) found that a large part of the metals was associated with the organic colloidal fraction, except for Zn, that associated mainly with the inorganic colloidal fraction. The majority of the metals was found in colloids smaller than 0.4 μm (Jensen et al. 1999).

The measurement of transport capacity of DOC in complex groundwater matrices is a matter of ongoing research not only for metal, but also for organic pollutants (Persson et al. 2010).

Boy et al. (2008) tested the hypothesis that the concentrations of elements in stream water depend on the depth of water flow through soil. Peak concentrations of Al and Mn have been associated with lateral flow (near-surface flow in saturated C and nutrient-rich topsoil). Greatest Na concentrations appeared during low baseflow conditions. This was not correlated with larger concentrations of elements in the soil-layer traversed by the flowing water, as all concentrations were larger in the topsoil. Different processes controlled the exports, e.g., weathering and oxidations of sulfides in the subsoil was responsible for the export of P, S, and Na, while OC probably played a role in the export from the topsoil.

The buffering effect of DOC–metal complexes could be lower in mountain shallow soils, leading to a larger indirect role in mountainous contaminated catchments. The DOC pool in shallow mountain soil is replenished after hydrologic events within days during summer and longer in the autumn because of lower temperatures (Halland and Mulder 2010). The pool of potential DOC and its replenishment rates depend on the season and are at maximum during the growing season. Under nonleaching conditions, the pool of potential DOC for export becomes constant, probably due to conversion to other forms or to consumption by microorganisms (Halland and Mulder 2010).

Metal inputs from groundwater to surface water may be important. Spinelly et al. (2002) reported that about 60% of the Co unknown sources of metals in a surface water system could be allocated to seepage from groundwater, but at most 4% in the case of Ag, Cd, Cu, Ni, and Zn. It is likely that such input may involve organic carbon–metal complexes only when the underground path is short (not the case in the cited study). Atekvana and Krishnamurthy (2004) studied the hyporheic ecotone near a landfill and found evidence for water enrichment in inorganic carbon attributable to the landfill.

15.4.7 Effects at Catchment Scale

Mixing of metal sources at basin scale is documented (Bird et al. 2010a, b), and part of it may be due to mechanisms influenced by OC. However, few studies have adequately investigated the migration of trace elements in the slope area hydrosystem (Miller and Miller 2007), and none, according to actual knowledge, looked for the role of OC in the export of metals from slope areas by hydrological processes. Also, the literature concerning the direct relationship between metals and organic carbon at basin scale is not very rich, but the correlation between the export of carbon and that of macro-nutrients is documented (for instance, most of the nitrogen exported from catchments with mature forests is in the organic form; Hedin et al. 1995). In this context, we attempt here to extract information from studies about the export of organic carbon and then from reported metals–organic carbon correlations.

In a pioneering study, Brooks et al. (1999) analyzed the pools of DOC available for export in catchments. Both surface water DOC concentrations and watershed DOC export were lower in areas where pyrite weathering resulted in lower soil pH, compared to neutral rocks weathering (Brooks et al. 1999). They also noticed important differences within catchments between land use types, and a correlation of the snow melt flushing of DOC and the heterotrophic activity in soils during winter. The mechanism of DOC export at catchment scale involved vertical preferential flow to the soil–bedrock interface and then lateral downslope flow, with a finite source of DOC in the organic horizon (Verseveld et al. 2008). Organic horizon leachate and transient groundwater were characterized by high DOC, deep groundwater was characterized by low DOC, and the mixing of deep groundwater and shallow transient groundwater was different at the hill-slope scale compared to the catchment scale (Verseveld et al. 2009).

In a unique and insightful study, Jardine et al. (2006) discussed the multiscale characteristics of organic carbon flow: "Common features across scales persist, including preferential flow, matrix diffusion, and geochemical reactions. However, new important phenomena, such as lateral subsurface flow, develop at the field scale and can significantly influence the movement of organic C through the soil profile. Storm intensity and duration also influence C dynamics and the extent of lateral flow." All these aspects are relevant for the role of OC in metals transport at various scales, although direct studies about carbon–metal interactions at multiple scales do not yet exist. The particularities of DOC export at site and small catchment scale depend on the characteristics of the hydrological event, especially on its intensity (Jardine et al. 2006). The area providing runoff changes seasonally, between storms, and during a rainfall event (Miller and Miller 2007). The intensity of the hydrological event controls the relative importance of vertical and subsurface flows and, because the vertical flow (via groundwater) allows a better filtration of organic carbon, controls the patterns of DOC export (Jardine et al. 2006), and presumably of the associated metals.

Gauthier (2009) studied the production of DOM and its fate in the catchment at scales from microcosm to catchment. During discharge events, DOM was mobilized from multiple sources, wetlands having an important role in the pattern generation. As wetlands function as buffer zones for element fluxes the export of OC during storms may be associated with an important export of metals.

The situation seems to be catchment specific, but possibilities for identifying general patterns exist to some extent. Luider et al. (2006) analyzed 581 low order streams and observed significant differences between biogeoclimatic zones and moisture subzones with respect to conductance, turbidity, pH, and DOC concentration. OC mean values and confidence intervals were generally larger in drier biogeoclimatic zones, and these zones accounted for 25% of the variation in DOC concentrations between catchments.

In another meta-analyses, Raymond and Saiers (2010) processed the data obtained in 30 small forested watersheds without wetlands. Eighty-six percent of the DOC was exported during storms and snowmelts, with the majority of the flux during the rising hydrograph. The highest DOC concentrations and flux were observed in warmer periods and when the preceding discharge was low. Such results underline the importance of timing in field studies estimating the indirect role of OC in metal mobility.

We illustrate now the literature explicitly approaching the influence of OC on metal export in catchments. The patterns of metal associated with OC in rivers as revealed by size fractionation methods (Contado et al. 2003) appear to be very heterogeneous. High Hg concentrations are associated with high DOC in catchments in late summer and early autumn presumably because of OM decomposition (Driscoll et al. 1994). Methylmercury is associated mainly with small size OC, while inorganic Hg is associated mainly with high molecular humic material. Mobilization and immobilization of Pb in catchments is also associated with the dynamic of the OM (Driscoll et al. 1994). The hydrogeochemistry of Mn and Fe in small boreal catchments depended on season, landscape structure (estimated by land cover and soil cover, with key parameter the number of wetlands), the size catchment, and the gradient of the catchment (Björkvald et al. 2008). The authors found that the temporal variation of Fe was likely related to varying hydrological pathways and that the wetland area and DOC were important for Fe concentrations, whereas silt deposits strongly influenced Mn concentrations (Björkvald et al. 2008). Based on their results, they recommend a landscape approach to the study of metal distribution in rivers. The Cd-binding capacity in stream water was due mainly to inorganic carbon, for Pb was predominantly due to OM, and for Cu to both inorganic and organic forms of Cu (Alberts et al. 1984). Park et al. (2007) analyze the climatic control on the export of DOM from a catchment and its influence on metal transport. DOM differed in quantity and quality along the river and depended at a sampling point on the timing of hydrological events, and was positively correlated with some of the trace elements analyzed such as As, Co, and Fe.

The above paragraph referred to stream water. But riparian areas are important as well in the mobility of metals. The sediment load of runoff is reduced when it crosses vegetation strips (transversal buffer zones). For instance, Abu-Zreig et al. (2004) measured efficiency in sediment trapping of 84% in vegetative strips compared to 25% in control. The major factor controlling sediment deposition was the width of the strip (up to 10 m in the mentioned study, then with no influence in the experimental conditions). Other factor influencing the efficiency were the incoming flow and the vegetation cover. The influence of sediment quality, in particular its content in organic matter, was not studied so far, according to our knowledge. To the extent that OM changes the properties of the particles relevant for sedimentation, one can expected to have an influence on the retention of the associated metals in transversal buffer strips. Vidon et al. (2010) reviewed the role of buffer systems, underlining the co-occurrence between the retention of OC and of other pollutants, including Hg. In the floodplain, U was found to be entrapped mainly by adsorption to OM in the water permeable sediment layer (Driscoll et al. 1994), but it can be easily remobilized (either from soil or from sediment) by acidification and introduction of complexing agents. The mobility of metals from overbank deposited sediments in floodplain contaminated by mining and smelter activities is crucial for downstream water quality (Gäbler 1997, Gäbler and Schneider 2000). The risk is relevant for both surface and underground water (Hürkamp et al. 2009). In a seminal article dedicated to the effect of DOM on the export of metals and As from contaminated floodplains, Kalbitz and Wennrich (1998) found that the concentrations of Cu, Cr, Hg, and As in the soil percolates was positively correlated with DOC, but Cd and Zn mobilization depended on pH, and not on DOC (the statistical relation between DOC and pH is not mentioned in the article, but presumably the contribution of these parameters to the multiple regression were not redundant by the chosen algorithm). DOM had no significance in alluvial soils with pH less than 4.5 (Kalbitz and Wennrich 1998). Grybos et al. (2007) identified four types of trace metal behavior in wetland soil: metals bound to OM and released by DOM (rare earth elements), metals bound only to Fe-oxyhydroxides and released by reductive dissolution (Co), metals bound both to organic matter and Fe-oxyhydroxides and released by DOM and reduction on ferric iron (Pb, Ni), and metals with unclear mechanism of release because reduction changes also their redox state and/or speciation (Cu, Cr, U and Th). Schulz-Zunkel and Krueger (2009) reviewed the dynamics of trace metals in the soils of Elbe floodplain and underlined the role of OM and its mineralization in the export of metals, with direct relevance for the Water Framework Directive. Iordache (2009) extensively characterized the role of lower Danube river floodplain in the longitudinal buffering of metal fluxes, pointing out the complex role of litter in mobilizing/immobilizing metals (see also Karavanova et al. 2006). In the same system, Neagoe and Iordache (2002) demonstrated that fast decomposition of detritus coupled with autumn floods play a key role in the export of metals (Fe, Mn, Zn, Cu, Pb, Cd) from alluvial islands in bioavailable forms, an aspect even more important taking into consideration that plants acted as pumps for metals from the soil to aboveground ecosystem parts (Neagoe et al. 2002). As a possible indirect proof of metals transport in organic form, l. Van Damme et al. (2010) identified Zn-humic acid and Zn-fulvic acid complexes in the overbank sediments of a river contaminated by mining and smelter using extended X-ray absorption fine structure spectroscopy. A recent review of trace metal behavior in estuarine and riverine floodplain soils is provided by Laing et al. (2009). They showed that the net effect of OM on metals mobility can be a decrease or an increase of the mobility depending on the situation (oxic/anoxic condition, other parameters).

The relative roles of upland and riparian areas in the hydrological export of DOC from catchments vary during a storm. The riparian runoff was larger on the rising than falling limb of the hydrograph, while the hillslope runoff was larger on the falling than rising limb of the hydrograph (McGlynn and McDonnel 2003). This once more underlines the importance of timing in studies for tracing the sources of metals exported associated with organic carbon from a catchment.

Mass balance and modeling of metals at catchment scale provide another line of evidence for the role of OC. In a recent study on metal stocks and fluxes at catchment scale, Huang et al. (2010) demonstrated positive and statistically significant correlations between Cu, Cr, and Ni concentrations and DOC concentrations in runoff. Shotbolt et al. (2008) estimated that the observed increase of DOC leaching from soils of a studied Nordic catchment could result in an increase in Pb export. The effect of OC on the retention of metals in catchments was evaluated also by modeling (Tipping et al. 2006). The retention of deposited metals strongly adsorbed in soils was 89–95% for Cu and 95–100% for Pb, while for metals which sorb weakly this depended on soil pH and varied from 5% for Ni to 57% for Zn (Tipping et al. 2006). For Ni, Cu, Zn, and Cd, the parameters describing the interaction with OM allowed a good agreement between observed and simulated metals in soil and surface water (Tipping et al. 2006).

The export from natural, noncontaminated systems also provides information about the role of OC at site scale. Sevel et al. (2009) concluded, by comparing the Cd budgets of two forest grown on soils with different pHs, that the forest on acid soil is a net source of Cd by soil leaching, with OC probably playing a role in this export (Boissier and Fontvieille 1993). However, their study was limited to the 0–90 cm layer. As the lower soil layers were not studied, this could lead to a homogenization of the percolating fluxes by intra-ecosystem buffering effects (Currie et al. 1996).

Finally, several words on the indirect role of OC in the atmospheric export of metals, with direct relevance for human health problems (e.g., Cheyran et al. 2000). The situation of organometal(loid) compounds and their mobility has been extensively reviewed by Thayer (2010). The atmospheric export of such compounds occurs especially in organic rich environment, such as landfills. The atmospheric export of organic forms of Hg, As, Sb, Bi, Se, and Te is documented.

Complementary sources of information in this respect are the reviews about capabilities of certain microbial groups to be involved in the volatilization of metals and metalloids (Meyer et al. 2008).

15.5 Conclusions

The number of pseudo-hierarchical levels needed to understand the influence of OC on metal mobility in catchments vary with the complexity of the catchment and include: soil aggregates and other small-scale soil subsystems, soil layer, soil column, site, slope area, elementary catchments, higher order catchments with transversal buffer zones (riparian areas), higher order catchments with longitudinal buffer zones (floodplains).

Depending on its quality, quantity, and physical phase, the OM can induce either the mobilization or the remobilization on of metals. The general concept of "organic matter" is too roughly defined to draw any conclusions on the mobility of metals from simple parameters characterizing it. It means that carbon biogeochemistry and the biogeochemistry of metals cannot be approached separately without an important loss of knowledge. From a different perspective, i.e., the metal effects on plants, Neagoe et al. (2011) arrived to a similar conclusion for the biogeochemistry of heavy metals, nitrogen, and phosphorus.

There is no extensive explicit knowledge about the role of OM in controlling the intensity of metal fluxes from environmental entities, whatever their scale, although the very small-scale mechanisms supporting their role in mobilization and immobilization of metals is rather clear. The buffering and remobilization of metal fluxes as a result of OM turnover across scales is even less understood.

Table 15.1 synthesizes the existing knowledge about the role of organic carbon (and of microorganisms, as revealed by the above analysis) in the mobility of metals and its effects on fluxes in contaminated sites at soil layer scale. For organic carbon, whose particulate forms are much larger than the microorganisms, there is available information also about its effects on fluxes occurring at site scale, categorized by the type of contaminated sites existing in a contaminated mining area, and at catchment scale (Table 15.2).

The nutrient spiraling concept in river biogeochemistry (Ensign and Doyle 2006) introduced the idea of spiraling length as the distance needed for an atom to pass from dissolved form to particulate form, to biological form and back to dissolved form (Newbold et al. 1981). In the multiscale framework proposed above for OM role in metal mobility by hydrological fluxes, specific hypotheses in metal biogeochemistry could be formulated with respect to the transient storage on metal spiraling at multiple-scale units of analyses in contaminated areas (soil column, underground flowpath, slope area flowpath, small catchment, and large catchment). Also, ecotoxicological hypotheses can be formulated with respect to the storage units about the effects of metal stock and retention time on the overall productivity in the storage units (primary productivity and productivity of microorganisms) and

Table 15.1 Roles of microorganisms and organic matter observable at soil aggregate scale and of their effects on fluxes of metals from soil layers

Soil layer	Roles	Role by immobilization metals	Role by mobilization of metals	Role by supporting the mobilization or immobilization of metals	Effects on mobility at larger scale
(Aggregates in) Soil layer relevant for plants	Microorganisms	K1 Biosorption, intracellular accumulation, biomineral formation, redox immobilization, metals sorption to biogenic minerals	K1 Chemolitotrophic leaching, chemoorganotrophic leaching, bioweathering, redox mobilization, methilation	K2 Organic matter decomposition, organic acid and syderophores exudates	K3 Transfer of metals to plants, and to lower soil layers
	Organic matter	K1 Immobilization in litter, immobilization in soil aggregates, chelates in fine pores	K1 Organochemical weathering, soluble chelates, organocolloids, free enzymatic degradation of immobile organic carbon	K1 Energy source for microorganisms, buffering of soil solution	
(Aggregates and particles in) Lower soil layers	Microorganisms	K2 Same as in upper layer	K2 Same as in upper layer	K2 Same as in upper layer	K4 Transfer of metals to lower soil layer or to subsoil
	Organic matter	K2 Immobilization in soil aggregates, chelates in fine pores	K2 Same as in upper layer	K2 Same as in upper layer	

Legend of the extent and coherence of existing knowledge: K1 = max to K5 = min

Table 15.2 Effects of organic matter on fluxes of metals observable at scales ranging from sites and buffer zones to catchments

Type of entity/effect	Immobilization	Mobilization	Effect at larger scale
Site with contaminated soil in the slope area	K2 Immobilization in litter, immobilization in soil aggregates	K3 Soluble chelates, organocolloids	K4 Transfer to subsoil and groundwater and then to surface water, transfer to surface water by lateral types of flows, transfer to floodplains, volatilization
Mining dump or tailing dam	K3 None, or same as above only in case of organic amendments		
Contaminated soil in the floodplain, contaminated stream ecotone	K3 Immobilization in litter, immobilization in soil aggregates		K4 Transfer to surface water during floods, transfer to groundwater, volatilization
Catchments	K5 Complex influence resulted from mixture of hydrological fluxes and from the retention and turnover of dissolved and particulated organic matter		K5 Export to higher order catchments

on functional parameters of the bacterial community (e.g., by comparison of soil columns with different trace elements biogeochemical parameters).

Acknowledgments This research was carried out within the Romanian Consortium for the Biogeochemistry of Trace Elements with financing from National University Research Council (CNCSIS) by projects 291 and 176 – 2007 (codes ID 1006 and 965) and within the frame of European project UMBRELLA (FP7 – ENV-2008 – 1 no. 226870). Special thanks go to two anonymous reviewers for the constructive criticism, which greatly improved the manuscript.

References

Abu-Zreig M, Rudra RP, Lalonde MN, Whiteley HR, Kaushik NK (2004) Experimental investigation of runoff reduction and sediment removal by vegetated filter strips. Hydrol Process 18:2029–2037

Alakangas L, Öhlander B (2006) Pilot-scale studies of different covers on unoxidised sulphide-rich tailings in Northern Sweden: the geochemistry of leachate waters. Mine Water Environ 25:171–183

Alberts JJ, Giesy JP, Evans DW (1984) Distribution of dissolved organic carbon and metal-binding capacity among ultrafilterable fractions isolated from selected surface waters of the Southeastern United States. Environ Geol Water Sci 6:91–101

Alkorta I, Hernández-Allica J, Becerril JM, Amezaga I, Albizu I, Anaindia M, Garbisu C (2004) Chelate-enhanced phytoremediation of soils polluted with heavy metals. Environ Sci Biotechnol 3:55–70

Amery F, Degryse F, Degeling W, Smolders E, Merckx R (2007) The copper-mobilizing-potential of dissolved organic matter in soils varies 10-fold depending on soil incubation and extraction procedures. Environ Sci Technol 41:2277–2281

Antoniadis V, Alloway BJ (2002) The role of dissolved organic carbon in the mobility of Cd, Ni, and Zn in sewage sludge-amended soils. Environ Pollut 117:515–521

Ashworth DJ, Alloway BL (2008) Influence of dissolved organic matter on the solubility of heavy metals in sewage-sludge0amended soils. Comm Soil Sci Plant Anal 39:538–550

Atekwana EA, Krishnamurthy RV (2004) Investigating landfill-impacted groundwater seepage into headwater streams using stable carbon isotopes, Hidrological Processes, 18(10):1915–1926

Banks MK, Schwab AP, Fleming GR, Hetrick BA (1994) Effects of plants and soil microflora on leaching of zinc from mine tailings. Chemosphere 29:1691–1699

Baun DL, Christensen TH (2004) Speciation of heavy metals in landfill leachate: a review. Waste Manag Res 22:3–23

Berg B, McCluagherty C (2008) Plant litter – decomposition, humus formation, carbon sequestration. Springer, Berlin

Besser JM, Rabeni CF (2009) Bioavailability and toxicity of metals leached from lead-mine tailings to aquatic invertebrates. Environ Toxicol Chem 6:879–890

Bird G, Brewer PA, Macklin MG, Nikolova M, Kotsev T, Mollov M, Swain C (2010a) Dispersal of contaminant metals in the mining-affected Danube and Maritsa Drainage Basins, Bulgaria, Eastern Europe. Water Air Soil Pollut 206:105–127

Bird G, Brewer PA, Macklin MG, Nikolova M, Kotsev T, Mollov M, Swain C (2010b) Quantifying sediment-associated metal dispersal using Pb isotopes: application of binary and multivariate mixing models at the catchment-scale. Environl Pollut 158:2158–2169

Björkvald L, Buffam I, Laudon H, Magnus Mörth C (2008) Hydrogeochemistry of Fe and Mn in small boreal streams: the role of seasonality landscape type and scale. Geochem Cosmochim Acta 72:2789–2804

Boissier JM, Fontvieille D (1993) Biodegradable dissolved organic carbon in seepage from two forest soils. Soil Biol Biochem 25:1257–1261

Bolan NS, Adriano DC, Luz M (2004) Dynamics and environmental significance of dissolved organic matter in soil. Super soil: 3rd Australian New Zeeland soils conference, 5–9 December 2004, Australia

Boy J., Valarezo C, Wilcke W (2008) Water flow paths in soil control element exports in an Andean tropical montane forest, Eur J Soil Sci, 59:1209–1227

Bronick CJ, Lal R (2005) Soil structure and management: a review. Geoderma 124:3–22

Brooks PD, McKnight DM, Bencala KE (1999) The relationship between soil heterotrophic activity, soil dissolved organic carbon (DOC) leachate, and catchment-scale DOC export in headwater catchments. Water Resour Res 35:1895–1902

Burckhard SR, Schwab AP, Banks MK (1994) The effects of organic acids on the leaching of heavy metals from mine tailings. J Hazard Mater 41:135–145

Carrillo-González R, Šimunek J, Sauve S, Adriano D (2006) Mechanisms and pathways of trace element mobility in soils. Adv Agron 91:111–178

Cheyran C, Lalere B, Donard OX (2000) Volatile metal and metalloid species (Pb, Hg, Se) in a European urban atmosphere (Bordeaux, France). Environ Sci Technol 34:27–32

Chow AT, Fengmao Guo F, Gao S, Breuer RS (2005) Trihalomethane formation potential of filter isolates of electrolyte-extractable soil organic carbon. J Environ Qual 34:1992–1997

Christensen TH, Kjeldsen P, Bjerg PL, Jensen DL, Christensen JB, Baun A, Albrechtsen HJ, Heron G (2001) Biogoechemistry of landfill leachate plumes. Appl Geochem 16:659–718

Citeau L, Lamy I, van Oort F, Elsass F (2003) Colloidal facilitated transfer of metals in soils under different land use. Colloid Surface Physicochem Eng Aspect 217:11–19

Contado C, Blo G, Conato C, Dondi F, Beckett R (2003) Experimental approaches for size-based metal speciation in rivers. J Environ Monit 5:845–851

Cornu JY, Denaix L, Schneider A, Pellerin S (2007) Temporal evolution of redox processes and free Cd dynamics in a metal-contaminated soil after rewetting. Chemosphere 79:306–314

Currie WS, Aber JD, McDowell WH, Boone RD, Magill AH (1996) Vertical transport of dissolved organic C and N under long-term N amendments in pine and hardwood forests. Biogeochemistry 35:471–505

Denaix L, Semlali RM, Douay F (2001) Dissolved and colloidal transport of Cd, Pb, and Zn in a silt loam soil affected by atmospheric industrial deposition. Environ Pollut 114:29–38
Degryse F, Smolders E, Parker E (2009) Partitioning of metals (Cd, Co, Cu, Ni, Pb, Zn) in soils: concepts, methodologies, prediction and applications – a review, Eur J Soil Sci. 60:590–612
Dessurealut-Rompre J, Nowack B, Schulin R, Tercier-Waeber ML, Luster J (2008) Metal solubility and speciation in the rhizosphere of *Lupinus albus* cluster roots. Environ Sci Technol 42:7146–7151
Driscoll CT, Otton JK, Iverfeld A (1994) Trace metals speciation and cycling. In Moldan B, Cerny J (eds) Biogeochemistry of small catchments: a tool for environmental research, vol 51. SCOPE Publications, John Willey & Sons Ltd, Chicester, pp 301–322
Dudal Y, Gérard F (2004) Accounting for natural organic matter in aqueous chemical equilibrium models: a review of the theories and applications. Earth Sci Rev 66:199–216
Dudal Y, Sevenier G, Dupont L, Guillon E (2005) Fate of the metal-binding soluble organic matter throughout the a soil profile. Soil Sci 170:707–715
Ensign SH, Doyle MW (2006) Nutrient spiralling in streams and river networks. J Geophys Res 111:G04009. doi:10.1029/2005JG000114, 13 p
Flores AN, Sola FM (2010) Evaluation of metal attenuation from mine Tailings in SE Spain (Sierra Almagreea): a soil-leaching column study. Mine Water Environ: on-line first. doi:10.1007/s10230-010-0099-z
Fontaine S, Barot S, Barré N, Mary B, Rumpel C (2007) Stability of organic carbon in deep soil layers controlled by fresh supply. Nature 450:277–281
Forsberg LS, Gustafsson JP, Kleja DB, Ledin S (2008) Leaching of metals from oxidizing sulphide mine tailings with and without sewage sludge application. Water Air Soil Pollut 194:331–341
Gäbler HE (1997) Mobility of heavy metals as a function of pH of samples from an overbank sediment profile contaminated by mining activities. J Geochem Explor 58:185–194
Gäbler HE, Schneider J (2000) Assessment of heavy metal contamination of floodplain soils due to mining and mineral processing in the Harz Mountains, Germany. Environ Geol 39:774–782
Gauthier A (2009) Production and fate of dissolved organic matter in ecosystems with low human impact. Ph.D. thesis, James Cook University
Goldberg S, Criscenti LJ, Turner DR, Davis JA, Cantrell KJ (2007) Adsorption-desorption precesses in subsurface reactive transport modeling. Vadose Zone J 6:407–435
Grybos M, Davranche M, Gruau G, Petitjean P (2007) Is trace metal release in wetland soils controlled by organic matter mobility or Fe-oxyhydroxides reduction? J Colloid Interf Sci 314:490–501
Guo X, Komnitsas K, Li D (2010) Correlation between herbaceous species and environmental variables at the abandoned Haizhou coal mining site. Environ Forensics 11:146–153
Gustafsson JP, Pechová P, Bergren D (2003) Modeling metal binding to soils: the role of natural organic matter. Environ Sci Technol 37:2767–2774
Haaland S, Mulder J (2010) Dissolved organic carbon concentrations in runoff from shallow heathland catchments: effects of frequent excessive leaching in summer and autumn, Biogeochemistry, 97:45–53
Haynes RJ, Beare MH (1997) Influence of six crop species on aggregate stability and some labile organic matter fractions. Soil Biol Biochem 29:1647–1653
Hedin LO, Armesto JJ, Johnson AH (1995) Patterns of nutrient loss from unpolluted, old-growth temperate forests: evaluation of biogeochemical theory. Ecology 76:493–509
Helvoort PJ, Griffioen J, Edmunds WM (2009) Occurrence and behavior of main inorganic pollutants in European groundwater. In: Quevauviller P, Fouillac A-M, Grath J, Ward R (eds) Groundwater monitoring. Willey, Chicester, pp 81–109
Huang JH, Ilgen G, Matzner E (2010) Fluxes and budgets of Cd, Zn, Cu, Cr and Ni in a remote forested catchment in Germany. Biogeochemistry 103(1–3):59–70
Hulshof AHM, Blowes DW, Gould DW (2009) Evaluation of situ layers for treatment of acid mine drainage: a field comparison. Water Res 40:1816–1826
Hürkamp K, Raab T, Völkel J (2009) Lead pollution of floodplain soils in a historic mining area-age, distribution and binding forms. Water Air Soil Poll 201:331–345

Iordache V (2009) Ecotoxicology of heavy metals in the Danube floodplain (in Romanian). Ars Docendi, Bucharest

Iordache V, Dumitru D, Constantinescu M (2004) Assessment and reduction of heavy metals input into agroecosystems of Romania. Mediul Inconjurator (Environ) 2:24–47

Iordache V, Neagoe A, Bergmann H, Kothe E, Buechel G (2006) Factors influencing the export of metals by leaching in bioremediation experiments. 23. Arbeitstagung in Jena, Lebensnotwendigkeit und Toxizität der Mengen-, Spuren- und Ultraspurenelemente, pp 288–295

Iordache V, Kothe E, Neagoe A, Gherghel F (2011) A conceptual framework for up-scaling ecological processes and application to ectomycorrhizal fungi. In: Rai M, Varma A (eds) Diversity and biotechnology of ectomycorrhiza. Springer, Berlin, Heidelberg, pp 255–299

Iordache V, Lăcătusu R, Scrădeanu D, Onete M, Jianu D, Bodescu F, Neagoe A, Purice D, Cobzaru I (2012) Contributions to the theoretical foundations of integrated modeling in biogeochemistry and their application in contaminated areas In: Kothe E, Varma A (eds) Bio-geo-interactions in contaminated soils. Springer, Berlin, Heidelberg

Jansen B, Nierop KGJ, Verstraten JM (2003) Mobility of Fe(II), Fe(III), and Al in acidic forest soils mediated by dissolved organic matter: influence of solution pH and metal/organic carbon ratios. Geoderma 114:323–340

Jansen B, Nierop KGJ, Verstraten JM (2005) Mechanisms controlling the mobility of dissolved organic matter, aluminium and iron in podzol B horizons. Eur J Soil Sci 56:537–550

Jardine PM, Mayes MA, Mulholland PJ, Hanson PJ, Traver JR, Luxmoore RJ, McCarthy JF, Wilson GV (2006) Vadose zone flow and transport of dissolved organic carbon at multiple scales in humid regimes. Vadose Zone J 5:140–152

Jensen DL, Ledin A, Christensen TH (1999) Speciation of heavy metals in landfill-leachate polluted groundwater. Water Res 33:2642–2650

Jianu D, Iordache V, Soare B, Petrescu L (2012) The role of mineralogy in the hazard potential of abandoned mine sites. In: Kothe E, Varma A (eds) Bio-geo-interactions in contaminated soils. Springer, Berlin, Heidelberg

Johnson DB, Hallberg KB (2005) Acid mine drainage remediation options: a review. Sci Total Environ 338:3–14

Kahle M, Kleber M, Jahn R (2004) Retention of dissolved organic matter by phyllosilicate and soil clay fractions in relation to mineral properties. Org Geochem 35:269–276

Kalbitz K, Wennrich R (1998) Mobilization of heavy metals and arsenic in polluted wetland soil and its dependence on dissolved organic matter. Sci Total Environ 209:27–39

Kalbitz K, Solinger S, Park JH, Michalzink B, Matzner E (2000) Controls on the dynamics of dissolved organic matter in soils: a review. Soil Sci 165:277–304

Karathanasis AD, Johnson DMC (2006) Subsurface transport of Cd, Cr, and Mo mediated by biosolid colloids. Sci Total Environ 354:157–169

Karavanova EI, Belyanina LA, Shapiro AD, Stepanov AA (2006) Effect of litters on the mobility of zinc, copper, manganese, and iron in the upper horizons of podzolic soils. Soil Chem 39:35–43

Kjeldsen P, Barlaz AM, Rooker AP, Baun A, Ledin A, Christensen TH (2002) Present and long-term composition of MSW landfill leachate: a review. Crit Rev Environ Sci Technol 32:297–336

Kjoller C, Postma D, Larsen F (2004) Groundwater acidification and the mobilization of trace metals in a sandy aquifer. Environ Sci Techol 38:2829–2835

Klein T, Niessner R (1998) Characterization of heavy metal containing seepage water colloids by flow FFF, ultrafiltration, ELISA and AAS. Mikrochim Acta 129:45–55

Klitzke S, Lang F (2007) Hydrophobicity of soil colloids and heavy metal mobilization: effects of dryinn. J Environ Qual 36:1187–1193

Klitzke S, Lang F (2009) Mobilization of soluble and dispersible lead, arsenic, and antimony in a polluted, organic-rich soil – effects of pH increase and counterion valency. J Environ Qual 38:933–939

Klitzke S, Lang F, Kaupenjohann M (2008) Increasing pH releases colloidal lead in a highly contaminated forest soil. Eur J Soil Sci 59:265–273

Kretzschman R, Schäfer (2005) Metal retention and transport on colloidal particles in the environment. Elements 1:205–210

Laing GD, Rinklebe J, Vandecasteele B, Meers E, Tack FMG (2009) Trace metal behavior in estuarine and riverine floodplin soils and sediments: a review. Sci Total Environ 407:3972–3985

Lal R (2009) Challenges and opportunities in soil organic matter research. Eur J Soil Sci 60:158–169

Lee MK, Saunders JA (2003) Effects of pH on metals precipitation and sorption: field bioremediation and geochemical modeling approaches. Vadose Zone J 2:177–185

Leug KT, Nandakumar K, Sreekumari K, Lee H, Trevors T (2007) Biodegradation and bioremediation of organic pollutants in soil. In: van Elsas JD, Jansson JK, Trevors JT (eds) Modern soil microbiology, vol 2. Tayor & Francis Group, Boca Raton, FL

Li Z, Zhou L (2010) Cadmium transport mediated by soil colloid and dissolved organic matter: a field study. J Environ Sci 22:106–115

Luider CD, Schere R, Jefferson Curtis P (2006) An empirical approach to predicting water quality in small strems of southern Britich Columbia using biogeoclimatic ecosystem classifications, BC Journal of Ecosystems and Management, 7(3):25–35

Luster J, Menon M, Hermle S, Schulin R, Gunthardt-Goerg MS, Nowack B (2008) Initial changes in refilled lysimeters built with metal polluted topsoil and acidic or calcareous subsoils as indicated by changes in drainage water composition. Water Air Soil Pollut 8:163–176

Marschner B, Kalbitz K (2003) Controls of bioavailability and biodegradability of dissolved organic matter in soils. Geoderma 113:211–235

McGlynn BL, McDonnel JJ (2003) Role of discrete landscape units in controlling catchment dissolved organic carbon dynamics. Water Resour Res 39:3–18

Mendez MO, Maier RM (2008) Phytoremediation of mine tailings in temperate and arid environments. Rev Environ Sci Biotechnol 7:47–59

Merckx R, Brans K, Smolders E (2001) Decomposition of dissolved organic carbon after soil drying and rewetting as an indicator of metal toxicity in soils. Soil Biol Biochem 33:235–240

Meyer J, Michalke K, Kouril T, Hensel R (2008) Volatilisation of metals and metalloids: an inherent feature of methanoarchaea? Syst Appl Microbiol 31:81–87

Mikutta R, Kleber M, Torn MS, Jhan R (2006) Stabilization of soil organic matter: association with minerals or chemical recalcitrance? Biogeochemistry 77:25–56

Miller JR, Miller SMO (2007) Contaminated rivers: a geomorphological-geochemical approach to site assessment and remediation. Springer, Dordrecht, 418 p. ISBN ISBN 1402052863; ISBN 9781402052866

Muller J (2000) Large scale field experiments on the mobility of heavy metals in ground water. Tracers and modeling in hydrology, IAHS Publication No. 262, pp 135–140

Nakamaru Y, Ishikawa N, Tagami K, Uchida S (2007) Role of soil organic matter in the mobility of radicesium in agricultural soils common in Japan. Colloid Surf 306:111–117

Neagoe A, Iordache V (2002) Preliminary assessment of the macrophytes role in the export of metals from Fundu Mare Island, Danube floodplain. In: Proceedings of the 5th international symposium on metal elements in environment, medicine and biology, pp 245–251

Neagoe A, Iordache V, Altorfer T (2002) Risk sources due to metals in the Danube floodplain. Mengen und Spuren Elemente 21:77–82

Neagoe A, Ebenå G, Carlsson E (2005) The effects of soil amendments on plant performance in an area affected by acid mine drainage. Chem Erde 65(S1):115–130

Neagoe A, Merten D, Iordache V, Buechel G (2009) The effect of bioremediation methods involving different degrees of soil disturbance on the export of metals by leaching and by plant uptake. Chem Erde 69:57–73

Neagoe A, Iordache V, Kothe E (2011) A concept of biogeochemical role and the challenge of upscaling the effects of arbuscular mycorrhizal fungi on metals mobility. In: Goltapeh EM, Danesh YR, Varma A (eds) Fungi as bioremediators, Soil Biology. Springer, Berlin, Heidelberg

Neff JC, Asner GP (2001) Dissolved organic carbon in terrestrial ecosystems: synthesis and model. Ecosystems 4:29–48

Newbold JD, Elwood JV, O'Neill RV, Winkle WV (1981) Measuring nutrient spiralling in streams. Can J Fish Aquat Sci 38:860–863

Nowack B (2005) Chelating agents and the environment. Environ Pollut 153:1–2

Nowack B, Rais D, Frey B, Menon M, Schulin R, Gunthardt-Goerg MS, Luster J (2006) Influence of metal contamination on soil parameters in a lysimeter experiment designed to evaluate phytostabilization by afforestations. Snow Landsc Res 80:201–211

Parat C, Denaix L, Lévêwue J, Chaussod R, Andreux F (2007) The organic carbon derived from sewage sludge as a key parameter determining the fate of trace metals. Chemosphere 69:636–643

Park JH, Lee JH, Kang SY, Kim SY (2007) Hydroclimatic controls on dissolved organic matter (DOM) characteristics and implications for trace metal transport in Hwangryong river watershed, Korea, during a summer monsoon period. Hydrol Process 21:3025–3034

Pédrot M, Dia A, Davranche M, Bouhnik-Le Coz M, Henin O, Gruau G (2008) Insights into colloid-mediated trace element release at the soil/water interface. J Colloid Interface Sci 355:187–197

Peppas A, Komnitsas K, Halikia I (2000) Use of organic covers for acid mine drainage control. Miner Eng 13(5):563–574

Persson L, Alsberg T, Odham G, Ledin A (2010) Measuring the pollutant transport capacity of dissolved organic matter in complex matrixes. Int J Environ Anal Chem 83:971–986

Praharaj T, Swai SP, Powell MA, Hart BR, Tripathy S (2002) Delineation of groundwater contamination around an ash pond: geochemical and GIS approach. Environ Int 27:631–638

Quails RG, Haines BL (1991) Geochemistry of dissolved organic nutrients in water percolating through a forest ecosystem. Soil Sci Soc Am J 55:1112–1123

Raymond PA, Saiers JE (2010) Event controlled DOC export from forested watersheds. Biogeochemistry 100:197–209

Robinson B, Schulin R, Nowack B, Roulier S, Menon M, Clothier B, Green S, Mills T (2006) Phytoremediation for the management of metal flux in contaminated sites. Snow Landsc Res 80:221–234

Rumpel C, Kögel-Knabner I (2010) Deep soil organic matter—a key but poorly understood component of terrestrial C cycle, Plant Soil, on-line first, DOI:10.1007/s11104-010-0391-5

Schulze WX, Gleixner G, Kaiser K, Guggenberger G, Mann M, Schulze ED (2005) A proteomic fingerprint of dissolved organic carbon and of soil particles. Oecologia 142:335–343

Schulz-Zunkel C, Krueger F (2009) Trace Metal Dynamics in Floodplain Soils of the River Elbe. J Environ Qual 38:1349–1362

Sen TK, Khilar KC (2006) Review on subsurface colloids and colloid-associated contaminant transport in saturated porous media. Adv Colloid Interface Sci 119:71–96

Seuntjens P, Nowack B, Schulin R (2004) Root-zone modeling of heavy metal uptake and leaching in the presence of organic ligands. Plant Soil 265:61–73

Sevel L, Hansen HCB, Raulund-Rasmussen KR (2009) Mass balance of cadmium in two contrasting oak forest ecosystems. J Environ Qual 38:93–102

Shotbolt LA, Rothwell JJ, Lawlor AJ (2008) A mass balance approach to quantifying Pb storage an fluxes in an upland catchment of the Peak District, north-central England, Earth Surface Processes and Landforms, 33:1721–1741

Smucker AJM, Park EJ, Dorner J, Horn R (2007) Soil micropore development and contributions to soluble carbon transport within macroaggregates. Vadose Zone J 6:282–290

Spinelli GA, Fissher AT, Wheat GC, Tryon MD, Brown KM, Flegal AR (2002) Groundwater seepage into northern San Francisco BayL Implications for dissolved metals budgest, Water Resources Research, 38:1107–1125

Stockdale A, Davison W, Zhang H (2009) Micro-scale biogeochemical heterogeneity in sediments: a review of available technology and observed evidence. Earth Sci Rev 92:81–97

Strobel BW, Borggaard OK, Hansen HCB, Andersen MK, Rasmussen KR (2005) Dissolved organic carbon and decreasing pH mobilize cadmium and copper in soil. Eur J Soil Sci 56:189–196

Tabak HH, Lens P, van Hullebusch ED, Dejonghe W (2005) Developments in bioremediation of soils and sediments polluted with metals and radionuclides – 1. Microbial processes and mechanisms affecting bioremedia- tion of metal contamination and influencing metal toxicity and transport. Rev Environ Sci Biotechnol 4:115–156

Tang XY, Weisbord N (2009) Colloid-facilitatcs transport of lead in natural discrete factures. Environ Pollut 157:2266–2274

Thayer JC (2010) Roles of organometal(loid) compounds in environmental cycles. Met Ions Life Sci 7:1–32

Tipping E, Lawlor AJ, Lofts S, Shotbolt L (2006) Simulating the long-term chemistry of an upland UK catchment: heavy metals. Environ Pollut 141:139–150

Uroz S, Calvaruso C, Turpault MP, Frey-Klett P (2009) Mineral weathering by bacteria: ecology, actors and mechanisms, Trends Microbiol 17:378–387

vad der Keur P, Iversen BV (2006) Uncertainty in soil physical data at river basin scale. Hydrol Earth Syst Sci 10:889–902

Van Damme A, Degryse F, Smolders E, Sarret G, Dewit J, Swennen R, Manceau A (2010) Zinc speciation in mining and smelter contaminated overbank sediments by EXAFS spectroscopy. Geochim Cosmochim Acta 74:3707–3720

van Gestel CAM (2008) Physico-chemical and biological parameters determine metal bioavailability in soil. Sci Total Environ 406:385–395

van Verseveld WJ, Mc Donnell JJ, Lajtha L (2008) A mechanistic assessment of nutrient flushing at the catchment scale. J Hydrol 358:268–287

van Verseveld WJ, Mc Donnell JJ, Lajtha L (2009) The role of hillslope hydrology in controlling nutrient loss. J Hydrol 367:177–187

Vangronsveld J, Herzig R, Weyens N, Boulet J, Adriaensen K, Ruttens A, Thewys T, Vassilev A, Meers E, Nehnajova E (2009) Phytoremediation of contaminated soils and groundwater lessons from the field. Environ Sci Pollut Res 16:765–794

Velde B, Barré P (2010) Soils, plants and clay minerals. Springer, Berlin

Vidon P, Allan C, Burns D, Duval N, Gurwick S, Inandar S, Lowrance R, Okay J, Scott D, Sebestyen S (2010) Hot sports and hot moments in riparian zones: potential for improved water quality management. J Am Water Resour Assoc 46:278–298

Wagai R, Mayer LM (2007) Sorptive stabilization of organic matter in soils by hydrous iron oxides. Geochim Cosmochim Acta 71:25–35

Wang B, Zhang H, Fan Z, Ju Y (2010) Compacted sewage sludge as a barrier for tailing impoundment. Environ Earth Sci 61:931–937

Warren LA, Haack EA (2001) Biogeochemical controls on metal behavior in freshwater environments. Earth Sci Rev 54:261–320

Wu J, Lu J, Chen T, He Z, Su Y, Yao X (2010) In situ biotreatment of acidic mine drainage using straw as sole substrate. Environ Earth Sci 60:421–429

Yin X, Gao B, Ma LQ, Saha UK, Sun H, Wang G (2010) Colloid- facilitated Pb transport in two shooting-range soils in Florida. J Hazard Mater 177:620–625

Zagury GJ, Kulnieks VI, Neculita CM (2006) Characterization and reactivity assessment of organic substrates for sulphate-reducing bacteria in acid mine drainage treatment. Chemosphere 64:944–954

Zhang M, Zhang H (2010) Co-transport of dissolved organic matter and heavy metals in soil induced by excessive phosphorus applications. J Environ Sci 22:598–606

Zhao Lu YL, Schulin R, Weng L, Nowack (2007) Coupled mobilization of dissolved organic matter and metals (Cu and Zn) in soil columns. Geochim Comsochim Acta 71:3407–3418

Zhou LX, Wong JWC (2003) Behavior of heavy metals in soil: effect of dissolved organic matter. In: Selim M, Kingery WL (eds) Geochemical and hydrological reactivity of heavy metals in soils. CRC, New York, pp 245–270

Zsolnay A (2003) Dissolved organic matter: artefacts, definitions, and functions. Geoderma 113:187–209

Chapter 16
Mycorrhizal-Based Phytostabilization of Zn–Pb Tailings: Lessons from the Trzebionka Mining Works (Southern Poland)

Katarzyna Turnau, Stefan Gawroński, Przemysław Ryszka, and Douglas Zook

16.1 Introduction

Because of its wide variety of mineral resources, southern Poland is the most industrialized region of the country. Ores rich in zinc and lead are especially common in the vicinity of Bytom, Olkusz, and Chrzanów. Economically important are metasomatic ores extracted from triassic dolomites (Strzyszcz 2003). The extraction of the metal ores dates back to the tenth century when surface ore concrete was used and to the twelfth century when ore mining had begun (Szuwarzyński 2000). Trzebionka Mining Works (ZG Trzebionka) is one of the companies extracting ores. It is located between Chrzanów and Trzebinia, sites of intensive industrial development over the last 100–150 years. Initially, only lead and silver were excavated, mostly by hand and selected visually. Much later, the excavated material was crushed and centrifuged in water within hydrocyclones. The ores were then collected from the bottom. This technology was not very efficient in that the wastes that were deposited in the form of heaps contained two times more precious metals than in the technologies which were developed later. On the other hand, the granularity of the material was 2–7 mm, and the wastes were not so easily moved by wind.

Since 1970, excavated ores have been subjected to flotation processes. The metals were collected from the top of the tanks. This type of technology was used

K. Turnau (✉) • P. Ryszka
Institute of Environmental Sciences, Jagiellonian University, Gronostajowa 7, Kraków 30-387, Poland
e-mail: katarzyna.turnau@uj.edu.pl

S. Gawroński
Institute of Botany, Jagiellonian University, Lubicz 46, Kraków 31-512, Poland
e-mail: stefangaw@gmail.com

D. Zook
Boston University, 875 Beacon Street, Boston, MA 02215, USA
e-mail: dzook@bu.edu

E. Kothe and A. Varma (eds.), *Bio-Geo Interactions in Metal-Contaminated Soils*, Soil Biology 31, DOI 10.1007/978-3-642-23327-2_16,

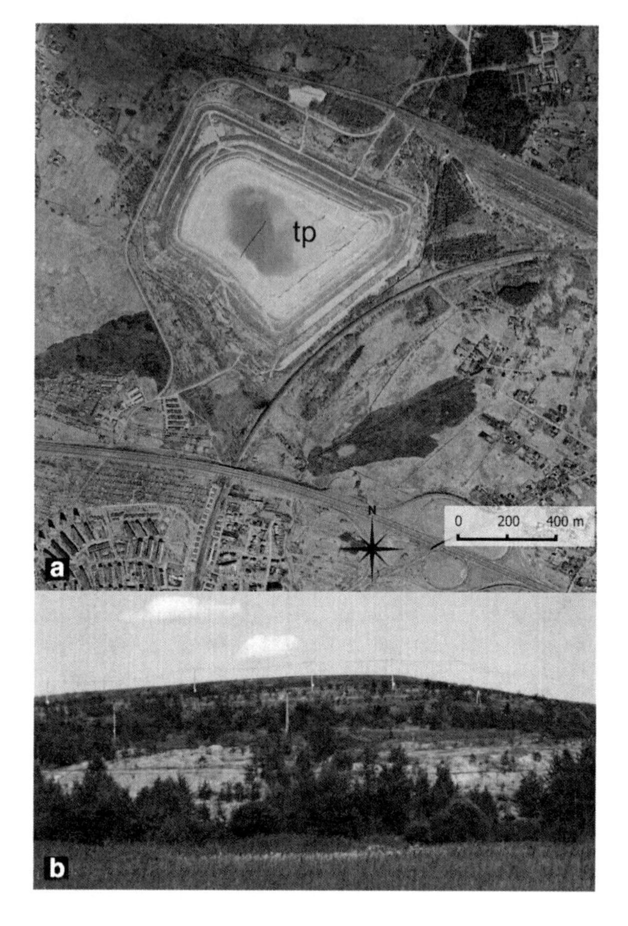

Fig. 16.1 Tailing pond of the ZG Trzebionka industrial site: (**a**) aerial view on the tailing pond (tp); (**b**) tailing pond as seen from ground surface

by ZG Trzebionka since the 1960s. Each year, 2.2 million tons of ore was extracted. This activity was terminated in July 2009 when the resources were exhausted. The waste material produced as a by-product of flotation was deposited 1 km from the ore extraction site and resulted in the formation of a 60-m high and 64-ha tailing area (Fig. 16.1). The solid fraction was used to form tailing ponds into which the liquid phase was dumped. The heavy-metal-rich wastes are subjected to wind and water erosion and thus pose a threat to the environment and human health. Although the pH of the waste ranges from 7 to 8 and the toxicity to living organisms is not as high as would be expected if the total content of the wastes was taken into account, nevertheless, the waste particles can move and increase the soil toxicity in the surrounding area, especially if soil pH is low. Soil monitoring that was carried out several times in the area surrounding the tailing, and the overall data leave no doubt that the tailing is the source of soil pollution by Cd (4–16 mg $\times$ kg^{-1}), Pb (200–400 mg $\times$ kg^{-1}), Zn (400–800 mg $\times$ kg^{-1}, Cu (20–40 mg $\times$ kg^{-1}), and Sr (40–80 mg $\times$ kg^{-1}) (all measured after extraction in HCl) (Lis and Pasieczna 1998), and increased concentrations of As, Co, Hg, Ni Ti, and V (vanadium) were detected. The introduction of vegetation on the outer slopes of the tailing was found to decrease slightly the extent of pollution outside the area of substrate deposition.

Future plans include leveling and revegetation of the tailing and subsequent use by the community as a recreation area. This goal is challenging, and the tailing locales still need continuous monitoring for potential threats to the biota. Here we summarize the research that has been carried out in this region since 1990.

16.2 Pb–Zn Tailing as a Substratum for Plant Growth

Postflotation wastes are extremely harsh substrates for plant growth and for any biological reclamation because they are composed of ca. 0.3 mm grains (Strzyszcz 1980, 2003). Low porosity results in unfavorable air–water conditions, restricted water infiltration during rainfall, and restricted water recharge by capillary rise from deeper layers during dry periods. These favor wind erosion in dry periods and water erosion during rainfall (Strzyszcz 1980, 2003). The slopes of the heaps are of similar type, although the grain sizes are often larger. The slopes of the heaps are also more susceptible to erosion and, therefore, the stabilization is even more difficult. The whole tailing was comparatively easy to manage while the extraction was carried out and water was pumped into the sedimentation tank. About 560,000 m^3 of water per year was being pumped inside, which was more than the atmospheric precipitation estimated for this area. Now, following the termination of the factory activity, the hydrological situation is expected to become even worse.

The chemical composition of the substrate is also unfavorable because the carbonate content is over 75%, with high concentrations of Ca^{2+}, SO_4^2, and low Na^+, K^+, Mg^{2+}, Cl^-, and HCO_3. No organic matter and very low P and N contents were found in the youngest part of the substrate. The heavy metal content of the substrate exceeds several times the values defined as tolerable in arable soils and contains 468 mg $\times$ kg^{-1} of total cadmium, 7,068 mg $\times$ kg^{-1} of total lead, and 53,303 mg $\times$ kg^{-1} of total zinc (Orłowska et al. 2005). Because of alkaline pH value, their availability is, however, not severely phytotoxic and appears to not be a major problem in reclamation of the wastes, given that plants can be fortified with several mechanisms to counteract toxic effects of metals. For example, these metals can be excluded by the plant as a result of chelation within fungal mycelium of mycorrhizal fungi, binding to root cell walls, and precipitation within the rhizosphere because of either root exudates or decreased net uptake (Lambers et al. 2008). Heavy metals that are taken up by plants can be detoxified by (1) complex organic molecules such as metallothioneins and phytochelatins that are synthesized by the organisms (Cobbett and Goldsbrough 2002; Hall 2002) and (2) heavy metal transporters, which are a broad group of different proteins such as CPx-ATPases for Cu or Cd, ABC-transporters for Cd transport into the vacuole, as well as ZIP transporters (ZRT- and IRT-related proteins for Fe or Zn), and Nramp transporters (Hall 2002). These mechanisms can attenuate toxic effects originating from (1) the generation of reactive oxygen species (ROS), e.g., by the Fenton reaction (e.g.,),

(2) the damaging of cellular components and interference with metabolic processes, and (3) the binding of heavy metals to SH groups of enzymes and inactivation of their catalytic domains as reviewed by Hall (2002).

Most plants occurring on Zn–Pb tailings (Appendix) are commonly found in other habitats and are classified as pseudometallophytes. The only plant that is characteristic for metal-rich soils is a calamine ecotype of *Silene vulgaris*, sometimes named *S. vulgaris* var. *humilis*. This ecotype is adapted to harsh conditions on the tailing and when compared to ecotypes growing on nonpolluted soil, it has smaller but thicker leaves, more abundant root hairs, a faster development to generative phase, 20% more flowers, and numerous seeds that are smaller. Such plants are more tolerant of drought and heavy metals such as Zn and Pb (Wierzbicka and Panufnik 1998).

16.3 Restoration Techniques Used in Poland Versus Spontaneous Succession

Phytostabilization is the only feasible strategy to reduce the toxic potential of industrial wastes (Pierzynski et al. 2000) because it minimizes wind and water erosion as well as the uptake of heavy metals by plants.

The tailing has the form of a steep hill. The oldest wastes are localized at the base while the younger layers are deposited progressively above. The restoration of the tailing includes covering it with a ca. 20-cm layer of humus. The use of soil without heavy metals is usually avoided as it is costly when the area is large. The thickness of the material varies from 20 cm to 2 m in places where the material slides down into local depressions as a result of water erosion. There are two important reasons for using this humus layer procedure. (1) Such a layer allows for easier establishment of the vegetation by arresting the factors which limit plant growth. (2) In this way, the transfer of heavy metals into the plant material will be diminished. In toto, almost 100 plant species were noted on the Trzebionka tailings (Appendix). Plants that are sown under restoration processes belong to selected cultivars of grasses such as *Festuca rubra*, *Dactylis glomerata*, *Poa pratensis*, *Lolium perenne*, and *Phleum pratense*, and they are enriched by *Medicago sativa*, *Trifolium pratense*, and *Melilotus albus*. Although populations of *L. perenne* covered up to 41% of the area after sowing the seeds for 2 years on old tailing, it was almost totally wiped out a few seasons later. Ten-year-old substrate was covered by *L. perenne* with ca. 0.05%, whereas others, such as *P. pratense* or *F. rubra*, formed a ca. 15% cover. All of them were almost totally out competed at older stages.

In 10–20 year old tailings, the substrate becomes re-exposed in large proportions because of the disappearance of the cover material and occasionally because of the accidental spill of flotation products from the destroyed tubes that lead the flotation pulp/liquid into the sedimentation tank. In such places, specimens of Caryophyllaceae (*Cerastium* spp., *Silene vulgaris*) and Brassicaceae (*Cardaminopsis arenosa*) appear. These are nonmycorrhizal plants, and because their biomass is small, they add little to the revegetation process.

Shrubs and trees are also introduced into the tailing under the revegetation program. Among them are native species such as *Betula pendula*, which is the most common; *Acer pseudoplatanus* and *Sorbus aucuparia* that are relatively common; *Hippophaë rhamnoides*, *Acer platanoides*, *Populus alba*, *P. tremula*, *Pinus sylvestris*, and *Tilia cordata* which are rare or appear locally; and *Alnus glutinosa* and *Larix decidua* which survive only sporadically. Exotic species introduced there include *Robinia pseudoaccacia*, the most common; *Caragana arborescens*, *Symphoricarpos albus*, relatively numerous; and *Deutzia* × *lemoinei*, *Eleagnus angustifolius*, *Philadelphus coronarius*, *Populus* × *canadensis*, *Physocarpus opulifolius*, *Sorbus aria* and *Spiraea douglasii*, *Cornus alba*, and *Padus serotina* which all become rare with time. Trees and shrubs are mostly located at the bottom of the tailing and their presence has a healthy influence on soil formation, shading, and the development of herbs.

The appearance of those plants that were not introduced by human beings but spread by natural succession is a slow process and starts from the oldest part at the base of the tailing. They can be assigned to taxa of dry habitats, sites which are poor in nutrients, meadows, and ruderal sites. The most abundant are thermophilic species recorded mostly among the class *Festuco-Brometea*. Among these, the most common were *Bromus inermis*, *Artemisia campestris*, and *Scabiosa ochroleuca*. The following are plants of dry, poor, sandy grasslands of the *Koelerio glaucae–Corynephoretea canescentis* class including *Festuca tenuifolia*, *Cardaminopsis arenosa*, *Rumex acetosella*, *Thymus serpyllum*, and *Corynephorus canescens* which form vegetation cover on 18% of the 20-year-old substrate. The following group of plants appears at the tailing with the humus/soil layer. The seeds originating from this layer belong to grassland and ruderal communities of *Molinio-Arrhenatheretea Stellarietea mediae* and *Artemisietea vulgaris*. Besides seeds, other propagules appear such as stolons of *Molinia caerulea*. This plant seems to be the most effective in establishing itself on the waste and the formation of dense root mats that decrease water erosion.

Plants characteristic for dry places form a dense cover on the older wastes and slowly colonize younger regions where the humus layer is already lost. They then meet plants that had arrived as seeds or other propagules, brought to the tailing with the humus. Particularly interesting are plants that are typical for moist areas (*Molinia caerulea*, *Carex flacca*, *Sanguisorba officinalis*, *Valeriana officinalis*, and *Phragmites australis*). They are often resistant to heavy metals due to increased content of Si in their tissues (Kabata-Pendias 2001), easily survive periodic changes of water level, and most of them develop mycorrhiza during dry periods. Despite the fact that water is a limiting factor at the tailing, they develop abundantly. However, most of these did not originate from seeds but from stolons. *Phragmites australis* and *Molinia caerulea* are the best example of these plants. Moreover, both are potentially mycorrhizal, and the stolons, mycorrhizal fungi and associating bacteria, are transferred to the industrial wastes within the soil used to cover the tailing slopes. Both plants develop seeds on the tailings, but germination and survival of seedlings are extremely low, thus showing the limits of the colonization area. In general, a limited number of species are able to develop from seeds on the bare tailings. This was highly visible during several unsuccessful experiments which centered on revegetation of the tailing with various plant species (Ryszka and Turnau 2007). As pointed out by Ahsan et al. (2007), seed germination is the most sensitive stage in heavy metal presence as a result of poorly developed defense mechanisms in the newly developed shoots and roots. Moreover, those that were able to germinate were easily killed by the lack of water during warm and dry periods.

16.4 Role of Mycorrhizal Fungi in Revegetation of ZG Trzebionka Tailings

Until recently, industries attempting restoration usually did not recognize that plants require appropriate below-ground ecosystems, especially at acutely challenging sites. Mycorrhizal fungi have been investigated at the ZG Trzebionka waste since 1990 with the main focus on arbuscular (AMF) and orchid mycorrhizae and only pilot studies on ectomycorrhiza. Mycorrhizal fungi are known to enhance root absorption area (Nye and Tinker 1977; Gianinazzi-Pearson and Gianinazzi 1983). The fungi provide access to nutrients and water which would otherwise not be accessible for plants (Jeffries et al. 2003; Pozo and Azcon-Aguilar 2007). The establishment of the vegetation and its survival under stress conditions is more likely if the mycorrhizal fungi are present (Smith et al. 1998). This is due not only to nutrient access but also to the enhancement and stabilization of the substrate structure from the formation of the hyphal net, a common feature of all types of mycorrhiza. Mycorrhizal fungi also commonly produce substances that bind or "glue" soil particles leading to aggregate formation and the improvement of soil structure quality (Jastrow et al. 1998; Rillig and Mummey 2006; Rillig et al. 2010). Moreover, AMF extraradical mycelium can accumulate 10–20 times more Cd than the roots of the host plant (Janouskova et al. 2006). The ability to chelate Cd, Cu, and Zn is now attributed to production of glomalin, identified as likely a 60-kDa heat shock protein homolog (Gadkar et al. 2006) which is localized mainly in outer layers of the mycelium cell wall (Purin and Rillig 2008). As pointed out by Joner and Leyval (1997), the efficiency of plant protection against heavy metals depends on the given AMF isolate, and this is presumably because of the different abilities of strains to immobilize metals within the mycelium. This was clearly visible in roots of *Fragaria vesca* collected from Trzebionka tailings. *Glomus mosseae* mycelium, as identified by molecular means, was the one that strongly accumulated heavy metals as shown by rhodizoniate test (Turnau et al. 2001).

The postflotation substrate freshly deposited on the tailing is devoid of arbuscular mycorrhizal fungi. There are two possibilities for the origin of AMF. Either the first propagules arrived with the soil or humus used to cover the industrial wastes, or they can be introduced by animals such as ants. The latter was observed for *Euphorbia cyparissias*, one of the first mycorrhizal plants which appeared on the tailings. The rhizosphere of this plant is abundantly visited by ants, with *Manica rubida* the most common (unpublished data). *E. cyparissias* is followed by other facultatively mycorrhizal grasses such as *Molinia caerulea*, *Agrostis gigantea*, *Bromus inermis*, *Calamagrostis epigejos*, and *Corynephorus canescens*. Grasses collected from zinc wastes are usually very strongly mycorrhizal (Ryszka and Turnau 2007). In some cases (e.g., *Festuca tenuifolia*), mycorrhizal colonization and arbuscular richness were two to five times higher than in natural stands (outside industrial areas). Mycorrhizal colonization of grasses varies considerably, but on industrial wastes the most successful were those developing abundant mycorrhizal colonization.

Research carried out on grasses (Newsham and Watkinson 1998) has shown that they respond differentially to mycorrhiza. The increase of growth following mycorrhization was observed in 37%, no reaction was found in another 46% of the species, and a negative effect was observed in 17% of species. This negative effect is probably caused by increased transfer of carbon compounds from the plant to a fungus, or by low photosynthetic activity, or by increased activity of the mycelium. Such a situation was shown in *Lolium perenne* (Buwalda and Goh 1982).

Not much is known about the precise effect of mycorrhization on those plants species that may be promising for phytostabilization of Zn wastes. Among cultivars of grasses that were possible to study under such conditions, two cultivars of *Festuca rubra* and one cultivar *Lolium perenne* showed increased growth, and one demonstrated a negative effect (*Poa pratensis* cv Alicja) (Ryszka 2006). However, the positive response to mycorrhiza may be visible in the second year of cultivation, e.g., with better survival of new seedlings (Dodd et al. 2002). As yet, this has not been verified. Although mycorrhizal colonization of grasses on zinc wastes was usually high, there were exceptions. The criteria used to determine mycorrhizal colonization and arbuscule richness showed very low levels or none in plants which occurred in places where the fresh liquid waste was spilled from accidentally broken tubes transporting the liquid waste to the sediment tanks. This shows that the waste is more harmful to plant colonization at the initial waste deposition stage.

The diversity of AMF was studied in zinc wastes of ZG Trzebionka. Nested polymerase chain reaction (PCR) with taxon-specific primers was used to identify the species *G. moseae*, *G. intraradices*, *G. claroideum*, *G. gerdemannii*, and *Paraglomus occultum* (Turnau et al. 2001). Furthermore, morphological analysis added *G. fasciculatum* and *G. aggregatum* to the list of AMF species in zinc wastes of Trzebionka Mining Company. Several, but not all, of these fungi are now available as monocultures. The roots of *Fragaria vesca* were used to evaluate which fungus was the most effective in mycorrhizal colonization. The most frequent fungus was *G. gerdemannii* which, however, was not obtained in trap cultures. Slightly less frequent were *G. claroideum* and *G. occultum*, while the least common were *G. intraradices* and *G. mosseae*. The easiest to obtain in trap cultures was *G. claroideum*. Experiments designed to compare different fungal species revealed that fungi such as *G. intraradices* and *G. clarum* isolated from industrial wastes were lower down on metal uptake into the roots and shoots of *P. lanceolata*, as compared to strains originating from soils of low heavy metal content (Orłowska et al. 2005). Moreover, depositions within the periplasmic space between the inner layer of the wall and plasmalemma were found in spores of *G. intraradices* isolated from the Trzebionka Pb–Zn wastes. The accumulation of heavy metals within these depositions was confirmed by energy dispersion spectroscopy (EDS) analysis coupled with the scanning electron microscope (SEM). The method was also used to estimate the percentage of the mycelia that show increased metal accumulation. In the case of *Euphorbia cyparissias* (Turnau 1998), about 80% of the intraradical mycelium showed an increase in heavy metals. However, simultaneously, the number of arbuscules is slightly lower than in roots which contain lower levels of these elements. On the other hand, *Fragaria vesca* roots from the Zn–Pb wastes have shown that increased metal accumulation is visible only in *G. mosseae*, which is one of the least efficient colonizers (below 10%) (Turnau et al. 2001). These roots were from the same industrial wastes that were selected for molecular studies concerning fungal identity. These results show that there are differences not only between fungi but also between plant species in its preference for symbiotic associations.

Under laboratory conditions, we mostly inoculated the plants with one fungal species, while under field conditions the plants are actually inhabited by several different strains or species. Studying the metal uptake by plants is one of the examples that must be carried out with a single species. It is well established that the response of the plant can vary depending upon the AMF isolate and species (Adjoud et al. 1996; Streitwolf-Engel et al. 2001; van der Heijden et al. 1998). AMF were already reported to decrease or increase HM uptake by plants (Leyval et al. 1997). Although these phenomena strongly depend on the selection of appropriate plant and fungal genotypes, there is mounting evidence that they can provide a potential for decreasing the health hazards which accompany edible plant production, and they can improve sustainable agriculture and phytoremediation technologies, including phytoextraction (Jurkiewicz et al. 2004; Turnau and Mesjasz-Przybylowicz 2003).

The research on heavy metal uptake by plants growing on zinc industrial wastes collected from Trzebionka was carried out using 15 cultivars of *Zea mays* (Jurkiewicz et al. 2004) and *P. lanceolata*. In the first case, the aims were to check whether (1) mycorrhizal colonization influences heavy metal uptake by maize, (2) there is a variability in uptake between plant cultivars, and (3) mycorrhiza development has any effect when EDTA – (a chelating agent known to mobilize heavy metals in soil) – is used. In this experiment, *Glomus intraradices* was chosen as a fungal symbiont. Analysis of mycorrhizal parameters indicated differences among the varieties, but between different treatments of the same variety, results were generally statistically insignificant. Although the EDTA treatment strongly decreased the activity of the fungal alkaline phosphatase (indicator of fungal viability), the treatment did not totally eliminate the arbuscular mycorrhizal fungi (AMF) from the soil. Modification of the appearance of AMF structures within roots of plants cultivated in EDTA-treated soil was observed. Among the heavy metals studied, the highest impact of the EDTA treatment on heavy metal uptake into shoots was found with Pb. In most case, the EDTA treatment significantly increased the Pb level in shoots of mycorrhizal plants, when compared with mycorrhizal, EDTA non-treated samples. Among samples treated with EDTA, mycorrhizal plants of six cultivars showed a higher Pb content in shoots than did the nonmycorrhizal ones. Significant differences in heavy metal content in plant material were demonstrated between the varieties tested.

Plant species can vary in their response to different AMF strains not only by metal uptake but also by survival rates. *Festuca rubra* survival was similar in association with different AMF species. *Plantago lanceolata* showed the highest survival on waste substrate when inoculated with *G. intraradices* and *G. geosporum*. The highest survival rate was observed in the presence of *G. etunicatum* and *G. claroideum*. Research with *Plantago lanceolata,* a plant highly responsive to mycorrhiza, examined the potential differences in Cd, Zn, and Pb uptake by plants inoculated with different AMF strains (Orłowska et al. 2005). The plants were grown in rhizo-boxes on substrate collected from zinc wastes similar to those from ZG Trzebionka. Heavy metal uptake clearly depended on the fungal strain used. Plants inoculated with AMF strains originating from soils not affected by heavy metals had higher metal concentrations in tissues than plants inoculated with strains from heavy-metal-polluted areas. However, few differences were found in the extent of mycorrhizal colonization and arbuscular richness between strains. This finding is significant because plants growing on zinc wastes often contain high metal content which can create the risk of metal transfer into the food web. Indeed, the industrial wastes are often visited by a wide range of wild animals. The introduction of the appropriate fungi can decrease metal uptake into the shoots while the available pool of metals could be stored within the root system. The effectiveness of introducing the selected fungus into the nonpolluted or moderately polluted soils could be low because of the presence of native fungi that are better adapted to the particular soil characteristics. However, zinc wastes at the start of the succession are often devoid of AMF and inoculation, therefore, is reasonable. Nevertheless, the practical application of the AMF awaits further experiments and studies.

16.5 Use of Legumes for Soil Bioremediation

Interactions between mycorrhizal fungi and rhizospheric microorganisms likely influence the soil bioremediation (Turnau et al. 2006). The legume (Fabaceae)–nitrogen-fixing bacteria symbiosis can play a key role. Legumes can be advantageous for phytoremediation of Trzebionka tailings because the substrate is poor in nutrients, especially nitrogen and phosphorus. Legumes are often the first colonizers of poor and degraded soils (Cardinale et al. 2010). The use of human-introduced fertilizers can have negative effects on plant diversity at the tailings, and such fertilizers would tend to leach out of the poorly absorbing substrate. Recently there has been an increased interest on this topic (Carrasco et al. 2005, Dashti et al. 2009, Khan et al. 2009, Lafuente et al 2010). Although legumes are thought to be tolerant to zinc, the Trzebionka tailings contain several other metals that can affect these plants. For example, As, Al, and Cu are known to reduce formation of root hairs and as a consequence reduce the formation of rhizobial infection (Kopittke et al. 2007; Pajuelo et al. 2008; Lafuente et al. 2010). Molecular studies showed that the expression of several nodulation genes can be blocked by any toxic metal or metalloid (Lafuente et al. 2010). Mycorrhizal fungal activity also influences both qualitatively and quantitatively the degree of bacterial activity and the overall functioning of the symbiosis (Barea et al. 2005). In the ZG Trzebionka tailings, legumes are represented by *Robinia pseudoaccacia*, which is an introduced plant that has established itself relatively well, *Vicia cracca*, *V. hirsuta*, *V tetrasperma*, *Lotus corniculatus*, *Medicago lupulina*, *Trifolium repens*, and *T. pratense*.

There were several attempts to introduce *Medicago sativa*, but this plant was not able to survive there. Therefore, an experiment under laboratory conditions on zinc waste and nonpolluted substrates was carried out to discover if the growth of alfalfa can be improved by the introduction of mycorrhizal fungi. Plants were either noninoculated or inoculated with AM fungi originating from industrial wastes or from area with low metal concentration. Although the extent of mycorrhizal colonization and arbuscule richness was slightly lower in plants cultivated on heavy-metal-rich substrates, the influence of mycorrhiza on plant development was more visible there. Plants grown on heavy-metal-rich substrates had much higher content of elements such as Pb, As, and Zn than those cultivated on control substrates (Tsimilli-Michael et al. 2008). Most heavy metals were found in higher concentrations in roots than in shoots. In shoots, a relatively high content of Pb, As, and Zn was observed. Fe was high both in roots and shoots. Plants inoculated with *G. clarum* originating from the tailing and cultivated on industrial wastes had the highest concentrations of potentially toxic heavy metals. These plants had also a high ratio (above 6) of Fe to Mn in shoots, while in roots all mycorrhizal and nonmycorrhizal plants cultivated on metal-rich substrates had a ratio in the range of 6.2–7.8. If this value is lower than 1.5, deficiency of Fe and toxicity of Mn is observed, and if it is above 2.5, the situation is the reverse, i.e., Fe is toxic and Mn is deficient (Kabata-Pendias and Pendias 2001). In this experiment, Fe/Mn was high in roots of M and NM plants which were cultivated on polluted soil. In shoots, the Fe/Mn ratio was high, particularly in the case of *G. clarum* samples from metal-rich substrates. The damaging effect represented by this ratio might be because of the destruction of indoleacetic acid auxin (IAA) by increased activity of IA oxidase, decreased activities of catalase, ascorbic acid oxidase, glutathione oxidase, and cytochrome C oxidase, lower ATP content and respiration rates (Foy et al. 1978). In addition to the high uptake of several heavy metals, this might be one of the explanations as to why *Medicago sativa* is not successfully introduced into zinc wastes. There, the fungal propagules are less abundant and the moisture conditions are far more inadequate than under experimental conditions. Under laboratory conditions, the plants survived, losing not more than 45% of their overall potential for photosynthesis (Tsimilli-Michael et al. 2008). This loss could be too high for *M. sativa* to survive on industrial wastes where the conditions are more extreme. This area needs further work, and careful selection of the appropriate symbiosis partners would be essential.

16.6 Use of Xerothermic Plants for Restoration

Among plant species that are relatively efficient in natural succession of ZG Trzebionka tailings, a wide group is characteristic for dry calcareous grasslands (Turnau 1998; Ryszka 2006). Both dry calcareous grasslands and industrial waste require that the plants are able to grow under water stress. Most plants inhabiting dry grasslands are strongly mycorrhiza dependent (van der Heijden 2002). Pilot studies that were carried out on the survival of species such as *Inula ensifolia* and *Aster amellus* under laboratory conditions showed that while plants grown on heavy-metal-poor soil do survive without mycorrhiza, they are not able to survive on metal-rich soils (unpublished data). Seeds germinated on polluted soil with a controlled water regime were able to survive, especially if arbuscular mycorrhizal fungi were supplied. Unfortunately, this was unsuccessful on the tailing due to sensitivity of the plants to drought periods. The first establishment of plants is a matter of chance, and introduction of the seeds or seedlings must be strictly followed by a relatively wet period. This was successful in 2003, when small seedlings of *Anemone sylvestris* and *Primula verris* fortified with AMF from xerothermic grasslands were introduced into Zn–Pb wastes. Both survived the following seasons, but they did not expand their population, albeit seed production was noted. This experiment drew our attention to other plants that proliferate in xerothermic grasslands. In such habitats clonal plants are the dominant growth form (Marshall and Price 1997). For example, *Hieracium pilosella* is a very common plant on the industrial wastes, flowering abundantly and forming numerous seeds. But its main way of spreading is clonal growth, and its rate of spreading depends on the number and distribution of ramets (genetically identical offspring) formed during growth (Harper 1977). As illustrated for other plants with this strategy, variation in clonal reproduction depends on abiotic resource heterogeneity and plant genotype (Kays and Harper 1974). The most common way of reproduction is by the formation of stolons or rhizomes, on which new ramets are formed (Jónsdóttir and Watson 1997). As shown by Streitwolf-Engel et al. (2001), different species of arbuscular mycorrhizal AMF fungi differentially alter clonal growth of *Prunella vulgaris,* a plant also originating from calcareous grasslands. The number of ramets produced by *P. vulgaris* and the effect of different AMF were more important than the effect of plant genotype on variation in size and ramet production. This subject is very promising when designing revegetation strategies for waste tailings. Certainly, *H. pilosella* would be a good candidate as a model plant, although its practical use would be limited due to its small biomass and due to the fact that it forms the small rosettes on bare ground only periodically under relatively good moisture conditions, and then disappears during dry periods. The new rosettes are formed from new buds developing from underground rhizomes. Survival is easier and longer if the ramets are formed in places where other plants such as *Festuca ovina* are present, because their abundant root systems hold moisture more efficiently. Their presence can be useful even in cases where the grass is already dead, for organic matter can efficiently hold water and can be the source of AMF propagules.

Despite the recent data outline, little is known about how individual plant species survive on wastes. Therefore, we engaged in research designed to introduce a wide group of plants from the calcareous grassland (Turnau et al. 2008). Plants were selected on the basis of floristic studies, and the seeds originated from dry calcareous grasslands or from private gardens where such plants were cultivated to obtain seeds. The seeds were germinated under laboratory conditions and introduced into soils with industrial wastes to adapt the seedlings to the presence of heavy metals. Plants of a few weeks were introduced into the tailing. The survival of most plant species in the first year was 100%. These included mycorrhizal plants such as *Melica transsilvanica*, *Bromus inermis*, *Agrostis capillaris*, *Agropyron intermedium*, *Brachypodium pinnatum*, *Cirsium pannonicum*, *Ononis arvensis*, and *Verbascum thapsus*. Nonmycorrhizal were *Echium vulgare* and *Carex* spp., although they were colonized by dark septated fungi (DSE). The comparison of plants growing on the tailing and those in xerothermic grasslands showed that in fact the overall photosynthetic potential (PI_{total}) of many of these plants was similar (Turnau et al. 2008). Despite this, only a few plant species actually formed seeds. Among the seed formers were nonmycorrhizal *Carex* sp. and mycorrhizal *Melica transsilvanica*, *Phleum boehmeri*, *Plantago media*, and *Verbascum thapsus*. Interestingly, only *V. thapsus* and *P. boehmeri* were able to spread due to production of seeds. *V. thapsus* is known as a biennial species producing seeds tolerant to various stresses. The seeds can survive up to 100 years in the soil. These plants had been growing and flowering during next few years. This experiment showed that there are many plants originating from grasslands that can survive under harsh conditions on the tailings. But a confounding fact is that the plants were planted in the location of the highest toxicity as a result of recent spills of the flotation liquid. In this location, the concentration of Cu, As, Fe, Cr, Ti, and Pb of the substratum was ca. two to four times higher than a few meters above where a similar substratum was present but outside the area of accidental spill of the flotation liquid. In the same area of the accidental spill, *Molinia caerulea* was introduced. Despite the fact that this plant was able to form dense tufts in 20-year-old substrate, here on the spill area it was not spreading at all. This shows clearly that the substrate must still be enhanced in some way. Further trials must be done to increase organic matter content of this substrate and not necessarily covering it with the soil layer. Expected increase of drought stress following termination of water release into the tailing pond will certainly decrease the ability of plant cultivars such as *Lolium perenne* to survive on the humus layer. Then the use of more tolerant plants will be necessary.

16.7 *Molinia caerulea* as a Colonizer of the Bare Substratum

Among various plants occurring on zinc wastes, one species was particularly noteworthy. *Molinia caerulea* was found to be an effective colonizer (Fig. 16.2). This perennial grass shows many advantages, which may be significant from a reclamation point of view. *M. caerulea* forms tussocks where accumulation of organic matter and humus layer formation can be observed. A dense cover of *Molinia* ensures phytostabilization of wastes because of its very robust root system, consisting of both thick and fibrous roots. These features also allow penetration into deeper substrate levels. *M. caerulea* is known to harbor a vast range of microorganisms, including free living nitrogen-fixing bacteria (diazotrophs) which are very important in nutrient-poor habitats (Hamelin et al. 2002), such as zinc wastes. Bacteria-driven nitrogen supply improves plant growth; *M. caerulea* has relatively high biomass among nonwoody plants colonizing the investigated wastes. Nutrients collected during the vegetative season are stored during winter in

Fig. 16.2 *Molinia caerulea* in Zn–Pb tailings of ZG Trzebionka Corporation: (**a**) dense patch at the border of erosion-affected portion of the wastes; (**b**) whole plant; (**c**) plantlets 1 year after introduction directly into barren substrate

swollen intercalary basal nodes. These bulb-shaped structures can release nutrients more than once in spring, allowing regrowth after drought periods. Once established, *Molinia* persists in increasing biomass and tussock diameter, which leads to changes in local soil conditions as erosive processes become less threatening and increased organic matter accumulation improves microbial activity. Furthermore, *M. caerulea* is among the plants which take up relatively few heavy metals as compared to other plants growing in the tailings. TXRF analysis of *M. caerulea* (unpublished data) show higher concentrations of Pb, As, and Zn, lower concentrations of Sr, Rb, Fe, Mn, and Ti, and no statistical differences in Cu and Ni concentrations in shoots in comparison with samples from heavy-metal-poor soil. Low concentration of heavy metals in grasses is probably the result of relatively small amount of polysaccharides that can bind Ca and heavy metals within intercellular spaces (Broadley et al. 2003). In general the concentration of Ca in grasses from polluted and unpolluted sites was relatively low (ca. 0.5%) in comparison with other plants inhabiting the tailings.

Seeds of *M. caerulea* can be spread a short distance by stalks falling in autumn and winter which form radial nets where organic matter (i.e., leaves) can be captured. The seeds, however, rarely develop into seedlings on the tailing due to drought or the presence of a pathogen, *Claviceps purpurea* (ascomycetous fungus), which fosters ergots instead of seeds. Although *C. purpurea* is known as a parasite of grasses (Shaw and Mantle 1980), the leaves of plants show numerous ergots characterized by more intensive green color, which suggests a positive role of the fungus on the vitality of the plant. These observations were supported by measurements of photosystem II performance with Handy PEA. Simultaneously, plants infected with *C. purpurea* showed lower values of mycorrhizal colonization, which could be due to secondary metabolites produced by *C. purpurea*.

Low seed germination and survival of seedlings were the bases for using plantlets obtained by tussock splitting in the experiment conducted in 2004 (Ryszka and Turnau 2007). It showed very high survival rates of plants that were introduced directly into bare zinc waste substrates. No treatments such as fertilization, watering, and introduction of mycorrhizal inocula were required, and plants showed great resistance against erosion.

Various organisms may have a selective advantage from the presence of *Molinia*. For example, ants seem to be well integrated. These insects build corridors and nests within the root system. As they rearrange the ecosystem in microscale, they can bring seeds of other plants (i.e., *Viola* spp.) into the nests, which can then result in increased plant species diversity. Ants are also able to spread spores of mycorrhizal fungi or fragments of root system, colonized by mycorrhizal fungi (Harinikumar and Bagyaraj 1994; Dauber et al 2008). All *Molinia* root samples collected on ZG Trzebionka zinc wastes showed the presence of arbuscular mycorrhizal fungi (AMF). AMF together with other symbionts might significantly contribute to improved nutrient supply in that extreme habitat. The data presented suggest that *Molinia* should be regarded as a "hot-spot" for diversity on zinc wastes. These results open new avenues for studying plant–microbial interactions on zinc wastes. Practical applications of native plants should take into account the multidimensional interrelationships between the wide array of organisms.

16.8 Metal Content of Plants

Plants that are used for phytostabilization of the tailings, which are rich in potentially toxic metals, should be chosen over those that are heavy metal excluders so as to avoid introduction of these elements into the food chain. To evaluate the phytostabilization potential of plants introduced into the tailing of ZG Trzebionka or being established during natural succession, total reflection X-ray fluorescence (TXRF) was employed (Turnau et al. 2010). Among 24 investigated plants, four grass species (*Melica transsilvanica*, *Bromus inermis*, and *Elymus hispidus*) and one legume (*Anthylis vulneraria*) were the most suitable for phytostabilization with relatively low concentrations of potentially toxic elements in the shoots. *Verbascum thapsus* accumulated high concentrations of As, Cr, Ni, Fe, Zn, and Y. In general, plants that had high concentrations of heavy metals usually were rich in Ca, suggesting a role of this element in detoxification. The use of TXRF allowed for comparisons of K concentration. The comparison with grassland specimens suggests that plant shoots from the tailings were more potassium-deficient than those from grasslands. This indicates the need for K-supplementation of the waste substrates. Through the TXRF analysis, discrepancies in heavy metal concentrations between green and brown leaves of plant species grown on industrial waste were noted. The analysis was performed only in two cases, involving *Echium vulgare* and *Verbascum thapsus*, when both types of leaves were present at the same time. *E. vulgare* had 2× higher concentration of Ti, 4× Fe and Mn, 5× Zn, 13× Pb, and much Cr with brown leaves compared with green ones. Similar data were found for *V. thapsus*, but, in addition, there were also several times higher concentrations of As and Cu. Similar phenomenon was already described for *Biscutella laevigata* growing on Zn–Pb wastes in Olkusz, Poland (Pielichowska and Wierzbicka 2004), and other plant species and habitats (Ernst et al. 1992). Such expression is likely a plant detoxification mechanism. Rascio (1977) first suggested this concerning *Biscutella laevigata* and *Anthyllis vulneraria*, both observed on zinc-mine deposits. He noted that both plants accumulate the highest amount of Zn in roots and less in leaves in spring. The metal is then translocated into the shoots in autumn, thus enabling elimination of toxic compounds which were taken up by the roots. The ability to accumulate heavy metals in plant roots is a known phenomenon; however, very few studies compared concentrations in brown and green leaves.

Populus spp. growing on metal-rich locations is known to accumulate up to 5,000 mg kg^{-1} of Zn (Strzyszcz 2003). Leaves of poplar growing on Trzebionka tailings had concentrations from 1,900 to 2,100 mg kg^{-1}. TXRF showed slight but significant differences between green and brown leaves with respect to Zn, although this is rather the effect of metal accumulation with time and not the mechanism as described above for *Biscutella laevigata* or *Verbascum thapsus*. The concentration of Zn was over four times higher in leaves from the tailings than from metal-poor soil, but the concentration is still in the range found in plants from sites outside the influence of industry. Interestingly, among the other heavy metals, only concentrations of Fe were two times higher in this grass than in plants from grasslands. Simultaneously, green and brown leaves differed strongly by three times lower K concentration, which is the result of K leakage during senescence (Turnau et al. 2010).

16.9 Orchid and Ectomycorrhiza

As we have shown, arbuscular mycorrhiza is the dominant symbiosis type at ZG Trzebionka tailings. But at least two other mycorrhizae, the orchid mycorrhiza and ectomycorrhiza, were also found. Recently, abundant populations of *Epipactis atropurpureum*, *E. helleborine*, and *Dactylorhiza majalis* have been observed on several industrial wastes in southern Poland, and all of them were strongly mycorrhizal. The mycelium colonizes root cortical cells, producing dense intracellular peletons. These peletons were shown to accumulate heavy metals such as Zn and Pb at high concentrations (Jurkiewicz et al. 2001). Orchids colonizing tailings rely on fungal mycelium to obtain necessary nutrients, particularly because their root system is not well developed. The orchids belong to metal excluders, and the fungus seems to play an important role in filtering out the heavy metals. The subject has low value for the restoration of the tailing, but because many orchids are rare plants, their presence in metal-rich habitats is of ecological conservation interest. Although orchids of Trzebionka Mining Works are carrying out photosynthesis, it is suspected that their fungal symbionts are shared with trees which associate with ectomycorrhiza. The orchids are usually growing close to trees, which also give shade to orchids.

On the Trzebionka substrate, the growth of such trees is slow and in many cases the herb layer underneath is poorly developed. Thus, the role of the trees in attenuation of wind erosion is not equivalent to grass cover. Zinc wastes of southern Poland were investigated concerning occurrence of macromycetes (Mleczko 2004), although no studies concerning this group of fungi exists from Trzebionka substrates. Very few fruiting bodies were found in this area thus far. Investigations carried out since 1990 show mostly saprobic fungi such as *Bovista* sp. and *Cyathus striatus*. This is contrary to observations carried out on other Zn wastes, e.g., in Bolesław where almost 70 species were recorded. Depending on the site, 40–70% species are known as ectomycorrhizal (Mleczko 2004). The most common fungi in this area belong to *Chroogomphus rutilus*, *Suillus luteus*, and *Rhizopogon roseolus* and form symbioses exclusively with pine. Pine is a common tree in this area, although it grows slowly. Pines of Trzebionka tailings are several times younger than those from Bolesław, which can influence the range of fungi that are forming symbiosis. A great deal of research needs to continue in this area, including more longitudinal, comprehensive studies.

16.10 Conclusion

Our data are based on a single industrial tailing. Conventional revegetation applied by site owner involves covering the tailings with a thin layer of humus and sowing seeds. So far there have been several experiments revealing weak points of the technology. Among these is the use of annual plants which has been shown not to be the best solution. Promising perennial plants are already known, but they need propagation to obtain suitable amounts of material. The metal content of the waste is not necessarily the most important problem in revegetation. Rather poor water-holding capacity, water/wind erosion, and mineral deficiency must also be evaluated and managed first.

The results clearly show the potential application of mycorrhizae in phytoremediation technologies; however, the direct transfer of the technology into a variety of wastes needs further work. Both plants and their symbionts must be selected individually for each tailing based on substrate features such as pH, water-holding capacity, and heavy metal bioavailability. Differences can also rely on climatic conditions. Although plants equipped with microbial symbionts may be able to grow directly on the tailings, their survival is much more likely if the tailing can be covered by the soil/substratum layer of better quality than by the tailing itself. Certainly, the development of rhizosphere consortia including AM fungi and bacteria is an important prerequisite for the successful site restoration.

Water-holding capacity can be improved, for example, by the use of hydrogels applied simultaneously with the inoculum. This saves the inoculum from removal by wind. The application of inorganic fertilizers should be replaced by compost or manure. The use of sewage sludges might be a good choice, but a corresponding increase of toxicity (e.g., Cu) may appear. Moreover, optimization of the fertilizer dosage must be performed with respect to the development of the subterranean microbial consortia. Restoration can also be enhanced by additional inoculations of appropriate symbiotic and saprophytic rhizosphere microorganisms isolated from HM soils.

Acknowledgment This chapter was written within the framework of the Project UMBRELLA (EU FP7 no 226870).

Appendix

Vascular plant species recorded on ZG Trzebionka zinc wastes. Phytosiociological units (according to Szafer 1966), followed by origin on wastes, are given in parentheses: Sm/Av – *Stellarietea mediae*/*Arrtemisietea vulgaris*; QF – *Querco-Fagetea*; MA – *Molinio-Arrhenatheretea*; FB – *Festuco-Brometea*; KgCc – *Koelerio glauce-Corynephoretea canescentis*; Vc – *Violetea calaminariae*; C – spontaneous colonization; S – sown within restoration practices.

Amaranthaceae: *Amaranthus retroflexus* (Sm/Av; C); Apiaceae: *Aegopodium podagraria* (QF; C), *Daucus carota* (FB; C), *Heracleum sphondylium* (MA; C), *Pastinaca sativa* (MA; C), *Peucedanum oreoselinum* (FB; C), *Pimpinella saxifraga* (MA; C); Artemisiaceae: *Artemisia campestris* (FB; C), *A. vulgaris* (Sm/Av; C); Asteraceae: *Anthemis arvensis* (Sm/Av; C), *Aster novi-belgii* (Sm/Av; C), *Bidens frondosa* (Sm/Av; C), *Cardus crispus* (Sm/Av; C), *Centaurea cyanus* (Sm/Av; C), *C. jacea* (MA; C), *Chamomilla recutita* (Sm/Av; C), *Cirsium arvense* (Sm/Av; C), *C. vulgare* (Sm/Av; C), *Conyza canadensis* (Sm/Av; C), *Crepis biennis* (MA; C), *Erigeron acris* (KgCc; C), *E. annuus* (Sm/Av; C), *Eupatorium cannabinum* (MA; C), *Gnaphalium uliginosum* (Sm/Av; C), *Helianthus tuberosus* (Sm/Av; C), *Hieracium bauhinii* (FB; C), *Hieracium pilosella* (KgCc; C), *H. piloselloides* (KgCc; C), *Hypocheris radicata* (Sm/Av; C), *Lactuca serriola* (Sm/Av; C), *Leontodon autumnalis* (MA; C), *L. hispidus* (MA; C), *Leucanthemum vulgare* (MA; C), *Matricaria maritima* subsp. *Inodora* (Sm/Av; C), *Picris hieracioides* (FB; C), *Rudbeckia bicolor* (Sm/Av; C), *R. laciniata* (Sm/Av; C), *Senecio jacobaea* (Sm/Av; C), *S. vernalis* (Sm/Av; C), *Solidago canadensis* (Sm/Av; C), *S. gigantea* (Sm/Av; C), *Sonchus arvensis* (Sm/Av; C), *S. asper* (Sm/Av; C), *Tanacetum vulgare* (Sm/Av; C), *Taraxacum officinale* (MA; C), *Tragopogon pratensis* (MA; C), *Tussilago farfara* (Sm/Av; C); Boraginaceae: *Echium vulgare* (FB; C), *Myosotis arvensis* (Sm/Av; C), *Myosoton aquaticum* (QF; C), *Symphytum officinale* (MA; C), *Veronica officinalis* (QF; C); Brassicaceae: *Armoracia rusticana* (Sm/Av; C), *Barbarea vulgaris* (Sm/Av; C), *Berteroa incana* (FB; C), *Cardaminopsis arenosa* (KgCc; C), *Lepidium campestre* (Sm/Av; C), *Rorippa sylvestris* (Sm/Av; C), *Sinapis arvensis* (Sm/Av; C); Campanulaceae: *Campanulla trachelium* (QF; C); Caryophyllaceae: *Arenaria serpyllifolia* (KgCc; C), *Cerastium holosteoides* (Sm/Av; C), *Dianthus carthusianorum* (FB; C), *Melandrium album* (Sm/Av; C), *Sclerantus perennis* (Sm/Av; C), *Silene vulgaris* var. *humillis* (Vc; C), *Viscaria vulgaris* (Sm/Av; C); Chenopodiaceae: *Chenopodium album* (Sm/Av; C), *C. polyspermum* (Sm/Av; C); Cistaceae: *Helianthemum nummularium* (FB; C); Convolvulaceae: *Convolvulus arvensis* (Sm/Av; C), *Calystegia sepium* (QF; C); Crassulaceae: *Sedum acre* (KgCc; C), *S. maximum* (FB; C); Cyperaceae: *Carex caryophyllea* (KgCc; C), *C. echinata* (FB; C), *C. flacca* (MA; C), *C. hirta* (MA; C); Dipsacaceae: *Dipsacus sylvestris* (Sm/Av; C), *Knaucia arvensis* (MA; C), *Scabiosa ochroleuca* (FB; C); Equiseataceae: *Equisetum variegatum* (MA; C), *E. arvense* (Sm/Av; C), *E. pratense* (Sm/Av; C); Euphorbiaceae: *Euphorbia cyparyssias* (FB; C), *E. esula* (Sm/Av; C); Fabaceae: *Anthyllis vulneraria* (FB; C), *Coronilla varia* (FB; C), *Lathyrus pratensis* (MA; C), *Lotus corniculatus* (MA; C), *Medicago falcata* (FB; C), *M. lupulina* (MA; C), *M. sativa* (Sm/Av; S), *Melilotus alba* (Sm/Av; S), *M. officinalis* (Sm/Av; C), *Ononis spinosa* (FB; C), *Trifolium arvense* (FB; C), *T. pratense* (MA; S), *T. repens* (MA; S), *Vicia cracca* (MA; C), *V. grandiflora* (Sm/Av; C), *V. hirsuta* (FB; C), *V. sativa* (Sm/Av; S), *V. tetrasperma* (FB; C); Geraniaceae: *Geranium pratense* (MA; C); Hypericaceae: *Hypericum perforatum* (FB; C); Lamiaceae: *Ajuga reptans* (QF; C), *Galeopsis speciosa* (Sm/Av; C), *Glechoma hederacea* (Sm/Av; C), *Mentha arvensis* (Sm/Av; C), *Origanum vulgare* (FB; C), *Stachys palustris* (Sm/Av; C), *Thymus pulegioides* (FB; C), *T. serpyllum* (KgCc; C); Linaceae: *Linum catharticum* (FB; C); Lythraceae: *Lythrum salicaria* (MA; C); Onagraceae: *Chamaenerion angustifolium* (Sm/Av; C), *Oenothera biennis* (Sm/Av; C); Orchidaceae: *Epipactis atrorubens* (FB; C), *E. helleborine* (QF; C); Oxalidaceae: *Oxalis dillenii* (Sm/Av; C); Papaveraceae: *Papaver argemone* (Sm/Av; C), *Papaver rhoeas* (Sm/Av; C); Plantaginaceae: *Plantago lanceolata* (MA; C), *P. major* (Sm/Av; C); Plumbaginaceae: *Armeria maritima* subsp. *elongata* (KgCc; C); Poaceae: *Agrostis capillaris* (FB; C), *A. gigantea* (MA; S), *Apera spica-venti* (Sm/Av; C), *Arrhenatherum elatius* (MA; C), *Avena sativa* (Sm/Av; C), *Avenula pubescens* (FB; C), *Bromus hordeaceus* (Sm/Av; C), *B. inermis* (FB; C), *Calamagrostis epigejos* (FB; C), *Corynephorus canescens* (KgCc; C), *Dactylis glomerata* (MA; S), *Deschampsia caespitosa* (MA; C), *Echinochloa crus-galli* (Sm/Av; C), *Elymus caninus* (Sm/Av; C), *E. repens* (Sm/Av; C), *Festuca arundinacea* (MA; S), *F. pratensis* (MA; S), *F. rubra* (MA; S), *F. tenuifolia* (KgCc; C), *F. trachyphylla* (FB; C), *Holcus lanatus* (MA; C), *Koeleria glauca* (KgCc; C), *Lolium multiflorum* (MA; S), *L. perenne* (MA; S), *Molinia caerulea* (MA; C), *Phalaris arundinacea* (MA; C), *Phleum pratense* (MA; S), *Phragmites australis* (MA; C), *Poa angustifolia* (FB; C), *P. compressa* (FB; C), *P. pratensis* (MA; S), *Setaria viridis* (Sm/Av; C); Polygonaceae: *Fallopia convolvulus* (Sm/Av; C), *Polygonum aviculare* (Sm/Av; C), *P. lapathifolium* subsp. *lapathifolium* (Sm/Av; C), *P. persicaria* (Sm/Av; C), *Rumex acetosa* (MA; C), *R. acetosella* (KgCc; C), *R. crispus* (Sm/Av; C); Primulaceae: *Lysimachia nummularia* (QF; C), *L. vulgaris* (MA; C); Ranunculaceae: *Consolida regalis* (Sm/Av; C), *Ranunculus acris* (MA; C), *R. repens* (MA; C); Rosaceae: *Achillea millefolium* (MA; C), *Agrimonia eupatoria* (FB; C), *Fragaria vesca* (FB; C), *Potentilla anserina* (Sm/Av; C), *P. argentea* (FB; C), *P. reptans* (MA; C), *Sanguisorba minor* (FB; C), *S. officinalis* (MA; C); Rubiaceae: *Galium molugo* (MA; C), *G. verum* (MA; C); Scrophulariaceae: *Chaenorhinum minus* (Sm/Av; C), *Linaria vulgaris* (FB; C), *Rhinanthus serotinus* (Sm/Av; C), *Scrophularia nodosa* (QF; C), *Verbascum lychnitis* (FB; C), *V. nigrum* (FB; C), *V. thapsus* (FB; C), *Euphrasia rostkoviana* (FB; C); Urticaceae: *Urtica dioica* (QF; C); Valerianaceae: *Valeriana officinalis* (MA; C); Violaceae: *Viola arvensis* (Sm/Av; C), *V. collina* (FB; C), *V. tricolor* (Sm/Av; C).

References

Adjoud D, Plenchette C, Halli-Hargas R, Lapeyrie F (1996) Response of 11 eucalyptus species to inoculation with three arbuscular mycorrhizal fungi. Mycorrhiza 6:129–135

Ahsan N, Lee DG, Lee SH, Kang KY, Lee JJ, Kim PJ, Yoon HS, Kim JS, Lee BH (2007) Excess copper induced physiological and proteomic changes in germinating rice seeds. Chemosphere 67:1182–93

Barea JM, Werner D, Azcon-Guilar C, Azcon R (2005) Interactions of Arbuscular Mycorrhiza and Nitrogen-Fixing Symbiosis in Sustainable Agriculture. In: Werner D, Newton W (eds) Nitrogen fixation in agriculture, forestry, ecology, and the environment. nitrogen fixation: origins, applications, and research progress, vol 4, pp 199–222

Broadley MR, Bowen HC, Cotterill HL, Hammond JP, Meacham MC, Mead A, White PJ (2003) Variation in the shoot calcium concentration of angiosperms. J Exp Bot 54:1431–1446

Buwalda JG, Goh KM (1982) Host fungus competition for carbon as a cause of growth depression in vesicular-arbuscular mycorrhizal ryegrass. Soil Biol Biochem 14:103–106

Cardinale M, Brusetti L, Lanza A, Orlando S, Daffonchio D, Puglia AM, Quatrini P (2010) Rehabilitation of Mediterranean anthropogenic soils using symbiotic wild legume shrubs: plant establishment and impact on the soil bacterial community structure. Appl Soil Ecol 46:1–8

Carrasco JA, Armario P, Pajuelo E, Burgos A, Caviedes MA, Lopez R (2005) Isolation and characterization of symbiotically effective *Rhizobium* resistant to arsenic and heavy metals after the toxic spill at the Aznalcollar pyrite mine. Soil Biol Biochem 37:1131–1140

Cobbett C, Goldsbrough P (2002) Phytochelatins and metallothioneis: roles in heavy metal detoxification and homeostatsis. Ann Rev Plant Physiol Plant Mol Biol 53:159–182

Dashti N, Khanafer M, El-Nemr I, Sorkhoh N, Radwan S (2009) The potential of oil utilizing bacterial consortia associated with legume root nodules for cleaning oily soil. Chemosphere 74:1354–9

Dauber J, Niechoj R, Baltruschat H, Wolters V (2008) Soil engineering ants increase grass root arbuscular mycorrhizal colonization. Biol Fert Soils 44:791–796

Dodd JC, Dougall TA, Clapp JP, Jeffries P (2002) The role of arbuscular mycorrhizal fungi in plant community establishment at Samphire Hoe, Kent, UK – the reclamation platform created during the building of the Channel tunnel between France and the UK. Biodivers Conserv 11:39–58

Ernst WHO, Verkleij JAC, Schat H (1992) Metal tolerance in plants. Acta Bot Neerl 41:229–248

Foy CD, Chaney RL, White MC (1978) The physiology of metal toxicity in plants. Ann Rev Plant Physiol 29:511–566

Gadkar V, Driver JD, Rillig MC (2006) A novel in vitro cultivation system to produce and isolate soluble factors released from hyphae of arbuscular mycorrhizal fungi. Biotechnol Lett 28:1071–1076

Gianinazzi-Pearson V, Gianinazzi S (1983) The physiology of vesicular-arbuscular mycorrhizal roots. Plant Soil 71:197–209

Hall JL (2002) Cellular mechanisms for heavy metal detoxification and tolerance. J Exp Bot 53:1–11

Hamelin J, Fromin N, Tarnawski S, Teyssier-Cuvelle S, Aragno M (2002) nifH gene diversity in the bacterial community associated with the rhizosphere of *Molinia coerulea*, an oligonitrophilic perennial grass. Environ Microbiol 4:477–481

Harinikumar KM, Bagyaraj DJ (1994) Potential of earthworms, ants, millipedes, and termites for dissemination of vesicular-arbuscular mycorrhizal fungi in soil. Biol Fert Soils 18:115–118

Harper JL (1977) Population biology of plants. Academic, London

Janouskova M, Pavlikova D, Vosatka M (2006) Potential contribution of arbuscular mycorrhiza to cadmium immobilisation in soil. Chemosphere 65:1959–1965

Jastrow JD, Miller RM, Lussenhop J (1998) Contributions of interacting biological mechanisms to soil aggregate atabilization in restored prairie. Soil Biol Biochem 30:905–916

Jeffries P, Gianinazzi S, Perotto S, Turnau K, Barea J-M (2003) The contribution of arbuscular mycorrhizal fungi in sustainable maintenance of plant health and soil fertility. Biol Fert Soils 37:1–16

Jónsdóttir IS, Watson MA (1997) Extensive physiological integration: an adaptive trait in resource poor environments? In: de Kroon H, van Groenendael J (eds) The ecology and evolution of clonal plants. Backhuys, Leiden, The Netherlands, pp 109–136

Joner EJ, Leyval C (1997) Uptake of ^{109}Cd by roots and hyphae of a *Glomus mosseae/Trifolium subterraneum* mycorrhiza from soil amended with high and low concentrations of cadmium. New Phytol 138:353–360

Jurkiewicz A, Orłowska E, Anielska T, Godzik B, Turnau K (2004) The influence of mycorrhiza and EDTA application on heavy metal uptake by different maize varieties. Acta Biol Cracov Bot 46:7–18

Jurkiewicz A, Turnau K, Mesjasz-Przybylowicz J, Przybylowicz W, Godzik B (2001) Heavy metal localization in mycorrhizas of *Epipactis atropurpureum* (Orchidaceae) from zinc wastes in Poland. Protoplasma 218:117–124

Kabata-Pendias A (2001) Trace elements in soils and plants, 3rd edn. CRC, Boca Raton, London, New York, Washington, D.C., pp. 413

Kays S, Harper JL (1974) The regulation of plant and tiller density in a grass sward. J Ecol 62:97–105

Khan MS, Zaidi A, Wani PA, Oves M (2009) Role of plant growth promoting rhizobacteria in the remediation of metal contaminated soils. Environ Chem Lett 7:1–19

Kopittke PM, Dart PJ, Menzies NW (2007) Toxic effects of low concentrations of Cu on nodulation of cowpea (*Vigna unguiculata*). Environ Pollut 145:309–315

Lafuente A, Pajuelo E, Caviedes MA, Rodriguez-Llorente ID (2010) Reduced nodulation in alfalfa induced by arsenic correlates with altered expression of early nodulins. J Plant Physiol 167:286–291

Lambers H, Chapin FS III, Pons TL (2008) Plant physiological ecology, 2nd edn. Springer, New York, pp 1–604

Leyval C, Turnau K, Haselwandter K (1997) Effect of heavy metal pollution on mycorrhizal colonization and function: physiological, ecological and applied aspects. Mycorrhiza 7:139–153

Lis J, Pasieczna A (1998) Kartografia geochemiczna obszarów zurbanizowanych i uprzemysłowionych. In: Kozłowski S (ed) Ochrona litosfery. Państwowy Instytut Geolologiczny, Warszawa, pp 248–252

Marshall C, Price EAC (1997) Sectoriality and its implications for physiological integration. In: de Kroon H, van Groenendael J (eds) The ecology and evolution of clonal plants. Backhuys, Leiden, The Netherlands, pp 79–107

Mleczko P (2004) Mycorrhizal and saprobic macrofungi of two zinc wastes in southern Poland. Acta Biol Cracov Bot 46:25–38

Newsham KK, Watkinson AR (1998) Arbuscular mycorrhizas and the population biology of grasses. In: Cheplick GP (ed) Population biology of grasses. Cambridge University Press, Cambridge, UK, pp 286–310

Nye PH, Tinker PB (1977) Solute movement in the soil-root system. Studies in ecology, vol 4. Blackwell Scientific, Oxford

Orłowska E, Jurkiewicz A, Anielska T, Godzik B, Turnau K (2005) Influence of different arbuscular mycorrhizal fungal (AMF) strains on heavy metal uptake by *Plantago lanceolata* L. Pol Bot Stud 19:65–72

Pajuelo E, Rodriguez-Llorente ID, Dary M, Palomeres AJ (2008) Toxic effects of arsenic on *Sinorhizobium-Medicago sativa* symbiotic interaction. Environ Pollut 154:2003–2011

Pielichowska M, Wierzbicka M (2004) The uptake and localization of cadmium by *Biscutella laevigata* – a cadmium hyperaccumulator. Acta Biol Cracov Bot 46:57–64

Pierzynski GM, Sims JT, Vance GF (2000) Soil and environmental quality, 2nd edn. CRC, Boca Raton, FL

Pozo M, Azcon-Aguilar C (2007) Unraveling mycorrhiza induced resistance. Curr Opin Plant Biol 10:393–398

Purin S, Rillig MC (2008) Immuno-cytolocalization of glomalin in the mycelium of the arbuscular mycorrhizal fungus *Glomus intraradices*. Soil Biol Biochem 40:1000–1003

Rascio N (1977) Metal accumulation by some plants growing on zinc-mine deposits. Oikos 29:250–253

Rillig MC, Mummey DL (2006) Mycorrhizas and soil structure. New Phytol 171:41–53

Rillig MC, Mardatin NF, Leifheit EF, Antunes PM (2010) Mycelium of arbuscular mycorrhizal fungi increases soil water repellency and is sufficient to maintain water-stable soil aggregates. Soil Biol Biochem 42:1189–1191
Ryszka P (2006) Mikoryza traw hałdy ZG Trzebionka. Ph.D. thesis, Faculty of Biology and Earth Sciences, Jagiellonian University, Kraków, Poland
Ryszka P, Turnau K (2007) Arbuscular mycorrhiza of introduced and native grasses colonizing zinc wastes: implications for restoration practices. Plant Soil 298:219–229
Shaw BI, Mantle PG (1980) Host infection by *Claviceps purpurea*. Trans Br Mycol Soc 75:77–90
Smith MR, Charvat I, Jacobson RL (1998) Arbuscular mycorrhizae promote establishment of prairie species in a tallgrass prairie restoration. Can J Bot 76:1947–1954
Streitwolf-Engel R, van Der Heijden MGA, Wiemken A, Sanders IR (2001) The ecological significance of arbuscular mycorrhizal fungal effects on clonal reproduction in plants. Ecology 82:2846–2859
Strzyszcz Z (1980) Właściwości fizyczne, fizyko-chemiczne i chemiczne odpadów poflotacyjnych rud cynku i ołowiu w aspekcie ich biologicznej rekultywacji (Physical, physical-chemical and chemical properties of wastes after flotation of zinc and lead ores with regard to their biological reclamation). Archiwum Ochrony Środowiska 3–4:19–50
Strzyszcz Z (2003) Some problems of the reclamation of waste heaps of zinc and lead ore exploitation in southern Poland. Z Geol Wiss, Berlin 31:167–173
Szafer W (ed) (1966) The vegetation of Poland. Pergamon, Oxford
Szuwarzyński M (2000) Zakłady Górnicze "Trzebionka" S.A. 1950–2000, Przedsiębiorstwo Doradztwa Technicznego "Kadra", Krakow, Poland
Tsimilli-Michael M, Turnau K, Ostachowicz B, Strasser RJ (2008) Effect of mycorrhiza on the photosynthetic performance of *Medicago sativa* L. cultivated on control and heavy rich substratum, studied in vivo with the JIP-test. COST 870 Meeting "From production to application of arbuscular mycorrhizal fungi in aqricultural systems: a multidisciplinary approach, 17–19 September 2008, Thessaloniki, Greece, pp 80–83
Turnau K (1998) Heavy metal uptake and arbuscular mycorrhiza development of *Euphorbia cyparissias* on zinc wastes in South Poland. Acta Soc Bot Pol 67(1):105–113
Turnau K, Anielska T, Ryszka P, Gawronski S, Ostachowicz B, Jurkiewicz A (2008) Establishment of arbuscular mycorrhizal plants originating from xerothermic grasslands on heavy metal rich industrial wastes – new solution for waste revegetation. Plant Soil 305:267–280
Turnau K, Jurkiewicz A, Lingua G, Barea JM, Gianinazzi-Pearson V (2006) Role of arbuscular mycorrhiza and associated microorganisms in phytoremediation of heavy metal-polluted sites. In: Prasad MNV, Sajwan KS, Naidu R (eds) Trace elements in the environment. Biogeochemistry, biotechnology, and bioremediation. Taylor & Francis, Boca Raton, pp 235–252
Turnau K, Mesjasz-Przybylowicz J (2003) Arbuscular mycorrhiza of *Berkheya coddii* and other Ni-hyperaccumulating members of Asteraceae from ultramafic soils in South Africa. Mycorrhiza 13:185–190
Turnau K, Ostachowicz B, Wojtczak G, Anielska T, Sobczyk Ł (2010) Metal uptake by xerothermic plants introduced into Zn-Pb industrial wastes. Plant Soil; DOI: 10.1007/s11104-010-0527-7
Turnau K, Ryszka P, Gianinazzi-Pearson V, van Tuinen D (2001) Identification of arbuscular mycorrhizal fungi in soils and roots of plants colonizing zinc wastes in Southern Poland. Mycorrhiza 10:169–174
van der Heijden MGA, Wiemken A, Sanders IR (1998) Different arbuscular mycorrhizal fungi alter coexistence and resource distribution between co-occurring plant. New Phytol 157:569–578
van der Heijden MGA (2002) Arbuscular mycorrhizal fungi as a determinant of plant diversity: in search for underlying mechanisms and general principles. In: van der Heijden MGA, Sanders IR (eds) Mycorrhizal ecology. Ecological studies, vol 157. Springer, Heidelberg, pp 243–265
Wierzbicka M, Panufnik D (1998) The adaptation of *Silene vulgaris* to growth on a calamine waste heap (S. Poland). Environ Pollut 101:415–426

Chapter 17
Bioremediation of Copper, Chromium and Cadmium by Actinomycetes from Contaminated Soils

María Julia Amoroso and Carlos Mauricio Abate

17.1 Introduction

Water and soil pollution has become a major concern in the word, as much of the population relies on ground water as its major source of drinking water as well as on soil as cultivable land. Heavy-metal contamination brings a potential health hazard that can cause metal toxicosis in animals and humans. Trace elements such as cadmium, copper, chromium and mercury are very toxic heavy metals and have been found in the human environment at increased concentrations, because a wide variety of industrial activities have accelerated the release of these metals at higher rates than natural geochemical cycling processes can tolerate.

Thus, chromium (Cr) is a naturally occurring element that is found in rocks, soil, plants, animals, volcanic dust, and gases. It can exist in several oxidation states, but the most stable and common forms in the environment are trivalent Cr(III) and hexavalent Cr(VI) species (Cheung and Gu 2007). Cr(III) is an essential nutrient required for normal sugar and fat metabolism (Cefalu and Hu 2004), whereas Cr (VI) is an established human carcinogen that can damage protein and cell structures, due to its strong oxidizing potential. Cr(VI) and its compounds are placed on the priority list of toxic chemicals of many countries including USA, UK and Canada (Hedgecott 1994; US NTP 2002). It is used in diverse industrial products and processes (Ryan et al. 2002), and it has been released to the environment via leakage, poor storage, or improper disposal practices (Viti et al. 2003).

Cr(VI) have long residence times in soils because of sorption to soil particles, which increases the potential risks for human exposure (Stewart et al. 2003). The plants growing on the metal contaminated sites are not suitable for use as animal or

M.J. Amoroso (✉) • C.M. Abate
Planta Piloto de Procesos Industriales y Microbiológicos (PROIMI). Avenida Belgrano y Pasaje Caseros, 4000 Tucumán, Argentina

Universidad Nacional de Tucumán, Ayacucho 471, 4000 Tucumán, Argentina
e-mail: amoroso@proimi.org.ar

E. Kothe and A. Varma (eds.), *Bio-Geo Interactions in Metal-Contaminated Soils*, Soil Biology 31, DOI 10.1007/978-3-642-23327-2_17,

human food because of the high metal accumulation in roots and leaves (Sinha and Gupta 2005).

Development of suitable methods for cleaning up such contaminated soils is an important topic to environmental recovery and protection. In particular, methods for remediation of Cr(VI)-contaminated soils must be performed.

Respect to copper, it is a very versatile heavy-metal and has a wealth of functions in biological systems, making it an essential requirement for all currently known life forms. However, the same chemistry that makes Cu essential also makes it a potent cytotoxin when Cu homeostatic controls fail (Georgopoulus et al. 2002). Copper cannot be destroyed and tends to be accumulated in soils, plants and animals, increasing their concentrations in the superior level of food chains (Georgopoulus et al. 2002). The many uses of copper in several applications lead to their wide distribution in soil, silt, waste and wastewater and to significant environmental problems that need to be addressed (Lloyd and Lovley 2001).

In relation to cadmium, it is a highly toxic heavy metal that is mainly used for Ni–Cd batteries and for pigments in plastics and glass, stabilizers for PVC, protective plating on steel and alloys (Wilson 1988; Scoullos et al. 2001) contributing to environmental pollution. It is also found in sewage sludge, from industrial areas, which is also used in agriculture as fertilizers at low concentrations. The average cadmium concentration in nonvolcanic soil ranges from 0.01 to 1 mg kg^{-1}, but in volcanic soil levels of up to 4.5 mg kg^{-1} have been found (Korte 1983). Cadmium is carcinogenic, embryotoxic, teratogenic and mutagenic (Sanders 1986), exhibiting both acute and chronic toxicity. Cadmium is a very mobile metal in soil, and can be transferred and accumulated in the roots, leaves and stems of plants (Mench et al. 1989).

A solution to the heavy-metal contamination would be the cleaning of polluted soils by conventional treatments used in industry: removal of the soil, deposition in land fill, or physico-chemical treatments, among others. The adsorption properties of mineral and organic amendments have been exploited (Bailey et al. 1999) but heavy-metal leakage can occur in the soil and with the water path, especially in humid climatic conditions (Antoniadis and Alloway 2001).

Bioremediation is particularly useful for large-scale application on heterogeneous environments, such as ground water, soil sludge and industrial waste (Boopathy 2000).

Different microorganisms such as bacteria and fungi, including yeasts, have a potential ability of metal biosorption; however, there is very few information on actinomycetes used for bioremediation of heavy metals in soils. These microorganisms constitute a significant component of the microbial population in most soils. Their metabolic diversity and particular growth characteristics, mycelial form and relatively rapid colonization of selective substrates, indicate them as well suited agents for bioremediation of metal and organic compounds. Amoroso et al. (1998) have reported that metal resistance and biosorption capability may be widespread among actinomycetes growing in contaminated environments. Later, Amoroso et al. (2001) reported Cr(VI) bioaccumulation by *Streptomyces* strains. Also, among actinomycete species, *Microbacterium*, *Arthrobacter* and *Streptomyces* show Cr(VI) reduction ability (Pattanapipitpaisal et al. 2001; Laxman and More 2002; Horton et al. 2006; Polti et al. 2008, 2010).

Copper resistance in actinobacteria has been described (Amoroso et al. 2001; Albarracín et al. 2005, 2008a; Schmidt et al. 2005). Also, Albarracín et al. (2010) studied the bioaugmentation soil microcosms experimentally polluted with copper using *Amycolatopsis tucumanensis* DSM 45259 and determined the ability of this strain to effectively diminish the bioavailability copper from soils.

Streptomyces sp. strains have been used for bioremediation of soil polluted with cadmium (Jézéquel and Lebeau 2008; Siñeriz et al. 2009); however, there is still not much information about the bioremediation of cadmium from the soils.

The aim of this chapter is to show all the obtained results related to the remediation of chromium, copper and cadmium by actinomycetes, isolated from contaminated Argentinean areas.

17.2 Cr(VI), Cu(II) and Cd(II)-Resistant Actinomycetes

17.2.1 Isolation and Selection

Because almost there is no information about high Cr(VI), Cu(II) and Cd(II) concentration reduction, Polti et al. (2007), Albarracín et al. (2005) and Siñeriz et al. (2009) worked in this subject. Forty-one actinomycete isolates were used for Cr(VI) resistance studies. Twenty-nine were originally isolated from the water reservoir El Cadillal (EC, noncontaminated area), nine from a waste water of a copper filter plant (CFP, contaminated area) and three from sugar cane plant (SC). Culture collection strains (CCS), ATCC 11048 (*Rhodococcus erythropoli*), ATCC 25603 (*Pilimelia terevasa*) and ATCC 15438 (*Streptomyces fradiae*) were used as negative Cr(VI) resistance control strains.

Sediment is an important sink and reservoir for copper. In pristine areas, sediment generally contains less than 50 mg g^{-1}; the level can reach several thousand mg g^{-1} in polluted areas (Georgopoulus et al. 2002). Copper concentration was measured in the sediment samples from CFP, the results obtained were 629 mg g^{-1} for the sample and only 30 mg g^{-1} from the EC water reservoir, Tucumán state, Argentina. Thirty-one actinomycetes were isolated from the noncontaminated area and 19 from the copper contaminated area.

To find actinomycete strains for cadmium bioremediation purposes, 46 strains were isolated from two contaminated places and one noncontaminated place; eight from the soils of a former uranium mine polluted with several heavy metals; and four from the Rio Hondo Dam, Tucumán state, Argentina, polluted by effluents from sugar mills, paper industry, mining, etc. Thirty-four colonies were isolated from the marine sediments from Ushuaia, considered nonpolluted. The unpolluted site was chosen to isolate actinomycetes, which could serve to look for presence of resistance mechanisms against cadmium in strains coming from pristine sites.

A qualitative screening assay was carried out in Petri dishes containing MM agar medium containing (in g L^{-1}): glucose, 10.0; L-asparagine, 0.5; K_2HPO_4, 0.5; $MgSO_4{\cdot}7H_2O$, 0.20; $FeSO_4{\cdot}7H_2O$, 0.01; agar 15 (Amoroso et al. 1998). Rectangular troughs were cut in the center of plates and filled with 5 mM filter – sterilized solution of Cr(VI) as $K_2Cr_2O_7$, $CuSO_4$ 16, 40 and 80 and 10 mg L^{-1} $CdCl_2$ solution. Isolates were inoculated by streaking of 100 μL of spore suspension perpendicularly to the troughs. Microbial growth was used as the qualitative parameter of metal resistance. This was systematized using two categories: heavy metal (Cr(VI), Cu(II) and Cd(II))-sensitive and heavy metal (Cr(VI), Cu(II) and Cd(II))-resistant strains. The inoculated plates were incubated at 30°C for 7 days. Growth controls were performed using medium without the heavy metals (Amoroso et al. 1998).

All the strains used in presence of Cr(VI), displayed resistance wards to 5 mM of Cr(VI). On the other hand, 89% from the CFP and only 62% from EC were found resistants (Polti et al. 2007). With respect to the results obtained in the presence of Cu(II), 100% of the isolated strains of the copper contaminated area (CFP) were resistant up to 80 mg L^{-1} of $CuSO_4$. On the contrary, actinomycete strains isolated from noncontaminated (EC) areas were more sensitive in copper amended media. Nevertheless, 19.4% of these strains were capable of growing with a concentration of 80 mg L^{-1} of $CuSO_4$. This could indicate that copper resistance mechanisms facultative exist in some cells (Albarracín et al. 2005).

In relation to test the Cd(II) resistance, primary qualitative screening showed that from the 46 colonies isolated from the polluted and nonpolluted sites, 26 were Cd(II) resistant and selected for the semi-quantitative screening; 18 from Ushuaia bay, 5 from Wismut and 3 from Rio Hondo (Siñeriz et al. 2009). Because the obtained results in the qualitative screening, 29 strains were selected for testing their resistance towards 3–20 mM Cr(VI). For this semi-quantitative screening of Cr(VI) resistance, 50 μL of $K_2Cr_2O_7$ solution at different Cr(VI) concentrations (3–20 mM) were used to fill the wells of Petri dishes culture media containing MM agar medium previously inoculated with 100 μL of spore suspension (1×10^9 CFU mL^{-1}) of the strain to be tested (Polti et al. 2007). The diameter of growth inhibition was measured after incubation at 30°C for 7 days. Strains were considered resistant when the inhibition diameter was less or equal than 7 mm, and nonresistant when it was higher than 7 mm (Amoroso et al. 1998).

Different levels of growth inhibition for the selected strains were found (Fig. 17.1a). All isolated strains from CFP showed no growth inhibition up to 5 mM Cr(VI), and only 14% of them could grow with 13 mM Cr(VI). However, 70% of EC isolates turned out to be sensitive to 3 mM Cr(VI), and 20% of them were able to grow at 13 mM of Cr(VI) (Fig. 17.1b). All isolated strains from SC could grow up to 17 mM Cr(VI). The strains resistant to 10 mM of Cr(VI) were: C16, C33, C35, C39 and C55 from EC; M3, M40 and M46 from CFP; and MC1, MC2 and MC3 from SC (Polti et al. 2007). These were selected for Cr(VI) toxicity assays.

For testing the resistance to copper, $CuSO_4$ solution at different concentrations (100–1,000 mg L^{-1}) was used to fill the wells of Petri dishes culture media previously inoculated with 100 mL of spore suspensions (1×10^9 CFU mL^{-1}) of the strain to be tested. The diameter of the growth inhibition was measured after

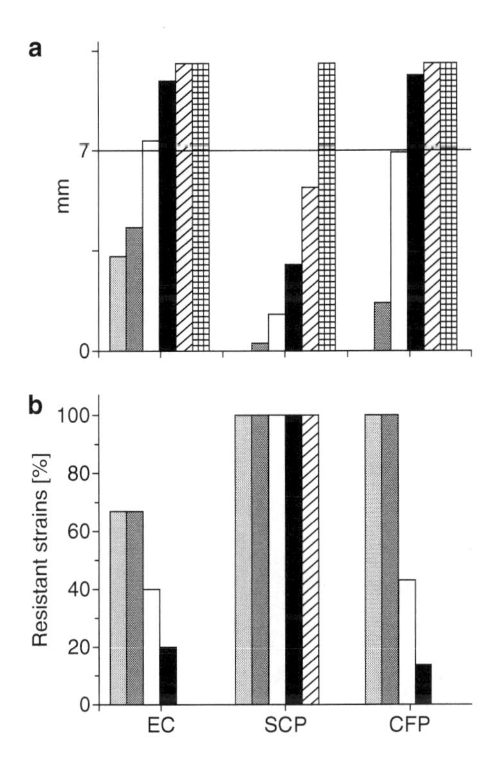

Fig. 17.1 Semi-quantitative resistance at 3 (▨), 5 (▨), 10 (□), 13 (■), 17 (▨) and 20 (▦) mM of Cr(VI); (**a**) measured as inhibition zone in mm. The *horizontal line* indicates the arbitrary limit of 7 mM Cr(VI) considered as resistant; (**b**) percentage of resistant strains from EC ($n = 15$), SCP ($n = 3$); CFP ($n = 7$) (Polti et al. 2007)

incubation at 30°C for 4 days. The strains were considered resistant using the same selection that for Cr(VI), as it was described previously (Fig. 17.2) (Albarracín et al. 2005).

The strains that could grow in the qualitative screening until 100 mg L^{-1} of Cd(II) was tested for Cd(II) toxicity on plates containing defined amounts of Cd. R25 strain could grow on plates with Cd(II) up to 5 mg L^{-1}, R22 to 2 mg L^{-1}, USH 1.9 until 10 mg L^{-1}, USH 3.15–6 mg L^{-1}, USH 1.12–2 mg L^{-1}, and F4 until 150 mg L^{-1} Cd(II). The control strains *Streptomyces lividans* TK24 and *S. coelicolor* A2 (3) could grow until 2 and 0 mg L^{-1} Cd(II), respectively (Siñeriz et. al. 2009).

17.2.2 Cr(VI), Cu(II) and Cd(II) Toxicity Assays

Spore suspensions (100 μL of 1×10^9 CFU mL^{-1}) of 11 selected chromium resistant strains, AB copper-resistant group, AB2C (the most sensitive strain) and F4 cadmium resistant strains were inoculated in liquid MM supplemented with

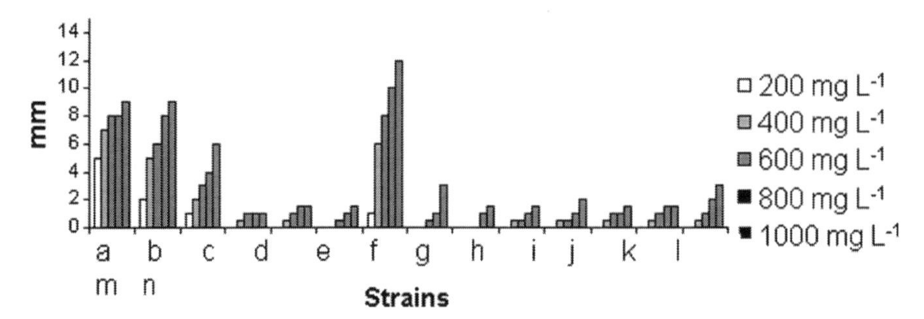

Fig. 17.2 Semi-quantitative resistance at 200, 400, 600, 800 and 1,000 mg L^{-1} of Cu(II) concentrations measured as inhibition zone in mm. The *horizontal line* indicates the arbitrary limit used to consider copper-resistant (*below*) and nonresistant (*up*) strains. AB-strains (d, e, f, g, h, I, j, k, l, m: AB0, AB2A, AB2B, AB2C, AB3, AB5A, AB5B, AB5C, AB5D, AB5E and AB5F) – from sediment of the drainage channel (CFP). C-strains (a, b, c: C16, C39 and C43) – from El Cadillal (EC) (Albarracín et al. 2005)

1 mM of Cr(VI) as $K_2Cr_2O_7$, 80 mg L^{-1} $CuSO_4$ and 8 mg L^{-1} $CdCl_2$, respectively. Cultures were incubated in an orbital shaker (220 rpm) at 30°C for 72 h. Biomass was collected by centrifugation at 12,000 g for 10 min; washed twice with distilled water and dried at 105°C until constant weight. Relative growth is expressed as percentage of that obtained in controls without heavy metals, which was taken as 100%.

Chromate-removing activity was estimated by measuring decrease of Cr(VI) concentration in the supernatant during the assay; using the Cr(VI)-specific colorimetric reagent 1,5-diphenylcarbazide (DPC), dissolved in acetone to a final concentration of 0.05% (APHA 1989). The absorbance was measured at 540 nm. The Cr(VI) concentration was estimated using a calibration curve (0–1 mM). Specific removal, defined as Cr(VI) removal (mg) per biomass (g), was also calculated. Uninoculated controls were performed to determine the Cr(VI) loss by the components of the culture medium (Polti et al. 2007).

Copper concentrations were determined by atomic absorption spectrometry (Perkin Elmer Aanalyst 100) in the supernatant and the biomass cultures. Total final biomass was first digested with concentrated nitric acid (Standard Methods, APHA 1992).

Cr(VI) resistance assays reported by Polti et al. (2007) were performed in a defined minimal medium (MM), because the supplemented metal does not form complexes with medium components and all the metal added is bioavailable (Amoroso et al. 2001). The studies on chromium tolerance, as published previously, were usually carried out on complex or organic media. Under those conditions, *Bacillus* species, isolated from tannery effluent, showed a minimal inhibitory concentration (MIC) of less than 1.92 mM (Srinath et al. 2002). Francisco et al. (2002) found a cultivable microbial community from activated sludge that were able to survive 2 mM Cr(VI) and Richards et al. (2002) found that *Frankia* strains Cr(VI) MIC were among 0.05 and 1.75 mM. A *Pseudomonas* strain isolated from a metal contaminated area was able to grow with 1 mM Cr(VI) in minimal medium, but not with 10 mM Cr(VI) (McLean et al. 2001). Laxman and More (2002) reported higher Cr(VI) tolerance in organic medium than in semi-synthetic medium, probably due to the binding of Cr to the organic constituents of the medium, which reduces toxicity.

Relative growth was less than 50% for strains C16, C33, C39 and C55, from EC and for M3 from CFP. Only one strain (C35) did not suffer inhibition of growth. No strain from SCP suffered an inhibition of more than 50% (Fig. 17.3a). Cr(VI) removal was prominent with more than 40% removal for strains from sugar cane plant (SCP). CFP strains varied with 10–45% removal. Strains from EC showed low (2–24%) removal ability (Fig. 17.3b).

The highest specific removal ability was seen with strain M3 (75.5 mg g^{-1} cells), and the lowest with C35 (1.5 mg g^{-1} cell). SCP strains showed values between 26.5 and 28.5 mg g^{-1} cells (Fig. 17.3c). A none spontaneous Cr(VI) removal was observed in MM medium without inoculum (Polti et al. 2007).

The growth expressed as dry weight (mg mL^{-1}) after cultivation in MM liquid medium supplemented with 80 mg L^{-1} Cu(II) was determined. Only the strains AB0, AB2A, AB5A and AB2B showed the greatest growth compared to the AB2C-sensitive strain (Fig. 17.4a). This experiment was carried out in triplicate. When copper residual was determined in the supernatant of the culture medium (Fig. 17.4b), the results revealed a diminution of 71% by AB0, 65% by AB2A, 27% AB5A and 23% by AB2B culture strains. Copper biosorption related to the cell growth (Fig. 17.4c) was also calculated and as expected, the two strains (ABO and AB2A) that had the lowest copper residual had the highest biosorption value. They look promising to study and perform copper bioremediation processes.

In relation to determine the toxicity of Cd(II), the strain F4 reaches the exponential phase on the fourth day on medium without Cd(II) (1,492 g L^{-1}) on days 6–8 the biomass is less, with 1,378 and 1,150 g L^{-1}, respectively, probably because of bacterial lyses. On the medium with 8 mg L^{-1} Cd(II)), less growth was observed with a maximum of 0.304 g L^{-1} on day 8. The growth on day 4 with Cd(II) was 20% (five times less) as compared to growth without the metal (Fig. 17.5) (Siñeriz et al. 2009).

17.3 Copper Soil Bioremediation

Copper resistance in actinobacteria has been also described (Amoroso et al. 2001; Schmidt et al. 2005; Albarracín et al. 2005, 2008a, b, 2010). However, there is not enough specific information on the application of actinomycetes to bioremediate copper polluted soil microcosms. *Streptomyces* sp. strains have been used for bioremediation of soil polluted with cadmium (Jézéquel et al. 2008), chromium (Polti et al. 2009) or plaguicides (Benimeli et al. 2008). *Amycolatopsis tucumanensis* DSM 45259T (also referred as ABO strain), isolated from copper polluted sediments, has been studied for its remarkable copper resistance as well as for its high bioaccumulation abilities (Albarracín et al. 2005, 2008a, 2010).

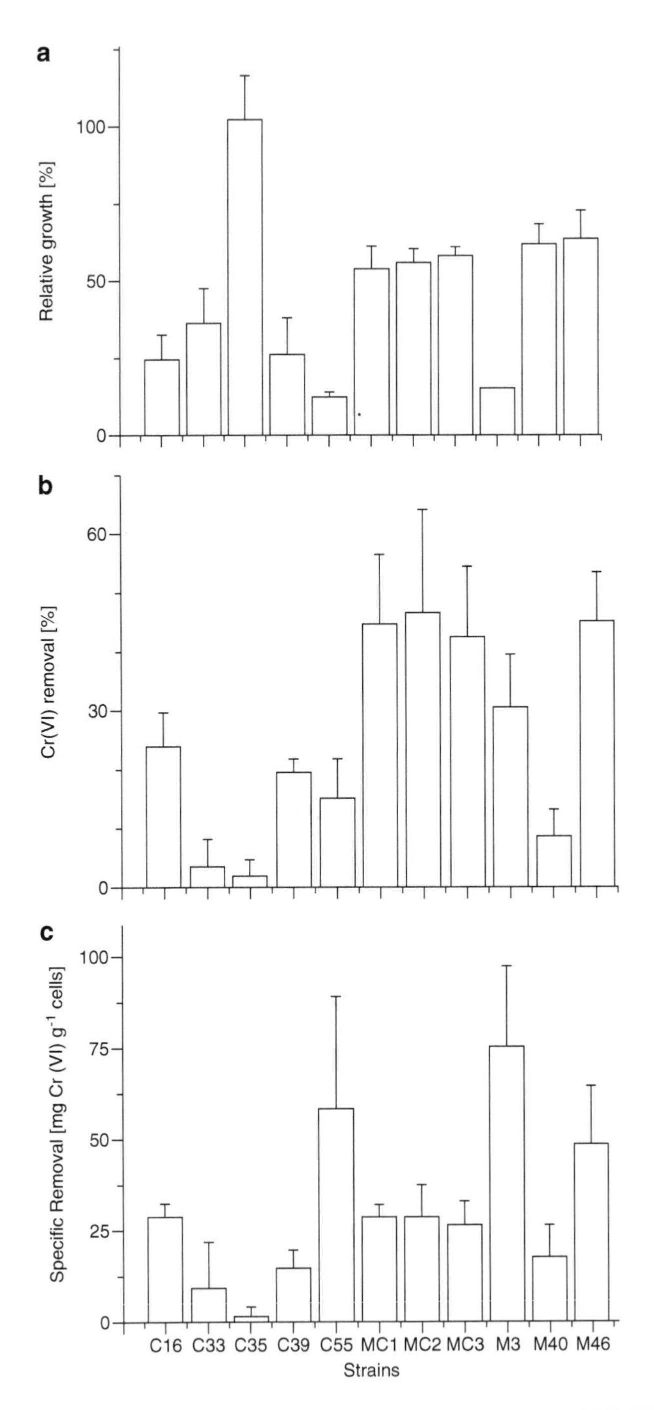

Fig. 17.3 (**a**) Relative growth in MM liquid medium supplemented with 1 mM Cr(VI), expressed as a percentage of growth obtained in controls without Cr(VI), which was taken as 100%; (**b**) Percentage of Cr(VI) removal after 3 days of growth; (**c**) Cr(VI) specific removal (mg g^{-1} cells) of 11 selected actinomycete strains (Polti et al. 2007)

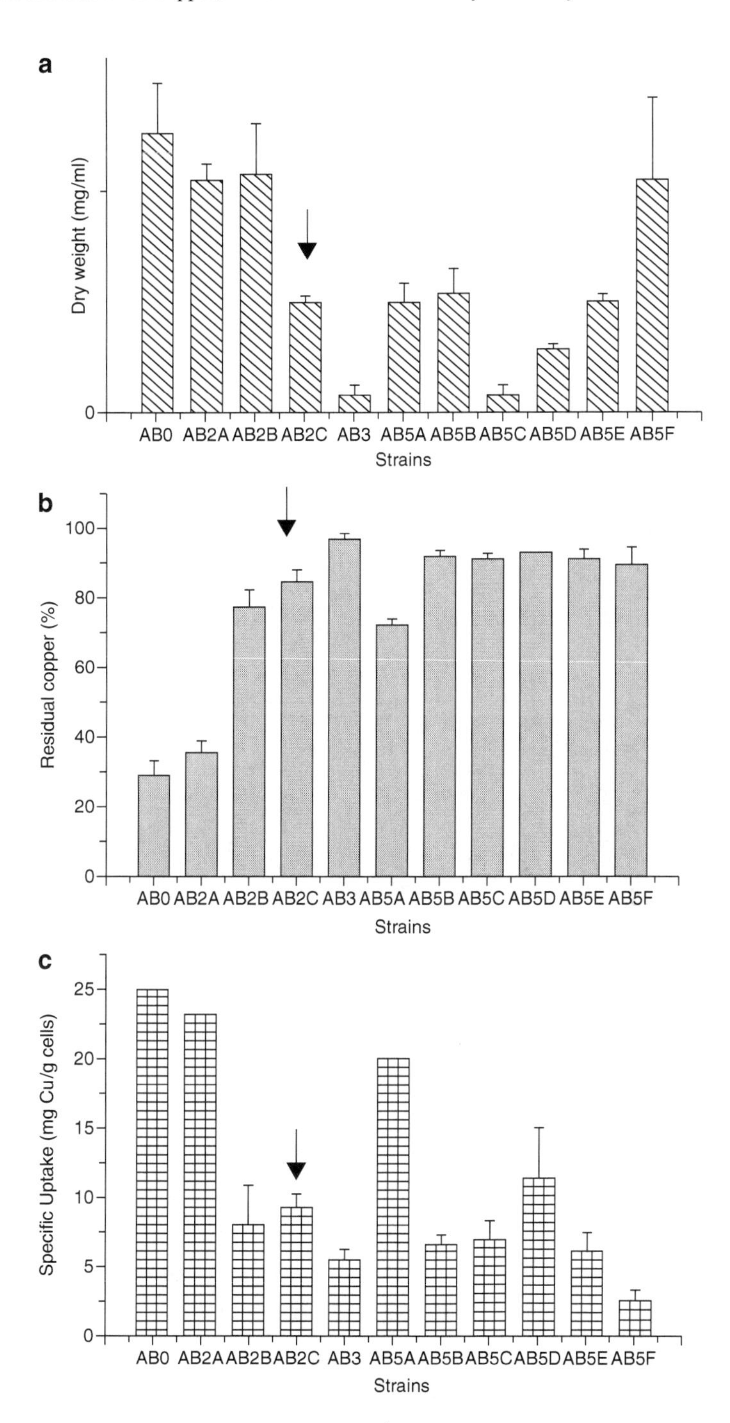

Fig. 17.4 (**a**) Biomass as dry weight (mg mL^{-1}) in MM liquid medium supplemented with 80 mg L^{-1} of $CuSO_4$; (**b**) residual copper concentration (mg L^{-1}) in the supernatant after 7 days of growth; (**c**) copper biosorption (mg of Cu g^{-1} of cells) of 11 selected actinomycete strains. The control sensitive strain, AB2C, is indicated with an *arrow* (k) (Albarracín et al. 2008a)

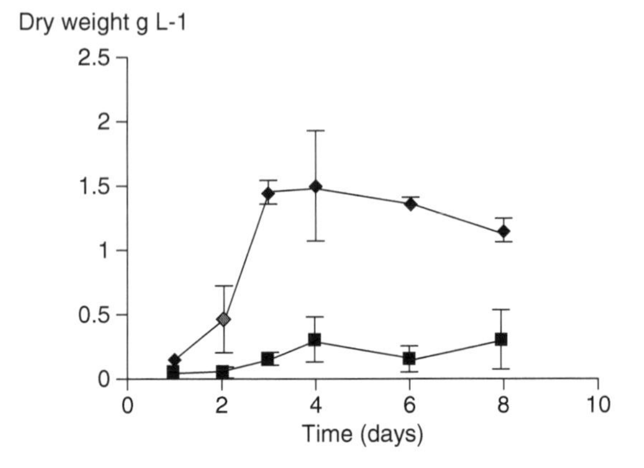

Fig. 17.5 *Streptomyces* sp. F4 growth in MM medium with 8 mg L^{-1} (◆) and without cadmium (■) measured as dry weight of biomass (Siñeriz et al. 2009)

However, there is not information on the application and metabolic activity in clean or polluted soil of any strain belonging to the *Amycolatopsis* genus. In this sense, the aim of this work was to bioaugment soil microcosms experimentally polluted with copper using *Amycolatopsis tucumanensis* DSM 45259^T and to study the ability of this strain to effectively diminish bioavailable copper from soils.

17.3.1 *Growth and Viability of* A. tucumanensis *in Sterile Soil Microcosms*

A. tucumanensis was inoculated in SM (soil medium) added with 80 mg of copper kg^{-1} of soil ($SM80_{b\ (biremediated)}$) in order to monitor its growth and viability in this environment. The growth was assessed after 7, 14, 21 and 28 days of incubation by counting CFU g^{-1} of soil. SM not experimentally polluted with copper was used as control ($SM20_b$).

A. tucumanensis displayed high colonization ability in both, $SM20_b$ and $SM80_b$. The maximal growth (2.5×10^9 CFU per g of soil) was obtained after 7 days of incubation in both cases (Fig. 17.6a). When a slower initial inoculum was used (10^6 CFU per g of soil), the same maximal population was obtained. A similar maximal growth efficiency was obtained for other actinobacteria strains applied to similar loamy soils; *Streptomyces* sp. M7, applied to lindane-polluted soil microcosms, reached 1.98×10^9 CFU per g of soil (Benimeli et al. 2008). Polti et al. (2009) studied the growth and Cr(IV) renoval ability of *Streptomyces* sp. MC1 in chromate-polluted soils which reached a maximal population of 6.6×10^9 CFU per g of soil.

Interestingly, *A. tucumanensis* did not show inhibition in its growth in $SM80_b$ compared to the control ($SM20_b$). Moreover, higher copper concentrations (300 mg kg^{-1}) in soil did not inhibit the strain growth (data not shown). This concurs with the results obtained for the same strain in soil extract and artificial soil medium.

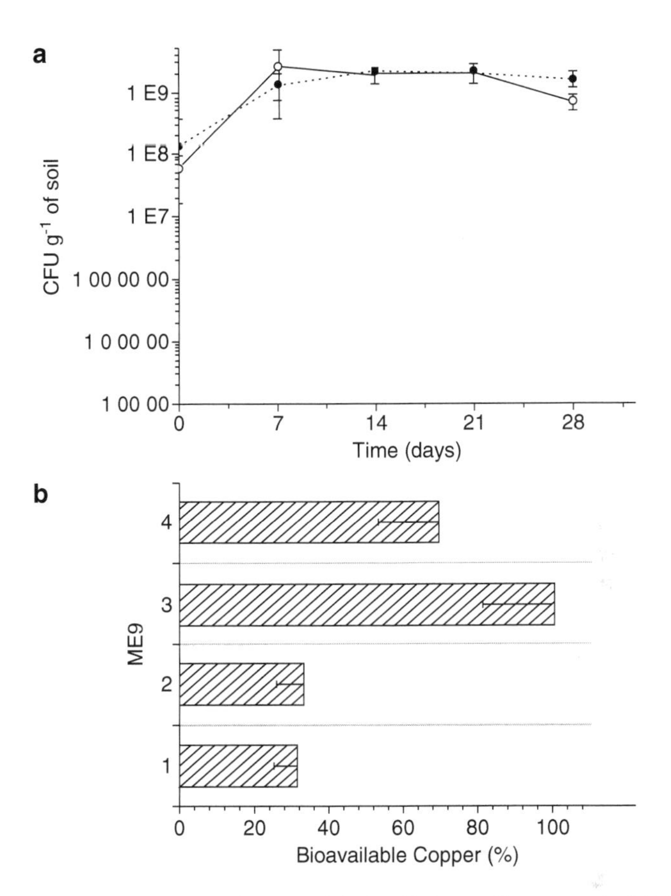

Fig. 17.6 (**a**) *A. tucumanensis* growth assessed during 28 days of incubation in SM20 (__o __) and SM80 (.....•.....). (**b**) Bioavailable copper measured in the soil solution. (1) SM20nb (2) SM20b (bioaugmented with *A. tucumanensis*) (3) SM80nb (4) SM80b (bioaugmented with *A. tucumanensis*). The values of the groups a, b and c are significantly different among them ($p < 0.05$) (Albarracín et al. 2008a)

17.3.2 Copper Removal Ability by A. Tucumanensis *in SM80*

Soil solution is the liquid mobile phase where most of the chemical and biological reactions of soil are produced. Its composition represents a dynamic index of the soil *status* and, in particular, the heavy-metal concentration on it represents an indicator of their bioavailability as it is the soil fraction upon which plants bioabsorb metals (Csillag et al. 1999). Taking this into account, it has been studied the copper bioinmobilization ability of *A. tucumanensis* on soil by assessing the bioavailable copper in the soil solution extracted from SM_b.

For $SM20_{nb\ (nonbioremediated)}$, the bioavailable copper measured in the soil solution was approximately 30% with respect to the one recorded in $SM80_{nb}$

(Fig. 17.6b). In addition, bioavailable copper for $SM20_{nb}$ (0.08 mg kg^{-1}) and in $SM80_b$ (0.24 mg kg^{-1}) was much more less than the total copper content in the soil (20 and 80 mg kg^{-1}, respectively). This same phenomenon was observed by Groudev et al. (2001) for a copper polluted soil, and it is not unexpected as it is known that most of the copper present in soil is complexed by humic and fulvic acids, and sorbed by arcilla and Mn and Fe oxides (Georgopoulus et al. 2002).

When the strain was applied to soil, no significant differences were observed between the values of bioavailable copper from $SM20_{nb}$ and $SM20_b$. On the contrary, a significant depletion of the bioavailable copper (31%) in $SM80_b$ was observed with respect to the total bioavailable copper present in $SM80_{nb}$ (Fig. 17.6b) demonstrating *A. tucumanensis* copper biosorption ability in a polluted soil.

To our knowledge, this work is the first to apply and to probe the colonization ability of an *Amycolatopsis* strain in soil microcosms and constitutes the first application of an *Amycolatopsis* strain on bioremediation of polluted soils.

17.4 Chromium Soil Bioremediation

However, there are only few studies on Cr(VI) bioreduction by actinomycetes and their potential for bioremediation processes in soil. The first report on Cr(VI) reduction by *Streptomyces* was by Das and Chandra (1990). Amoroso et al. (2001) reported on Cr(VI) bioaccumulation by *Streptomyces* strains, whereas Laxman and More (2002) determined Cr(VI) reduction by *Streptomyces griseus*. In a previous work, our research group determined Cr(VI) removal by nine *Streptomyces*. However, *Streptomyces* sp. MC1, isolated from sugar cane, was selected because it showed significant growth and capacity to remove Cr(VI) in liquid minimal medium (Polti et al. 2007), abilities that may be useful for bioremediation processes.

17.4.1 Growth of Streptomyces *sp. MC1 in Soil Samples; Determination of Cr(VI) Bioavailability*

In order to determine the ability of *Streptomyces* sp. MC1 to grow on and reduce Cr (VI) assays with sterilized and nonsterilized soil sample (SS A) were carried out. *Streptomyces* sp. MC1 did not show growth inhibition in presence of 50 mg kg^{-1} of Cr(VI) (Fig. 17.7) (Polti et al. 2008). The Cr(VI) concentration reduced from 50 to 5 mg kg^{-1} in SS A inoculated with *Streptomyces* sp. MC1 after 3 weeks of incubation, whether the soil was initially sterilized or not. Moreover, analyses revealed that Cr was not present in its hexavalent form, indicating that it was metabolically reduced by *Streptomyces* sp. MC1. Only a slight decrease in Cr(VI)

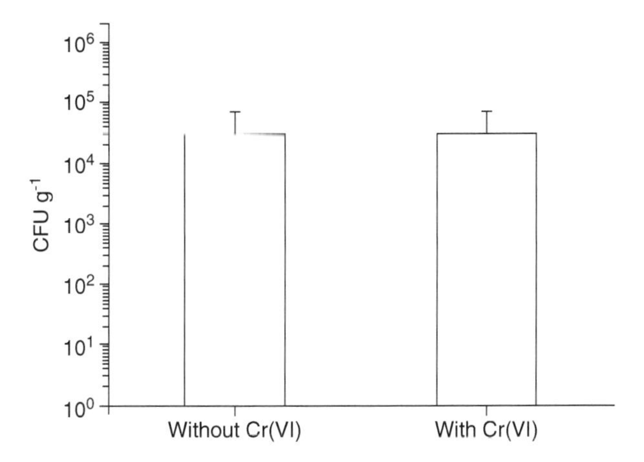

Fig. 17.7 Growth of *Streptomyces* sp. MC1 in SS A without Cr(VI) and after addition of 50 mg kg^{-1} of Cr(VI) after 3 weeks of incubation at 30°C (Polti et al. 2008)

concentration (2%) was observed in noninoculated samples, indicating that SS A probably does not contain significant amounts of any substance or microorganism able to spontaneously reduce Cr(VI) under the experimental conditions. The autochthonous culturable microflora was 3,107 CFU g^{-1} at the beginning and 2,107 CFU g^{-1} of soil at the end of the essay. Finally, growth of *Streptomyces* sp. MC1 was studied in sterilized SS B without and with 50 mg kg^{-1} of Cr(VI) for 4 weeks (Fig. 17.8) and Cr(VI) reduction was simultaneously determined. No growth inhibition was observed when *Streptomyces* sp. MC1 was grown with Cr (VI). After 1 week of incubation, Cr(VI) was reduced from 50 to 5 mg kg^{-1} during the exponential growth phase. Once Cr(VI) was reduced by *Streptomyces* sp. MC1, no reoxidation took place. Because Cr(VI) concentrations in sterile pots were practically constant, reduction of Cr(VI) corresponded to *Streptomyces* sp. MC1 activity. Our results also demonstrate the absence of active soil compounds able to reduce Cr(VI) under the current assay conditions (Fig. 17.8).

To date there exist no other studies about Cr(VI) removal by actinomycetes from soil samples without addition of substrates and at normal humidity level. Our results show for the first time how a *Streptomyces* strain was able to remove 90% of Cr(VI) from a soil sample supplemented with 50 mg kg^{-1} Cr(VI) after 7 days of incubation. It is also important to point out that this *Streptomyces* strain was able to remove Cr(VI) from nonsterile soil and practically to the same degree. This last result suggests that bioremediation activity by *Streptomyces* sp. MC1 is not inhibited by the natural soil microbial flora. Previous studies determined that greatest removal efficiency was observed in soil supplemented with an organic substrate (Vainshtein et al. 2003). Organically modified soil in pots reduced Cr(VI) in ground water from 1 mg L^{-1} to less than 50 μg L^{-1}. Cifuentes et al. (1996) demonstrated that indigenous microbes in a shaker flask supplemented with yeast extract aerobically reduced. Approximately 100 mg L^{-1} of Cr(VI) to Cr(III) after 15 days at room temperature. Desjardin et al. (2003) observed up to 99% reduction

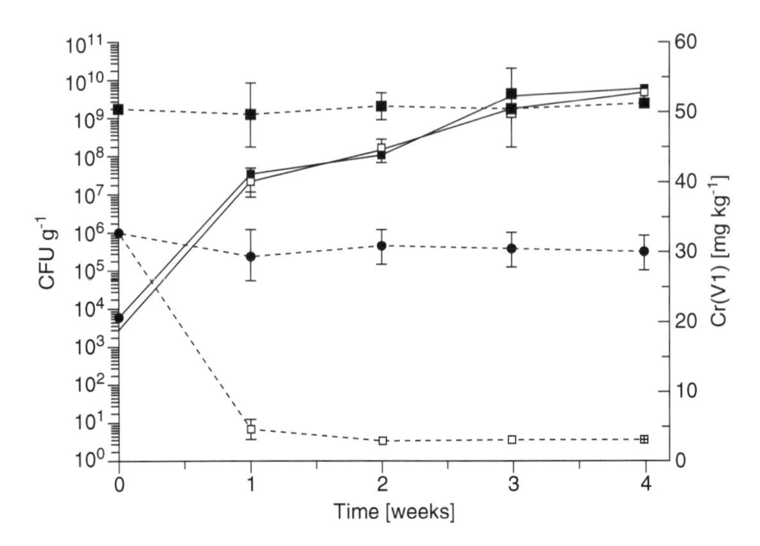

Fig. 17.8 Effect of inoculation of *Streptomyces* sp. MC1 in sterile soil sample SS B during 4 weeks of incubation in pots. Growth without (—■—) and with Cr(VI) (50 mg kg^{-1}) (—□—) revealed by colony-forming units per gram (CFU g^{-1}); concentration of residual Cr(VI) in inoculated (- - □- -) and noninoculated (- - ●- -) soil; total Cr in inoculated soil (- - ■- -) (Polti et al. 2008)

of 180 mg L^{-1} Cr(VI) by supernatants of *Streptomyces* NH50 after 1 month of incubation, using 2.5 g of nonsterile contaminated soil in 25 ml of liquid medium. Cr(VI) reduction by *Streptomyces* sp. MC1 represents a useful mechanism for bioremediation in soil samples and biotechnological processes could be improved. *Streptomyces* sp. MC1 could not only be applied to solid systems such as soil but also to liquid or semi-liquid systems such as waste water or slurry. Particularly important is the fact that the system is environment friendly. Studies at field scale are required for future applications of *Streptomyces* sp. MC1 to bioremediation processes.

References

Albarracín VH, Amoroso MJ, Abate CM (2005) Isolation and characterization of indigenous copper resistant actinomycete strains. Chem Erde Geochem 65:145–156

Albarracín VH, Avila AL, Amoroso MJ, Abate CM (2008a) Copper removal ability by *Streptomyces* strains with dissimilar growth patterns and endowed with cupric reductase activity. FEMS Microbiol Lett 288:141–148

Albarracín VH, Winik B, Kothe E, Amoroso MJ, Abate CM (2008b) Copper bioaccumulation by the actinobacterium *Amycolatopsis* sp. AB0. Basic Microbiol 48:323–330

Albarracín VH, Amoroso MJ, Abate CM (2010) Bioaugmentation of copper polluted soil microcosms with *Amycolatopsis tucumanensis* to diminish phytoavailable copper for Zea mays plants. Chemosphere 79:131–137

Amoroso MJ, Castro GR, Carlino FJ, Romero NC (1998) Screening of heavy metal-tolerant actinomycetes isolated from the Salí River. J Gen Appl Microbiol 44:129–132

Amoroso MJ, Castro GR, Durán A, Peraud O, Oliver G, Hill RT (2001) Chromium accumulation by two *Streptomyces* sp isolated from riverine sediments. J Ind Microbiol Biotechnol 26:210–215

Antoniadis V, Alloway BJ (2001) Availability of Cd, Ni and Zn to ryegrass in sewage sludge-treated soils at different temperatures. Water Air Soil Pollut 132:201–214

Bailey SE, Olin TJ, Bricka RM, Adrian DD (1999) A review of potentially low-cost sorbents for heavy metals. Water Res 33:2469–2479

Benimeli CS, Fuentes MS, Abate CM, Amoroso MJ (2008) Bioremediation of lindane-contaminated soil by *Streptomyces* sp. M7 and its effects on Zea mays growth. Int Biodeter Biodegr 61:233–239

Boopathy R (2000) Factors limiting bioremediation technologies. Bioresour Technol 74:63–67

Cefalu WT, Hu FB (2004) Role of chromium in human health and in diabetes. Diabetes Care 27:2741–2751

Cheung KH, Gu JD (2007) Mechanism of hexavalent chromium detoxification by microorganisms and bioremediation application potential: a review. Int Biodeter Biodegr 59:8–15

Cifuentes FR, Lindemann WC, Barton LL (1996) Chromium sorption and reduction in soil with implications to bioremediation. Soil Sci 161:233–241

Csillag J, Pártay G, Lukács A, Bujtás K, Németh T (1999) Extraction of soil solution for environmental analysis. Int J Environ Anal Chem 74:305–324

Das S, Chandra AL (1990) Chromate reduction in *Streptomyces*. Experientia 46:731–733

Desjardin V, Bayard R, Lejeune P, Gourdon R (2003) Utilisation of supernatants of pure cultures of *Streptomyces thermocarboxydus* NH50 to reduce chromium toxicity and mobility in contaminated soils. Water Air Soil Pollut 3:153–160

Francisco R, Alpoim MC, Morais PV (2002) Diversity of chromium-resistant and -reducing bacteria in a chromium-contaminated activated sludge. J Appl Microbiol 5:837–43

Georgopoulus PG, Roy A, Opiekun RE, Yonone-Lioy MJ, Lioy PJ (2002) Environmental dynamics and human exposure to copper. In: Georgopoulus PG, Roy A, Opiekun RE, Yonone-Lioy MJ, Lioy PJ (eds) Environmental dynamics and human exposure issues, vol 1. International Copper Association, New York, USA, pp 10–23

Groudev SN, Spasova II, Georgiev PS (2001) In situ bioremediation of soils contaminated with radioactive elements and toxic heavy metals. Int J Miner Process 62:301–308

Hedgecott S (1994) Prioritization and standards for hazardous chemicals. In: Calow P (ed) Handbook of ecotoxicology. Blackwell, Oxford, UK, pp 378–382

Horton RN, Apel WA, Thompson VS, Sheridan PP (2006) Low temperature reduction of hexavalent chromium by a microbial enrichment consortium and a novel strain of *Arthrobacter aurescens*. BMC Microbiol 6:5–12

Jézéquel K, Lebeau T (2008) Soil bioaugmentation by free and immobilized bacteria to reduce potentially phytoavailable cadmium. Bioresour Technol 4:690–698

Korte F (1983) Ecotoxicology of cadmium: general overview. Ecotoxicol Environ Saf 7:3–8, Cadmium-environmental aspects. Environmental Health Criteria 135:1992

Laxman SR, More S (2002) Reduction of hexavalent chromium by *Streptomyces* griseus. Miner Eng 15:831–837

Lloyd JR, Lovley DR (2001) Microbial detoxification of metals and radionuclides. Curr Opin Biotechnol 12:248–253

McLean J, Terry J, Beveridge TJ (2001) Chromate Reduction by a Pseudomonad Isolated from a Site Contaminated with Chromated Copper Arsenate. Appl Environ Microbiol 3:1076–1084

Mench M, Tancogne J, Gomez A, Juste C (1989) Cadmium bioavailability to *Nicotiana tabacum* L., *Nicotiana rustica* L., and *Zea mays* L. grown in soil amended with cadmium nitrate. Biol Fertil Soils 8:48–53

Pattanapipitpaisal P, Brown NL, Macaskie LE (2001) Chromate reduction by *Microbacterium liquefaciens* immobilised in polyvinyl alcohol. Biotechnol Lett 23:61–65

Polti MA, Amoroso MJ, Abate CM (2007) Chromium(VI) resistance and removal by actinomycete strains isolated from sediments. Chemosphere 67:660–667

Polti MA, Amoroso MJ, Abate CM (2010) Chromate reductase activity in *Streptomyces* sp. MC1. J. Gen. Appl. Microbiol. 56:11–18

Polti MA, García RO, Amoroso MJ, Abate CM (2008) Bioremediation of chromium(VI) contaminated soil by *Streptomyces* sp. MC1. J Basic Microbiol 49:285–292

Polti MA, García R, Amoroso MJ, Abate CM (2009) Chromium (VI) soil bioremediation by *Streptomyces* sp. MC1. J Basic Microbiol 49:285–292

Richards JW, Krumholz GD, Chval MS, Tisa LS (2002) Heavy metal resistance patterns of Frankia strains. Appl Environ Microbiol 2:923–927

Ryan MP, Williams DE, Chater RJ, Hutton BM, McPhail DS (2002) Why stainless steel corrodes. Nature 415:770–774

Sanders CL (1986) Toxicological aspects of energy production. MacMillian, New York, pp 158–162

Scoullos M, Vonkeman G, Thornton I, Makuch Z (2001) Mercury, cadmium, lead: handbook for sustainable heavy metals policy and regulation. Kluwer, Dordrecht, Nordic Council of Ministers

Schmidt A, Haferburg G, Siñeriz M, Merten D, Buchel G, Kothe E (2005) Heavy metal resistance mechanisms in actinobacteria for survival in AMD contaminated soils. Chem Erde Geochem 65:131–144

Sinha S, Gupta AK (2005) Translocation of metals from fly ash amended soil in the plant of *Sesbania cannabina* L. Ritz: effect on antioxidants. Chemosphere 61:1204–1214

Siñeriz ML, Kothe E, Abate CM (2009) Cadmium biosorption by *Streptomyces* sp. F4 isolated from former uranium mine. J Basic Microbiol 49:55–62

Stewart MA, Jardine PM, Brandt BC, Barnett MO (2003) Effects of contaminant concentration, aging, and soil properties on the bioaccessibility of Cr(III) and Cr(VI) in soil. Soil Sediment Contam 12:1–21

Srinath T, Verma T, Ramteke PW (2002) Chromium (VI) biosorption and bioaccumulation by chromate resistant bacteria. Chemosphere. 4:427–435

US NTP (2002) United States National Toxicology Program. The Report on Carcinogens, Tenth Edition. U.S. Department of Health and Human Services, National Institute of Health Public Health Service, National Institute of Environmental Health Sciences. (http://ehp.niehs.nih.gov/roc/toc10.html)

Viti C, Pace A, Giovànnetti L (2003) Characterization of Cr(VI)-resistant bacteria isolated from chromium-contaminated soil by tannery activity. Curr Microbiol 46:1–5

Vainshtein M, Kuschk P, Mattusch J, Vatsourina A, Wiessner A (2003) Model experiments on the microbial removal of chromium from contaminated groundwater. Water Res 37:1401–1405

Wilson DN (1988) Cadmium – market trends and influences. In: Cadmium 87. Proceedings of the 6th International Cadmium Conference, London, Cadmium Association, pp 9–16

Chapter 18
Bioremediation and Heavy Metal Uptake: Microbial Approaches at Field Scale

Frank Schindler, Matthias Gube, and Erika Kothe

18.1 Introduction

Pollution of the biosphere with toxic heavy metals is a widespread ecological problem resulting from anthropogenic activities such as fossil fuel burning, ore mining and smeltering, industrial and municipal waste disposal, and agricultural activities (Nriagu 1979; Adriano 2001; Kratz and Schnug 2006). In Western Europe alone, about 300,000 sites have been contaminated with heavy metals (Gade 2000; McGrath et al. 2006). The retention time of metals in soil is thousands of years because, unlike organic pollutants, metals are not degraded biologically. They rather are transformed from one oxidation state or organic complex into another and therefore persist in soil (Gisbert et al. 2003). However, the mobility of (heavy) metals may change, resulting in wash-out into ground and surface waters or uptake into plants via microbial physiological processes. While the major metal contamination is specific for each site, most operations will lead to multimetal contamination, which in most cases include Cd, Pb, and Cr among the pollutants. Since these three metals may be viewed as models for different behavior of heavy metals, we will focus on these three metals as soil contaminants in this chapter.

The mobility of metals in soil is a key determinant for distribution to other compartments, including the biosphere. Thus, metal chemistry has to be viewed with respect to solubility of complexes in order to evaluate distribution and the potential for uptake into biomass. Cd is one of the most toxic metals with effects on soil biology, plant metabolism, and animal health. It forms metal–organic complexes with a solubility significantly lower than that of other heavy metals leading to enrichment and high stability. In soil, Cd is strongly affected by pH, but in acidic soils, organic matter and sesquioxides may largely control the solubility (Forbes et al. 1976; Gadde and Laitinen 1974; Street et al. 1977). In contrast, Pb is

F. Schindler • M. Gube • E. Kothe (✉)
Institute of Microbiology, Friedrich-Schiller-Universität, Neugasse 25, 07743 Jena, Germany
e-mail: erika.kothe@uni-jena.de

E. Kothe and A. Varma (eds.), *Bio-Geo Interactions in Metal-Contaminated Soils*, Soil Biology 31, DOI 10.1007/978-3-642-23327-2_18,

very immobile in soils due to specific absorption mainly into insoluble metal–organic complexes. The third metal considered here, Cr, is one of the most widely used metals in industry and mainly is present as chromate soluble under oxidizing conditions. In soil, it is very immobile. Cr depresses plant growth, decreases net photosynthesis, and increases the vapor pressure difference (Davies et al. 2001).

Cd, Cr and Pb uptake into plants varies widely. Their toxicity may lead to typical stress symptoms including chlorosis, wilting and stunted growth, but even healthy plants may contain high metal contents. If metal ions are entering the root, still the root-to-shoot transfer may be limited (Scheffer et al. 2008). To calculate the capacity of a plant to extract heavy metals into the biomass, a transfer coefficient is used which indicates the ratio of a metal in plant tissue divided by the soil content. The resulting plant contents then can be evaluated with respect to ecologically critical concentrations of metals (Table 18.1).

The transfer coefficients usually are based on the soil total content of a metal. However, only a portion of all metal is bioavailable to the plant root, namely the water soluble or easily extractable fraction, which can enter a plant with the root water uptake. Since the same fraction of a heavy metal can be regarded most harmful, due to potential wash-out into ground and surface waters or transfer into the food chain, it may be advisable to differentiate between the transfer coefficient based on total contents and a new factor describing the transfer with regard to the bioavailable content of a given metal in soil. This bioavailable fraction is the amount of metal directly accessible to microorganisms such as bacteria, achaea, and fungi, as well as to higher organisms such as plants.

Phytoremediation is a fast developing technology using green plants to clean up the environment from pollutants in a cost-effective and noninvasive way as a substitute to the conventional, engineering-based remediation methods. Conventional techniques for reclamation of metal contaminated soils are expensive and destroy soil structure. They may be even environmentally harmful by allowing for dust emission during excavation and transport of contaminated soils. In contrast, bioremediation offers an economical alternative to remediate contaminated areas in situ. As a strategy for bioremediation, phytoextraction is applied where harvestable above-ground biomass is used as sink, which is then composted, fermented or burned to further concentrate metals (Dushenkov et al. 1995, 1997; Salt et al. 1995). In optimal cases, this can even be linked to phytomining. Alternatively, phytostabilization with sequestration of metals in soil and root zone is applied with the effect that shoot biomass can be used for applications such as energy production without harm (see also Haferburg et al. Chap. 2).

Table 18.1 Transfer coefficient soil–plant, normal and critic concentrations of heavy metals in plants (in [mg kg^{-1} dry matter]) after Scheffer et al. (2008)

Metal	Transfer coefficient	Normal in plants	Critical for plant growth	Critical for animal feed
Cd	0.03–10	<0.1–1	5–10	0.5–1
Pb	<0.5	1–5	10–20	10–30
Cr	<0.5	0.1–1	1–2	50–3,000

The strong interactions of plant roots with the rhizosphere bacterial and fungal communities allow for a large impact of microbes on phytoremediation (Baum et al. 2006; Rivera-Becerril et al. 2002; Wang et al. 2005). Although in situ application is the ultimate goal of bioremediation studies, open land experiments using microbial biomass have been less than often performed. This is probably due to the increased complexity of environmental parameters in field trials. Yet, even seemingly basic features, such as element uptake, may differ between laboratory and field scale (Baldrian 2003).

Soil polluted with heavy metal has, in addition, successfully been treated with fungi, termed mycoremediation, followed by leaching of the pollutants (Humar et al. 2004). Numerous fungal species abundantly produce fruiting bodies on contaminated sites into which many species incorporate considerable amounts of elements, which can surpass soil contents (García et al. 1998; Gast et al. 1988; Kalač and Svoboda 2000; Svoboda et al. 2000). For use of fungi in mycoremediation approaches, natural hyperaccumulator species of fungi need to be identified and evaluated.

18.2 Field-Scale Experiments at a Former Uranium Leaching Heap Site

In the process of uranium mining at the former Ronneburg district, Germany, low-grade silicate minerals with an uranium content below 300 g/ton and a high pyrite content were dumped since 1972 at the former clay mining area Gessen. The 28.7 ha large and 45 m high Gessen heap was leached with uranium-rich, acidic pit waters (pH 2.7–2.8) and later with sulfuric acid. The low pH mobilization was used to precipitate uranium yellow cake from the collected leachate. After removal of the primary contamination source, sandy-silty soil was applied to a depth of approximately 40 cm. By capillary rise and horizontal infiltration, this soil now contains heavy metals at concentrations above threshold values. The study area forms a shallow valley in E-NE direction (70° true bearing). Soil and groundwater characteristics, and residual contamination of the area were analyzed (Carlsson and Büchel 2005; Grawunder et al. 2009). At this site, a 2,500 m^2 test field was established in 2004, with three fields of 12 m × 12 m, each divided into nine subplots of 4 m x 4 m (Büchel et al. 2005). On the experimental plots, investigations of microbially assisted phytoremediation, microbially assisted are performed (Neagoe et al. 2005). To improve plant growth, one of the fields was amended with about 5 cm compost, which was plowed into a depth of approximately 20 cm. A second field was amended with topsoil in a similar way, while the third field was left untreated. On each of these three fields, nine subplots were either inoculated with different microorganisms or left unamended to examine the microbial influence on plant growth and bioremediation. We used microbial inoculation in order to increase metal mobility in the soil with the idea that a better performance of phytoextraction might be reached.

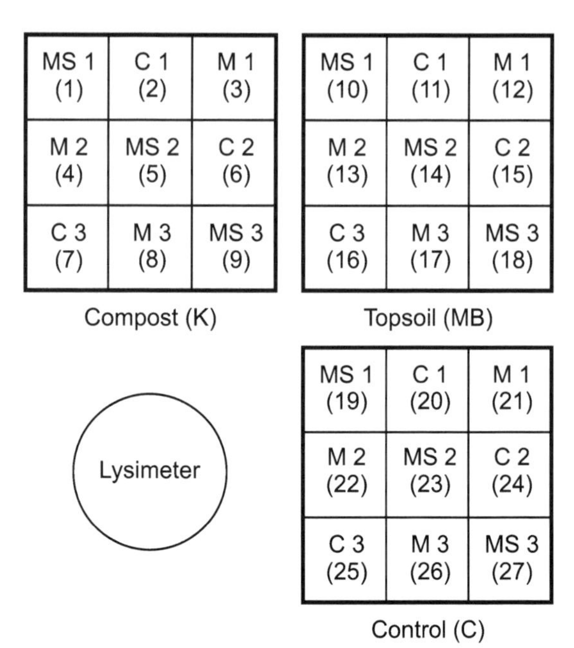

Fig. 18.1 The inoculation design of the test field site

The microbial inoculations were performed in a Latin square design, with three control subplots, three subplots inoculated with *Glomus* and three with both, *Glomus* and *Streptomyces* (Fig. 18.1).

The total concentration of Cd, Pb and Cr in the test field soils range from 0.8 to 47 μg/g. While the content of Cd is almost constant in the plots, Pb and Cr total contents vary considerably. The Cd content of 0.9 μg/g in the substrate is below action limits as defined by German law. The mean total content of all substrates for Pb is 37.34 μg/g, with slightly higher contents in the topsoil and compost plots. These values are well within common levels for Germany. The average Cr contents for all plots is 47.79 μg/g, being slightly increased in comparison with the values indicated for normal soils (Zeien and Brümmer 1989).

18.3 Microbially Enhanced Phytoextraction at the Field Site

18.3.1 Inoculation Influence on Bioavailability of Cd, Pb and Cr

As introduced above, not the total contents, but rather bioavailable fractions should be used to define microbially enhanced phytoremediation. Microbes are able to change the redox status and alter the material metals are absorbed to. Thus, monitoring of different associations is a basic necessity to predict future metal behavior. To define different fractions of metals, sequential extraction was performed (Zeien and Brümmer 1989). It is applicable to soil with low $CaCO_3$-content or $CaCO_3$-free soils (Grawunder et al. 2009), which is the case for our test field site. The heavy metals are divided into seven fractions: (F1) mobile, (F2) specifically adsorbed, (F3) bound to Mn-oxides, (F4) bound to organic matter, (F5) bound to amorphous Fe-oxides, (F6) bound to crystalline Fe-oxides, and (F7) the residual fraction. The residual fraction contains on average about 55% of the total Cd.

This again shows the necessity to investigate different fractions when considering phytoremediation (Fig. 18.2). Still, Cd is readily available in the substrate, since about 20–25% of the total content is found in the mobile fraction. Enhanced Cd content in the organic fraction (F4) is visible only for the compost amended plot. With inoculation, more Cd is found in the organic fraction with 20% compared to 0.7% in the control subplots. Thus, the addition of organic matter has decreased mobility and hence potential for phytoextraction. We found low contents of bioavailable Cd in the compost and the nonamended control. The potentially mobile pool was 33–38% of total content. The inoculation had almost no effect on bioavailability. Only in the double inoculated topsoil subplots, the bioavailable Cd level decreased. Thus, the microbial inoculation rather decreased Cd mobility, which should allow for lower phytotoxicity. Thus, inoculation would rather be applicable for phytostabilization than for phytoextraction. Pb was largely immobile in all three soils with a tendency to remove some Pb into the organic matter. Inoculation led to a slight increase of Pb in the bioavailable fractions, indicating a potential for microbial enhancement of phytoextraction. Still, Pb was always inhomogenously distributed. In all soil samples, Cr was present at high proportions, with 70–80%, in the crystalline Fe bound and the residual fractions. The bioavailable fraction was less than 1% of the total content; inoculation with fungi and the double inoculation with fungi and bacteria lowered the bioavailable contents to amounts below those of the uncontaminated control.

A correlation analysis using Pearson index revealed a weak negative correlation between Cd and Pb, as well as between Pb and Cr. However, no correlation between Cd and Cr bioavailable fractions was visible. The correlation can most likely be attributed to coprecipitation or adsorption processes (Gherghel 2009; Krpata et al. 2008; Mleczko 2004; Hui et al. 2011).

18.3.2 Metal Uptake into Plant Biomass

From the bioavailable pool, the metal may be taken up into plant biomass. However, different plants are known to perform differently, specifically with respect to root–shoot transfer. The test field has been planted with sunflower (*Helianthus annuus*) for four subsequent years. The results shown thus are reflecting a trial in four growth periods. For the analyses of plant contents, five shoots and roots were dried and metal load was determined. Above-ground biomass (Fig. 18.3) and roots (Fig. 18.4) were examined separately.

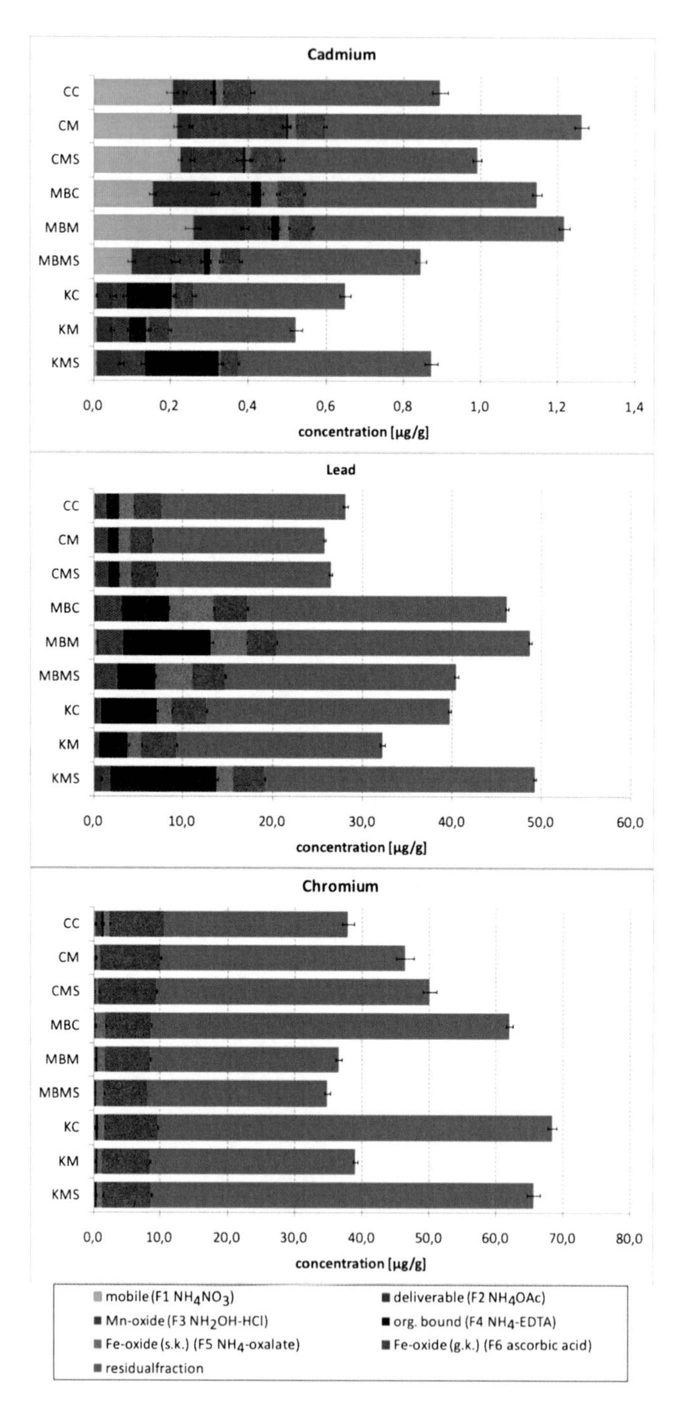

Fig. 18.2 Test field soil contents of Cd, Pb, and Cr by sequential extraction after Zeien and Brümmer (1989)

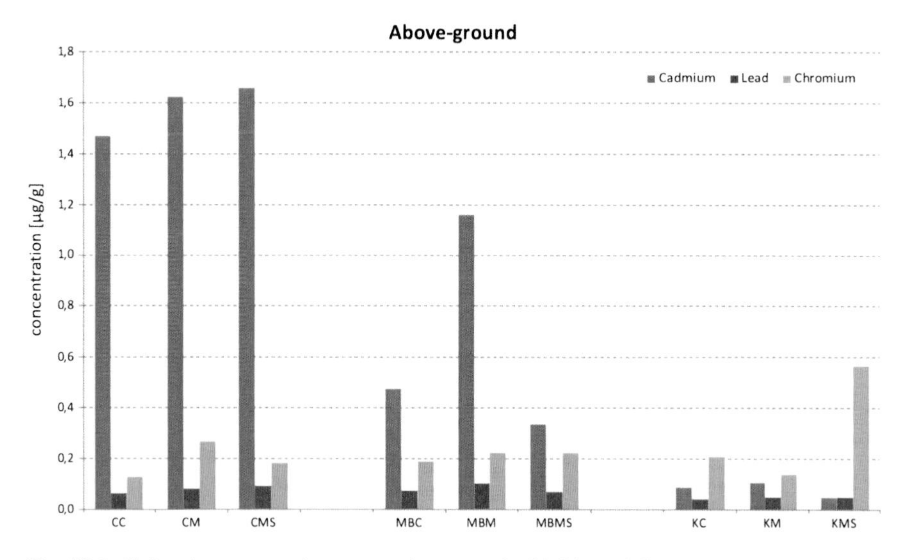

Fig. 18.3 *Helianthus annuus* above-ground contents in Cd, Pb, and Cr

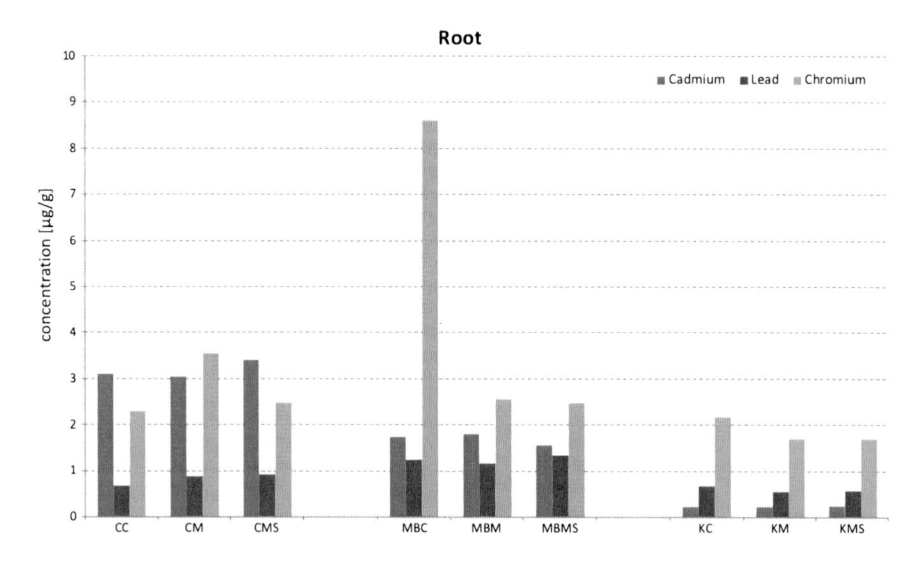

Fig. 18.4 Cd, Pb, and Cr contents of *Helianthus annuus* roots

With respect to Cd, increased plant contents were found with amounts about 12 times higher than the average plant (Scheffer et al. 2008). On the compost plot, plant Cd concentrations were comparatively low (Fig. 18.3).

Most Cd was found in the roots, where it is transported via apoplastic transport due to its high mobility (Kabata-Pendias and Pendias 2001). Microbial enhancement of bioavailability is indicated in both higher root and shoot contents. Thus, microbial inoculation seems favorable for phytoextraction strategies. While Pb contents generally were very low, as expected due to the low bioavailability (Hemkes et al. 1983; Karamanos et al. 1976; Korcak and Fanning 1985), the compost amended soil showed even lower Pb concentrations in plant biomass, visible for both above- and below-ground biomass. The inoculation with mycorrhizal fungi and bacteria increased plant Pb uptake. Plant Cr contents were fivefold enhanced compared to standard values (Scheffer et al. 2002); an influence of inoculation was not visible. This is easily explained by the low solubility. However, uptake in the above-ground tissues was slightly increased with mycorrhizal fungi inoculation. In recent studies, it has been shown that mycorrhization with *Glomus intraradices* enhanced growth, gas exchange, Cr uptake, and Cr tolerance in sunflower (Davies et al. 2001).

18.3.3 Transfer Coefficients with Regard to the Bioavailable Fractions of Metals

The transfer coefficients for the individual treatment types and elements were calculated from the bioavailable metal contents in the soil (Fig. 18.5).

Since Cd was present in relatively high bioavailable concentrations in the nonamended control substrate, Cd uptake and transfer coefficient were high on that plot with a further increase following inoculation. Cd uptake was in a range, which might be critical for plant growth (Scheffer et al. 2002). However, judging from our results for Cd, an increasing of the soil organic matter by amendments should be avoided while applying microbial enhancement. In contrast, transfer

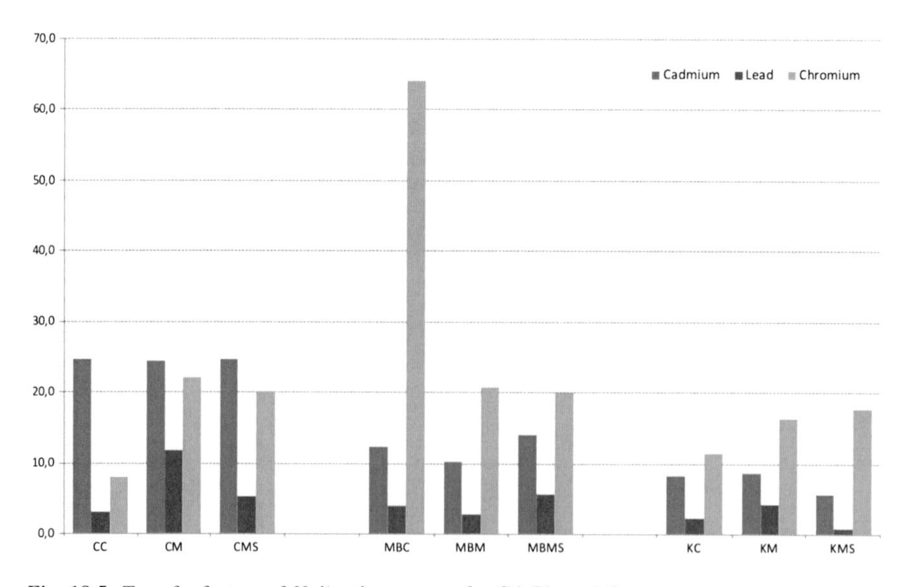

Fig. 18.5 Transfer factors of *Helianthus annuus* for Cd, Pb, and Cr

coefficients were relatively small for Pb, which does not allow for recommendations on microbially assisted phytoextraction. The results rather implicate phytostabilization approaches for Pb. Although Cr accumulation was low, an extraction could be shown, especially in the inoculated subplots. This again indicates microbial enhancement as a feasible strategy for this metal.

18.4 Mycoremediation

18.4.1 Diversity of Fungal Fruiting Bodies at the Test Field Site

We collected fruiting bodies from the vicinity of our test site (Table 18.2). In our study area, species composition of mycorrhizal fungi matched several descriptions of comparable successional stages associated with *Betula* and *Pinus* (Dighton et al. 1986; Fleming 1985; Shaw et al. 2003; Shaw and Lankey 1994). While early stage mycorrhizal species (Deacon et al. 1983), especially *Thelephora terrestris*, *Laccaria proxima*, and *Paxillus involutus*, seem to prevail in high abundancies for a relatively long time, some later stage mycorrhizal species, such as *Amanita muscaria*, are still absent from the studied site. A study on ectomycorrhizal diversity close to our test field site, on the former cone heaps near Ronneburg (Staudenrausch et al. 2005) revealed an interesting species composition, with both early and late colonizers abundant. Differences in pedological and hydrological characteristics to our site are evident and the *Betula* stands were older. Still, both results indicate a generally longer dominance of early colonizer species in these disturbed habitats. Differences in species composition between contaminated and control sites have been shown in another investigation in the same mining area and elsewhere even for later successional stages (Gherghel 2009; Krpata et al. 2008; Mleczko 2004). However, early colonizers have only sometimes been shown dominant at contaminated sites (Hui et al. 2011). This indicates extended, but finally still limited niche availability for such species.

To analyze ectomycorrhizal ecotypes, reference is often taken to their exploration type (Agerer 2001) in terms of rhizomorph functioning. They could function for avoiding heterogeneously distributed hot spots in the area, while still making nutrients available to both fungus and host tree. Exploration types doubtlessly influence mycelial biomass and surface, and thus can be linked to nutrient as well as heavy-metal uptake. Unfortunately, exploration types have not been assigned to saprobic species, which constitute the majority of our sampled species. It has been suggested that long distance exploration type species should be more prevalent in areas with lower root density (Peay et al. 2011). While this may be valid for forest edges, we cannot find a dominance of long distance types among the relatively loose stands of young trees at the test field, where no transition to woodland exists (Table 18.2).

Table 18.2 Species found in the vicinity of the test field site; ecolgy and exploration types

Species	2007	2008	2009	2010	Ecology[a]	Exploration type[b]
Agaricus campestris		x	x		s	Unknown
Agaricus semotus		x	x		s	Unknown
Bovista plumbea	x	x	x	x	s	Unknown
Cladonia sp.	x	x	x	x	l	Unknown
Claviceps purpurea			x		p	Unknown
Clitocybe dealbata	x	x	x		s	Unknown
Conocybe pulchella		x			s	Unknown
Coprinus comatus	x			x	s	Unknown
Cyathus olla	x	x	x		s	Unknown
Entoloma sericeum	x	x	x		s	Unknown
Erysiphe trifolii	x	x	x	x	p	Unknown
Hebeloma crustuliniforme				x	em	sd
Inocybe similis	x	x			em/lm	sd
Inocybe splendens				x	em/lm	sd
Laccaria proxima	x	x	x	x	em	mds
Lacrymaria lacrymabunda	x		x		s	Unknown
Lactarius rufus		x		x	lm	c
Lactarius torminosus	x	x	x	x	em/lm	c
Leccinum scabrum	x	x	x	x	lm	ld
Leccinum versipelle	x		x	x	lm	ld
Lepista sordida		x			s	Unknown
Leucoagaricus leucothites	x	x	x		s	Unknown
Lycoperdon lividum	x			x	s	Unknown
Lycoperdon pratense	x	x	x	x	s	Unknown
Marasmius oreades	x	x			s	Unknown
Melanoleuca brevipes	x		x		s	Unknown
Paxillus involutus	x	x	x	x	em	ld
Peziza sp.			x		s	Unknown
Pisolithus tinctorius		x	x	x	em	ld
Psathyrella candolleana	x	x			s	Unknown
Puccinia poarum	x	x	x	x	p	Unknown
Rickenella fibula	x		x		p	Unknown
Russula aeruginea	x		x		lm	c
Russula exalbicans		x			lm	c
Stropharia coronilla	x	x	x		s	Unknown
Suillus bovinus		x	x		em	ld
Suillus luteus	x	x	x		em/lm	ld
Thelephora penicillata					em	mds
Thelephora terrestris	x	x	x		em	mds
Tricholoma fulvum	x	x	x		lm	mdf

[a]*S* saprobic, *L* lichen, *EM* early stage mycorrhiza, *LM* late-stage mycorhiza, *P* parasitic
[b]*c* contact, *ld* long distance, *mds* medium distance smooth, *mdf* medium distance fringe

Unfortunately, literature concerning succession of saprobic open land species is comparatively scarce (Arnolds 1988; Lange 1982, 1984, 1991). Still, it is evident here that species composition is not hugely different from similar habitats not influenced by heavy metals. Nitrophilic species (Lange 1982, 1991), such as *Marasmius oreades*, *Lycoperdon pretense*, *Leucoagaricus leucothites*, and certain *Agaricus* species are especially abundant, indicating the influence of the surrounding agriculture. However, certain parts of the study area are rather dominated by nitrophobous (Lange 1982, 1991) species, such as *Rickenella fibula* and *Entoloma sericeum*. These species accompany more open vegetation dominated by mosses and lichens rather than herbs. This indicates a heterogenous distribution of nitrogen. On the test field plots, fruiting bodies of nitrophilic species are becoming more abundant with the addition of nutrient-rich substrates. While on the control plot fruiting bodies were never observed, addition of top soil led to occasional growth of *Agaricus semotus*, which is known from humus-rich habitats (Vellinga 2001). The compost plot yielded abundant fruiting bodies of *Leucoagaricus leucothites* and *Cyathus olla*, both clearly nitrophilic species.

In summary, species composition at our test field site reflects both the state of plant succession and the influence of habitat conditions, both natural and anthropogenic. While the contamination has minor effects on species composition, nitrogen availability and flora composition seem to show a greater influence.

18.4.2 Cd, Cr, and Pb Accumulation in Fungi at the Test Field Site

To judge the potential use of species from the test field site for mycoremediation, heavy-metal transfer coefficients of mushrooms need to be defined, which has rather rarely been performed (Colpaert et al. 2000). Again, bioavailability should also be taken into account. To our knowledge, metal uptake into fruiting bodies of *Thelephora terrestris*, *Stropharia coronilla*, *Lepista sordida*, and *Melanoleuca brevipes* were tested for the first time. Cd uptake in our fungal samples varied considerably, both among species (from 0.056 μg/g in *Pisolithus tinctorius* to 26.64 μg/g in a *Leucoagaricus leucothites* sample) and within species (with *Leucoagaricus leucothites* samples ranging from 0.102 to 26.64 μg/g). At a total soil value of 0.48 μg/g in average, transfer coefficients range from 0.19 to 55.5. When the bioavailable Cd at 0–24 μg/g is considered, accumulation ranges from 0 to 111. Eleven of the 21 sampled species generally accumulated Cd over the total soil content, namely *Agaricus campestris*, *Agaricus semotus*, *Bovista plumbea*, *Leccinum scabrum*, *Lepista sordida*, *Lycoperdon pratense*, *Melanoleuca brevipes*, *Peziza* sp., *Psathyrella candolleana*, *Stropharia coronilla*, and *Tricholoma fulvum*. Additionally, *Laccaria proxima*, *Lactarius rufus*, and *Thelephora terrestris* always accumulated at least above the bioavailable content. Only *Pisolthus tinctorius* shows no Cd accumulation at all. Of the remaining species, some samples exceed at least bioavailable soil values. In a review of metal contents in fruiting bodies (Kalač and Svoboda 2000), Cd and Pb values have been compared for a range of species. While a very broad range of values for both elements is evident, many patterns can be confirmed with our values. Our measurements of Cd within *Agaricus campestris* and *Agaricus semotus* are far below the most extreme values of 300 μg/g for *Agaricus* species. Still, average uptake of related species is relatively high, especially not only in *Leucoagaricus leucothites*, but also *Bovista plumbea* and *Lycoperdon pratense*. The close relationship of these puffball species with Agaricaceae has only recently been revealed (Gube 2009; Hibbett et al. 1997; Vellinga 2004), and similar uptake characteristics might be attributed to phylogenetically conserved mechanisms (Stijve et al. 2001). High values for *Leccinum* species are also confirmed. The high Cd uptake potential of *Melanoleuca brevipes*, *Psathyrella candolleana* and *Tricholoma fulvum* was not known before, but as yet is based on single specimens. Surprisingly low are the values for *Thelephora terrestris*, which greatly accumulates both other metals tested (see below).

For Cr, great variance of uptake is evident, ranging from 0.048 μg/g in a *Bovista plumbea* sample to 165.33 μg/g in a *Thelephora terrestris* sample, corresponding to a 3,374-fold higher Cr content in the latter. Cr accumulation ranges from 0.002 to 6.08 for total soil values, and from 0.15 to 501 for the bioavailable contents. The largest variation within a species is also observed with *Thelephora terrestris*, spanning from 9.7 to 165.33 μg/g. Total soil contents of Cr average 27.19 μg/g, which is only occasionally surpassed by *Thelephora terrestris* fruiting bodies. Concerning the bioavailable fractions at 0.33 μg/g, *Laccaria proxima*, *Leccinum versipelle*, *Lepista sordida*, *Melanoleuca brevipes*, *Peziza* sp., *Pisolithus tinctorius*, *Psathyrella candolleana*, *Stropharia coronilla*, and *Tricholoma fulvum* also accumulate Cr. Only *Bovista plumbea* was never found accumulating Cr above at least the bioavailable soil content. Although this metal is extremely harmful for health and environment (Cervantes et al. 2001; Cohen et al. 1993), Cr contents in fungal fruiting bodies are recorded relatively rarely (Malinowska et al. 2004; Michelot et al. 1998; Radulescu et al. 2010; Sarikurkcu et al. 2010). Among these references, the highest value measured was 16.4 μg/g for *Amanita caesarea*. While our *Thelephora terrestris* samples surpass this value up to tenfold, all other species stay below 6 μg/g. Like for Cd and Pb, uptake of Cr varies greatly within species in our data and also in literature. Apart from *Thelephora*, soil inhabiting Ascomycete fruiting bodies might have a potential for higher accumulation. This is evident in samples of *Morchella esculenta* and *Helvella crispa* (Michelot et al. 1998), and our *Peziza* sample, which have relatively high values, always above 4 μg/g.

Pb uptake also has a wide range among species, from 0.039 μg/g in a *Lactarius torminosus* sample to 15.53 μg/g in a *Leucoagaricus leucothites* sample. Corresponding accumulation factors span from 0.019 to 0.76 for the total soil content, and 0.16 to 62.12 for the bioavailable contents. In *Leucoagaricus leucothites*, the within-species uptake range is especially high, from 0.099 to 15.53 μg/g, equaling a 157-fold difference. However, the high-Pb sample is exceptional among samples of that species, which otherwise show an average uptake of 0.303 μg/g lead. The highest mean uptake is evident in *Thelephora terrestris*, where it averages 4.57 μg/g. Total soil content, at 20.36 μg/g, is higher than that of any sample, while *Agaricus semotus*, *Bovista plumbea*, *Leccinum versipelle*, *Lepista sordida*, *Lycoperdon pratense*, *Melanoleuca brevipes*, *Peziza* sp., *Pisolithus tinctorius*, *Psathyrella candolleana*, *Stropharia coronilla,* and *Thelephora terrestris* always surpass the bioavailable Pb content at 0.25 μg/g. Both *Lactarius* species and *T. fulvum* never exceeded that value. Like for Cd, our values for Pb are well below the most extreme records of 300 μg/g (Kalač and Svoboda 2000), but again representatives of the genera designated as accumulators there showed higher uptake also in our samples. In accordance with this, *Lycoperdon pratense* and *Lepista sordida* are among the high accumulating species, along with other Agaricaceae, *Leccinum*, and especially *Thelephora terrestris*.

Our results underline the relevance of fungi for element cycling, especially at contaminated habitats. Their partly immense metal uptake capacities and resistances suggest an important role through modification of soil element mobility and distribution of heavy metals through the food chain. It also makes use of fungal fruiting bodies for mycoremediation approaches attractive.

18.4.3 *Feasibility Study of* Leucoagaricus leucothites *Mycoremediation in Conjunction with Microbial Inoculation*

Leucoagaricus leucothites abundantly produced fruiting bodies on the compost amended plot of our test field site in 2008 and 2009. Thus, we compared metal uptake characteristics for samples from each subplot (Fig. 18.6). On subplots C2 and C3, fruiting bodies were only present in 2008, while none were produced on subplot M1 in 2008 or 2009. In 2010, no fruiting of *Leucoagaricus leucothites* occurred in the area. Productivity on average slightly decreased in 2009, but variation among

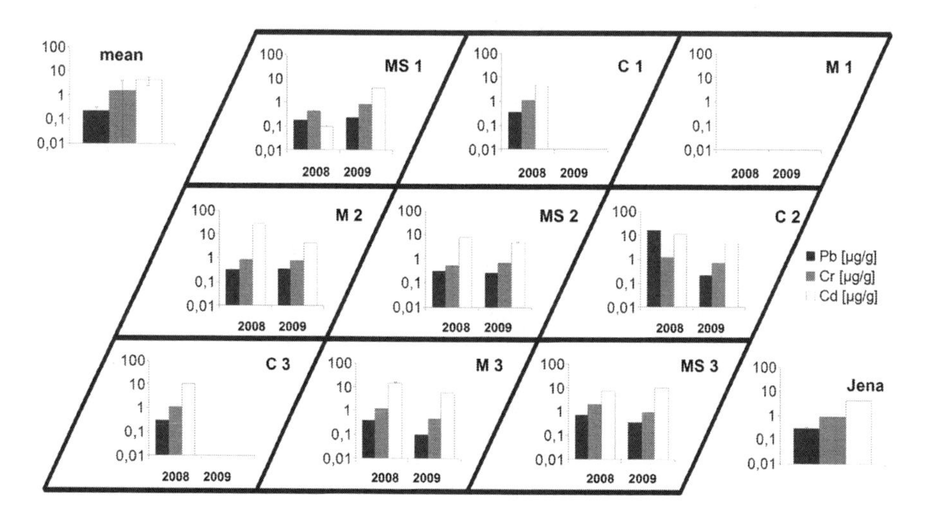

Fig. 18.6 Uptake of Cd, Cr, and Pb into fruiting bodies of *Leucoagaricus leucothites* in 2008 and 2009

absolute values within subplots was quite high. Thus, yearly differences are not significant. Apart from an exceptionally high Pb values on the C2 subplot in 2008, and a very low Cd uptake on the MS1 plot in 2008, uptake of the three tested metals remained within the same order of magnitude. Uptake occurred in a similar pattern, even in fruiting bodies outside the test field. Exceptions are the C2 and MS1 plots in 2008. Correlation, indicated by Pearson's r, was always higher than 0.98. Additionally, Cr and Pb uptake were correlated in these samples (Pearson's $r = 0.88$). Assuming that translocation to fruiting bodies is about equal within one mycelium (Bazała et al. 2008; Brunnert and Zadražil 1983), this could be interpreted with an increasing dominance of one clonal mycelium over competitors of the same species. However, comparisons of uptake patterns with *Leucoagaricus leucothites* samples from other areas, contaminated and uncontaminated with heavy metals, shows patterns similar to the majority of the subplot patterns. Other possible reasons for differences in uptake are the age of the fruiting bodies (Thomet et al. 1999), which was not uniform in the samples, or irregularities in metal ion mobility at the sample site, overriding balancing mycelial transport.

Cd, Pb, and Cr contents have been tested before in *Leucoagaricus leucothites* (Doğan et al. 2006; Sarikurkcu et al. 2010; Seeger 1978; Tüzen et al. 2007; Yilmaz et al. 2003). Other *Leucoagaricus* species were also analyzed, namely *L. nympharum*, *L. cinerascens,* and *L. barssii* as *L. macrorhizus* (Işiloğlu et al. 2001; Tüzen et al. 2007). All show similar characteristics for Cd, Pb, and Cr. Reported Cd contents range from 0.73 to 11.6 μg/g (mean 2.4 μg/g), Pb contents from 0.73 to 16.5 μg/g (mean 5.21 μg/g) and Cr contents 3.4–12 μg/g (mean 7.7 μg/g). In the case of Pb and Cr, this exceeds values of our analysis, while Cd uptake is slightly lower in comparison. While our sampling area is not specifically rich in the tested elements, soil contents were not measured in the literature records, so accumulation factors cannot be compared. High Pb levels were to be expected in the soil accommodating some Turkish roadside samples; as leaded gasoline was used at least until 1996 in Turkey (Karakaya et al. 1996).

Considering all this, *Leucoagaricus leucothites* is a close to ideal model fungus to analyze the potential of mycoremediation by fruiting bodies. It is a fast-growing saprobic species and fruits abundantly on our testing site. It is furthermore related to commercially cultivated species, therefore cultivation protocols should not be too hard to establish. Most importantly, it has the potential to accumulate certain heavy metals to a significant degree, possibly allowing for broader scale application. However, the range of uptake varies greatly, which might be due to so far unknown biotic and/or abiotic factors. To understand the impact of these factors on metal uptake, a more profound knowledge of the genetic regulation of the metal responsive factors of this species seems desirable.

18.5 Conclusions

In our study, we tested two soil amendments, addition of low amounts of compost or topsoil. Especially the compost addition changed the mobility of the tested metals, followed by topsoil amendment, while metals were mainly bioavailable in the nonamended control field. With higher soil organic matter contents, more heavy metals were bound to the organic fraction, probably as insoluble complexes. Especially for the immobile Pb (Svoboda et al. 2000), accumulation in the organic material was seen (Korcak and Fanning 1985; Dighton et al. 1986; Fleming 1985; Shaw et al. 2003), which would limit bioremediation. However, some of the Pb was remobilized following inoculation, showing enhanced applicability for Pb phytoextraction. Cd is particularly well suited for bioremediation or phytoextraction due to its high proportion in the bioavailable phase. The amendments can minimize the bioavailable fractions of heavy metals, thus lowering wash-out. Inoculation with fungi and bacteria clearly can increase this effect. We thus have performed an proof-of-principle for successful microbial enhanced bioremediation.

Plant–microbe associations can be exploited to remediate heavy-metal-contaminated soils. By promoting plant growth, microorganisms can aid sustainable biomass production and improve plant establishment (Agerer 2001). Improved yields also aid phytoextraction, as the total content removed is dependent on specific uptake as well as total biomass produced. Microbially enhanced phytoremediation is an emerging technology with potential for the future with little imminent risk to health or the environment. However, more field studies are needed to fully exploit the potentials of this methodology, as clearly seen in our study for both inoculation for stimulated phytoextraction and mycoremediation.

Due to their unique abilities to cope with heavy metals, fungi play a prominent role in metal affected soil. Although the major influence on fungal species diversity at the test site can be attributed to other natural and anthropogenic influences, contamination may be indicated by a shift in diversity toward species typical for earlier succession stages. Uptake of Cd, Cr, and Pb into fruiting bodies did confirm their relevance for element cycling. The considerable enrichment abilities of certain species suggest an application in biomonitoring and bioprospection, and most importantly mycoremediation by fungal fruiting bodies. The common, fast-growing mushroom *Leucoagaricus leucothites* was designated as an ideal model to analyze the potential of such applications.

Acknowledgments The authors are grateful to FSU Jena Applied Geology staff especially D. Merten, U. Buhler, I. Kampand and G. Weinzierl for laboratory assistance, and to S. Formann for help in collecting fruiting bodies. The work was financially supported by the excellence graduate school JSMC and DFG GRK1257.

References

Adriano DC (2001) Trace elements in terrestrial environments: biogeochemistry, bioavailability, and risks of metals, 2nd edn. Springer, New York

Agerer R (2001) Exploration types of ectomycorrhizae – a proposal to classify ectomycorrhizal mycelial systems according to their patterns of differentiation and putative ecological importance. Mycorrhiza 11:107–114

Arnolds E (1988) Dynamics of macrofungi in 2 moist heathlands in Drenthe, the Netherlands. Acta Bot Neerlandica 37:291–305

Baldrian P (2003) Interactions of heavy metals with white-rot fungi. Enzyme Microb Technol 32:78–91

Baum C, Hrynkiewicz K, Leinweber P, Meissner R (2006) Heavy-metal mobilization and uptake by mycorrhizal and nonmycorrhizal willows (Salix x dasyclados). J Plant Nutrition Soil Sci 169:516–522

Bazała MA, Gołda K, Bystrzejewska-Piotrowska G (2008) Transport of radiocesium in mycelium and its translocation to fruitbodies of a saprophytic macromycete. J Environ Radioact 99:1200–1202

Brunnert H, Zadražil F (1983) The translocation of mercury and cadmium into the fruiting bodies of 6 higher fungi – a comparative-study on species specificity in 5 lignocellulolytic fungi and the cultivated mushroom agaricus-bisporus. Eur J Appl Microbiol Biotechnol 17:358–364

Büchel G, Bergmann H, Ebena G, Kothe E (2005) Geomicrobiology in remediation of mine waste. Chemie Erde-Geochemistry 65S1:1–5

Carlsson E, Büchel G (2005) Screening of residual contamination at a former uranium heap leaching site, Thuringia, Germany. Chemie Erde-Geochemistry 65S1:75–95

Cervantes C, Campos-García J, Devars S, Gutiérrez-Corona F, Loza-Tavera H, Torres-Guzmán JC, Moreno-Sánchez R (2001) Interactions of chromium with microorganisms and plants. FEMS Microbiol Rev 25:335–347

Cohen MD, Kargacin B, Klein CB, Costa M (1993) Mechanisms of chromium carcinogenicity and toxicity. Crit Rev Toxicol 23:255–281

Colpaert JV, Vanden Koornhuyse P, Adriaensen K, Van Gronsveld J (2000) Genetic variation and heavy metal tolerance in the ectomycorrhizal basidiomycete *Suillus luteus*. New Phytol 147:367–379

Davies FT, Puryear JD, Newton RJ, Egilla JN, Grossi JAS (2001) Mycorrhizal fungi enhance accumulation and tolerance of chromium in sunflower (*Helianthus annuus*). J Plant Physiol 158:777–786

Deacon JW, Donaldson SJ, Last FT (1983) Sequences and interactions of mycorrhizal fungi on birch. Plant Soil 71:257–262

Dighton J, Poskitt JM, Howard DM (1986) Changes in occurrence of basidiomycete fruit bodies during forest stand development - with specific reference to mycorrhizal species. Trans Br Mycol Soc 87:163–171

Doğan HH, Şanda MA, Uyanöz R, Öztürk C, Çetin Ü (2006) Contents of metals in some wild mushrooms. Biol Trace Elem Res 110:79–94

Dushenkov S, Vasudev D, Kapulnik Y, Gleba D, Fleisher D, Ting KC, Ensley B (1997) Removal of uranium from water using terrestrial plants. Environ Sci Technol 31:3468–3474

Dushenkov V, Kumar PBAN, Motto H, Raskin I (1995) Rhizofiltration – the use of plants to remove heavy-metals from aqueous streams. Environ Sci Technol 29:1239–1245

Fleming LV (1985) Experimental-study of sequences of ectomycorrhizal fungi on birch (Betula Sp) seedling root systems. Soil Biol Biochem 17:591–600

Forbes EA, Posner AM, Quirk JP (1976) Specific adsorption of divalent Cd, Co, Cu, Pb, and Zn on goethite. J Soil Sci 27:154–166

Gadde RR, Laitinen HA (1974) Studies of heavy-metal sorption by hydrous oxides. Abstracts of papers of the American Chemical Society, p 142

Gade LH (2000) Highly polar metal – metal bonds in "early-late" heterodimetallic complexes. Ang Chem-Int Ed 39:2659–2678

García MA, Alonso J, Fernández MI, Melgar MJ (1998) Lead content in edible wild mushrooms in northwest Spain as indicator of environmental contamination. Arch Environ Contam Toxicol 34:330–335

Gast CH, Jansen E, Bierling J, Haanstra L (1988) Heavy-metals in mushrooms and their relationship with soil characteristics. Chemosphere 17:789–799

Gherghel F (2009) Identification and characterization of *Quercus robur* ectomycorrhiza in relation to heavy metal contamination. Dissertation, Friedrich-Schiller-Universität Jena, Jena

Gisbert C, Ros R, De Haro A, Walker DJ, Bernal MP, Serrano R, Navarro-Avino J (2003) A plant genetically modified that accumulates Pb is especially promising for phytoremediation. Biochem Biophys Res Commun 303:440–445

Grawunder A, Lonschinski M, Merten D, Büchel G (2009) Distribution and bonding of residual contamination in glacial sediments at the former uranium mining leaching heap of Gessen/ Thuringia, Germany. Chemie Erde-Geochemistry 69:5–19

Gube M (2009) Ontogeny and phylogeny of gasteroid members of Agaricaceae (Basidiomycetes). Dissertation, Friedrich-Schiller-Universität Jena, Jena

Hemkes OJ, Kemp A, Vanbroekhoven LW (1983) Effects of applications of sewage-sludge and fertilizer nitrogen on cadmium and lead contents of grass. Neth J Agric Sci 31:227–232

Hibbett DS, Pine EM, Langer E, Langer G, Donoghue MJ (1997) Evolution of gilled mushrooms and puffballs inferred from ribosomal DNA sequences. Proc Natl Acad Sci USA 94: 12002–12006

Hui N, Jumpponen A, Niskanen T, Liimatainen K, Jones KL, Koivula T, Romantschuk M, Strömmer R (2011) EcM fungal community structure, but not diversity, altered in a Pb-contaminated shooting range in a boreal coniferous forest site in Southern Finland. FEMS Microbiol Ecol 76:121–132

Humar M, Pohleven F, Šentjurc M (2004) Effect of oxalic, acetic acid, and ammonia on leaching of Cr and Cu from preserved wood. Wood Sci Technol 37:463–473

Işiloğlu M, Merdivan M, Yilmaz F (2001) Heavy metal contents in some macrofungi collected in the northwestern part of Turkey. Arch Environ Contam Toxicol 41:1–7

Kabata-Pendias A, Pendias H (2001) Trace elements in soils and plants, 3rd edn. CRC, Boca Raton, FL

Kalač P, Svoboda L (2000) A review of trace element concentrations in edible mushrooms. Food Chem 69:273–281

Karakaya A, Ilko M, Ulusu T, Akal N, Isimer A, Karakaya AE (1996) Lead levels in deciduous teeth of children from urban and suburban regions in Ankara (Turkey). Bull Environ Contam Toxicol 56:16–20

Karamanos RE, Bettany JR, Stewart JWB (1976) Uptake of native and applied lead by alfalfa and bromegrass from soil. Can J Soil Sci 56:485–494

Korcak RF, Fanning DS (1985) Availability of applied heavy-metals as a function of type of soil material and metal source. Soil Sci 140:23–34

Kratz S, Schnug E (2006) Rock phosphates and P fertilizers as sources of U contamination in agricultural soils. In: Merkel BJ, Hasche-Berger A (eds) Uranium in the environment. Springer, Berlin, pp 57–67

Krpata D, Peintner U, Langer I, Fitz WJ, Schweiger P (2008) Ectomycorrhizal communities associated with *Populus tremula* growing on a heavy metal contaminated site. Mycol Res 112:1069–1079

Lange M (1982) Fleshy fungi in grass fields. Dependence on fertilization, grass species, and age of field. Nord J Bot 2:131–143

Lange M (1984) Fleshy fungi in grass fields. 2. Precipitation and fructification. Nord J Bot 4:491–501

Lange M (1991) Fleshy fungi of grass fields.3. Reaction to different fertilizers and to age of grass turf – periodicity of fruiting. Nord J Bot 11:359–368

Malinowska E, Szefer P, Falandysz J (2004) Metals bioaccumulation by bay bolete, *Xerocomus badius*, from selected sites in Poland. Food Chem 84:405–416

McGrath SP, Lombi E, Gray CW, Caille N, Dunham SJ, Zhao FJ (2006) Field evaluation of Cd and Zn phytoextraction potential by the hyperaccumulators *Thlaspi caerulescens* and *Arabidopsis halleri*. Environ Pollut 141:115–125

Michelot D, Siobud E, Doré JC, Viel C, Poirier F (1998) Update on metal content profiles in mushrooms – toxicological implications and tentative approach to the mechanisms of bioaccumulation. Toxicon 36:1997–2012

Mleczko P (2004) Mycorrhizal and saprobic macrofungi of two zinc mines in southern Poland. Acta Biol Cracoviensa Series Botanica 46:25–38

Neagoe A, Ebena G, Carlsson E (2005) The effect of soil amendments on plant performance in an area affected by acid mine drainage. Chemie Erde-Geochemistry 65:115–129

Nriagu JO (1979) Global inventory of natural and anthropogenic emissions of trace-metals to the atmosphere. Nature 279:409–411

Peay KG, Kennedy PG, Bruns TD (2011) Rethinking ectomycorrhizal succession: are root density and hyphal exploration types drivers of spatial and temporal zonation? Fungal Ecol 4:233–230

Radulescu C, Stihi C, Busuioc G, Gheboianu AI, Popescu IV (2010) Studies concerning heavy metals bioaccumulation of wild edible mushrooms from industrial area by using spectrometric techniques. Bull Environ Contam Toxicol 84:641–646

Rivera-Becerril F, Calantzis C, Turnau K, Caussanel JP, Belimov AA, Gianinazzi S, Strasser RJ, Gianinazzi-Pearson V (2002) Cadmium accumulation and buffering of cadmium-induced stress by arbuscular mycorrhiza in three *Pisum sativum* L. genotypes. J Exp Bot 53:1177–1185

Salt DE, Blaylock M, Kumar NPBA, Dushenkov V, Ensley BD, Chet I, Raskin I (1995) Phytoremediation: a novel strategy for the removal of toxic metals from the environment using plants. Nat Biotech 13:468–474

Sarikurkcu C, Tepe B, Semiz DK, Solak MH (2010) Evaluation of metal concentration and antioxidant activity of three edible mushrooms from Mugla, Turkey. Food Chem Toxicol 48:1230–1233

Scheffer F, Schachtschabel P, Blume H-P, Scheffer S (2002) Lehrbuch der Bodenkunde. 15. Aufl./ edn. Spektrum, Akad. Verl., Heidelberg

Scheffer F, Schachtschabel P, Blume H-P, Scheffer S (2008) Lehrbuch der Bodenkunde. 15. Aufl., [Nachdr.]/edn. Spektrum, Akad. Verl., Heidelberg

Seeger R (1978) Cadmium in mushrooms. Zeitschrift für Lebensmitteluntersuchung und -forschung 166:23–34

Shaw PJA, Kibby G, Mayes J (2003) Effects of thinning treatment on an ectomycorrhizal succession under Scots pine. Mycol Res 107:317–328

Shaw PJA, Lankey K (1994) Studies on the scots pine mycorrhizal fruitbody succession. Mycologist 8:173–175

Staudenrausch S, Kaldorf M, Renker C, Luis P, Buscot F (2005) Diversity of the ectomycorrhiza community at a uranium mining heap. Biol Fertil Soils 41:439–446

Stijve T, Andrey D, Goessler W, Guinberteau J, Dupuy G (2001) Étude comparative des métaux lourds et d'autres éléments traces dans Gyrophragmium dunalii et dans les agarics jaunissants de la section Arvenses. Bulletin trimestriel de la Société mycologique de France 117:133–144

Street JJ, Lindsay WL, Sabey BR (1977) Solubility and plant uptake of cadmium in soils amended with cadmium and sewage sludge. J Environ Qual 6:72–77

Svoboda L, Zimmermannová K, Kalač P (2000) Concentrations of mercury, cadmium, lead and copper in fruiting bodies of edible mushrooms in an emission area of a copper smelter and a mercury smelter. Sci Total Environ 246:61–67

Thomet U, Vogel E, Krähenbühl U (1999) The uptake of cadmium and zinc by mycelia and their accumulation in mycelia and fruiting bodies of edible mushrooms. Eur Food Res Technol 209:317–324

Tüzen M, Sesli E, Soylak M (2007) Trace element levels of mushroom species from East Black Sea region of Turkey. Food Control 18:806–810

Vellinga EC (2001) Leucoagaricus. In: Noordeloos ME, Kuyper ThW, Vellinga EC (eds) Flora Agaricina Neerlandica, vol 5. Agaricaceae. A.A. Balkema, Lisse, Abingdon, Exton, Tokyo, pp 85–108

Vellinga EC (2004) Genera in the family Agaricaceae: evidence from nrITS and nrLSU sequences. Mycol Res 108:354–377

Wang FY, Lin XG, Yin R (2005) Heavy metal uptake by arbuscular mycorrhizas of *Elsholtzia splendens* and the potential for phytoremediation of contaminated soil. Plant Soil 269:225–232

Yilmaz F, Işiloğlu M, Merdivan M (2003) Heavy metal levels in some macrofungi. Turk J Bot 27:45–56

Zeien H, Brümmer GW (1989) Chemische Extraktion zur Bestimmung von Schwermetallbindungsformen in Böden. Mitt Dtsch Bodenkdl Ges 59:505–515

Chapter 19
Contributions to the Theoretical Foundations of Integrated Modeling in Biogeochemistry and Their Application in Contaminated Areas

V. Iordache, R. Lăcătuşu, D. Scrădeanu, M. Onete, S. Ion, I. Cobzaru, A. Neagoe, F. Bodescu, D. Jianu, and D. Purice

19.1 Introduction

This chapter is a continuation and development of the ideas introduced in a review of integrated modeling in metals biogeochemistry (Iordache et al. 2009). We attempt here to make operational steps toward the research direction identified there. Two types of integrated models have been identified in the mentioned review: models integrating biological and abiotic processes, and models integrating processes of the same type (biological or abiotic) occurring at different scales.

Integrated modeling efforts are not specific only to metals biogeochemistry, but also to all other areas characterizing from different perspectives the planetary productive systems (ecological, socioeconomic, or socioecological). One can

V. Iordache (✉) • A. Neagoe • F. Bodescu
Research Centre for Ecological Services (CESEC), Faculty of Biology, University of Bucharest, Spl. Independentei 91-95, Bucharest 050089, Romania
e-mail: virgil.iordache@g.unibuc.ro

R. Lăcătuşu
Research Institute for Soil Science and Agrochemistry (ICPA), Bucharest, Romania

D. Scrădeanu
Laboratory of Hydrogeology, Faculty of Geology and Geophysics, University of Bucharest, Bucharest, Romania

M. Onete • I. Cobzaru • D. Purice
Institute of Biology Bucharest, Romanian Academy, Bucharest, Romania

S. Ion
Institute of Mathematical Statistics and Applied Mathematics, Romanian Academy, Bucharest, Romania

D. Jianu
Lythos Research Center, Faculty of Geology, University of Bucharest, Bucharest, Romania

E. Kothe and A. Varma (eds.), *Bio-Geo Interactions in Metal-Contaminated Soils*, Soil Biology 31, DOI 10.1007/978-3-642-23327-2_19,

even say that integration is underdeveloped in metal biogeochemistry. But fully covering all approaches for integrated modeling in environmental sciences in order to extract lessons to learn is a matter of writing a book, rather than a chapter. However, a screen of several examples from the literature can easily reveal the main trends.

What is at stake is to use knowledge (and models) developed in different disciplines in order to solve more complex problems, cross-cutting the disciplinary fields. There are two strategical trends: one is to react to the existence of many individual models developed disciplinary and to try ad-hoc integrations without paying much attention to relevance of the integrated model for the real coupled processes. The second is more proactive and suppose the development of new research areas conceived from the start as interdisciplinary.

The first (reactive) strategy is illustrated by the platforms/frameworks for linking existing models (reviewed by Argent et al. 2006). An example of software framework for integrating reusable components describing hydrological processes are Branger et al. (2010), one of a software platform for integrating existing models is Kraft et al. (2010), and an example of technical aspects raised by the integration software are illustrated by the work of Schmitz et al. (2011).

For the second, proactive strategy examples are the development of hydropedology and ecohydrology. Lin (2003) is a promoter of the integration between pedology, soil physics, and hydrology within a new disciplinary field hydropedology. His review uniquely points out the problems related to linking data and scale specific to separate disciplines, and is followed by conceptual developments (Lin 2010a, b) aimed at catalyzing the development of the new discipline. No consequences for integrated modeling in practice seem to have been generated by this effort by now, excepting for the estimation of pedotransfer functions (Pachepsky et al. 2006), suggesting that classic hydrological approaches are still dominant (Kohne et al. 2009). But, the accent of Lin and coworkers on data and concepts coherentization is a better strategy than hurrying up for integration of old models within and between disciplines. Following general recommendations for the evolution of hydrology ("cross-disciplinary integration must become a primary characteristic of hydrologic research" – Wagener et al. 2010), ecohydrology is an effervescent field for modeling coupled processes on a realistic base, from basic data, not by integrating previously developed disciplinary models (Hwang et al. 2009). The scale dependence of the studied processes in ecohydrology is in the top of research priorities in this field (Thompson et al. 2011).

The above two examples are bilateral integrations. How far can one go with this trend? Are there conceptual limitations for integrated many disciplines, as many as needed in biogeochemistry for dealing with biological and abiotic processes at large scale? Such limitations have been signaled within each discipline. For instance, besides technical aspects precluding integration and model performance in hydrology (Buytaert et al. 2008; Kavetski and Klark 2011), there are also general conceptual problems and attempts to solve them by alternatives to the current approaches (Schaefli et al. 2011 in hydrology), but have been not systematically characterized for large-scale ecological systems.

The extreme form of integration would be a holistic one (Odum 1995). Even more ambitious is the holistic research and integrated modeling of socioecological systems (Seppelt 2003). In this line of thinking, there is a recognized need for a holistic, integrated approach in metal biogeochemistry and in ecotoxicology (Matyssek et al. 2006; Breure et al. 2008). But, the accent of the holistic approach is more on the biological part of the ecosystem (methodology of systems identification Pahl-Vostl 1995), and less on the heterogeneity of the abiotic part. Most of the abiotic part of the holistically conceived ecosystems remain unstructured in systems ecology description, and are put under general headings such as biotope, or hydro-geomorphic unit. These abiotic features are characterized in detail by other scientific disciplines, such as geomorphology, soil science, hydrology, and hydrogeology.

In this chapter, we explore the following questions:

1. Is there an optimal level of integration for understanding the biogeochemical role of abiotic and biological objects, larger than the bilateral interdisciplinary integrations exemplified above, but smaller than the holistic integration?
2. If there is such optimal level of integration, to what extent is it practically possible to produce such integrated models?

The ideas introduced here have roots in previously published theoretical work (Lăcătuşu and Iordache 2008; Iordache et al. 2009, 2010a, b; Iordache 2010). They have been developed in the frame of biogeochemical interdisciplinary research programs (Neagoe 2007a, b; Petrescu 2007; Kothe 2009) in order to conceptually support the modeling of effects of local management measures in contaminated sites on processes occurring downstream in the contaminated catchments.

19.2 Basic Characterization of Productive Entities

We first introduce a concept of complex object based on properties following Ryan (2007), as different from the concept of complex object in standard system analysis. The new concept will be used later in describing the general form of a productive entity.

The nested hierarchical spatial partitioning is the most convenient approach in modeling spatial processes. Band et al. (2000) use this for watersheds and include one-dimensional ecosystem models in the patches. Also, Thompson et al. (2011) recently express their confidence in the usefulness of emergent properties in understanding the relationships across processes and scales in catchments. Mathematical modeling in standard systems analyses paradigm requires as a first step systems identification, i.e., the definition of boundary, components, and interactions. For a spatial model, one has to choose a spatial extent (spatial scope) and a granularity (spatial resolution), and the scope is assumed to be the size of the investigation area (Seppelt 2003). Then this is said to be the scale of the modeled object. While in the case of time scale, it is accepted that a model can include slow and fast processes as they are in reality, the system boundary is *defined* by the modeler and associated for spatial models with one scale. This kind of approach leads naturally to nested hierarchies of environmental objects by the simple association of the discretization units of space and to the emergence of *new* and irreducible properties at each hierarchical level in terms of new processes characterizing that level, an idea difficult to accept if the higher level system results exclusively from lower level systems. Such a model can then be regionalized by changing the model parameters and eventually the model structure in function of the spatial location of the discretization unit, and simulating the model in each grid cell. Alternatively, a spatially explicit model, which allows exchange of matter between the spatial units can be built (Seppelt 2003). In both cases, no explicit coupling between entities of the same or of different scales is conceptualized. This single scale raster cell is assumed to represent an ecosystem and the aggregated cells a landscape.

Ryan (2007) notices that using emergent hierarchies to give account for emergence is circular and propose to define emergence by the extent (scope) of the system whose emergent properties are observed. This spatial extent can be wider or narrower depending on the scale of observation. Instead of starting the individuation of the object by delineating the boundaries of the system, one starts with the set of properties and finds the scales of observation at which those properties emerge. Then the complex entity is characterized by a multitude of scales of observation, its boundaries being the reunion of the boundaries of the subentities characterized by properties observable at certain scales. This view developed from general physics perspective is convergent with the idea that at least some environmental objects are multiscale not only in time, but also in space (developmental systems, Iordache et al. 2011). From this perspective, a model "flattened" at a single scale is always a stiff system *sensu* Seppelt (2003). The variability in space of the state variables within the classic space–time boundaries of the system may reflect the ontologic diversity of its subsystems, and ignoring this is simply not realistic enough for scientific purposes. The main advantage of this concept of complex object is that allows an objective, independent from the researcher, allocation of spatial scales to the environmental entity. The ontological structure of the system is ensured and not sacrificed from the start in order to reduce the dimensionality of the model.

We assume that a procedure for producing a model of a complex entity such as above involves three steps. Step 1: define extratheoretical objects at scales xi, step 2: measure/observe properties at scales xi, step 3: define/model complex multiscale theoretical environmental object at scales xi by lawful connection between measured/observed properties.

Now we can introduce a concept of productive entity developed elsewhere based on a reconstruction of Darwin's "Origin of species," the first book dealing implicitly with the ecological (nature's) productivity (simplified from Iordache 2010). A productive entity "*i*" is a system of the following form:

(P_i, I_i, G_i, I_j, G_j, M_i^{rel}, M_j^{rel}, S, M^{ob}) where properties which should be characterized using an observation model, theoretically independent from the structural model S, are:

P_i, a property or a set of properties describing the biological production of the entity i, with i from 1 to n, where n is the populations size and $n \geq 2$ (biomass, or number of descendants)

I_i, a set of observable properties at space–time scales smaller than the maximal scale of the organism i (parts, genes, etc.)

G_i, a set of properties observable at the maximal scale of the organism

I_j, sets of observable properties at space–time scales smaller than the maximal scale of the organisms j, with j from 1 to n excepting for i

G_j, sets of properties observable at the maximal scale of the organisms j, with j from 1 to n excepting for i

M_i^{rel}, a set of relational properties between the organism i and its environment (the "subjective" environment of the organism i characterized from the perspective of the scientific observer)

M_j^{rel}, sets of relational properties between the organism j and their environment, with j from 1 to n excepting for i, the "subjective" environment of each organism from the perspective of the scientific observer

M^{ob}, intrinsic and relational properties (different from M^{rel}) of the environment, the "objective" environment of the organisms from the perspective of the scientific observer

S is the structure of the system of properties described above. S is characterizable by a structural model decomposable into a production law L expressed by an unknown mathematical function (submodel, "subjectivistic model" in Iordache et al. 2009) of the form $P_i = L\ (I_i, G_i, I_j, G_j, M_i^{rel}, M_j^{rel})$ and one or many coupled structural models of environmental entities characterized by the properties M_i^{rel}, M_j^{rel} and M^{ob} ("objectivistic models" in Iordache et al. 2009). For productive objects without competition the production law reduces to $P_i = L\ (I_i, G_i, M_i^{rel})$. The coupling between production submodels and objectivistic models occurs at the level of M^{rel}. The coupling between productive submodels takes place at the level of P and M^{rel} (P or a property structurally linked to P – I or G – in a productive submodel is M^{rel} in another productive submodel).

The potential unit of selection is a system (P_i, I_i, G_i, M_i^{rel}, S) describing the development of the organisms (developmental system, DS, Iordache et al. 2011, as an extension of the homonymous concept of Oyama, e.g., Oyama et al. 2000). For selection to occur, the following conditions should be fulfilled.

The condition of finite lifetime: the lifetime of the potential units of selection is finite.

The condition of coupling: M_i^{rel} and M_j^{rel} are not decoupled in space–time (i.e., with independent dynamic) at the existence time scale of the units of selection (i.e., either these properties characterize environmental objects which from the perspective of the observer – by M^{ob} – are the same for all organisms, or are characterize parts of a complex environmental object, which are in causal relation at the existence time scale of the units of selection).

The condition of scarcity: to have scarcity in the productive object (for the reason of space we will not develop this here; in short, it is a condition linked to the functional relationship between P and M^{rel} such as to have "struggle for existence").

The condition of variability: to have variability of the values of I_i and/or G_i and/or M_i^{rel} in such a way that P_i would be different (i.e., the fitness would be different). This condition is needed for sorting the units of selection by P_i.

In the absence of the conditions for selection and by eliminating the condition the $n \geq 2$ one has the general structure of a productive object. The relevance of this structure for the biogeochemistry of metals comes from the fact that metals are in this abstract model both at the level of I properties (internal resources or toxicants for the organism), at the level of M^{rel} properties (external resources or potential toxicants for the organism), and at the level of M^{ob} properties (part of the structure of the objective environment linking the organisms, where transport takes place).

This model of productive objects implicit in the "Origin of species" cannot be studied operationally as a whole because of its extreme complexity. But Darwin work (in the interpretation associated with the reconstructed model) puts the bases for a research program of biology. The multiscale system of coupled productive systems in the ecosphere can be studied only by disciplinary fragmentation and discretization. Current systems biology tries, for instance, only to relate I and G properties, while biogeochemistry works in investigating the circulation of elements only with G, M^{rel} and M^{ob} properties. Other environmental disciplines such as hydrology focus only on some of the M^{ob} properties, or on some of the G, M^{rel} and M^{ob} properties in the case of more integrated developments such as ecohydrology. The way of discretizing the physical space, which includes the productive object controls the solution to the "ontology" problem much studied by computer scientist from the database and knowledge base design perspectives (Sui and Maggio 1999; Pundt and Bishr 2002), and the way of representing the real productive system model is linked to the problem of relational data bases in GIS (Cova et al. 2002). The discretization in space–time for research purposes is needed both for well delimited objects and for fields (e.g., mobile masses), although it is appropriate for describing the environmental reality only in the first case. In the case of fields, the partitioning into discrete pieces is needed for accommodating the finite computing environment (Bian 2007, a source nicely reviewing the object-oriented representation of environmental phenomena).

The scales of the productive entity are the space–time scopes and the minimal associated resolutions at which the properties can be observed. Lifetime is the scope in time of an entity, whether biotic or abiotic. Characterizing the entity only by properties (without reference to scales) is not enough for meaningful individuation of the productive entity. Modeling the productive systems makes physical sense only within its range of ST scales, so the relevant mathematical properties of the model are only those characterized within these scales. A productive entity is multiscale by its different properties. The maximal such scale is *the* scale of this complex object. The same ideas apply to other complex objects. For instance, *the scale of an organism* is the scope of the physical space–time needed by an individual to develop its lifecycle, a function of both the size of the organism and its

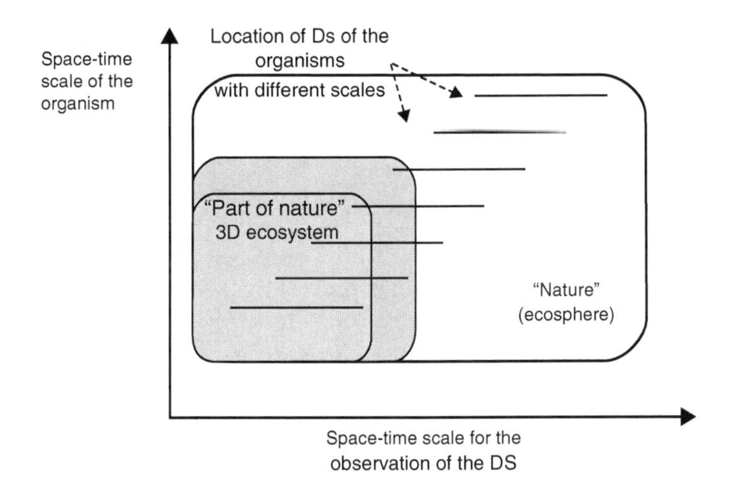

Fig. 19.1 Relationiships between three-dimensional (3D) ecosystems in space (or "parts of nature" in Darwinian terms) and the space–time scale of organisms and of their developmental systems (DSs). The point is that always some parts of the DSs will be out 3D physical space delineated by convention as ecosystem for management reasons. The DSs in this situation at one ecosystem scale (hierarchical level) are apparent structural emergent properties at higher level in the hierarchy of ecosystems. See Iordache et al. (2011) for other representations supporting the same idea

mobility in space, and the minimal resolution needed to observe the characteristic processes of the lifecycle (Iordache et al. 2010). It can be dynamic along the lifecycle length.

Due to the multiscale character of the DS, the discretization of the physical space of the ecosphere in nested three-dimensional units (nested hierarchies of ecosystems) will always cut the continuum of scales in such a way that some of the DS will have parts outside the ecosystem (Fig. 19.1). In Iordache et al. (2011), we have analyzed this situation working only with the biological objects from the DS and demonstrated that the obtained hierarchy of ecosystems is not a true one (with true emergent properties at each level), but a pseudo-hierarchy, with new structural subsystems at each pseudohierachical level not arising from the interaction of the parts, and with the possibility to cut the hierarchies across scales on subjective grounds (according to human interests).

Two important consequences result from this discussion with respect to the need for integrated modeling. The first one is that there is no need to study the function of the whole three-dimensionally delimited ecosystem in order to understand the role of a population of certain organisms in that ecosystem. This is important when at stake is to understand the role of a service production unit (Luck et al. 2003), and not all the mechanisms supporting the biogeochemical services provided by the ecosystem. A simpler, with minimal complexity, homomorphic model can be built according to the structure of the DSs of those organisms [see, for example, ectomycorrhizal fungi in Iordache et al. (2011), and arbuscular mycorrhizal fungi in Neagoe et al. (2011)]. The second is that it makes little sense to study the circulation of one element (e.g., a heavy metal) in the DSs identified by the homomorphic model separately from the circulation of other elements playing the role of resources or toxicants (e.g., in the case of plants from the circulation of nitrogen or phosphorus), because all of them influence the productivity of the DSs. Based on these ideas in Neagoe et al. (2011), a concept of biogeochemical role of organisms has been proposed, and a concept of integrated biogeochemistry. In other chapters of this book, (Neagoe et al. 2012; Jianu et al. 2012) the role of abiotic objects (organic matter, minerals) is investigated in the same framework.

Next in this chapter we further develop these ideas by looking at two aspects. In part X.3, we look at the way of coupling biological and abiotic objects resulted from the discretization in space–time of the productive objects at multiple scales. At stake here is how to produce a homomorphic model of a productive object with the general form introduced above [note that in Iordache et al. (2011), we limited the methodology of homomorphic model construction to the biological compartments without taking into detailed analyses the abiotic objects]. In part X.4, we synthesize from the literature the processes of different scales occurring in two common types of management units in contaminated areas, sites (from smallest scale environmental process up to 10^5 m^2) and regions (processes from 10^5 to 10^{10} m^2). In the same part, we screen the ontological assumptions of soil science and of catchment hydrology, and the possibilities of coupling between theoretical objects of these disciplines and selected groups of organisms. In the end of the chapter (part X.5), we extract several applied consequences of this theoretical framework for the risk assessment of contaminated areas.

19.3 Theoretical Framework for Coupling Environmental Entities in Order to Characterize Scale-Specific Processes in Productive Entities

Once one has discretized the multiscale productive system into environmental objects with well-defined positions in space–time (either specific to the observation model, or to disciplinary theories independent of the observation model), its reconstruction depends on how the interactions (the coupling mechanisms) between the environmental objects are conceptualized.

A *coupling mechanism* is the process of interaction between at least two entities explaining a pattern observed in one or all entities (for example, either in the entity with smaller substance turnover rate – distribution pattern of metals in soil as a result of the interaction with atmosphere, or in the entity with larger turnover rate: in distribution pattern of metals in organisms as a result of the interaction with soil). A coupling mechanism suppose the existence of at least two scales of observation: that of each entity involved in coupling, and that of the complex entity resulted from coupling. A coupling mechanism in nature takes place by an exchange of substance and energy between the coupled entities, i.e., is a causal mechanism. However, there is no need for a speak in terms of bottom-up and top-down causation, as the multiscale nature of the complex systems does not entail necessarily true hierarchies of systems (Iordache et al. 2011).

The *scale of coupling* is the scale of the coupling mechanism, is implicitly the *scale of the process*, and is the space–time scope and minimal resolution at which the interactions of two subentities with the same or different scales of a complex system can be observed. The ontological condition for coupling, and implicitly for the complex entity, is the existence of a causal relation (a process) between the coupled entities at the scale of coupling. Not everything is coupled to everything because of the limited lifetime of the entities and the fact that causal connections occur in time. Processes cannot interact, only distinct entities by processes. A *pattern* is the distribution of a measurable variable in space or in time within the scope of the coupling between two entities, and is caused by this coupling. Resolution is the minimal fragmentation of the scope at which a variable should be measured in order to observe the pattern.

The *scales of the external entities with value* (EV, characterized by M^{rel} properties in the general model of productive entities) for an organism are the scales of coupling between an organism and other abiotic or biotic entities with positive or negative value for it. The scale of the organism is the maximal scale at which one should estimate the objective (from the observer perspective) distribution in space–time of the entities with values for an organism. Figure 19.2 shows a general representation of a pseudohierarchy of environmental entities.

The *scale-specific mechanisms of metals mobility* are the coupling mechanisms between environmental entities involving fluxes of metals and causing the generation of patterns in the space–time distribution of metals. Within this framework, it does not make sense to speak of scale-specific processes of metal mobility in general, but only within multiscale entities, when we are looking for explanations

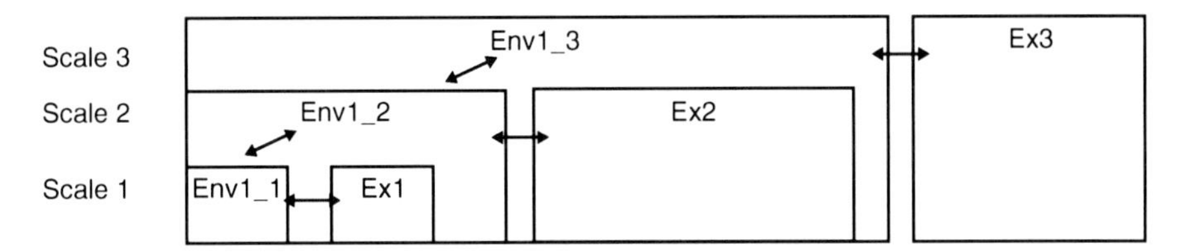

Fig. 19.2 General representation of a pseudohierarchy of environmental entities (Env) showing the coupling mechanism with external entities (acronym Ex, *continuous arrows*) at each scale, which generate scale-specific patterns, and implicitly the partial decoupling (*dashed arrows*) of the larger scale patterns from smaller scale mechanisms and associated patterns. The environmental systems Env1 from scales 1 to 3 (e.g., soils, biological systems, hydro-systems, ecosystems, socioeconomic systems) are not a true hierarchy of nested systems, as usually approached in a systems analyses context, but only a pseudohierarchy including in their structure new entities at each scale (usually conceptualized as external driving "forces", or forcing functions in a nested hierarchy paradigm). In order to observe the complex environmental entity Env-Ex, one needs a scope of observation larger than the scope associated with the scale of Env or of Ex taken separately

of a mechanismic type. The fact that such processes are reported in the literature separately by discipline means only that disciplinary knowledge is waiting for integration.

Conceptualizing coupling mechanisms is trivial in abstract terms, but is far from trivial when one works with entities defined and characterized in scientific disciplines with different traditions. This is the case of biogeochemistry. In this case, conceptualizing the coupling involves a translation between the entities assumed in each discipline in terms of scopes and generally accepted (transdisciplinary) measurable properties. The first problem in modeling the coupling mechanisms or in up-scaling and downscaling is to conceptualize relationships between entities developed under different scientific paradigms. The external entities represented in Fig. 19.1 may not be of the same type or may be even complex systems (think at the influence of bedrock geochemistry on the patterns of metals in soil, in comparison with metal patterns generated by solid phase inputs from industry, by atmospheric phase inputs from industry, or by atmospheric inputs modulated by different vegetation roughness covering the soil). The external entities may be from the perspective of the observer of a larger scale than the coupling scale (think, for instance, at the coupling between a mycorrhizal fungi and a plant).

The patterns in one entity resulted by the coupling with external entities at the coupling scale is dependent to some extent also on smaller scale processes occurring in the entity manifesting that (target) pattern. In this sense, there is a partial internal decoupling of the target scale pattern from smaller scale internal processes due to the external coupling at the target scale.

By *role* of a variable in the mobility of elements, we understand the causal influence of a variable in a coupling mechanism in producing the outcome of a process involving it. The *role* of a subsystem of the coupled entities (characterized by a variable) is specific to the coupling scale. The fluxes resulted from this role at the coupling scale propagate to larger or smaller scales (these effects at distance could eventually be labeled as indirect roles). Neagoe et al. (2012) explore the role of organic carbon in the mobility of metals, and Jianu et al. (2012) characterize the role of mineralogy.

One important use of the above framework, and actually the research problem which pushed us to its development, is in modeling the effects of local processes at large distance in space–time (Fig. 19.3). From the existence of large distance effects, it results that an indirect role of a variable in metal mobility can be not only across scales, but also at the same scale if the effect is localized at distance in space–time.

19.4 Scale-Specific Processes in Metal Biogeochemistry

In order to build models with the general structure presented in Fig. 19.3, a first step is to have an idea about the scale specific to various processes involving environmental entities. Based on a literature review and our own expertise, Tables 19.1 and

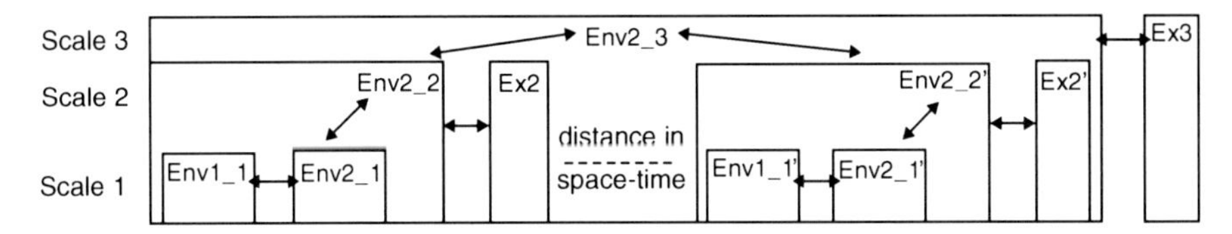

Fig. 19.3 General representation of the structure of a model in an integrated approach in order to assess the effects of local processes involving metals at one space–time location on other local processes at large distance in space–time (designate by '). Env1_1 can be a tailing-dam or a mining dump or a polluted soil, or a complex soil-vegetation entity. Env2 (scale 1–3) can be hydro-systems, Ex2 can be geomorphological, pedologic and microclimatic entities in the landscape, Ex3 geologic and regional climatic features of a larger catchment, and the model could attempt to predict the effects of local phyto-remediation of a contaminated site in the slope area of a small catchment on late and distant bioaccumulation of metals in crops in an agricultural floodplain site. The coupling at the same scale and between scales is based on variables characterizing the coupled entities. Successive up-scaling and downscaling of some variables are needed in order to predict ST large-distance effects of local changes

19.2 show a synthesis of the site (from smallest scale environmental process up to 10^5 m^2) and region (10^5–10^{10} m^2) specific processes of metal mobility by various pathways. The larger literature body supporting this synthesis is not presented here for reasons of space, but some aspects will be detailed in part X.5.

These processes generate patterns in metal distribution at different scales, and what we have to do is to separate by modeling and in situ observation the patterns generate by the processes of interest from the patterns generated by other processes. For instance, the contamination of the floodplain in the vicinity of a smelter can occur both by hydrological processes and by atmospheric deposition, and we have to separate the effect of sedimentation during floods from the effect of particles deposition.

From a different perspective, because multiscale experiments are a useful strategy in designing the remediation of sites contaminated with metals we performed, we also underlined the specific processes which can be investigated by such experiments at each scale (Table 19.3).

In the sense of this chapter, biogeochemistry is interdisciplinary and deals with the mobility of elements in three-dimensional (physical) volumes (sites, regions, etc.) hosting complex systems formed by coupled entities fully or partially located in that physical volume. Such clearly delineated in space entities are management units providing local resources and services and having large distance effects on the local resources and services production in other sites. The processes specific to this approach result from the multiple coupling between entities occurring in a three-dimensional physical space volume. Each simple couple of entities support phase specific, multiphase abiotic or biological processes. Thus, what is specific to interdisciplinary biogeochemistry is the complexity of the processes, and not the scale. We define then a scale-specific process in metal biogeochemistry as the complex process resulting from the multiple coupling of the entities involved in processes occurring at a certain scale in a three-dimensional physical space.

Table 19.1 Site-specific processes involved in metals mobility

Scale	Transport pathway of metals	Mechanism
Part of soil column 10^{-8} to 10^{-4} m^2	Various	Chemical and microbiological weathering
Part of soil column 10^{-4} m^3	Biological	Microbiological direct and indirect (by organic carbon) immobilization/mobilization for hydrological fluxes
Rhizosphere	Biological	Microbiological direct and indirect (by organic carbon) immobilization/mobilization for plants
Soil column 10^{-8} to 10^0 m^2	Hydrological	Diffusion and dispersion
Soil column 10^0 m^2	Various	Other biological weathering (by plants, invertebrates)
Soil column 10^0 m^2	Hydrological	Colloidal transport
Soil column 10^0 m^2	Hydrological	Soluble transport
Soil column 10^0 m^2	Hydrological	Soluble complexes transport
Soil column 10^0 m^2	Hydrological	Preferential flow (vertical)
Soil column 10^0 m^2	Biological	Bioaccumulation in soil invertebrates with low mobility
Bioaccumulation area 10^{-2} to 10^4 m^2	Biological	Plant uptake (bioaccumulation in plants)
Field 10^3–10^4 m^2	Hydrological	Unsaturated (preferential) flow (to groundwater)
Slope area 10^2–10^4 m^2	Hydrological	Infiltration excess overland flow (dissolved and particulate)
Slope area 10^3–10^5 m^2	Hydrological	Retention in and remobilization from transversal buffer zones
Bioaccumulation area 10^3–10^5 m^2	Biological	Bioaccumulation in mobile epigeous invertebrates
Large slope area 10^3–10^6 m^2	*Hydrological*	*Saturation excess overland flow (dissolved and particulate)*
Large slope area 10^4–10^8 m^2	*Hydrological*	*Subsurface storm flow (lateral flow)*

The italic entries indicate processes crossing the site-region scale boundary

The complexity *of the model* can further increase by coupling besides processes at the same scale, also processes between scales. This is the most interesting case (coupling both at the same and between scales, Fig. 19.2), because this provides an operational approach for studying the effects of local management actions in space–time at large distance.

In order to make operational this general framework for contaminated sites management, one needs to compare the ontologies of soil science, hydrology, and population ecology to identify scale-specific patterns of metals distribution reported in each discipline, and to infer the coupling mechanisms between disciplinary entities in the cases where one or more variables explicitly or implicitly common are reported to be involved in the generation of the patterns. Fully following these steps is beyond the objective of this text, but we will illustrate the approach in the next chapter.

Table 19.2 Region specific processes involved in metals mobility

Scale	Transport pathway of metals	Mechanism
Region 10^4–10^{10} m^2	*Hydrological*	*Groundwater flow in different types of aquifers*
Region 10^4–10^8 m^2	*Atmospheric*	*Dry and wet deposition from local sources*
Bioaccumulation area 10^4–10^8 m^2	Biological	Bioaccumulation in mammals and in nonmigratory birds
1st order catchment 10^5–10^6 m^2	Hydrological	Retention in and remobilization from transversal buffer zones
2nd–6th order catchment 10^6–10^8 m^2	Hydrological	Interactions between types of hydrological flows
Region of 10^6–10^7 m^2	Various	Soil catena formation
Large order catchment 10^7–10^9	Hydrological	Retention in and remobilization from longitudinal buffer zones (floodplains)
Region of 10^6–10^7 m^2	Atmospheric	Volatilization
Bioaccumulation area 10^5–10^{12} m^2	*Biological*	*Bioaccumulation in migratory birds*
Region 10^8–10^{11} m^2	*Various*	*Zonal soil formation*
Region 10^9–10^{12} m^2	*Atmospheric*	*Dry and wet deposition from distant sources*

The italic entries indicate processes crossing the site-region scale boundary and the upper scale boundary of regions

Table 19.3 Processes involved in metals mobility investigated in experiments at different scales

Name of the system and usual scales	Environmental complex system studied at these scales	Processes, fluxes, effects studied/control variables
Pot 10^{-2} m^2	Soil + plants	Exploration by root, bioaccumulation/ microorganisms, organic carbon, level and spatial structure of amendments
Lysimeter 10^{-1} to 10^0 m^2	Soil + plants + small-scale hydro-system	Same as in pots + leaching, internal redistribution, net outputs/same as in pots + soil structure, hydraulic conductivity, humidity, redox potential on profile
Plot 4×10^0–10^2 m^2	Soil + plants + larger scale hydro-system + other organisms	Same as in pots + effects of heterogeneity in space, margin effects, other processes due to external entities (consumers, runoff, etc.)/same as in pots + variables for external entities

19.5 Contributions to a Comparative Ontology of Environmental Sciences

19.5.1 Soil Science

From soil science, we refer here only to aspects related to soil classification and mapping. Soils are described for their classification by soil profiles, so the scale of observation is that of a soil column of the order of 1 m^2. The concept of soil column refers to a three-dimensional physical object organized in horizons and having one or more properties, with no equivalent in other scientific disciplines (e.g., gleyc, vermic). These properties are derived from a combination of characteristics indicating soil-forming processes (European Commission 2005). The characteristics comprise single observable or measurable variables. Their relation to the properties is to a consistent extent tacit knowledge of the soil expert (Scull et al. 2003), and for this reason difficult to include in an explicit knowledge base useful for integrated interdisciplinary modeling. Moreover, there are country specific classification schemes, the official classification scheme of the International Union of Soil Science serving only for comparing national soil classification systems (European Commission 2005). It is not explicit in soil science theory if a soil unit mapped at a small map scale is composed (in a systemic sense) of the soil units mapped at a larger map scale in the same area. One cannot identify emergent properties of the soil type mapped at smaller scale compared to the soil types mapped in the same area at larger scale. In a review paper, Scull et al. (2003) goes up to stating that "the soil landscape is continuous and is not composed of distinct individual soil types." Soil classification is in this view only a way to summing up the information obtained about soil columns, the "pedons" sharply represented would not have a strong ontological status as a spatial object.

If this is the case, then what is real in the case of soil? First of all, of course, the soil layer as an integral object extending in space. Another possibility, pursued to some extent in soil science, is to interpret soil as an ecosystem. In this case, the minimal scale ecosystems could be a soil column hosting plants, and larger scale soil ecosystems would go up to the surface needed for supporting a population of large-scale soil organisms (e.g., mobile animals). This approach would work well for biological properties of the soil, and can lead to soil pseudohierarchies with biological apparently emerging properties (actually resulting from the coupling of the soil parts with organisms of different scales). But predicting soil abiotic properties would much more effective by assuming a continuous distribution of these properties in space. Soil columns, such as organisms, seem to be unique and classified only methodologically in the same type; the real entities are soil columns, not the types of soil columns. A better prediction of soil abiotic properties can be done either by refining the soil classification procedure (Qi et al. 2006), or by attempting to detect the space–time distribution patterns of abiotic properties at various scales (Lin et al. 2005, Scull 2010). Patterns of measurable characteristics (pedochemical, geochemical) can be observed by grid sampling or other sampling methods. In this approach, soil is treated as physical entity, not as an ecosystem.

The patterns of metal distribution in soil reflect internal processes (resulted from coupling between internal entities) and external processes (resulted from coupling of soil with external entities). The coupling of soil with external entities leads to apparent soil pseudohierarchies (Fig. 19.1) with patterns of abiotic parameters as apparently emerging properties. It is interesting to note that the soil manifest separate biotic and abiotic pseudohierarchies. This leads us to the theoretical hypotheses that the pseudohierarchy concept is but one way to reduce the complexity of the environment, more complicated and appropriate that nested hierarchies, but no more than a concept. Studying its mathematical properties is a research direction.

Internal redistribution processes in soil will be discussed in Neagoe et al. (2012) and Jianu et al. (2012) in terms of the role of organic carbon and mineralogy in the mobility of metals in soil. These internal processes influence the hydrological export of metals, the export by wind and the biological export.

External coupling processes of soil involves the interaction with atmosphere (liquid, particulate and colloidal phases; wet and dry deposition), with hydrosystems (liquid, particulate and colloidal phases (deposition and resuspension), and with organisms (input by organic matter, output by bioaccumulation processes). In terrestrial contaminated areas, the soil has a key role for the investigation of the patterns resulted from external coupling because its turnover time is much lower than that of the biological, hydrological, and atmospheric entities, but higher enough to allow the detection of patterns as a result of these interactions at the scale of human observation. Soil is a more reactive entity than bedrock, for instance, with much higher reactive surfaces because of its fragmentary structure. The reactive inorganic parts in the soil are minerals and organic carbon, whose importance is underlined by their use in soil classification. The soil as physical multiphase entity (solid, liquid, and gaseous) links entities with value for plants and for soil organisms, and provides variables also in hydro-system model (for vertical and runoff flows, etc.). The internal (within soil) distribution of variables is a key for understanding the coupling between plants, soil organisms, and hydro-systems.

The scale of the externally patterns differs with the coupling mechanism: solid waste deposition, bedrock weathering, differential deposition of atmospheric pollutants around industrial sources (gradients and barrier effects), and differential sedimentation from surface hydrological fluxes (lateral and longitudinal buffer zones). Other soil characteristics are controlled by topography, climate, etc., with their different scale of coupling.

Rodriguez et al. (2008) studied the patterns of metal distribution in agricultural soils. They found that the heavy metal concentration is influenced by bedrock composition and dynamics at all the spatial scales, while human activities had a clear effect only at the short- and medium-range scale of variation. There were differences between metals: Cu, Pb, and Zn (and secondary Cd) were associated with agricultural practices (at the short-range scale of variation), whereas Hg variation at the short and medium scale of variation was related to atmospheric deposition (Rodriguez et al. 2008). The multiscale structure of metals in soils is studied also by Xiaoni et al. (2010), who found scale-dependent variability of metals reflecting the existence of pollution hot spots. Lăcătuşu and coworkers extensively studied the distribution of microelements in Romanian soils not only at small resolution on large surfaces, but also at large resolution in contaminated areas (Răuţă et al. 1995; Lăcătuşu et al. 1996). In certain areas, a natural high background of metal concentration due to geologic bed led to important ecotoxicological and human health problems (Lăcătuşu et al. 1993). The distribution patterns of elements concentrations studied by Galan et al. (2008) have been controlled by the lithology and geochemistry nature of bedrock and by the occurrence of metallogenic belts in the studied region. However, not always the natural background in mining areas leads to large concentrations of metals in the soil because of the differences between deep and surface geochemistry (Lăcatusu et al. 2009); review of the geochemical background concepts and of their application is provided by Galuszka (2007). Meirvenne and Meklit (2010) analyzed a data set of 14,674 copper and 12,441 cadmium observations in the topsoil of more the 10,000 km^2 area and identified regional patterns and potential causes (smelters of effects of First World War). However, they eliminated outliers, not being interested in small-scale patterns of distribution (Meklit et al. 2009). By now, we have given examples of patterns only for total metals. The problem is complicated by the factors controlling the availability of metals, such as soil humidity, organic carbon, pH, porosity (Keur and Iversen 2006; Scrădeanu et al. 2010a, b). We will not develop these aspects here, but one should consider also these patterns of distribution when evaluating the mobility of metals at site and region scale.

19.5.2 Hydrology

From this disciplinary field, we refer here only on some aspects of surface hydrology, mainly catchment hydrology. The entities assumed in hydrology are hydro-systems, thus entities restricted to the aquatic phases, considered separated from their solid environment, which however controls water movements. The concept of hydro-system includes implicitly external ontological assumptions by the relations of the water with entities determining the types of flows at different scales. In the area of hydrology the distribution across scales of the processes is well documented (Bloeschl and Sivapalan 1995; Bloeschl 2001). Clear is also the fact that in the water quality modeling currently there are no models operating at multiple space–time scales, the integration of processes occurring at different scale being possible only by including separate models in common GIS-type platform (Srivastava et al. 2007). This is due to the coupling between processes of different scale generating nonlinearities in the system (Beven 2006). In this context, integrating the knowledge about processes occurring at different scales and places is a priority in hydrology (Bloeschl 2006). However, this cannot be done restricting the integration only to hydrological processes, because the hydrologic processes "emergent" at different scales are controlled by new types of not hydrological entities occurring at those scales (slope areas, geomorphological patterns, geological structures, etc.). Relationships with external biological entities are assumed also by their influence on water flow or chemical parameters (for instance, by microorganisms clogging of pores, or consumption of chemical substances, or vegetation influence of water distribution in soil by evapotranspiration, or of preferential flow in former roots spaces, or increase of the roughness of the surfaces over which the water flows). The study of such coupled ecological–hydrological processes generated an entire research subfield, the ecohydrology (Manfreda et al. 2010).

The sources concerning metal mobility by hydrologic fluxes are numerous, and it cannot be our purpose here to fully cover this body of knowledge, but only give some examples. A hot area in hydrology is the influence of upstream on downstream processes, the fluvial connectivity, especially because of its jurisdictional consequences under water quality regulations (Caruso and Haynes 2010). At catchment level, there is a coupling between upstream and downstream elements in terms of fluxes, there is a partial and decreasing with increasing stream order coupling, but in terms of patterns of distribution in space–time there is not a simply observable coupling. Alexander et al. (2007) found, for instance, that first-order headwaters contribute approximately 70% of the mean-annual water volume and 65% of the nitrogen flux in second-order streams. Their contributions to mean water volume and nitrogen flux decline only marginally to about 55 and 40% in fourth- and higher-order rivers that include navigable waters and their tributaries (Alexander et al. 2007). As for patterns, Saunders et al. (2004) found a decoupling of upland and stream inorganic nutrient patterns at base-flow conditions independent of the season. The longitudinal pattern of the total copper distributions in a large system reflected the balance of flushing, sources, and losses (Chadwick et al. 2004). From an applied perspective, metals dispersal at large catchment scale is a hot problem in the light of Water Framework Directive (Bird et al. 2010). Freeman et al. (2007), also, point out the key ecological role played by headwater streams on downstream processes, in terms of water-mediated transport of substance and energy.

Many factors influence the transit time of elements from upstream to downstream parts of the catchment: intra-catchment topographic variability, catchment area, path length, and slope gradient the aspect of the land surface, factors that influence the local energy budget (latitude of the study site, the ratio between direct and diffuse solar radiation, prevailing winds, and shading due to distant topography) (McGuire and McDonnell 2006; Broxton et al. 2009). A key role in the transit is played by the riparian area. Understanding of the co-occurrence of hot spots and moments for contaminants in riparian systems is essential for designing the management strategies of pollutant removal at the catchment scale (Vidon et al. 2010). At the other extreme of the range of scales, the daily patterns of elements distribution in water are controlled by temperature and the activity of organisms (Tercier-Waeber et al. 2009). For instance, the diel cycle of Fe is controlled by photoreduction and reoxidation processes (Gammons et al. 2005) inducing a diel cycle of other toxicologically important elements. Daily temperature-dependent adsorption onto actively precipitating hydrous Fe and Al oxides can occur, because the adsorption of metal cations onto oxide surfaces is endothermic (Chapin et al. 2007). In a multiscale study of catchments impacted by mining (monthly, daily, and bi-hourly), Nagorski et al. (2003) observed that monthly changes were dominated by snowmelt and precipitation dynamics, that on the daily scale, post-rain surges in some solute and particulate concentrations were similar to those of early spring runoff flushing characteristics on the monthly scale, and on the bi-hourly scale, a diel cycling for pH, dissolved oxygen, water temperature, dissolved inorganic carbon, total suspended sediment, and some total recoverable metals occurred. They found that short-term (daily and bi-hourly) variations of some geochemical parameters covered large proportions of the variations found on a much longer term (monthly) time scale (Nagorski et al. 2003). A review of remote in situ voltammetric techniques to characterize the biogeochemical cycling of trace metals in aquatic systems is made by Tercier-Waeber and Taillefert (2008). Micro multiscale studies of parameters such as porosity (Stockdale et al. 2009) in sediments and soils with appropriate micro-scale technologies (Viollier et al. 2003) are essential for understanding the mechanisms supporting the mobility and bioaccumulation of metals beyond a black-box approach.

In what concerns the groundwater, most of the leachate contamination plumes do not exceed the width of the landfill (Christensen et al. 2001). Metal transport in groundwater at large scale is increased as the sediment permeability increased, and the groundwater pH decreased (Simpson et al. 2004). Acidic groundwater (pH from 3 to 3.5) transported high concentrations of Pb, Cd, Zn, Cu, and Fe at a former battery recycling plant (Lee and Saunders 2003). Seepage of acidic groundwater at the base of tailing dams is common (Lottermoser and Ashley 2005).

The small watershed approach is a valuable tool for understanding the transport and cycling of metals (review in Driscoll et al. 1994, with case studies for Pb, U, Pb, and Al). The biogeochemical controls on metal behavior in aqueous environments involve complex linkages of biological (mainly microbiological) and geochemical processes occurring at microscopic and macroscopic scales (Warren and Haack 2001). Seasonal and spatial variations in metal concentrations and pH were found in a stream at a restored copper mine site (Bambic et al. 2006). Mechanistic studies of the interactions between groundwater and surface water fluxes of metals are, however, very few. Fritz and Arntzen (2007) study the flux of uranium in the hyporheic zone and found an important influence of the river level by changing the groundwater discharge and by dilution of the uranium in groundwater. The discharge of contaminant from groundwater by seepage areas was spatially variable depending on river valley morphology, and in some cases a dilution by uncontaminated groundwater occurred at the seepage point (Fryar et al. 2000).

19.5.3 Ecophysiology

The general ontological status of organisms and the case of ectomycorrhizal fungi have been characterized elsewhere (Iordache et al. 2011). Here, we point out the situation of three types of organisms with larger scale: plants, carabids, and birds. We investigate the possibility of characterizing an apparently simple process, the bioaccumulation of metals. These organisms have been selected as representative for the complexity of this problem (estimating the role of organisms in the circulation of metals). What is at stake is how to model the bioaccumulation of metals in such organisms and the toxicological (at individual level) and ecotoxicological (at population level) effects and how to build integrated transport – bioaccumulation models for metals. In order to do this, one needs basic information concerning the scale (in the sense introduced above) of the organisms, and that of the populations (their location in space–time).

Although for plants their scale seems to be very easy to characterize, the situation is not as straightforward as it seems to be. The bioaccumulation from the soil takes from the volume occupied by the roots systems, so one needs to have knowledge about this volume, the heterogeneity of metals distribution in this volume and the root growth strategy. Only the heterogeneity occurring at the scale of the plant affects plant uptake and the leaching of trace elements (Robinson et al. 2006). Very often such detailed information for the species occurring in contaminated sites is not available, so field observations are needed. Deeply penetrating roots such those of trees complicate the problem of correctly estimating the average concentration of metals in the volume of soil explored by the roots. To do this an estimation of metals distribution in the full soil profile is needed, beside characterizing their horizontal heterogeneity. But an even more complicated situation is that of the clonal plants, and this is the case that we want to develop here in order to illustrate the difficulty of computing even apparently simple bioaccumulation factors.

Almost all plants have modular structure (Harper 1977; White 1989). Plants with vegetative spread (clonals) grow realizing sets of shoot or roots units named ramets. Every ramet can survive alone once established in the proper habitat. The modularity combines the transport of recourses among ramets allowing plant to survive in more places in the same time (Oborny 1994; Oborny et al. 2000). It is considered that the ability of connected ramets is their specialization for the usage of local abundant resources, permitting the plants to have great flexibility in adjusting their relative abilities in usage of different resources. Producing different units of shoots and roots, the plants (clonal individuals, synonym with genets) allocate different ratio of biomass for light acquisition for photosynthesis versus water and mineral resources acquisition (Stuefer 1997). From the physiologic point of view, clonal plant acts systemic, the resources are transported from parts of the plant living in resources-rich area to parts of the plant living in resources-poor area. This specialization is based on clonal integration named "labor division" (Charpentier and Stuefer 1999). Clonal plants consists of possible interconnected (via stolons or rhizomes) and repetitive units (ramets). Clonal plants have two reproductive strategies: sexual reproduction via seeds, and clonal propagation via the development of vegetative offshoots from the parents (Canullo and Falinska 2003).

Isozyme analysis of perennial stoloniferous *Glechoma hederacea* (Widdén et al. 1994) in Sweden collected from natural population showed that one genotype might cover 20 m length and 10 m width surface. At *Hylocomium splendens*, the isozyme electrophoresis could identify 25 genotypes among total 75 ramets; most genets (clonal individuals) occurred in one plot (10 cm $\times$ 10 cm), two genets occurred in more than one plot inside of every site (10 m $\times$ 10 m) and four genets occurred in more than one site (separated by 500–1,000 m). The conclusion is that one genet might occupy a surface bigger than 1 km^2. The physiological integration is another problem, but at least in the case of tree clonal plants this can occur over very large areas. Anyway, for clonal plants what is characteristic is that the integration is at the scale larger than that of separately observed ramets.

The morphological plasticity in plants is very important, and in clonal ones even more important. In natural patchy heterogeneous environment, the plasticity in growth form may increase the probability that ramets are placed in the more favorable microhabitats within such environment. In *Carex flacca,* the depth of the shoot below soil surface may vary between 0.2 and 5 m, and rhizome length is separated in three classes: < 5 mm (0), 5–49 mm (5) and ≥ 50 mm (50). This evidences that stolon internode and rhizome ("spacer") lengths may be extremely variable. Tamm et al. (2002) highlighted the ramet lifespan (years) of many species (Table 19.4) and their vegetative mobility (mm/year) in natural wooded meadows. For example, *Fragaria vesca* has 2 years max ramet lifespan with 465 mm/year max vegetative mobility for stolons and 6 years max ramet lifespan with 15 mm/ year max vegetative mobility for rhizomes. *Rubus caesius* has 2 years max ramet lifespan for stolons and rhizomes with 595 mm/year max vegetative mobility for stolons and 40 mm/year max vegetative mobility for rhizomes. The internodes that separate the individual clonal units often prove highly responsive to environmental conditions (Oborny 1994).

The "foraging" rule is that the plant tend to "escape" from adverse conditions and to "remain" in favorable ones. Metal heterogeneity in soil may provide such adverse conditions. The plant morphology responds to site quality, meaning the environmental variability induced morphological variability within clone, and this reduced the degree of intraclonal competition. Spatial and temporal effect cannot be separated, thus, proper characterization of the environment exactly on the scale of clonal growth is of special importance. Spatial patterning of plant populations and the relative roles in intra- and inter-specific interactions play an important role in plant adaptation, species interactions (Silvertown and Charlesworth 2001) and community dynamics (Morin 2005). The availability of resources often changes considerably over a broad spatial distance and at different time scale. Clonal species are more likely to explore environmental heterogeneity than nonclonal plants species due to their spatial distribution and the longevity of the genet. Storage of resources is very common phenomenon in clonal plants. Experiencing temporal and/or spatial heterogeneous environments, the resource storage may serve

Table 19.4 Maximum vegetative mobility (mm/year) of some clonal plants from natural wooded meadows (according to Tamm et al. 2002)

Species/life span	1	2	3	4	5	6	7	8	9
Achillea millefolium	202								
Aegopodium podagraria					645				
Agrostis stolonifera	135								
Anemone ranunculoides	32								
Arrhenatherum elatius	16								
Brachypodium pinnatum	129								
Briza media				161					
Calamagrostis epigeios		275							
Carex tomentosa		190							
Festuca arundinacea			90						
Festuca pratensis					130				
Festuca rubra		260							
Filipendula vulgaris						19			
Galium mollugo	263								
Galium verum	190								
Geum rivale								25	
Helianthemum nummularium	210								
Hypericum maculatum	152								
Hypochaeris maculata							12		
Leucanthemum vulgare			83						
Lotus corniculatus		70							
Medicago lupulina		30							
Melica nutans	136								
Pimpinella saxifraga						13			
Plantago lanceolata									4
Plantago media			3						
Poa angustifolia		190							
Polygala amarella	75								
Polygonatum odoratum	48								
Potentilla erecta					10				
Primula veris							6		
Prunella vulgaris	144								
Ranunculus acris		10							
Stachys sylvatica	350								
Trifolium pratense			30						
Veronica chamaedrys	578								
Veronica officinalis	160								
Vicia cracca	300								
Vicia sepium	420								
Viola mirabilis			72						

1–9 are the maximum ramet life spans of the species (years). The soil surface and volume relevant for bioaccumulation of metals are constantly increasing each year with species and environmental dependent rates ecological functions in similar ways to temporal and spatial heterogeneity, which are partially different from nonclonal species (Suzuki and Stuefer 1999).

Almost 400 plant taxa are classified as heavy metal hyperaccumulators, part of which is clonal. Studies have been developed in the last years regarding population gene flow patterns and the structure of genetic diversity. *Arabidopsis halleri*, a close wild relative of *A. thaliana*, is a clonal insect-pollinated herb tolerant to heavy metals (Zn, Pb, Cd) and a hyperaccumulator of Zn and Cd. Clonal spread occur only at short distance (<1 m). Metallicolous populations occurring in polluted areas are of particular interest. The contaminated site act like an island in noncontaminated areas with a very high heavy metals concentration in the soil, meaning there are very strong ecological constrains. *Arabidopsis halleri* (Brassicaceae) is used as a model species because is closely related to *A. thaliana* for which a vast array of genomic tools and molecular markers is available (Van Rossum et al. 2004).

Clonal spread was reported to be higher under harsh ecological conditions (Falińska 1998). Heavy metals significantly decreased the performance of *Potentilla anserina* (Saikkonen et al. 1998) measured as number of ramets, total vegetative biomass, and number of flowers, limiting the growth of the species. Hyperaccumulator plants have greater requirements for the metal of interest is given to the fact that hyperaccumulator plants allocate more roots into soil patches containing potentially phytotoxic metal concentration (Liu et al. 2010). Metal concentration (Zn or Cd) affected both root and shoot biomass in *Sedum alfredii*. The experimental studies showed that when the same amount of Zn/Cd was provided heterogeneously rather than homogenously, the plants allocated approximately 90% of root biomass to metal rich patches.

We will explore now the problem of assessing the bioaccumulation in mobile organisms, namely in carabids and birds. Carabids are epigeous invertebrate able to inspect favorable and unfavorable microhabitats. Knowledge about their habitat preferences, a mapping of the habitats at their scale and knowledge about their movement and feeding patterns is needed. The situation is complicated when the species have flying capabilities.

Carabids are mobile organisms as adults and less mobile, even immobile in immature stages. Their mobility is determined by extrinsic factors (abiotic and biotic conditions) and intrinsic factors (fiziological state, age, etc.). Their movement increases almost linearly with temperature and starvation (Raworth and Choi 2001). The movement pattern is randomized in a favorable environment and with a precise direction in unfavorable conditions (Lys an Nentwig 1991). At flying species, the fly occurs only for young individuals (Meijer 1974). Observations about carabids movement are mainly done on populations, not on individuals. The surface covered by a population range from 1 ha to several square km. The home range of a population may change with the time of observation, as the species are freely moving (Loreau and Nolf 1993). It is not clear the scale of a carabid (i.e., its foraging area). One can make an idea about this by looking at change of the location in time. Using radio-transmitters mounting on large carabids, Riecken and Raths (1996) found that in 12 h an individual of *Carabus coriaceus* moved 51.25 m, and in 17 days 191.1 m. This technique is limited to species with large individuals.

A *Pterostichus cupreus* moved in 1 month 250 m (Thiele 1977). In other cases, the distance varied between meters to tenths of meters in a day (Lys and Nentwig 1991; Joyce et al. 1999). But a starved *Calosoma sycophanta* moved 3.7 km in 3 days (Thiele 1977). From this information, one can conclude that modeling or even computing a realistic bioaccumulation factor of metals in contaminated areas for carabids is practically impossible with the current knowledge about their ecology. The risk assessment in case of such species can be made based on structural differences between the contaminated and not contaminated landscapes, and not on a description of their role in metals mobility (which is not to say that this does not exist, but only that we are not ready to understand it because basic population ecology is not characterized with enough accuracy).

In the end of this part, we assess the possibility to model the bioaccumulation in birds. There is in the literature a cellular automata model for assessing the bioaccumulation of metals in birds (Cormont et al. 2005), suggesting that this might be feasible. From our research of the communities in mining areas (Ampoi and Geoagiu river catchments, Romania) and from the literature, we established the list of the present or potential species in contaminated areas (not exhaustive, but large enough to give an idea about the problem), and then inspected the literature for the scale of these organisms. Results are presented in Table 19.5 (spatial scales of individual organisms, not of populations).

One can notice two important aspects: most of the species are migratory, and the home range of birds potentially present in the contaminated areas varies one hectare to tenths of square kilometers. The species with the smaller home range, which could feed mainly in the contaminated area, are migratory, so they are not exposed continuously to the flux of metals. From these data, it results that in principle a risk assessment for birds based on metals mobility from the contaminated area can be done in principle, but exact knowledge about the migratory routes and/or about the use of habitats in the home range are needed.

19.6 Consequences for Risk Assessment

In the risk assessment of contaminated sites, there is the assumption that the larger the distance from the contaminated sites, the smaller is the contamination. However, as a result of the coupling between of large-scale processes (like atmospheric dispersion or surface water transport) and local scale processes one can have high contamination areas at large distance from the source (Fig. 19.4), so-called "hot spots." As an illustration, we present an inventory of such situations empirically proved in Romania sites studied within national and international projects (Table 19.6).

Although the lifetime of these hot spots can be short (low retention time of metals from geomorphological processes perspective), it is long enough from the perspective of the organisms and humans coupling to the contaminated systems. Finding the hot spot by standard monitoring means is not feasible as a solution

Table 19.5 Spatial scale of several bird species occurring in Romanian contaminated areas (mining dumps, tailing dams)

Taxon	Ph	Trophic niche	Specific spatial scale	Source
Accipiter gentilis	S	Raptor	12–863 ha	Rutz (2006), Squires and Kennedy (2006)
Corvus corax	S	Opportunist	700 ha	Rösner et al. (2005)
Dendrocopos major	S	Insectivorous	4.5–5 ha	Pavlik (1999)
Emberiza citrinella	S	Opportunist	10 ha	Golawski and Dombrowski (2002)
Garrulus glandarius	S	Opportunist	42.5–358.8 ha	Patterson et al. (1991)
Parus caeuleus	S	Insectivorous	11 km	Pinowski (1987)
Parus major	S	Insectivorous	3.3 km	Pinowski (1987)
Passer domesticus	S	Opportunist	59.2 km	Pinowski (1987)
Passer montanus	S	Opportunist	26 km	Pinowski (1987)
Pica pica	S	Opportunist	3–13 ha	Birkhead et al. (1986)
Buteo buteo	PM	Raptor	123–454 ha	Sim et al. (2001), Rodriguez et al. (2010)
Fringilla coelebs	PM	Opportunist	1,000 m from singing territory	Haila et al. (1989)
Fringilla coelebs	PM	Opportunist	27 individuals/ha	Mikkonen (1985)
Sturnus vulgaris	PM	Opportunist	40 ha	Paton et al. (2005)
Turdus merula	PM	Insectivorous	50–400 m from nest	Khokhlova (2009)
Cuculus canorus	M	Insectivorous	32.7–314.6 ha	Vogl (2004)
Lanius collurio	M	Insectivorous	1.5 ha	Golawski and Golawska (2008)
Phoenicurus ochruros	M	Insectivorous	2 ha	Personal estimation Cobzaru
Saxicola torquata	M	Insectivorous	2 ha	Personal estimation Cobzaru
Upupa epops	M	Insectivorous	3 ha	Personal estimation Cobzaru
Anthus trivialis	MD	Insectivorous	1–5 ha	Burton (2007), Moga et al. (2009)
Delichon urbica	MD	Insectivorous	400 from the colony	Personal estimation Cobzaru
Hirundo rustica	MD	Insectivorous	400 from the colony	Snapp (1976)
Phylloscopus trochilus	MD	Insectivorous	1.5 ha	Personal estimation Cobzaru
Sylvia atricapilla	MD	Insectivorous	1.3 ha	Shaeffer and Barkov (2004)
Sylvia curruca	MD	Insectivorous	1.3 ha	Personal estimation Cobzaru

Ph phenology, *S* sedentary, *PM* partial migratory, *M* migratory, *MD* migratory at large distance

because of the high monitoring resolution needed for large areas. The solution developed is to produce models coupled across scales able to identify the probable location of such contaminated areas. The general structure of the modeling approach was introduced in Fig. 19.2. The preliminary phase of contaminated sites risk assessment (aiming at the characterization of pathways of metals to receptors) should include models able to detect distant hotspot if parts of the

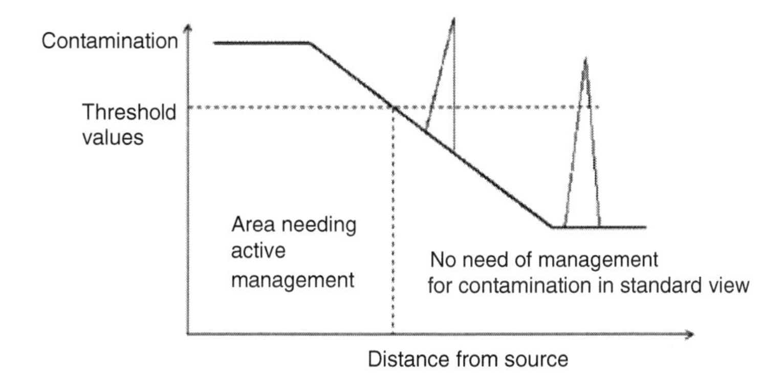

Fig. 19.4 The current assumption in risk assessment and the problem of hot spots at distance

environmental costs are not to be externalized from the eventual remediation project. Starting from this, a new concept of hazard assessment (the first phase of risk assessment) is introduced in another chapter of this book (Jianu et al. 2012F).

19.7 Conclusions

Metal biogeochemistry is only one side of the research of productive systems. There is an optimal complexity of the integrated models in metal biogeochemistry. A minimal complex homomorphic model can be built according to the structure of the developmental systems included in the productive entity, allowing for structural integration of several environmental entities. A complementary functional integration results from the observation that it makes little sense to study the circulation of one element (e.g., a heavy metal) in the population of developmental systems (DSs) identified by the homomorphic model separately from the circulation of other elements playing the role of resources or toxicants (e.g., macronutrients), because all of them influence the productivity of the DSs.

Metal mobility results from the interaction (coupling) of environmental entities at a multitude of scales. These interactions generate patterns of metal distribution. The environmental entities are frequently studied within different scientific disciplines, so understanding the coupled biogeochemical processes involved in the mobility of metals is a matter of interdisciplinary research and integrated modeling. There is not a "site" (ecosystem) or "region" (landscape) scale specific for the mobility of metals. This kind of speaking refers only to a simple modularization of the environment for management needs. A scientifically based management should consider the coupling between processes occurring at different scales because of the effects at distance generated by processes at the local scale of the contaminated site.

To answer the questions formulated in the introduction, there is an optimal level of integration for understanding the biogeochemical role of abiotic and biological

Table 19.6 Examples of hot spots contaminated with metals in Romania caused by the coupling of processes with different scales

Source of metals	Large-scale process	Distance to "hot spot"	Local scale process at distance (in "receptor area")		Location in Romania
			1	2	
Batteries factory	Atmospheric dispersion	2–3 km	Forest barrier effect	–	Pantelimon (NEFERAL/ Acumulatorul)
Smelter		2–5 km	Runoff	Transversal particles buffering	(geomorphology + plants)
Ampoi – Zlatna					
Smelter		2–4 km	Runoff	Longitudinal particles buffering	(geomorphology + plants)
Ampoi – Zlatna					
Smelter	4–5 km	Runoff	Longitudinal buffering	V. Viilor – Copsa Mica	
Mining dump	Surface water transport	12 km	Longitudinal buffering	–	Ampoi – Hg mining dumps to floodplain
Mining dump + tailing dams + polluted soil		25–40 km	Longitudinal buffering	–	Ampoi – various sources to floodplain
Acid mine drainage		10–15 km	Groundwater recharge in karstic NATURA 2000 area	–	Geoagiu – mine to downstream groundwater Ardeu

objects, and the practical possibility to produce realistic integrated models is still strongly limited by the available knowledge base produced by environmental scientific disciplines.

Acknowledgments This research was done in the Romanian Consortium for the Biogeochemistry of Trace Elements with financing from National University Research Council (CNCSIS) by project 291/2007 MECOTER, from National Center for the Management of Projects (CNMP) by projects 31012/2007 FITORISC, 31043/2007 PECOTOX, 52175/2008 METAGRO, and in the international consortium of the project UMBRELLA, FP7-ENV-2008-1 no. 226870. Special thanks go to the anonymous reviewers for their constructive criticism which improved the quality of the manuscript.

References

Alexander RB, Boyer EW, Smith RA, Schwarz GE, Moore RB (2007) The role of headwater streams in downstream water quality. J Am Water Resour Assoc 43:41–59

Argent RMA, Voinov A, Maxwell T, Cuddy SM, Rahman JM, Seaton S, Vertessy RA, Braddock RD (2006) Comparing modeling frameworks: a workshop approach. Environ Model Softw 21:895–910

Bambic DG, Alpers CN, Green PG, Fanelli E, Silk WK (2006) Seasonal and spatial patterns of metals at a restored copper mine site I. Stream copper and zinc. Environ Pollut 114:774–782

Band LE, Tague CL, Brun SE, Tenenbaum DE, Fernandes RA (2000) Modeling watershed as spatial object hierarchies: structure and dynamics. Transact GIS 4:181–196

Beven K (2006) Searching for the Holy Grail of scientific hydrology. Hydrol Earth Syst Sci 10:609–618

Bian L (2007) Object-oriented representation of environmental phenomena: is everything best represented as an object? Ann Assoc Am Geograph 97:267–281

Bird G, Brewer PA, Macklin MG (2010) Management of the Danube drainage basin: implications of contaminant-metal dispersal for the implications of the EU Water Framework Directive. Int J River Basin Manag 8:63–78

Birkhead TR, Eden SF, Clarkson K, Goodburn S, Pellat J (1986) Social organisation of a population of magpies *Pica pica*. Ardea 74:59–68

Bloeschl G (2001) Scaling in hydrology. Hydrol Process 15:709–711

Bloeschl G (2006) Hydrological syntheses – across processes, places and scales. Water Resour Res 42:W03S02

Bloeschl G, Sivapalan M (1995) Scale issues in hydrological modeling – review. Hydrol Process 9:251–290

Branger F, Braud I, Debionne S, Viallet P, Dehotin J, Henine H, Nedelec Y, Anquetin S (2010) Towards multi-scale integrated hydrological models using the LIQUID framework: overview of concepts and first application examples. Environ Model Softw 25:1672–1681

Breure AM, Groot M, Eijsackers HJP (2008) System oriented ecotoxicological research: which way to go? Sci Total Environ 406:530–536

Broxton PD, Troch PA, Lyon SW (2009) On the role of aspect to quantify water transit times in small mountainous catchments. Water Resour Res 45:W08427

Burton NHK (2007) Influences of restock age and habitat patchiness on Tree Pipits Anthus trivialis breeding in Breckland pine plantations. Ibis 149S2:193–204

Buytaert W, Reusser D, Krause S, Renaud J-P (2008) Why can't we do better than Topmodel. Hydrol Process 22:4175–4179

Canullo R, Falinska K (2003) Ecologia vegetale. La struttura gerarchica della vegetatione, Liguori Editore Napoli Italia

Caruso BS, Haynes J (2010) Connectivity and jurisdictional issues for rocky mountains and great plains aquatic resources. Wetlands, doi:10.1007/s13157-010-0084-0

Chadwick DB, Zirino A, Rivera-Duarte I, Katz CN, Blake AC (2004) Modeling the mass balance and fate of copper in San Diego Bay. Limnol Oceanogr 49:255–266

Chapin TP, Nimick DA, Gammous CH, Wanty RB (2007) Diel cycling on zinc in a stream impacted by acid rock drainage: initial results from a new in situ Zn analyzer. Environ Monit Assess 133:161–167

Charpentier A, Stuefer JF (1999) Functional specialization of ramets in Scirpus maritimus – splitting the tasks of sexual reproduction, vegetative growth, and resource storage. Plant Ecol 141:129–136

Christensen TH, Kjeldsen P, Bjerg PL, Jensen DL, Christensen JB, Baun A, Albrechtsen HJ, Heron G (2001) Biogoechemistry of landfill leachate plumes. Appl Geochem 16:659–718

Cormont A, Baveco JM, van den Brink NV, 2005, Effects of spatial foraging behaviour on risks of contaminants for wildlife. Alterra-rapport 1369, Wageningen

Cova JT, Goodchild FM (2002) Extending geographical representation to include fields of spatial objects, Int. j. geographical information science, 16:509–532

Driscoll CT, Otton JK, Iverfeld A (1994) Trace metals speciation and cycling. In: Moldan B, Cerny J (eds) Biogeochemistry of small catchments: a tool for environmental research, vol 51. SCOPE, Wiley, New York, pp 301–322

European Commission (2005) Soil Atlas of Europe, European Soil Bureau Network Office for Official Publications of the Eur Communities, p 128

Falińska K (1998) The framework for plant population biology. In: Falińska K (ed) Plant population biology and vegetation processes. W. Szafer Institute of Botany, Polish Academy of Science, Krakow, pp 10–42

Freeman MC, Pringle CM, Jackson CR (2007) Hydrologic connectivity and the contribution of stream headwaters to ecological integrity at regional scales. J Am Water Resour Assoc 43:5–14

Fritz BG, Arntzen EV (2007) Effect of rapidly changing river stage on uranium flux through the hyporheic zone. Ground Water 45:753–760

Fryar AE, Wallin EJ, Brown DL (2000) Spatial and temporal variability in seepage between a contaminated aquifer and tributaries to the Ohio River, Groundwater Monitoring and REmediation, 20:129–146

Galan E, Fernández-Caliani JC, Gonzalez I, Aparicio P, Romero A (2008) Influence of geological setting on geochemical baselines of trace elements in soils: application to soils of South–West Spain. J Geochem Explor 98:89–106

Galuszka A (2007) A review of geochemical background concepts and an example using data from Poland. Environ Geol 52:861–870

Gammons CH, Nimick DA, Parker SR, Cleasby TE, McCleskey RB (2005) Diel behavior of iron and other heavy metals in a mountain stream with acidic to neural pH: Fisher Creek, Montana, USA. Geochim Cosmochim Acta 69:2505–2516

Golawski A, Dombrowski A (2002) Habitat use of Yellowhammers Emberiza citrinella, Ortolan Buntings E. hortulana, and Corn Buntings Miliaria calandra in farmland of east-central Poland. Ornis Fennica 79

Golawski A, Golawska S (2008) Habitat preference in territories of the red-backed shrike Lanius collurio and their food richness in an extensive agriculture landscape. Acta Zool Acad Sci Hung 54:89–97

Haila Y, Hanuski IH, Raivio S (1989) Methodology for studying the minimum habitat requierments of forest birds. Ann Zool Fenn 26:173–180

Harper JL (1977) Population biology of plants. Academic, London

Hwang T, Band L, Hales TC (2009) Ecosystem processes at the watershed scale: extending optimality theory from plot to catchement. Water Resour Res 45:W11425

Iordache V (2010) The reconstruction of productive objects in Darwin's "Origin of species". In: Flonta M, Staicu L, Iordache V (eds) "Darwin, the evolution of species and evolutionary thinking" (in Romanian). Pelican, Giurgiu, Romania, pp 75–108

Iordache V, Ion S, Pohoaţă A (2009) Integrated modeling of metals biogeochemistry: potential and limits. Chem Erde Geochem 69:125–169
Iordache V, Kothe E, Neagoe A, Gherghel F (2011) A conceptual framework for up-scaling ecological processes and application to ectomycorrhizal fungi. In: Rai M, Varma A (eds) Diversity and biotechnology of ectomycorrhiza. Springer, Berlin, Heidelberg
Iordache V, Onete M, Paucă M, Oromulu L, Honciuc V, Purice D, Cobzaru I, Gomoiu I, Neagoe A (2010) Biological communities in mining areas: scale dependent patterns, organisms' potential as bioindicators, and native plants for remediation. In: Proceedings 7th European Conference on Ecological Restoration, Avignon, France, 23-27/08/2010
Jianu D, Iordache V, Soare B, Petrescu L, Neagoe A (2012) The role of mineralogy in the hazard potential of abandoned mine sites. In: Kothe E, Varma A (eds) Bio-geo-interactions in contaminated soils, Springer, Berlin, Heidelberg
Joyce KA, Holland JM, Doncaster CP (1999) Influences of hedgerow intersections and gaps on the movement of carabid beetles. Bull Entomol Res 89:523–531
Kavetski D, Klark MP (2011) Numerical troubles in conceptual hydrology: approximations, absurdities, and impact on hypothesis testing. Hydrol Process 25:661–670
vad der Keur P, Iversen BV (2006) Uncertainty in soil physical data at river basin scale. Hydrol Earth Syst Sci Discuss 3:1281–1313
Khokhlova TY (2009) Juvenile moult and spatial behaviour of first-year Blackbirds Turdus merula on the northeast edge of the range. Avian Ecol Behav 15:1–22
Kohne JM, Kohne S, Simunek J (2009) A review of model applications for structured soils: a water flow and tracer transport. J Contaminant Hydrol 104:4–35
Kothe E (2009) Using microbes for the regulation of heavy metals mobility at ecosystem and landscape scale. Research project. http://www.umbrella.uni-jena.de
Kraft P, Multsch S, Vache KB, Frede H-G, Breuer L (2010) Using phyton as a coupling platform for integrated catchement models. Adv Geosci 27:51–56
Lăcătuşu R, Iordache V (2008) Integrated modeling of geochemical and biogeochemical processes across space-time scales. In: Proceedings of the 8th International symposium on "Metal elements in environment, medicine and biology", Eurobit Publishing House, Timişoara, pp 41–54
Lăcătuşu R, Răuţă C, Cârstea S, Ghelase I (1996) Soil-plant-man relationships in heavy metal polluted areas in Romania. Appl Geochem 11:105–107
Lăcătuşu R, Răuţă C, Grigore C, Cârstea S, Ghelase I (1993) Geochemical and biogeochemical features of areas of haematuria in cattle in the Eastern Carpathian Mountains of Romania. Environ Geochem Health 15:3–13
Lăcatusu R, Citu G, Aston J, Lungu M, Lacatusu AR (2009) Heavy metals soil pollution state in relation to potential future mining activities in the Rosia Montana Area. Carp J Earth Environ Sci 4:39–50
Lee MK, Saunders JA (2003) Effects of pH on metals precipitation and sorption: field bioremediation and geochemical modeling approaches. Vadose Zone J 2:177–185
Lin H (2003) Hydropedology: bridging disciplines, scales, and data. Vadose Zone J 2:1–11
Lin H (2010a) Earth's critical zone and hydropedology: concepts, characteristics, and advances. Hydrol Earth Syst Sci 14:25–45
Lin H (2010b) Linking principles of soil formation and flow regimes. J Hydrol 393:3–19
Lin H, Wheeler D, Bell J, Wilding L (2005) Assessment of soil spatial variability at multiple scales. Ecol Modell 182:271–290
Liu P, Tang Y, Du R, Yang H, Wu Q, Qiu R (2010) Root foraging for zinc and cadmium requirement in the Zn/Cd hyperaccumulator plant Sedum alfredii. Plant Soil 327:365–375
Loreau M, Nolf C-L (1993) Occupation of space by carabid beetle Abax ater. Acta Oecol 14:247–258
Lottermoser BG, Ashley PM (2005) Tailings dam seepage at the rehabilitated Mary Kathleen uranium mine, Australia. J Geochem Explor 85:119–137
Luck GW, Daily GC, Ehrlich PR (2003) Population diversity and ecosystem services. Trends Ecol Evol 18:331–336

Lys J-A, Nentwig W (1991) Surface activity of carabid beetles inhabiting cereal fields. Pedobiologia 35:129–138

Manfreda S, Smettem K, Iacobellis V, Montaldo N, Sivapalan M (2010) Coupled ecological-hydrological processes. Echohydrology 3:131–132

Matyssek R, Schulin R, Günthardt-Goerg MS (2006) Metal fluxes and stresses in terrestrial ecosystems: synopsis towards holistic understanding. Snow Landsc Res 80:139–148

McGuire KJ, McDonnell JJ (2006) A review and evaluation of catchment transit time modeling. J Hydrol 330:543–563

Meijer J (1974) A comparative study of the immigration of carabids (Coleoptera, Carabidae) into a new polder. Oecologia 16:185–208

Van Meirvenne M, Meklit T (2010) Geostatistical simulation for the assessment of regional soil pollution. Geograph Anal 42:121–135

Meklit T, Van Meirvenne M, Verstraete S, Bonroy J, Tack F (2009) Combining marginal and spatial outliers identification to optimize the mapping of the regional geochemical baseline concentration of soil heavy metals. Geoderma 148:413–420

Mikkonen AV (1985) Establishment of breeding territory of the chaffinch, Fringilla coelebs, and the brambling, F. montifringilla, in northen Finland. Ann Zool Fenn 22:137–156

Moga CI, Hartel T, Ollerer K (2009) Ancient oak wood-pasture as a habitat for the endangered tree pipit Anthus trivialis. Biologia 64:1011–1015

Morin PJ (2005) Community ecology. Blackwell, Oxford, UK

Nagorski SA, Moore JN, McKinnon TE, Smith DB (2003) Scale-dependent temoral variations in stream water geochemistry. Environ Sci Technol 37:859–864

Neagoe A (coord) (2007a) Mathematical modeling of key mechanisms involved in metals mobility in terrestrial and wetlands ecosystems, Research project, University of Bucharest. http://www.mecoter.cesec.ro

Neagoe A (coord) (2007b) Procedure for risk assessment and phytoremediation of contaminated areas, Research project, University of Bucharest. http://www.fitorisc.cesec.ro

Neagoe A, Iordache V, Farcasanu I (2012) The role of organic matter in the mobility of metals in contaminated sites. In: Kothe E, Varma A (eds) Bio-geo-interactions in contaminated soils. Springer, Berlin, Heidelberg

Neagoe A, Iordache V, Kothe E (2011) A concept of biogeochemical role and the challenge of up-scaling the effects of arbuscular mycorrhizal fungi on metals mobility. In: Ebrahim Mohammadi Goltapeh, Younes Rezaee Danesh, Ajit Varma (eds) Fungi as bioremediators. Soil biology, Springer, Berlin, in press

Oborny B (1994) Spacer length in clonal plants and the efficiency of resource capture in heterogenous environments: a Monte Carlo simulation. Folia Geobot Phytotaxonomica 29:139–158

Oborny B, Kun Á, Czárán T, Bokros S (2000) The effect of clonal integration on plant competition mosaic habitat space. Ecology 81:3291–3304

Odum HT (1995) Energy systems concepts and self-organisation: a rebuttal. Oecologia 104:518–522

Oyama S, Griffiths PE, Gray RD (eds) (2000) Cycles of contingency: developmental systems and evolution. MIT, Cambridge, MA

Pachepsky YA, Rawls WJ, Lin HS (2006) Hydropedology and pedotransfer functions. Geoderma 131:308–316

Pahl-Vostl C (1995) The dynamic nature of ecosystems. Wiley, New York

Paton DC, Sinclair RG, Bentz CM (2005) Ecology and management of the common starling (sturnus vulgaris) in the McLaren Vale Region. Final report to Grape and Wine research & Development Corporation

Patterson IJ, Cavallini P, Rolando A (1991) Density, range size and diet of the European Jay Garrulus glandarius in the Maremma Natural Park, Tuscany, Italy in Summer and Autumn

Petrescu L (coord) (2007) Procedure for the economic assessment of stable pollutants retention in river systems. Research project, University of Bucharest. http://www.pecotox.cesec.ro

Pinowski J (1987) Experimental studies on the dispersal of young tree sparrows. Ardea 55:241–248
Pundt H, Bishr Y (2002) Domain ontologies for data sharing – an example from environmental monitoring using field GIS
Qi F, Zhu A-X, Harrower M, Burt JE (2006) Fuzzy soil mapping based on prototype category theory. Geoderma 136:774–787
Răuţă C, Lăcătuşu R, Cârstea S (1995) Heavy metal pollution in Romania. In Salomons W, Forstner U, Mader P (eds) Heavy metals: problems and solutions, Springer, Berlin, pp 359–372
Raworth DA, Choi M-Y (2001) Determining numbers of active carabid beetles per unit area from pitfall-trap data. Entomol Exp Appl 98:95–108
Riecken U, Raths U (1996) Use of radio telemetry for studying dispersal and habitat use of Carabus coriaceus L. Ann Zool Fenn 33:109–116
Robinson B, Schulin R, Nowack B, Roulier S, Menon M, Clothier B, Green S, Mills T (2006) Phytoremediation for the management of metal flux in contaminated sites. Snow Landsc Res 80:221–234
Rodriguez B, Siverio F, Rodriguez A, Siverio M, Hernandez JJ (2010) Figuerola J, Density, habitat selection and breeding biology of Common Buzzards *Buteo buteo* in an insular environment. Bird Study 57:75–83
Rodriguez JA, Nanos N, Grau JM, Gil L, Arias ML (2008) Multiscale analysis of heavy metal contents in Spanish agricultural topsoils. Chemosphere 70:1085–1096
Rösner S, Selva N, Müller T, Pugacewicz E, Laudet F (2005) Raven *Corvus corax* ecology in a primeval temperate forest. W: Corvids of Poland.Red. In: Jerzak L, et al. (eds) Bogucki Wyd. Nauk., Poznań, pp 385–405
Rutz C (2006) Home range size, habitat use, activity patterns and hunting behaviour of urban-breeding Northern Goshawks *Accipiter gentilis*. Ardea 94:185–202
Ryan A (2007) Emergence is coupled to scope, not to level. Complexity 13:67–77
Saikkonen K, Koivunen S, Vuorisalo T, Mutikainen P (1998) Interactive effects of pollination and heavy metals on resource allocation in Potentilla anserina L. Ecology 79:1620–1629
Saunders TJ, McClain ME, Llerena CA (2004) The biogeochemistry of surface and subsurface runoff in a small montane catchment of the Peruvian Amazon. Hydrol Process 20:2549–2562
Schaefli B, Harman CJ, Sivapalan M, Schymansky SJ (2011) Hydrologic predictions in a changing environment: behavioral modeling. Hydrol Earth Syst Sci 15:635–646
Schmitz O, Karssebnerg D, de Jong K, de Kok J-L (2011) Constructing integrated models: a scheduler to execute coupled components. In: Geertman S, Reinhardt W, Toppen F (eds), ISBN: 978-90-816960-1-2, In: Proceedings of AGILE 2011, The 14th AGILE International Conference on Geographic Information Science, Advancing Geoinformation Science for a Changing World. http://plone.itc.nl/agile_old/Conference/2011-utrecht/contents/pdf/shortpapers/sp_98.pdf. Accessed July 2011
Scrădeanu D, Ioane D, Chitea F, Tevi G (2010) The model moisture distribution in heterogenius soil under complexe hydrological conditions. Geophysical research abstracts 12: EGU2010-12863-1
Scrădeanu D, Mezincescu M, Pagnejer M (2010) The monitoring and control of space-time moisture distribution in unsaturated soil with a sensitive informatic system. Geophys Res abstracts, 12, EGU2010-12900
Scull P (2010) A top-down approach to the state factor paradigm for use in macroscale soil analysis. Ann Assoc Am Geograph 100:1–12
Scull P, Franklin J., Chadwick OA, McArthur D (2003) Predictive soil mapping: a review, Progress in Physical Geography 27: 171–197
Seppelt R (2003) Computer-based environmental management. Willey-VCH, Weinheim
Silvertown J, Charlesworth D (2001) Introduction to plant population biology, 4th edn. Blackwell, Oxford, UK
Sim IMW, Cross AV, Lamacraft DL, Pain DJ (2001) Correlates of common Buzzard Buteo buteo density and breeding success in the West Midlands. Bird Study 48(3):317–329

Simpson LS, Maher EJ, Jolley DF (2004) Process controlling metal transport and retention as metal-contaminated ground water efflux through estuarine sediments, Chemosphere, 56:821–831
Snapp BD (1976) Colonial breeding in the barn swallow (*Hirundo rustica*) and its adaptive significance. Condor 78:341–480
Squires JR, Kennedy PL (2006) Northern Goshawk ecology: an assessment of current knowledge and information needs for conservation and management. Stud J Avian Biol 31:8–62
Srivastava P, Migliaccio KW, Šimůnek J (2007) Landscape models for simulating water quality at point, field, and watershed scales. Transact ASABE 50:1683–1693
Stockdale A, Davison W, Zhang H (2009) Micro-scale biogeochemical heterogeneity in sediments: a review of available technology and observed evidence. Earth Sci Rev 92:81–97
Stuefer JF (1997) Division of labour in clonal plants? On the response of stoloniferous herbs to environmental heterogeneity, Ph.D. Thesis, Utrecht University
Sui DZ, Maggio RC (1999) Integrating GIS with hydrological modelling: practices, problems, and prospects, Computers. Environ Urban Syst 23:33–51
Suzuki J-I, Stuefer JF (1999) On the ecological and evolutionary significance of storage in clonal plants. Plant Species Biol 14:11–17
Tamm A, Kull K, Sammul M (2002) Classifying clonal growth forms based on vegetative mobility and ramet longevity: a whole community analysis. Evol Ecol 15:383–401
Tercier-Waeber ML, Hezard T, Masson M, Schafer J (2009) In situ monitoring of the diurnal cycling of dynamic metal species in a stream under contrasting photobenthic biofilm activity and hydrological conditions. Environ Sci Technol 43:7137–9244
Tercier-Waeber ML, Taillefert M (2008) Remote in situ voltammetric techniques to characterize the biogeochemical cycling of trace metals in aquatic systems. J Environ Monit 10:30–54
Thiele HU (1977) Carabid beetles in their environment. Springer, Berlin
Thompson SE, Harman CJ, Troch PA, Brooks PD, Sivapalan M (2011) Spatial scale dependence of ecohydrology mediated water balance partitioning: a syntheses framework for catchement ecohydrology. Water Res 47: W00J03
Van Rossum F, Bonnin I, Fenart S, Pauwels M, Petit D, Saumitou-Laprade P (2004) Spatial genetic structure within metallicolous population of *Arabidopsis halleri*, a clonal, self-incompatible and heavy-metal-tolerant species. Mol Ecol 13:2959–2967
Vidon P, Allan C, Burns D, Duval N, Gurwick S, Inandar S, Lowrance R, Okay J, Scott D, Sebestyen S (2010) Hot sports and hot moments in riparian zones: potential for improved water quality management. J Am Water Resour Assoc 46:278–298
Viollier E, Rabouille C, Apitz SE, Breuer E, Chaillou G, Dedieu K, Furukawa Y, Grenz C, Hall P, Janssen F, Morford JL, Poggiale JC, Roberts S, Shimmield T, Taillefert M, Tengberg A, Wenzhofer F, Witte U (2003) Benthic biogeochemistry: state of the art technologies and guidelines for the future of in situ survey. J Exp Mar Biol Ecol 285–286:5–31
Vogl W (2004) Habitat and space use of european cuckoo females during the egg laying period. Behaviour 141:881–898
Wagener T, Sivapalan M, Troch PA, McGlynn BL, Harman CJ, Gupta HV, Kumar P, Rao PSC, Basu NB, Wilson JS (2010) The future of hydrology: an evolving science for a changing world. Water Resour Res 46:W05301
Warren LA, Haack EA (2001) Biogeochemical controls on metal behavior in freshwater environments. Earth Sci Rev 54:261–320
White J (1989) The plant as a metapopulation. Ann Rev Ecol Syst 10:109–145
Widdén B, Cronberg N, Widdén M (1994) Genotypic diversity, molecular markers and spatial distribution of genets in clonal plants – a literature survey. Folia Geobot Phytotax 29:245–263
Xiaoni H, Hong L, Danfeng S, Liandi Z, Baoguo L (2010) Multi-scale spatial structure of heavy metals in agricultural soils in Beijing. Environ Monit Assess 164:605–616

Index

E. Kothe and A. Varma (eds.), *Bio-Geo Interactions in Metal-Contaminated Soils*, Soil Biology 31, DOI 10.1007/978-3-642-23327-2,

N

O